[illegible]gress in Mathematics

Volume 20[illegible]

Series Editors

[illegible]

[illegible]

A. Weinstein

Progress in Mathematics
Volume 207

Yiming Long

Index Theory for Symplectic Paths with Applications

Springer Basel AG

Author:

Yiming Long
Nankai Institute of Mathematics
Nankai University
Tianjin 300071
People's Republic of China
e-mail: longym@nankai.edu.cn

2000 Mathematics Subject Classification 34C25, 58E05, 70H03, 70H05, 70H12

A CIP catalogue record for this book is available from the Library of Congress, Washington D.C., USA

Deutsche Bibliothek Cataloging-in-Publication Data

Long, Yiming:
Index theory for symplectic paths with applications / Yiming Long. – Basel ; Boston ; Berlin : Birkhäuser, 2002
(Progress in mathematics ; Vol. 207)
ISBN 978-3-0348-9466-1 ISBN 978-3-0348-8175-3 (eBook)
DOI 10.1007/978-3-0348-8175-3

Originally published by Birkhäuser Verlag in 2002
Softcover reprint of the hardcover 1st edition 2002

Printed on acid-free paper produced of chlorine-free pulp. TCF ∞

9 8 7 6 5 4 3 2 1

www.birkhauser-science.com

Contents

Preface

This book is based upon my monograph *Index Theory for Hamiltonian Systems with Applications* published in 1993 in Chinese, and my notes for lectures and courses given at Nankai University, Brigham Young University, ICTP-Trieste, and the Institute of Mathematics of Academia Sinica during the last ten years. The aim of this book is twofold:

(1) to give an introduction to the index theory for symplectic matrix paths and its iteration theory, which form a basis for the Morse theoretical study on Hamiltonian systems, and to give applications of this theory to periodic boundary value problems of nonlinear Hamiltonian systems. Here the iteration theory means the index theory of iterations of periodic solutions and symplectic matrix paths.
(2) to serve as a reference book on these topics.

There are many different ways to introduce the index theory for symplectic paths in order to establish Morse type index theory of Hamiltonian systems. In this book, I have chosen a relatively elementary way, i.e., the homotopy classification method of symplectic matrix paths. It depends only on linear algebra, point set topology, and certain basic parts of linear functional analysis. I have tried to make this part of the book self-contained and at the same time include all of the major results on these topics so that researchers and students interested in them can read it without substantial difficulties, and can learn the main results in this area for their possible applications.

In this book, I have introduced numerous examples and more elementary cases to indicate the intuitive ideas behind the theorems. For example, the $\mathbf{R}^3$-cylindrical coordinate representation of Sp(2) and proofs of many results at the Sp(2) level are given, and they yield suggestive models and intuitive ideas for the general case. I hope this will help readers to understand the material.

It is my great pleasure to take this opportunity to thank Professor Paul Rabinowitz, who introduced me into this amazing area eighteen years ago, and whose perspective on this field has greatly influenced my research interest, for his constant encouragement and helps.

It is also my great pleasure to acknowledge my debt to the following mathematicians who have ever encouraged or helped me in various ways: Antonio Ambrosetti, Kung-Ching Chang, Shing-shen Chern, Han-Ying Deng, Ivar Ekeland, Helmut Hofer, Zixin Hou, Guoding Hu, Yu-Min Huang, John Mather, Jürgen Moser, Yongbin Ruan, Gang Tian, Alan Weinstein, Zhihong Xia, and Eduard Zehnder. I would also like to thank the National Natural Science Foundation ofChina, the Cheung Kong Scholars Programme of the Education Ministry of China and Li Ka Shing Foundation of Hong Kong, the Mathematics Center of the Education Ministry of China, the Research Fund for the Doctoral Program of Higher Education of the Education Ministry of China, the 973 Program of the Science and Technology Ministry of China, the City Education Committee of Tianjin, S.S. Chern Foundation, and the Hong Kong Qiu Shi Science and Technology Foundation for their support during my preparation of this book.

Nankai University, Tianjin
October 2001

Yiming Long

Introduction

The variational structure of Hamiltonian systems was observed more than one hundred years ago. But because of its degeneracy, caused by strong indefiniteness, until the 1970s it had been thought to be useless in the theoretical study of existence of solutions of Hamiltonian systems. It was P. Rabinowitz who, in his pioneering work [Rab2] of 1978, first demonstrated that this degenerate variational principle can be efficiently used in the existence study of periodic solutions of Hamiltonian systems. Since then, a great number of research papers have appeared in this and related areas, and many aspects of critical point theory have been applied to the variational study of Hamiltonian systems.

On the other hand, in the study of closed geodesics on Riemannian manifolds, M. Morse in the 1930s successfully developed his index theory. The iteration theory of the Morse index for closed geodesics was developed by R. Bott in his celebrated work [Bot1] of 1956. Based upon their works, many interesting and deep results on closed geodesics have been obtained via the Morse index theory.

Therefore it is natural to consider the problem of applying the Morse theoretical method to problems involving periodic solutions of Hamiltonian systems. In this consideration, a major difficulty is that the functionals corresponding to Hamiltonian systems are strongly indefinite, which causes the related Morse indices to be always infinite. It is important to find certain finite representations of these infinite Morse indices. Since 1980, two different index theories for Hamiltonian systems have appeared. One index theory was developed by I. Ekeland in the 1980s for convex Hamiltonian systems. A beautiful systematic treatment of this index theory was given in 1990 in his celebrated book '*Convexity Methods in Hamiltonian Mechanics*' ([Eke3]). The other index theory was defined for periodic solutions of general Hamiltonian systems without any non-degeneracy assumptions, and even more generally for any paths in the group of symplectic matrices. The current book is devoted to this index and its iteration theory as well as their applications.

Historically, as far as the author knows, the classification and an index theory for linear Hamiltonian systems with continuous symmetric periodic coefficients began with the work of H. Amann and E. Zehnder in their paper [AmZ1] of 1980. They established the corresponding index theory for systems with constant coefficients. After that many mathematicians worked on this problem. We refer to the papers [Cha1] of K.C. Chang in 1981 and [LiL1] of S. Li and J.Q. Liu in 1989 for example.

Since the fundamental solution of a general linear Hamiltonian system with continuous symmetric periodic coefficients is a path in the symplectic matrix group $\mathrm{Sp}(2n)$ starting from the identity, to be sufficiently flexible such an index theory needs to be generalized to any such symplectic paths. Here the symplectic group is defined by

$$\mathrm{Sp}(2n) = \{M \in \mathrm{GL}(\mathbf{R}^{2n}) \mid M^T J M = J\},$$

where $J = \begin{pmatrix} 0 & -I \\ I & 0 \end{pmatrix}$, I is the identity matrix on $\mathbf{R}^n$, and M^T denotes the transpose of M. For $\tau > 0$, we define the set of symplectic matrix paths by

$$\mathcal{P}_\tau(2n) = \{\gamma \in C([0,\tau], \mathrm{Sp}(2n)) \mid \gamma(0) = I\}.$$

Because Sp($2n$) is homeomorphic to the product of the unit circle and a simply connected space, a path $\gamma \in \mathcal{P}_\tau(2n)$ rotates naturally in Sp($2n$) along this unit circle. The point here is to find a way to count this rotation so that the rotation number represents intrinsically the corresponding Morse index of the related Hamiltonian system. For periodic boundary value problems of Hamiltonian systems, because of this consideration, we call a path $\gamma \in \mathcal{P}_\tau(2n)$ *degenerate* if 1 is an eigenvalue of $\gamma(\tau)$, and *non-degenerate* otherwise.

In their celebrated paper [CoZ2] of 1984, C. Conley and E. Zehnder defined an index theory for any non-degenerate path in Sp($2n$) with $n \geq 2$, i.e. the so called Conley–Zehnder index theory. This index theory was further defined for the non-degenerate paths in Sp(2) by E. Zehnder and the author in [LZe1] of 1990. The index theory for degenerate linear Hamiltonian systems was defined by C. Viterbo in [Vit4] and the author in [Lon5] of 1990 independently. Then in [Lon13] of 1997, this index theory was further extended to any degenerate path in $\mathcal{P}_\tau(2n)$ and an axiomatic characterization of the index theory was given.

In 1990, D. Salamon and E. Zehnder gave the first correct proof in [SaZ1] of the simply connectedness of the subset $\mathrm{Sp}(2n)_1^*$ in Sp($2n$), where $\mathrm{Sp}(2n)_1^*$ contains all matrices in Sp($2n$) possessing no eigenvalue equal to 1. This fact was used in the definition of the index theory in [CoZ1] for the non-degenerate case.

In this book, we denote this index by

$$(i_1(\gamma), \nu_1(\gamma)) \in \mathbf{Z} \times \{0, 1, \ldots, 2n\}, \quad \forall \gamma \in \mathcal{P}_\tau(2n).$$

In fact, there are infinitely many ways to define index theories for paths of symplectic matrices. A definition of the index theory for symplectic paths is meaningful if it can be applied to different problems. Here the crucial point for our index theory is its relation with the Morse index theory of general Hamiltonian systems, i.e., Theorem 6.1.1 below in this book with $\omega = 1$, which was proved by C. Conley, E. Zehnder, and the author in [CoZ2], [LZe1], [Lon5], and [Lon13]. For Lagrangian systems on $\mathbf{R}^n$, it was proved in [Vit1] by C. Viterbo that the Morse index of the corresponding functional coincides with our index theory for symplectic paths (cf. Theorem 7.3.1 below by a proof of T. An and the author in [LoA1]). Via such theorems, our index theory forms a natural basis of the Morse theoretical study on general nonlinear Hamiltonian systems.

Suppose c is a closed geodesic on a compact Riemannian manifold (M, g). Denote the sequence of Morse indices and nullities of the iterations c^m of c by $\{(i(c^m), \nu(c^m))\}$, where $c^m(t) = c(mt)$ for any $t \in \mathbf{R}$. In his pioneering work [Bot1] in 1956, R. Bott studied properties of this sequence, and established the now so-called Bott formula for iterations, i.e.,

$$i(c^m) = \sum_{\omega^m=1} \Lambda(\omega), \quad \nu(c^m) = \sum_{\omega^m=1} N(\omega),$$

where for a fixed closed geodesic c on (M, g), Λ and N are the Bott index functions for second order differential systems which are defined on the unit circle and take non-negative integer values. Since Bott's work, various iteration properties of different Morse and Maslov type index theories were studied by [Edw1] of H. Edwards in 1964, [Kli1] of W. Klingenberg in 1978, [BTZ1] of W. Ballmann, G. Thobergsson, and W. Ziller in 1982, [Eke1]–[Eke3] of I. Ekeland in 1984–1990, [Vit3] of C. Viterbo in 1989, and others. Index iteration theories were established in these works for closed geodesics (cf. [Bot1], [BTZ1], [Kli1]), convex Hamiltonian systems (cf. [Eke1]–[Eke3]), and star-shaped Hamiltonian systems under certain non-degeneracy conditions (cf. [Vit3]). Based upon these iteration theories, many interesting results on multiplicity, stability, and other problems of geometrically distinct closed geodesics on Riemannian manifolds and closed characteristics of convex Hamiltonian systems appeared later. Some unsuccessful attempts appeared earlier than 1990 on establishment of a complete iteration theory for the general index of symplectic paths. Note also that J. Robbin and D. Salamon in [RoS1] of 1993 extended the definition of the index of Conley and Zehnder to the degenerate case too so that a catenation iteration property holds. But their extension in the degenerate case is different from those of [Lon5], [Lon13], and [Vit1] (cf. Corollary 3.1 of [LZu1]).

Similarly, in the study on the minimality of period, multiplicity, and stability of geometrically distinct periodic solution orbits of nonlinear Hamiltonian systems, the author was led to the study of iteration theory of the above mentioned index theory for symplectic paths. For our applications, for a symplectic matrix path $\gamma \in \mathcal{P}_\tau(2n)$ we face two types of problems:

⟨1⟩ knowing the end point $\gamma(\tau)$ of a path $\gamma \in \mathcal{P}_\tau(2n)$, the initial index $(i_1(\gamma), \nu_1(\gamma))$, and the iteration time m, we want to find the index $i_1(\gamma^m)$ of the m-th iterated path γ^m;

⟨2⟩ knowing the end point $\gamma(\tau)$ of a path $\gamma \in \mathcal{P}_\tau(2n)$, the initial index $(i_1(\gamma), \nu_1(\gamma))$, and the index $(i_1(\gamma^m), \nu_1(\gamma^m))$ of the m-th iterated path γ^m, we want to find the iteration time m.

To solve these problems, the first natural consideration is to generalize Bott formulae to the index theory for general symplectic paths. To the author's knowledge, for this purpose, such a hopefully correct study of iteration theory for the index of any symplectic path in this full generality started from [DoL1] of Di Dong and the author published in 1997. A Nankai Institute preprint of this paper appeared in 1993. The basis of this iteration theory was established in [Lon16] of 1999, where a generalized index function theory $(i_\omega(\gamma), \nu_\omega(\gamma))$ parameterized by ω in the unit circle $\mathbf{U}$ of the complex plane $\mathbf{C}$ for every symplectic matrix path γ started from the identity matrix was introduced, a Bott-type iteration formula was proved, and the mean index was defined. This Bott-type formula is given in Theorem 9.2.1 of this book, i.e., for any $\tau > 0$, $\gamma \in \mathcal{P}_\tau(2n)$, $z \in \mathbf{U}$, and $m \in \mathbf{N}$,

$$i_z(\gamma, m) = \sum_{\omega^m = z} i_\omega(\gamma), \tag{1}$$

$$\nu_z(\gamma, m) = \sum_{\omega^m = z} \nu_\omega(\gamma). \tag{2}$$

To further solve the problems $\langle 1 \rangle$ and $\langle 2 \rangle$, we need to go beyond the Bott-type formula (1)–(2). For a given path γ it is natural to consider constructing a new path η in $\mathcal{P}_\tau(2n)$ so that

$$i_1(\gamma^m) = i_1(\eta^m), \quad \nu_1(\gamma^m) = \nu_1(\eta^m), \quad \forall m \in \mathbf{N}, \tag{3}$$

and that $(i_1(\eta^m), \nu_1(\eta^m))$ is easy enough to compute. This leads to finding a homotopy $\delta : [0, 1] \times [0, \tau] \to \mathrm{Sp}(2n)$ from γ to η in $\mathcal{P}_\tau(2n)$ and keeping the end points of this homotopy always in a certain suitably chosen subset Γ of $\mathrm{Sp}(2n)$, as shown in Figure 1 below, so that (3) always holds.

The classical way to determine the set Γ before our study was to set

$$\Gamma = [\gamma(\tau)] \equiv \{P^{-1}\gamma(\tau)P \mid, P \in \mathrm{Sp}(2n)\},$$

the conjugate set of $\gamma(\tau)$ in $\mathrm{Sp}(2n)$. This means that only symplectic coordinate changes are allowed in computations. Simplification of the computation of the index is then limited to those paths whose end points are symplectic normal form matrices. When the dimension of such normal forms are large, the computations are still rather complicated.

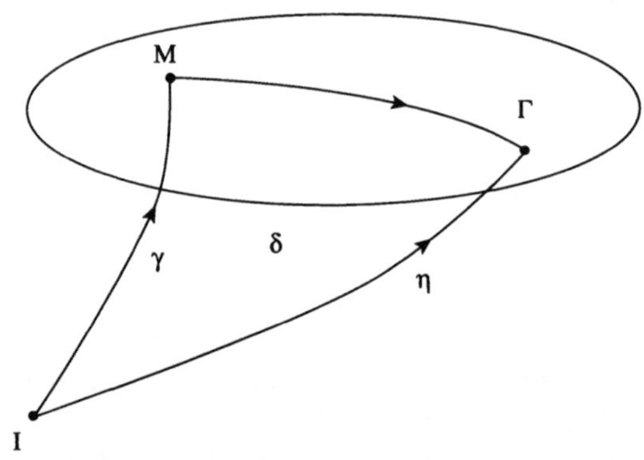

Figure 1: *The homotopy from γ to η and the subset Γ*

In order to simplify the computations as much as possible, we try to find the subset Γ of $\mathrm{Sp}(2n)$ as big as possible so that (3) still holds. By the Bott-type formula (1)–(2), to make (3) always hold it is necessary and sufficient to ensure that all the index functions $i_\omega(\delta(s,\cdot))$ do not change for all $s \in [0,1]$ whenever ω is a root of unit and an eigenvalue of $\gamma(\tau)$. Then by stability properties of Morse indices of symmetric matrices (Lemma 6.1.3 in this book) and the fact that roots of unit are dense in $\mathbf{U}$, it is necessary and sufficient to ensure that all the nullity functions $\nu_\omega(\delta(s,\tau))$ do not change for all $s \in [0,1]$ whenever ω is an eigenvalue of $\gamma(\tau)$ on $\mathbf{U}$. Therefore in [Lon16] we let the subset Γ be the path connected component $\Omega^0(M)$ containing $M = \gamma(\tau)$ (cf. Definition 1.8.5 below) of the set

$$\begin{aligned}\Omega(M) = \{N \in \mathrm{Sp}(2n) \mid \sigma(N) \cap \mathbf{U} = \sigma(M) \cap \mathbf{U}, \text{ and} \\ \nu_\lambda(N) = \nu_\lambda(M)\ \forall \lambda \in \sigma(M) \cap \mathbf{U}\}\end{aligned} \tag{4}$$

and call $\Omega^0(M)$ the **homotopy component** of M in $\mathrm{Sp}(2n)$. Clearly for any given path $\gamma \in \mathcal{P}_\tau(2n)$, $\Omega^0(\gamma(\tau))$ is the largest set for end points of any homotopy δ of γ so that (3) holds with η replaced by $\delta(s,\cdot)$. Note that $[\gamma(\tau)]$ is a proper subset of $\Omega^0(\gamma(\tau))$ in general.

Within $\Omega^0(\gamma(\tau))$, we then establish a theorem (Theorem 1.8.10 below) to decompose $\gamma(\tau)$ into products of certain special 2×2 and 4×4 symplectic normal form matrices, which we call **basic normal forms** in Definition 1.8.9. Correspondingly by the homotopy invariance and symplectic additivity of the index theory, the computations in (3) are reduced to those of iterations of paths in Sp(2) or Sp(4) whose end points are basic normal form matrices. Then this idea is used to derive an algebraic characterization of the topologically defined splitting numbers for symplectic matrices in [Lon16] and to establish the precise iteration formulae of indices of symplectic paths in [Lon18]. Based upon these results, various new iteration inequalities and iteration properties of the index theory are established. The work [Lon16] is based upon the earlier works [DoL1], [LoD1], and [HaL1] of Di Dong, Jianlong Han, and the author on normal forms of symplectic matrices, and [Lon15] on structures of ω-subsets of symplectic groups. Motivated by the precise iteration formulae of [Lon18], in a recent paper [LZu3] of Chaofeng Zhu and the author, the abstract index iteration formula of any symplectic path was established. On the other hand in [DoL1] by D. Dong and the author, [LLo1] of 1997 and [LLo3] of 2000 by Chun-gen Liu and the author, and the recent paper [LZu3], various index iteration inequalities were established for any symplectic paths. Especially in the recent [LZu3], the first result on the common index iteration property of any finite family of symplectic paths with positive mean indices, which we call the **common index jump theorem,** was established. All these results form the iteration theory of the index theory for symplectic matrix paths and general Hamiltonian systems without any non-degeneracy conditions. Note that our results on the iteration theory unify and generalize the above mentioned known results obtained before 1990.

By applying this index and its iteration theory to studies on the existence, multiplicity, and stability problems of periodic solutions for nonlinear Hamiltonian systems, we have obtained many interesting results. I believe that this index and its iteration theory possess great potential in solving many other related problems, and thus it is worthy to write such a book to introduce them to more people.

Note that in 1965, V.P. Maslov introduced his celebrated index theory for loops in Lagrangian Grassmanians (cf. [MaF1]). A solid mathematical foundation for this index theory was established by V.I. Arnold in his famous paper [Arn1] of 1967. The main differences between our index theory and the Maslov index are threefold:

(A) The Maslov index is only defined for loops and curves with ends not on the singular cycle. But in our case, the index is also defined for paths with end points on the singular cycle, and the role played by end points of symplectic paths is rather crucial in our applications.

(B) In the case of the Maslov index, the relation theorem with Morse index is only established for Lagrangian systems, a special class of Hamiltonian systems under certain non-degeneracy conditions. But in our case, the relation of the index theory with Morse index of general Hamiltonian systems without any non-degeneracy conditions is established.

(C) In our case, the iteration theory is completely established. To the author's knowledge, such an iteration theory seems to be unknown for the Maslov index theory.

Next a brief introduction is given to the contents of this book. The first four parts of the book are devoted to establish and to study our index and its iteration theory. The last part is devoted to their applications to periodic solution problems of nonlinear Hamiltonian systems.

Part I contains Chapters 1 and 2. This part studies the algebraic and topological aspects of symplectic matrices and groups.

In Chapter 1, the algebraic properties of symplectic matrices are studied. Here the focus is on the eigenvalues, Krein types, and normal forms of symplectic matrices. Based upon the study of the perturbations of normal forms, homotopy components and basic normal forms of symplectic matrices are introduced. Especially a new concept, the ultimate type of ω for M is defined for any $\omega \in \mathbf{U}$ and $M \in \mathrm{Sp}(2n)$. These concepts are fundamental for the study of our index and its iteration theory.

In Chapter 2, the topological aspects of the symplectic group and its ω subsets are studied. Here for any $\omega \in \mathbf{U}$, the ω-singular subset $\mathrm{Sp}(2n)_\omega^0$ of the symplectic group $\mathrm{Sp}(2n)$ consists of all symplectic matrices which possess ω as an eigenvalue, and the ω-regular subset $\mathrm{Sp}(2n)_\omega^*$ is its complement in $\mathrm{Sp}(2n)$. Here the most important topological properties are: $\mathrm{Sp}(2n)$ is homeomorphic to a product of $\mathbf{U}$ and a simply connected topological space; $\mathrm{Sp}(2n)_\omega^*$ contains precisely two path-connected components and is simply connected in $\mathrm{Sp}(2n)$; rotation paths are transversal to $\mathrm{Sp}(2n)_\omega^0$; and the regular part of $\mathrm{Sp}(2n)_\omega^0$ is orientable. These properties form the basis for our index function and its iteration theory.

Part II contains Chapters 3 and 4. In Chapter 3, classical results on canonical transformations and generating functions are introduced. Chapter 4 is devoted to the variational structure of Hamiltonian systems. Here two of the most popular finite dimensional approximation methods are introduced, the Galerkin approximation method and the saddle point reduction method. More attention is put on the saddle point reduction method which gives precise information about the index theory later. This part also makes necessary preparations for our applications in Part V.

Part III contains Chapters 5 to 7, in which an index function theory for symplectic paths is defined and its properties are studied. We call a path $\gamma \in \mathcal{P}_\tau(2n)$ to be *ω-degenerate* if $\gamma(\tau) \in \mathrm{Sp}(2n)^0_\omega$, or *$\omega$-nondegenerate* otherwise. In Chapter 5, the index function theory parametrized by $\omega \in \mathbf{U}$ is established for ω-nondegenerate symplectic paths by a homotopy classification for them (cf. Definition 5.2.8), and for any ω-degenerate symplectic path by a perturbation to a special nearby ω-nondegenerate path which takes the minimal index value in a certain sense (cf. Definition 5.4.2 and Theorem 6.1.8 below) based on our studies in Part I. For any $\omega \in \mathbf{U}$, this index function theory assigns a pair of integers to any $\gamma \in \mathcal{P}_\tau(2n)$ which is denoted by

$$(i_\omega(\gamma), \nu_\omega(\gamma)) \in \mathbf{Z} \times \{0, 1, \ldots, 2n\}.$$

More precisely, the nullity is defined by

$$\nu_\omega(\gamma) = \dim_{\mathbf{C}} \ker_{\mathbf{C}}(\gamma(\tau) - \omega I).$$

Note that γ is ω-degenerate if and only if $\nu_\omega(\gamma) > 0$. The rotation index $i_\omega(\gamma)$ can be viewed as the intersection number of the joint path $\gamma * \xi$ and $\mathrm{Sp}(2n)^0_\omega$:

$$i_\omega(\gamma) = [\mathrm{Sp}(2n)^0_\omega : \gamma * \xi],$$

when γ is ω-nondegenerate, where the path $\xi : [0, \tau] \to \mathrm{Sp}(2n)$ is defined by

$$\xi(t) = \mathrm{diag}\left(2 - \frac{t}{\tau}, \ldots, 2 - \frac{t}{\tau}, \left(2 - \frac{t}{\tau}\right)^{-1}, \ldots, \left(2 - \frac{t}{\tau}\right)^{-1}\right), \quad \forall t \in [0, \tau].$$

When γ is ω-degenerate, we prove in Section 6.1 that $i_\omega(\gamma)$ defined by the perturbation method can also be given by

$$i_\omega(\gamma) = \sup_{U \in \mathcal{N}(\gamma)} \inf\{i_\omega(\beta) \mid \beta \in U, \beta(\tau) \in \mathrm{Sp}(2n)^*_\omega\},$$

where $\mathcal{N}(\gamma)$ is the set of all open neighborhoods of γ in $\mathcal{P}_\tau(2n)$. Note that when $\omega = 1$, the index function is precisely that defined by Conley, Zehnder, Viterbo, and the author.

In Chapter 6, the relation between the index of symplectic paths and the Morse index of the truncated direct variational functional for Hamiltonian systems is established via a saddle point reduction method. An axiomatic characterization of the index function theory is given together with some further important properties of the index theory. This result shows that this index function theory is completely characterized by the homotopy invariance, symplectic additivity and its definition on paths in Sp(2) uniquely.

In Chapter 7, the relation between the index of symplectic paths and other Morse indices is established, including those for the Galerkin approximation method, second order Hamiltonian systems, and general Lagrangian systems.

Part IV, which contains Chapters 8 to 11, is devoted to the study of the index iteration theory. This index iteration theory works also for iterations of closed geodesics on Riemannian manifolds as proved in Chapter 12.

Let $\mathbf{N}$ denote the set of natural numbers. For any path $\gamma \in \mathcal{P}_\tau(2n)$ and $m \in \mathbf{N}$, we define the m-th iteration $\gamma^m \in \mathcal{P}_{m\tau}(2n)$ of γ by

$$\gamma^m(t) = \gamma(t - j\tau)\gamma(\tau)^j, \quad \text{for } j\tau \le t \le (j+1)\tau,\ j = 0, 1, \ldots, m-1.$$

We denote the index and nullity of γ^m by

$$(i(\gamma, m), \nu(\gamma, m)) = (i_1(\gamma^m), \nu_1(\gamma^m)).$$

In Chapter 8, the study of the index for iterations of any symplectic paths is carried out for paths in Sp(2) via the $\mathbf{R}^3$-cylindrical coordinate representation of Sp(2), then for hyperbolic and elliptic paths in Sp($2n$). Based on the study of homotopy components in Section 1.8, this yields the precise iteration formula (Theorem 8.3.1) of the index theory for any symplectic path $\gamma \in \mathcal{P}_\tau(2n)$ in terms of the basic norm form decomposition (cf. Theorem 1.8.10) of $\gamma(\tau)$, $(i(\gamma, 1), \nu(\gamma, 1))$, and the iteration time m.

In Chapter 9, the **splitting numbers** $S_M^{\pm}(\omega)$ of a symplectic matrix M at $\omega \in \mathbf{U}$ are defined topologically by

$$S_M^{\pm}(\omega) = \lim_{\epsilon \to 0^+} i_{\omega \exp(\pm\sqrt{-1}\epsilon)}(\gamma) - i_\omega(\gamma),$$

via any $\gamma \in \mathcal{P}_\tau(2n)$ satisfying $\gamma(\tau) = M$. Then it is proved in Theorem 9.1.7 that the splitting numbers of M at ω coincide with the algebraically defined ultimate type of ω for M in Section 1.8. In this chapter, the Bott-type iteration formulae (Theorem 9.2.1) for any symplectic path is also proved. This formula gives

$$i(\gamma, m) = \sum_{\omega^m = 1} i_\omega(\gamma), \quad \nu(\gamma, m) = \sum_{\omega^m = 1} \nu_\omega(\gamma), \quad \forall m \in \mathbf{N}.$$

It yields especially that the **mean index** of any symplectic path $\gamma \in \mathcal{P}_\tau(2n)$ defined below is always a finite real number:

$$\hat{i}(\gamma, 1) \equiv \lim_{m \to \infty} \frac{i(\gamma, m)}{m} \in \mathbf{R}.$$

Based upon these studies, the abstract precise iteration formula (Theorem 9.3.1) for any symplectic path $\gamma : [0, \tau] \to \mathrm{Sp}(2n)$ in terms of splitting numbers of $\gamma(\tau)$, $i(\gamma, 1)$, and the iteration time m is established. This theorem of C. Zhu and the author [LZu3] yields: *for any path $\gamma \in \mathcal{P}_\tau(2n)$, let $M = \gamma(\tau)$. Then for any $m \in \mathbf{N}$*

we have

$$\begin{aligned} i(\gamma,m) &= m(i(\gamma,1)+S_M^+(1)-C(M)) \\ &\quad +2\sum_{\theta\in(0,2\pi)} E\left(\frac{m\theta}{2\pi}\right) S_M^-(e^{\sqrt{-1}\theta}) - (S_M^+(1)+C(M)), \end{aligned} \tag{5}$$

where $C(M)=\sum_{0<\theta<2\pi} S_M^-(e^{\sqrt{-1}\theta})$ *and* $E(a)=\min\{k\in\mathbf{Z}\,|\,k\ge a\}$ *for* $a\in\mathbf{R}$.

In Chapter 10, based upon results of Chapters 8 and 9, various iteration inequalities of the index theory for any symplectic paths are established. For example, for any $\gamma\in\mathcal{P}_\tau(2n)$ and $m\in\mathbf{N}$, the following typical iteration inequalities always hold.

Estimate via mean index (by C. Liu-Y. Long [LLo1, LLo3]):

$$m\hat{i}(\gamma,1)-n\le i(\gamma,m)\le m\hat{i}(\gamma,1)+n-\nu(\gamma,m). \tag{6}$$

Estimate via initial index (by C. Liu-Y. Long [LLo3]):

$$\begin{aligned} m(i(\gamma,1)+\nu(\gamma,1)-n)+n-\nu(\gamma,1) &\le i(\gamma,m) \\ \le m(i(\gamma,1)+n)-n-(\nu(\gamma,m)-\nu(\gamma,1)). \end{aligned} \tag{7}$$

Successive index estimates (by Y. Long-C. Zhu [LZu3]):

$$\begin{aligned} \nu(\gamma,m)-\frac{e(\gamma(\tau))}{2} &\le i(\gamma,m+1)-i(\gamma,m)-i(\gamma,1) \\ &\le \nu(\gamma,1)-\nu(\gamma,m+1)+\frac{e(\gamma(\tau))}{2}. \end{aligned} \tag{8}$$

Here we define $e(M)$ to be the total multiplicity of eigenvalues of a symplectic matrix M on $\mathbf{U}$ and call it the **elliptic height** of M.

A consequence of the iteration inequality (7) together with the necessary and sufficient conditions for any equality in (7) to hold for some m is a new proof of the following theorem of [DoL1] on *controlling the iteration time m via indices*: *for any* $\gamma\in\mathcal{P}_\tau(2n)$ *and* $m\in\mathbf{N}$,

$$\begin{cases} i(\gamma,m)\le n+1, \\ i(\gamma,1)\ge n, \\ \nu(\gamma,1)\ge 1, \end{cases} \quad\Rightarrow\quad m=1. \tag{9}$$

In Chapter 11, the common index jump theorem of C. Zhu and the author [LZu3] as a common property of any given finite family of symplectic paths $\gamma_j\in\mathcal{P}_{\tau_j}(2n)$ with $1\le j\le q$ is proved. For any $\gamma:[0,\tau]\to\mathrm{Sp}(2n)$, its m**-th index jump** $\mathcal{G}_m(\gamma)$ is defined to be the open interval

$$\mathcal{G}_m(\gamma)=(i(\gamma,m)+\nu(\gamma,m)-1,\, i(\gamma,m+2)). \tag{10}$$

Suppose all these paths possess positive mean indices and large enough initial indices:

$$\hat{i}(\gamma_j,1)>0,\quad i(\gamma,1)\ge n,\qquad 1\le j\le q.$$

Then there are infinitely many possibilities that these index jumps match together nicely and reach their own largest sizes. More precisely, there exist infinitely many positive integer tuples $(N, m_1, \ldots, m_q) \in \mathbf{N}^{q+1}$ such that

$$\emptyset \neq [2N - \kappa_1, 2N + \kappa_2] \subset \bigcap_{j=1}^{q} \mathcal{G}_{2m_j-1}(\gamma_j), \tag{11}$$

for some positive constants κ_1 and κ_2 independent of j given by

$$\begin{aligned} \kappa_1 &= \min_{1\le j\le q} (i(\gamma_j, 1) + 2S^+_{\gamma_j(\tau_j)}(1) - \nu(\gamma_j, 1)), \\ \kappa_2 &= \min_{1\le j\le q} i(\gamma_j, 1) - 1. \end{aligned}$$

In order to prove this theorem, we need to complete two steps: to make each index jump to be as big as possible, and to make their largest sizes happen simultaneously to guarantee the existence of a non-empty common intersection interval among them and the size of this interval to be as large as possible. By the term $E(\frac{m\theta}{2\pi})$ in the abstract iteration formula (5), such proofs are reduced to a dynamics problem in a torus. Then this problem is solved by properties of closed additive subgroups of tori.

In Chapter 12, the classical Morse index theory of closed geodesics on Riemannian manifolds is briefly reviewed. Then it is proved that the splitting numbers for closed geodesics coincides completely with the splitting numbers of Definition 9.1.4 for symplectic matrices. This then proves that the iteration theory we have established in Chapters 8 to 11 for symplectic matrix paths works completely for the Morse index theory of closed geodesics.

Part V contains Chapters 13 to 15. In this part, three important problems on periodic solutions of nonlinear Hamiltonian systems are studied as applications of our index and its iteration theory.

Let $T > 0$ and suppose x is a non-constant T-periodic solution of the nonlinear Hamiltonian system

$$\dot{x} = JH'(x). \tag{12}$$

Suppose the minimal period of x is $\tau = T/k$ for some $k \in \mathbf{N}$. Denote by γ_x the fundamental solution of the linearized Hamiltonian system at x:

$$\dot{y} = JH''(x(t))y. \tag{13}$$

Then $\gamma_x \in \mathcal{P}_\tau(2n)$, and is called the **associated symplectic path** of x. Note that

$$\gamma_x|_{[0,m\tau]} = (\gamma_x|_{[0,\tau]})^m.$$

The iterated index of x is defined by

$$(i(x, m), \nu(x, m)) = (i_1(\gamma_x|_{[0,m\tau]}), \nu_1(\gamma_x|_{[0,m\tau]})), \quad \forall m \in \mathbf{N}.$$

This yields a way to study iterations x^m of x for $m \in \mathbf{N}$ defined by

$$x^m(t) = x(t - j\tau), \qquad j\tau \le t \le (j+1)\tau,\ j = 0, 1, \ldots, m-1.$$

In [Rab2] of 1978, P. Rabinowitz posed a conjecture on whether the Hamiltonian system possesses periodic solutions with prescribed minimal period when the Hamiltonian function satisfies his superquadratic conditions. This conjecture is studied in Chapter 13 as an application of the index iteration theory. Note that for a non-constant τ-periodic solution x of (12), the condition on the nullity in (9) always holds because the system (12) is autonomous. Thus the above estimate (9) in Chapter 10 can be rephrased as follows:

Theorem A *For any non-constant τ-periodic solution x of* (12), *denote its minimal period by τ/m for some $m \in \mathbf{N}$. Suppose*

$$i(x|_{[0,\tau]}, 1) \le n+1, \quad n \le i(x|_{[0,\tau/m]}, 1). \tag{14}$$

Then $m = 1$, i.e., τ is the minimal period of x.

Here the first estimate in (14) holds if x is obtained by minimax or minimization methods, and the second estimate in (14) holds if the Hamiltonian function H is convex in a certain weak sense along the orbit of x. This result was proved by D. Dong and the author in [DoL1] of 1997. It reveals the intrinsic relationship between the minimal period of a periodic solution and its indices, and unifies all the results obtained so far on Rabinowitz's conjecture under various convexity conditions.

In 1984, C. Conley stated a conjecture on whether the Poincaré map of any 1-periodic time dependent Hamiltonian system defined on the standard torus T^{2n} always possesses infinitely many periodic points which are produced by contractible periodic solutions of the corresponding Hamiltonian system on T^{2n}. A celebrated progress on this conjecture was made by D. Salamon and E. Zehnder in [SaZ2] of 1992 for a large class of symplectic manifolds (cf. Theorem 14.6.1 below). Restricted to the standard torus, Conley's conjecture was proved under the condition that every contractible solution of the Hamiltonian system has at least one Floquet multiplier not equal to 1. This result answers partially this Conley conjecture. So far the Conley conjecture is still open and seems far from being completely understood.

In Chapter 14, we study the Lagrangian system version of this conjecture. The main theorem proved in this chapter is the following:

Theorem B *The Poincaré map of a Lagrangian system on the standard n-dimensional torus T^n always possesses infinitely many periodic points which are produced by contractible integer periodic solutions of this system on T^n, provided the Lagrangian function satisfies the Legendre condition (L1) in Chapter 14 and is at least C^3.*

This result was proved in [Lon17] of 2000. Here the above mentioned estimate (6) is one of the crucial ingredients in the proof. By this estimate, a global homological injection map can be constructed which maps a certain non-trivial local critical group to a nontrivial homology class $[\sigma]$ in a global homology group, if the number of contractible integer periodic solutions of the related Hamiltonian system is finite. But on the other hand, by a technique of V. Bangert and W. Klingenberg in [BaK1], it is shown that this homology class $[\sigma]$ must be trivial globally. This contradiction then yields the conclusion of Theorem B.

In Chapter 15, closed characteristics on compact strictly convex C^2 hypersurfaces in $\mathbf{R}^{2n}$ are studied. Denote the set of all such hypersurfaces by $\mathcal{H}(2n)$. Let $\Sigma \in \mathcal{H}(2n)$. For $x \in \Sigma$, let $N_\Sigma(x)$ be the outward normal unit vector at x of Σ. We consider the problem of finding $\tau > 0$ and a C^1 curve $x : [0, \tau] \to \mathbf{R}^{2n}$ such that

$$\begin{cases} \dot{x}(t) = JN_\Sigma(x(t)), \quad x(t) \in \Sigma, \quad \forall t \in \mathbf{R}, \\ x(\tau) = x(0). \end{cases} \tag{15}$$

A solution (τ, x) of the problem (15) is called a *closed characteristic* on Σ. Two closed characteristics (τ, x) and (σ, y) are *geometrically distinct*, if $x(\mathbf{R}) \neq y(\mathbf{R})$. We denote by $\mathcal{J}(\Sigma)$ and $\tilde{\mathcal{J}}(\Sigma)$ the set of all closed characteristics (τ, x) on Σ with τ being the minimal period of x and the set of all geometrically distinct ones respectively. For $(\tau, x) \in \mathcal{J}(\Sigma)$, we denote by $[(\tau, x)]$ the set of all elements in $\mathcal{J}(\Sigma)$ which are geometrically the same as (τ, x). ${}^\#A$ denotes the total number of elements in a set A. This problem has been studied for more than 100 years. A long standing conjecture on the multiplicity of closed characteristics is whether

$${}^\#\tilde{\mathcal{J}}(\Sigma) \geq n, \qquad \forall\, \Sigma \in \mathcal{H}(2n).$$

In Chapter 15, the following answers to this conjecture are given via a new shape invariant $\varrho_n(\Sigma)$ defined by (15.4.1) below for any $\Sigma \in \mathcal{H}(2n)$. This $\varrho_n(\Sigma)$ takes values in $\mathbf{N} \cup \{+\infty\}$:

Theorem C *There holds*

$${}^\#\tilde{\mathcal{J}}(\Sigma) \geq \varrho_n(\Sigma), \quad \forall \Sigma \in \mathcal{H}(2n). \tag{16}$$

Especially a lower bound of $\varrho_n(\Sigma)$ *is given by*

$$\varrho_n(\Sigma) \geq \left[\frac{n}{2}\right] + 1, \quad \forall \Sigma \in \mathcal{H}(2n), \tag{17}$$

where $[a] = \max\{k \in \mathbf{Z} \mid k \leq a\}$ *for any* $a \in \mathbf{R}$.

Moreover, for a $\Sigma \in \mathcal{H}(2n)$, *if all the closed characteristics on* Σ *are non-degenerate, then*

$$\varrho_n(\Sigma) \geq n, \tag{18}$$

or for a $\Sigma \in \mathcal{H}(2n)$, *if* Σ *is symmetric with respect to the origin, i.e.,* $x \in \Sigma$ *implies* $-x \in \Sigma$, *then*

$${}^\#\tilde{\mathcal{J}}(\Sigma) \geq n. \tag{19}$$

A long standing conjecture on the stability of closed characteristics is whether there always exists at least an elliptic closed characteristic on any $\Sigma \in \mathcal{H}(2n)$. Here a $(\tau, x) \in \mathcal{J}(\Sigma)$ is elliptic if its all Floquet multipliers are on $\mathbf{U}$ when (τ, x) is viewed as a periodic solution of a corresponding Hamiltonian system. The following partial results on this conjecture are given in Chapter 15:

Theorem D *For $\Sigma \in \mathcal{H}(2n)$, suppose*

$$^{\#}\tilde{\mathcal{J}}(\Sigma) < +\infty. \tag{20}$$

Then there exists at least an elliptic closed characteristic on Σ, and there exists at least $\varrho_n(\Sigma) - 1$ ($\geq [\frac{n}{2}]$) geometrically distinct closed characteristics on Σ possessing irrational mean indices.

Moreover, suppose

$$n \geq 2, \qquad ^{\#}\tilde{\mathcal{J}}(\Sigma) \leq 2(\varrho_n(\Sigma) - 1). \tag{21}$$

Then there exist at least two elliptic elements in $\tilde{\mathcal{J}}(\Sigma)$.

All the above results in Theorems C and D on the multiplicity and stability of closed characteristics were proved by Y. Long-C. Zhu in [LZu3] recently except the result (19) when Σ is symmetric with respect to the origin, which was proved by C. Liu-Y. Long-C. Zhu in [LLZ1]. More historical remarks on the problem (15) can be found in Chapter 15.

Note that each closed characteristic $[(\tau, x)]$ on Σ corresponds to infinitely many critical values in the whole critical value sequence $\{c_k\}$ of the related dual action functional as studied by I. Ekeland in Section V.3 of [Eke3]. But it is very difficult to extract information from such subsequences of $\{c_k\}$. In order to count the number of closed characteristics on Σ, in stead of studying $\{c_k\}$, our method is to study how the index intervals of iterated closed characteristics cover the set of integers $2\mathbf{N} - 2 + n$. In the proof of the multiplicity claims (16)–(18) of Theorem C, this method is realized via the following most important ingredients, the common index jump theorem (11) and the following observations:

$$\begin{aligned} ^{\#}\tilde{\mathcal{J}}(\Sigma) &\geq q \\ &\geq {}^{\#}((2\mathbf{N} - 2 + n) \cap \cap_{j=1}^{q} \mathcal{G}_{2m_j-1}(\gamma_{x_j})) \\ &\geq {}^{\#}((2\mathbf{N} - 2 + n) \cap [2N - \kappa_1, 2N + \kappa_2]) \\ &\geq \varrho_n(\Sigma), \end{aligned} \tag{22}$$

where q is the number of infinitely variationally visible closed characteristics on Σ given by Definition 15.3.3. Our proof of the second inequality in (22) uses the Lyusternik–Schnirelman theoretical argument, i.e., the Fadell–Rabinowitz S^1-cohomology index theory, and those of the last two inequalities in (22) use the Morse theoretical approach. Here the common index jump theorem picks up enough points of $2\mathbf{N} - 2 + n$ in the interval

$$[2N - \kappa_1, 2N + \kappa_2] \subset \cap_{j=1}^{q} \mathcal{G}_{2m_j-1}(\gamma_{x_j}),$$

and this number yields a lower bound for ${}^{\#}\tilde{\mathcal{J}}(\Sigma)$. Because the invariant $\varrho_n(\Sigma)$ is defined via the initial indices and splitting numbers of closed characteristics on Σ, our previous studies on the index theory produce the estimates (16)–(18) on the number ${}^{\#}\tilde{\mathcal{J}}(\Sigma)$.

The estimate (19) on ${}^{\#}\tilde{\mathcal{J}}(\Sigma)$ for any symmetric hypersurface Σ in $\mathcal{H}(2n)$ depends on the common index jump theorem and two separate lower bound estimates on the numbers of two different families of closed characteristics. One family contains the symmetric orbits, i.e., those $[(\tau, x)]$'s which satisfy $x(\mathbf{R}) = -x(\mathbf{R})$. The other one contains asymmetric orbits, i.e., those $[(\tau, x)]$'s which satisfy $x(\mathbf{R}) \cap -x(\mathbf{R}) = \emptyset$. The sum of the two lower bound estimates is at least n.

The proofs for the stability results depend further on an enhanced version of the iteration estimate (8) on the elliptic height. This result relies also on the common index jump theorem which yields precise iterated index information about the solutions found by variational methods.

At the end of Chapter 15, a result in [DDE1] of G.F. Dell'Antonio, B. D'Onofrio, and I. Ekeland proved in 1992 is also introduced, which shows that there is always an elliptic closed characteristic on any $\Sigma \in \mathcal{H}(2n)$ provided it is symmetric with respect to the origin.

For the background material on critical point theory and convex Hamiltonian systems needed for this book, readers are refereed to the excellent books [Cha6] of K.C. Chang, [Eke3] of I. Ekeland, [Gho2] of N. Ghoussoub, and [Rab7] of P. Rabinowitz. This allows the part of this book on applications to be kept at a reasonable size, and results in less of an overlap with other books.

In each section of this book, theorems, definitions, propositions et al are numbered together sequentially. Within the same section they are refereed to by this number. If it is in another section, or another chapter, the corresponding section number as well as the chapter number are also recalled. For example, Theorem 1 means the first theorem in the same section. Theorem 2.1 means Theorem 1 in Section 2 of the same chapter, and Theorem 3.2.1 means Theorem 1 in Section 2 of Chapter 3. Equations are numbered in a similar way. Note that equations in this Introduction are numbered as in Section 0 of Chapter 0 in the text below. For example, equation (16) in this Introduction is denoted by equation (0.0.16) when it is quoted in the text of this book.

At the end of each section and the Notes and Comments of the last three chapters, historical remarks and further comments on results mentioned in the text are made. These remarks may not be complete, because of the limitation of the author's knowledge and the main interest of the book.

Part I
The Symplectic Group Sp(2*n*)

Chapter 1

Algebraic aspects

In the first three sections of this chapter, we study linear symplectic algebra, especially the algebraic properties of symplectic matrices, their eigenvalues and the Krein type of eigenvalues on the unit circle in the complex plane. In Sections 4 to 7, we derive normal forms of symplectic matrices according to their eigenvalues. In Sections 8 and 9, we introduce the homotopy component, basic normal forms, and the ultimate type of symplectic matrices, and study their basic properties.

Systematic studies of symplectic matrices, symplectic groups, and linear Hamiltonian systems were done by M. Krein, I. Gelfand, V. Lidskii, and J. Moser in the 1950s independently for the stability problem. These works are summarized in the book [YaS1] of V. Yakubovich and V. Starzhinski.

The basic properties of symplectic matrices described in Sections 1 and 3 can be found in the pioneering works [GeL1] of I. Gelfand and V. Lidskii and [Mos1] of J. Moser, and the books [Eke3] of I. Ekeland, [MoZ1] of J. Moser and E. Zehnder, [Wey1] of H. Weyl, and [YaS1] of V. Yakubovich and V. Starzhinski. Studies of the Krein type of eigenvalues of symplectic matrices on $\mathbf{U}$ in Section 3 are taken from [YaS1] and [Eke3] respectively. Here I restrict myself to only what is needed for our later usage in this book. For other properties of the Krein type numbers we refer to the books [YaS1] and [Eke3].

1.1 Symplectic matrices

In this section, we shall define the symplectic matrix, and study its basic algebraic properties. In this book, we denote by $\mathbf{N}$, $\mathbf{Z}$, $\mathbf{R}$, and $\mathbf{C}$ the sets of all natural, integral, real and complex numbers respectively. By $\mathbf{U}$ we denote the unit circle in the complex plane $\mathbf{C}$. For a given matrix M with entries in $\mathbf{R}$ or $\mathbf{C}$, we denote the transpose of M by M^T, and the complex conjugate of M^T by M^*. For any $n \in \mathbf{N}$ and any field K, denote by K^n the linear product vector space formed by all the column vectors of the form $x = (x_1, \ldots, x_n)^T$ with $x_i \in K$. Let $\mathcal{L}(K^n)$ be the group of all $n \times n$ matrices with entries in the field K under the usual matrix multiplication. In this book, we are only interested in two cases of the real number field $K = \mathbf{R}$ and the complex number field $K = \mathbf{C}$. Identifying with the n^2-dimensional topological product space K^{n^2}, the group $\mathcal{L}(K^n)$ becomes a topological group. Any of its subgroups, equipped with the induced topology, also becomes a topological group. Denote by $\mathcal{L}_s(K^n)$ the symmetric matrix subgroup of $\mathcal{L}(K^n)$. As usual, the orthogonal group $O(n, \mathbf{R})$ is the subgroup of $\mathcal{L}(\mathbf{R}^n)$ formed by those elements M satisfying $MM^T = I$. The unitary group $U(n, \mathbf{C})$ is the subgroup of $\mathcal{L}(\mathbf{C}^n)$ formed by those elements M satisfying $MM^* = I$.

Definition 1 A matrix $M \in \mathcal{L}(\mathbf{R}^{2n})$ is a **symplectic matrix**, if

$$M^T J M = J, \tag{1}$$

where $J = \begin{pmatrix} 0 & -I_n \\ I_n & 0 \end{pmatrix}$ is called the standard symplectic matrix on $\mathbf{R}^{2n}$. $I_n \in \mathcal{L}(\mathbf{R}^n)$ is the identity matrix. When there is no confusion we shall omit the subindex n. We denote the zero matrix by 0. All the $2n \times 2n$ symplectic matrices form a subgroup of $\mathcal{L}(\mathbf{R}^{2n})$, which we denote by $\mathrm{Sp}(2n, \mathbf{R})$ and call it the symplectic group. When there is no confusion, we simply denote it by $\mathrm{Sp}(2n)$ in this book.

Denote the inner product in $\mathbf{R}^{2n}$ by $a \cdot b$. Then a matrix $M \in \mathcal{L}(\mathbf{R}^{2n})$ is symplectic, if and only if

$$JMa \cdot Mb = Ja \cdot b, \qquad \forall\, a, b \in \mathbf{R}^{2n}.$$

From the definition, every symplectic matrix satisfies $(\det M)^2 = 1$. Especially, all the symplectic matrices are non-singular. Later $\det M = 1$ will be further proved.

Consider a 2×2 real matrix

$$M = \begin{pmatrix} a & b \\ c & d \end{pmatrix}.$$

By Definition 1, the matrix M is symplectic if and only if $ad - bc = 1$. This result has the following generalization.

Lemma 2 *Suppose a $2n \times 2n$ matrix M has the square block form*

$$M = \begin{pmatrix} A & B \\ C & D \end{pmatrix}.$$

1° *The matrix M is symplectic if and only if both $A^T C$ and $B^T D$ are symmetric, and $A^T D - C^T B = I$.*

2° *Especially when $C = 0$, the matrix M is symplectic if and only if A is non-singular, and M can be represented in the form*

$$M = \begin{pmatrix} A & 0 \\ 0 & (A^T)^{-1} \end{pmatrix} \begin{pmatrix} I & E \\ 0 & I \end{pmatrix},$$

where E is a symmetric matrix.

Proof: 1° follows from the definition. From 1° we get $A^T D = I$. Let $E = A^{-1} B$ from which we obtain 2°. □

Similar to the polar decomposition of complex numbers, symplectic matrices have also their polar decompositions. The following lemma yields this result.

Lemma 3 *Any symplectic matrix $M \in \mathrm{Sp}(2n)$ can be represented in the form*

$$M = AU, \tag{2}$$

where A is a symmetric symplectic positive definite matrix, U is a symplectic orthogonal matrix, and they are uniquely determined by M.

Proof: Let $A = \sqrt{MM^T}$. Then A is symmetric and non-negative definite. Since M is non-singular, A is positive definite. Let $U = A^{-1}M$. Then (2) holds, and from

$$UU^T = A^{-1}MM^TA^{-1} = A^{-1}A^2A^{-1} = I,$$

we obtain that U is orthogonal.

If M possesses two polar decompositions $M = A_1U_1 = A_2U_2$, then $U_1^TA_1 = U_2^TA_2$. Thus

$$A_1^2 = A_1U_1U_1^TA_1 = A_2U_2U_2^TA_2 = A_2^2,$$

and then $A_1 = A_2$. This implies $U_1 = U_2$ and proves the uniqueness of the decomposition (2).

By the definition of a symplectic matrix, $M = J^{-1}(M^T)^{-1}J$. Thus by (2), we obtain

$$M = J^{-1}A^{-1}(U^T)^{-1}J = J^{-1}A^{-1}JJ^{-1}(U^T)^{-1}J.$$

By the uniqueness of the decomposition (2) just proved we obtain $A = J^{-1}A^{-1}J$ and $U = J^{-1}(U^T)^{-1}J$. Thus both A and U are symplectic. □

It is well known that every positive definite symmetric matrix P can be uniquely represented by

$$P = \exp(M),$$

where M is a real symmetric matrix, and

$$\exp(M) = I + M + \frac{1}{2!}M^2 + \frac{1}{3!}M^3 + \dots,$$

M is called the logarithm matrix of P. Especially the following result holds for symplectic matrices.

Lemma 4 *A positive definite symmetric matrix $P \in \mathcal{L}(\mathbf{R}^{2n})$ is symplectic, if and only if it has the form*

$$P = \exp(M),$$

where $M = \begin{pmatrix} A & B \\ B & -A \end{pmatrix}$, A and B are symmetric $n \times n$ matrices.

Proof: *Necessity*. Since $\exp(M) = P$ is symplectic,

$$\exp(M) = J^{-1}(\exp(M))^{-1}J = \exp(-J^{-1}MJ).$$

Since $-J^{-1}MJ$ is real symmetric, by the uniqueness of the above representation we obtain

$$M = -J^{-1}MJ. \tag{3}$$

Setting $M = \begin{pmatrix} A & B \\ C & D \end{pmatrix}$ in (3) yields $D = -A$ and $B = C$. On the other hand, by the symmetry of M, we obtain $B^T = C^T = B$ and $A^T = A$. The necessity is proved.

Sufficiency. $P = \exp(M)$, $M = \begin{pmatrix} A & B \\ B & -A \end{pmatrix}$ together with the symmetries of A and B imply (3) and the symmetry of M. Thus P is symplectic. □

Lemma 5 1° *An orthogonal matrix U is symplectic, if and only if it has the form*

$$U = \begin{pmatrix} A & -B \\ B & A \end{pmatrix}, \tag{4}$$

where A^TB is symmetric, and $A^TA + B^TB = I$.

2° *These conditions are also the necessary and sufficient conditions such that $A - \sqrt{-1}B$ and $A + \sqrt{-1}B$ are unitary matrices.*

Proof: 1° *Sufficiency*. This follows by direct computations.

Necessity. By the given conditions it holds that $U^TKU = K$, where $K = I$ or J. Thus it also holds for $K = I + \sqrt{-1}J$. Let

$$U = \begin{pmatrix} A & B \\ C & D \end{pmatrix}.$$

Then

$$\begin{pmatrix} A^T & C^T \\ B^T & D^T \end{pmatrix}\begin{pmatrix} I & \sqrt{-1}I \\ -\sqrt{-1}I & I \end{pmatrix}\begin{pmatrix} A & B \\ C & D \end{pmatrix} = \begin{pmatrix} I & \sqrt{-1}I \\ -\sqrt{-1}I & I \end{pmatrix}.$$

Comparing both sides we obtain

$$E^*E = F^*F = I, \qquad E^*F = \sqrt{-1}I, \tag{5}$$

where $E = A + \sqrt{-1}C$, $F = B + \sqrt{-1}D$, E^* denotes the complex conjugate of E^T. This implies

$$E^*(E + \sqrt{-1}F) = 0.$$

By (5) we obtain that E^* is non-singular. Thus $E + \sqrt{-1}F = 0$. This proves $D = A$ and $B = -C$.

The symmetry of A^TB and $A^TA + B^TB = I$ follow from the orthogonality of U immediately.

The proof of 2° is similar, and is left to the readers. □

The above lemma establishes actually an isomorphism between groups

$$\mathrm{Sp}(2n, \mathbf{R}) \bigcap O(2n, \mathbf{R}) \cong U(n, \mathbf{C}).$$

Note that in the sense of the topological groups, this is also a homeomorphism.

Define the $2n \times 2n$ complex matrix

$$T = \begin{pmatrix} I & \sqrt{-1}I \\ I & -\sqrt{-1}I \end{pmatrix}.$$

Then the matrix U in Lemma 5 is transformed to

$$TUT^{-1} = \begin{pmatrix} A + \sqrt{-1}B & 0 \\ 0 & A - \sqrt{-1}B \end{pmatrix}. \tag{6}$$

Lemma 6 *For any symplectic matrix M,*

$$\det M = 1. \tag{7}$$

Proof: By Lemma 3, the matrix M has its polar decomposition $M = PU$, where P is symplectic symmetric positive definite and U is symplectic orthogonal. Then

$$\det M = \det P \det U.$$

By (6) we obtain

$$\det U = \det(TUT^{-1}) = \det[(A + \sqrt{-1}B)(A + \sqrt{-1}B)^*] = 1,$$

where $A^* = A$, $B^* = B$, and $A + \sqrt{-1}B$ are unitary. Since P is symplectic, $(\det P)^2 = 1$. But P is positive definite, then $\det P = 1$. Therefore (7) holds. □

Our above proof shows that the determinant of every symplectic matrix is 1. More importantly we shall see in Chapter 2 that following this discussion the global topological structure of the symplectic group can be obtained.

1.2 Symplectic spaces

In this section we briefly introduce the theory of symplectic spaces.

Definition 1 Suppose V is a real vector space of dimension m, and ω is an antisymmetric nondegenerate bilinear form defined on V, i.e., ω satisfies

$$\omega(u, v) = -\omega(v, u), \qquad \forall u, v \in V, \tag{1}$$

$$\omega(u, v) = 0 \quad \text{for some } u \in V \text{ and any } v \in V \text{ implies } u = 0. \tag{2}$$

Then ω is called a **symplectic form** defined on V, and (V, ω) is a **symplectic space**.

Let (V, ω) be a symplectic space. $u, v \in V$ are **symplectically orthogonal** and denoted by $u \perp v$, if $\omega(u, v) = 0$. For $A, B \subset V$, we define $A \perp B$ if $u \perp v$ for all $u \in A$ and $v \in B$. By the skew-symmetry of ω, it holds that $v \perp v$ for every $v \in V$. Let E be a linear subspace of V. Define

$$E^{\perp} = \{u \in V \mid \omega(u, v) = 0, \forall v \in E\}. \tag{3}$$

Then $E^{\perp}$ is a linear subspace of V, and

$$(E^{\perp})^{\perp} = E.$$

By the non-degeneracy of ω,

$$\dim E + \dim E^{\perp} = \dim V. \tag{4}$$

Definition 2 A linear subspace E of the symplectic space (V, ω) is

isotropic,	if	$E \subset E^{\perp}$,
coisotropic,	if	$E^{\perp} \subset E$,
Lagrangian,	if	$E = E^{\perp}$,
symplectic,	if	$E \cap E^{\perp} = \{0\}$.

If E is a subspace of (V, ω), then the restriction of ω on E is still a symplectic form if and only if $E \cap E^{\perp} = \{0\}$. This is the reason why such an E is called a symplectic subspace of (V, ω). Clearly, every 1-dimensional subspace of (V, ω) is isotropic.

Lemma 3 *Let (V, ω) be a symplectic space. Then it must be even dimensional. Denote its dimension by $2n$. Then there exists a base $\{v_1, \ldots, v_{2n}\}$ such that*

$$\omega(v_i, v_{n+i}) = 1, \qquad \text{for} \quad 1 \le i \le n, \tag{5}$$

$$\omega(v_i, v_j) = 0, \qquad \text{otherwise}. \tag{6}$$

The base satisfying these conditions is called a symplectic base of (V, ω).

Proof: Fix a non-zero vector $v_1 \in V$. By the non-degeneracy of ω, there exists a vector $v_{n+1} \in V$ such that $\omega(v_1, v_{n+1}) = 1$. Thus v_1 and v_{n+1} are linearly independent. If $m \equiv \dim V \le 2$, the lemma is proved. If $m > 2$, let $E = \operatorname{span}\{v_1, v_{n+1}\}$. Then both of E and $E^{\perp}$ are symplectic subspaces of (V, ω). Since $\dim E^{\perp} = m - 2 > 0$, by an induction argument on the dimension the proof is complete. □

The most popular example is the so-called **standard symplectic space** $(\mathbf{R}^{2n}, \omega)$ with ω defined by the standard symplectic matrix J introduced in Section 1.1:

$$\omega(u, v) = Ju \cdot v, \qquad \forall u, v \in \mathbf{R}^{2n}. \tag{7}$$

Here ω is the standard symplectic form on $\mathbf{R}^{2n}$. In terms of the notation in the exterior algebra, if we introduce the coordinates $(p_1, \ldots, p_n, q_1, \ldots, q_n)$ on $\mathbf{R}^{2n}$, the symplectic form ω can be represented by

$$\omega = dp \wedge dq = \sum_{i=1}^{n} dp_i \wedge dq_i. \tag{8}$$

Let $\{v_1, \ldots, v_{2n}\}$ be a symplectic base of $(\mathbf{R}^{2n}, \omega)$. For any x and $y \in \mathbf{R}^{2n}$, let

$$x = \sum_{i=1}^{2n} x_i v_i \equiv (x_1, \ldots, x_{2n})^T,$$

$$y = \sum_{i=1}^{2n} y_i v_i \equiv (y_1, \ldots, y_{2n})^T.$$

Then

$$\omega(x, y) = Jx \cdot y.$$

Especially,

$$v_i^T J v_{n+i} = J v_{n+i} \cdot v_i = -\omega(v_i, v_{n+i}) = -1, \quad \text{for} \quad 1 \le i \le n, \tag{9}$$

$$v_i^T J v_j = J v_j \cdot v_i = -\omega(v_i, v_j) = 0, \quad \text{otherwise.} \tag{10}$$

Corollary 4 *Let (V_1, ω_1) and (V_2, ω_2) be two symplectic spaces with the same dimension $2n$. Then there exists an invertible linear map $T : V_1 \to V_2$ such that*

$$\omega_2(Tu, Tv) = \omega_1(u, v), \quad \forall u, v \in V_1. \tag{11}$$

Proof: For $i = 1$ and 2, let $(v_1^i, \ldots, v_{2n}^i)$ be a symplectic base of (V_i, ω_i). Define

$$T v_k^1 = v_k^2, \qquad \forall k = 1, \ldots, 2n.$$

Then extending T linearly to the whole V_1 completes the proof. □

In the sense of this corollary, all the symplectic spaces of the same dimension are equivalent. Therefore the study on the general symplectic spaces can be reduced to that on $(\mathbf{R}^{2n}, \omega)$ with no loss of generality.

1.3 Eigenvalues of symplectic matrices

Denote by $\mathbf{U} = \{z \in \mathbf{C} \mid |z| = 1\}$ the unit circle on the complex plane $\mathbf{C}$. For any $n \times n$ matrix (real or complex), as usual we denote by $\sigma(M)$ the spectrum of M, i.e., the set of all $\lambda \in \mathbf{C}$ such that

$$\det(M - \lambda I) = 0.$$

Suppose $\lambda \in \sigma(M)$. The **geometric multiplicity** of $\lambda \in \sigma(M)$ is defined to be $\dim_{\mathbf{C}} \ker_{\mathbf{C}}(M - \lambda I)$. Note that when M is real and $\lambda \in \mathbf{R}$,

$$\dim_{\mathbf{C}} \ker_{\mathbf{C}}(M - \lambda I) = \dim_{\mathbf{R}} \ker_{\mathbf{R}}(M - \lambda I).$$

Denote the complex root vector space of M belonging to λ by

$$E_\lambda(M) = \cup_{k \geq 1} \ker_{\mathbf{C}}(M - \lambda I)^k \subset \mathbf{C}^{2n}. \tag{1}$$

The **algebraic multiplicity** of $\lambda \in \sigma(M)$ is defined to be $\dim_{\mathbf{C}} E_\lambda(M)$.

Fix any $M \in \mathrm{Sp}(2n)$. In order to determine eigenvalues of M, for $\lambda \in \mathbf{C}$ we consider its characteristic polynomial

$$f_M(\lambda) = \det(M - \lambda I). \tag{2}$$

This is a polynomial of degree $2n$ in λ with real coefficients. Since M is symplectic, we have

$$f_M(\lambda) = \det M \det(I - \lambda J^{-1} M^T J) = \det(I - \lambda M^T).$$

Thus

$$f_M(\lambda) = \lambda^{2n} f_M(\lambda^{-1}). \tag{3}$$

Therefore $f_M(\lambda)$ can be written in the symmetric form

$$f_M(\lambda) = \lambda^n \sum_{k=0}^{n} a_k(\lambda^k + \lambda^{-k}), \tag{4}$$

where $a_n = 1$, $a_k = \overline{a}_k$, $k = 0, 1, \ldots, n$, $\overline{a}_k$ denotes the complex conjugate of a_k.

From these derivations we obtain

Theorem 1 1° *If* $\lambda \in \sigma(M)$ *and* $\mathrm{Im}\lambda \cdot (|\lambda|^2 - 1) \neq 0$, *then* λ, $\overline{\lambda}$, λ^{-1}, *and* $(\overline{\lambda})^{-1}$ *belong to* $\sigma(M)$, *and possess the same geometric and algebraic multiplicities.*

2° *If* $\lambda \in \sigma(M)$ *and* $\lambda \in \mathbf{R}\backslash\{\pm 1\}$, *then* λ, $\lambda^{-1} \in \sigma(M)$, *and possess the same geometric and algebraic multiplicities.*

3° *If* $\lambda \in \sigma(M)$ *and* $|\lambda| = 1$, $\lambda^2 \neq 1$, *then* λ, $\overline{\lambda} \in \sigma(M)$ *and possess the same geometric and algebraic multiplicities.*

4° *If* $+1$ *or* $-1 \in \sigma(M)$, *then its algebraic multiplicity is even.*

Proof: 1° to 3° follow from (3) and (4). We prove 4° next. From 1°–3° we obtain that the total algebraic multiplicity of all eigenvalues in $\sigma(M)$ not equal to ± 1 is an even integer, say $2k$. Thus the total algebraic multiplicity of $+1$ and -1 is $2(n-k)$. Since $\det M = 1$, the algebraic multiplicity of -1 and then that of $+1$ must be even. □

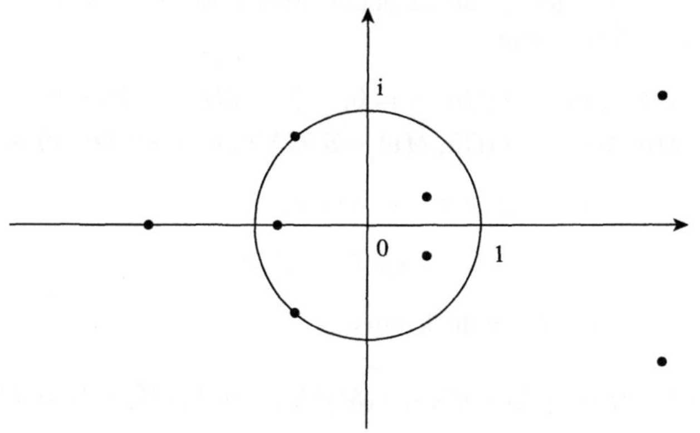

Figure 1: *The distribution of eigenvalues of symplectic matrices*

Figure 1 gives a description of the distribution of eigenvalues of symplectic matrices.

Following M. Krein, we define

$$G = \sqrt{-1}J \in \mathcal{L}(\mathbf{C}^{2n}). \tag{5}$$

Denote the inner product in $\mathbf{C}^k$ for any $k \in \mathbf{N}$ by $\langle \cdot, \cdot \rangle$. With the aid of the matrix G, the **Krein form** is defined by

$$\langle Gx, y \rangle \qquad \forall\, x, y \in \mathbf{C}^{2n}. \tag{6}$$

Recall that a subspace $E \subset \mathbf{C}^{2n}$ is an **invariant subspace** of M if $ME \subset E$. Then the above $E_\lambda(M)$ is invariant under M. An invariant subspace E is **irreducible**, if it is not a direct sum of two non-trivial invariant subspaces. Then any irreducible invariant subspace is contained in some $E_\lambda(M)$ and every $E_\lambda(M)$ is a direct sum of irreducible invariant subspaces. An eigenvalue λ of M is called **semi-simple**, if each irreducible invariant subspace of $E_\lambda(M)$ is one dimensional, i.e., all the elementary divisors belonging to λ of the characteristic polynomial of M are simple.

Lemma 2 *For $M \in \mathrm{Sp}(2n)$, suppose λ and $\mu \in \sigma(M)$, and $\lambda\bar{\mu} \neq 1$. Then $E_\lambda(M)$ and $E_\mu(M)$ are G-orthogonal.*

Proof: Let $u \in E_\lambda(M)$ and $v \in E_\mu(M)$. By definition there exist k and $h \in \mathbf{N}$ such that $(M - \lambda I)^k u = 0$ and $(M - \mu I)^h v = 0$. We continue the proof by induction on $k + h$.

If $k + h = 2$, then $k = h = 1$ and $Mu = \lambda u$ and $Mv = \mu v$. Thus,

$$\langle (\lambda\bar{\mu} - 1)Gu, v \rangle = \langle \sqrt{-1}JMu, Mv \rangle - \langle \sqrt{-1}Ju, v \rangle = 0.$$

Thus Lemma 2 holds in this case.

Let $m \geq 3$, and suppose the lemma holds for any k and $h \in \mathbf{N}$ such that $k + h < m$. Let $\hat{u} = (M - \lambda I)u$ and $\hat{v} = (M - \mu I)v$. Then $(M - \lambda I)^{k-1}\hat{u} = 0$ and

$(M-\mu I)^{h-1}\hat{v}=0$. Thus by the induction assumption, $\langle G\hat{u}, v\rangle = 0$, $\langle Gu, \hat{v}\rangle = 0$, and $\langle G\hat{u}, \hat{v}\rangle = 0$. They yield

$$\langle GMu, v\rangle - \lambda\langle Gu, v\rangle = 0, \quad \langle Gu, Mv\rangle - \overline{\mu}\langle Gu, v\rangle = 0,$$
$$\langle GMu, Mv\rangle - \lambda\langle Gu, Mv\rangle - \overline{\mu}\langle GMu, v\rangle + \lambda\overline{\mu}\langle Gu, v\rangle = 0.$$

Plugging the first two into the third, we obtain

$$(1-\lambda\overline{\mu})\langle Gu, v\rangle = 0.$$

By the fact $\lambda\overline{\mu}\neq 1$ we obtain the lemma. □

Corollary 3 *For $M\in \mathrm{Sp}(2n)$, if $\lambda\in\sigma(M)\backslash\mathbf{U}$, then $E_\lambda(M)$ is G-isotropic, i.e.,*

$$\langle Gx, y\rangle = 0 \qquad \forall\, x, y\in E_\lambda(M).$$

By Lemma 2, the complex space $\mathbf{C}^{2n}$ possesses a G-orthogonal decomposition

$$\mathbf{C}^{2n} = (\oplus_{\lambda\in\sigma(M)\cap\mathbf{U}} E_\lambda(M)) \oplus F(M), \tag{7}$$

where $F(M)$ is the direct sum of those $E_\lambda(M)s$ with $\lambda\in\sigma(M)\backslash\mathbf{U}$.

Corollary 4 *For $M\in \mathrm{Sp}(2n)$, if $\lambda\in\sigma(M)\cap\mathbf{U}$, then the restriction of G to $E_\lambda(M)$ is non-degenerate, i.e., for $x\in E_\lambda(M)$,* $\langle Gx, y\rangle = 0$ for all $y\in E_\lambda(M)$ implies $x=0$.

Proof: By Lemma 2 and (7), such an x must belong to the kernel of G on $\mathbf{C}^{2n}$. Thus $x=0$ by the non-degeneracy of G on $\mathbf{C}^{2n}$. □

Corollary 5 *For $M\in \mathrm{Sp}(2n)$, if $\lambda\in\sigma(M)\cap\mathbf{U}$ is not semi-simple, then there exists an eigenvector x belonging to λ of M which is G-isotropic, i.e., $\langle Gx, x\rangle = 0$.*

Proof: Since λ is not semi-simple, using one Jordan block belonging to λ of M, we obtain an eigenvector x and a root vector y in $E_\lambda(M)$ such that $My=\lambda y + x$. Thus

$$\langle Gx, y\rangle = \langle GMx, My\rangle = \langle \lambda Gx, \lambda y + x\rangle = |\lambda|^2\langle Gx, y\rangle + \lambda\langle Gx, x\rangle.$$

Thus $\langle Gx, x\rangle = 0$. □

Definition 6 Given $M\in \mathrm{Sp}(2n)$, suppose $\lambda\in\sigma(M)\cap\mathbf{U}$. By Corollary 4, the restriction $G|_{E_\lambda(M)}$ of G to $E_\lambda(M)$ is non-degenerate. Denote the total multiplicities of positive and negative eigenvalues of $G|_{E_\lambda(M)}$ by p and q respectively. The integer pair (p, q) is called the **Krein type number** of $\lambda\in\sigma(M)$. If $q=0$, λ is **Krein-positive**. If $p=0$, λ is **Krein-negative**. If both p and q are positive, λ is of **mixed kind**.

Note that by direct computation, $e^{\sqrt{-1}\theta}$ and $e^{-\sqrt{-1}\theta}$ are eigenvalues of the matrix $R(\theta) = \begin{pmatrix}\cos\theta & -\sin\theta\\ \sin\theta & \cos\theta\end{pmatrix}$ with Krein type $(0,1)$ and $(1,0)$ respectively for $\theta\in(0,\pi)\cup(\pi,2\pi)$.

Lemma 7 *For any M and $P \in \mathrm{Sp}(2n)$, suppose $\lambda \in \sigma(M) \cap \mathbf{U}$. Then the algebraic and geometric multiplicities and the Krein-type number of λ as an eigenvalue of $P^{-1}MP$ are the same as those of M.*

Proof: By direct computations. □

Lemma 8 *For $M \in \mathrm{Sp}(2n)$, if $\lambda \in \sigma(M) \cap \mathbf{U}$ has Krein type (p, q), then $\bar{\lambda}$ has Krein type (q, p). Specially, if 1 or $-1 \in \sigma(M)$, it must have Krein type (p, p) for some $p \in \mathbf{N}$.*

Proof: Note that for any $x \in \mathbf{C}^{2n}$, the real part of $\langle Jx, x\rangle$ is zero. Thus

$$\langle G\bar{x}, \bar{x}\rangle = \sqrt{-1}\langle J\bar{x}, \bar{x}\rangle = -\langle Gx, x\rangle. \tag{8}$$

We continue our study in two cases.

Case 1 $\lambda \in \sigma(M) \cap (\mathbf{U}\setminus\mathbf{R})$. Let $\{x_1, \ldots, x_k\}$ be a G-orthogonal base of $E_\lambda(M)$. Then $\{\overline{x_1}, \ldots, \overline{x_k}\}$ form a G-orthogonal base of $E_{\bar{\lambda}}(M)$. Then from (8) the lemma follows in this case.

Case 2 $\lambda = \pm 1$. Then the corresponding $E_\lambda(M)$ is $2d$-dimensional for some $d \in \mathbf{N}$ and real. Denote the Krein type belonging to the eigenvalue λ of M by (p, q) with $p + q = 2d$. Then the same argument as in Case 1 shows that $p = q$. □

Theorem 9 *For $M \in \mathrm{Sp}(2n)$, suppose $\omega \in \sigma(M) \cap \mathbf{U}$ is a p-fold Krein definite eigenvalue. Then*

1° *ω must be semi-simple,*

2° *there exists $\epsilon > 0$ and $\delta > 0$ such that any $N \in \mathrm{Sp}(2n)$ satisfying $\|N - M\| \le \delta$, any eigenvalue λ of N satisfying $|\lambda - \omega| \le \epsilon$ must locate on $\mathbf{U}$, and be semi-simple and of the same Krein sign.*

Proof: By the hypothesis, we have $\omega \in \mathbf{U}\setminus\mathbf{R}$.

1° Since ω is Krein definite, by Corollary 5, it must be semi-simple.

2° Without loss of generality, we suppose ω has Krein-type number $(p, 0)$. If the claim does not hold, there exists a sequence of matrices M_i and eigenvalues $\lambda_i \in \sigma(M_i)$ such that $M_i \to M$, $\lambda_i \to \omega$, and one of the following three cases happens: $|\lambda_i| \ne 1$, λ_i is not semi-simple, or λ_i is Krein negative. By Corollaries 3 and 5, there exists $x_i \in E_{\lambda_i}(M)$ such that $\langle Gx_i, x_i\rangle \le 0$ and $|x_i| = 1$. By passing to a subsequence of $\{x_i\}$, we can assume $x_i \to x$. This yields $Mx = \omega x$, $|x| = 1$, and $\langle Gx, x\rangle \le 0$. This contradicts the Krein positive definiteness of ω as an eigenvalue of M.

□

It is well known that $\sigma(U) \subset \mathbf{U}$ for any unitary matrix $U \in \mathcal{L}(\mathbf{C}^n)$. The following result on the spectrum of symplectic orthogonal matrices is based on the study of D. Salamon and E. Zehnder in the third paragraph on p.1316 of [SaZ2].

Theorem 10 *Suppose* $M \in \mathrm{Sp}(2n) \cap O(2n)$ *with* $M = \begin{pmatrix} A & -B \\ B & A \end{pmatrix}$ *as in Lemma* 1.5. *Then*

$$\sigma(M) = \sigma(A - \sqrt{-1}B) \cup \sigma(A + \sqrt{-1}B) \subset \mathbf{U}. \tag{9}$$

Moreover, $\lambda \in \sigma(M)$ *possesses Krein type* (p, q) *if and only if* $\bar{\lambda} \in \sigma(M)$ *possesses Krein type* (q, p), *if and only if* λ *and* $\bar{\lambda} \in \sigma(A - \sqrt{-1}B)$ *possess multiplicities* p *and* q *respectively, if and only if* λ *and* $\bar{\lambda} \in \sigma(A + \sqrt{-1}B)$ *possess multiplicities* q *and* p *respectively.*

Proof: We denote for short by (u, v) the vector $(u^T, v^T)^T$ in the following. Fix such an M. By an induction argument, for any $\xi \in \mathbf{C}^n$ and $k \in \mathbf{N}$ we obtain

$$\begin{aligned} &(M - \lambda I)^k(-\sqrt{-1}\xi, \xi) \\ &\quad = (-\sqrt{-1}(A - \sqrt{-1}B - \lambda I)^k\xi, (A - \sqrt{-1}B - \lambda I)^k\xi) \end{aligned} \tag{10}$$

$$\begin{aligned} &(M - \bar{\lambda} I)^k(\sqrt{-1}\xi, \xi) \\ &\quad = (\sqrt{-1}(A + \sqrt{-1}B - \bar{\lambda} I)^k\xi, (A + \sqrt{-1}B - \bar{\lambda} I)^k\xi). \end{aligned} \tag{11}$$

This especially proves (9).

By direct computation, we obtain

$$Q_{\pm} \equiv \ker_{\mathbf{C}}(\mp\sqrt{-1}I - J) = \{(\mp\sqrt{-1}\xi, \xi) \mid \xi \in \mathbf{C}^n\}. \tag{12}$$

Note that the Krein form $G = \sqrt{-1}J$ is positive definite on Q_+ and negative definite on Q_-. Note also $\mathbf{C}^{2n} = Q_+ \oplus Q_-$. From $JM = MJ$, we obtain $MQ_{\pm} \subset Q_{\pm}$. This implies $MQ_{\pm} = Q_{\pm}$.

For $\lambda \in \sigma(M)$ let $E_\lambda(M)$ be the root vector space of M in $\mathbf{C}^{2n}$ belonging to the eigenvalue λ. Let $(u, v) \in E_\lambda(M) \subset \mathbf{C}^{2n}$, and

$$\xi = \frac{1}{2}(u - \sqrt{-1}v), \qquad \eta = \frac{-\sqrt{-1}}{2}(u + \sqrt{-1}v).$$

Then it holds that

$$\begin{aligned} &(u, v) = (-\sqrt{-1}\xi, \xi) + (\sqrt{-1}\eta, \eta) \in Q_+ + Q_-, \\ &(-\sqrt{-1}\xi, \xi) \quad \text{and} \quad (\sqrt{-1}\eta, \eta) \in E_\lambda(M). \end{aligned}$$

Let

$$E_\lambda^{\pm}(M) = E_\lambda(M) \cap Q_{\pm}. \tag{13}$$

Then

$$E_\lambda(M) = E_\lambda^+(M) \oplus E_\lambda^-(M). \tag{14}$$

Together with (10)–(11) and Lemma 8, we obtain the second conclusion of the theorem. □

Note that by our above proof, suppose $\lambda \in \sigma(M)$ possesses Krein type (p, q). Then by (12)–(14),

$$\dim_{\mathbf{C}} E_\lambda^+(M) = p, \quad \dim_{\mathbf{C}} E_\lambda^-(M) = q. \tag{15}$$

Theorem 11 (I. Ekeland, Proposition I.3.2 of [Eke3]). *For $t_0 \in \mathbf{R}$ and $\epsilon_0 > 0$, suppose $\gamma \in C^1([t_0 - \epsilon_0, t_0 + \epsilon_0], \mathrm{Sp}(2n))$ satisfying that $B(t) = -J\dot{\gamma}(t)\gamma(t)^{-1}$ is positive definite, and $\gamma(t_0)$ possesses a Krein-negative eigenvalue $\omega = e^{\sqrt{-1}\varphi} \in \mathbf{U}$ with eigenspace E_ω. Then*

1° *there exist small enough $\epsilon \in (0, \epsilon_0)$ and $\delta > 0$ such that for $t \in (t_0 - \epsilon, t_0 + \epsilon)$ the matrix $\gamma(t)$ possesses at least one eigenvalue $e^{\sqrt{-1}\theta} \in \mathbf{U}$ with $|\theta - \varphi| < \delta$, which are all Krein-negative and their total multiplicity is $\dim E_\omega$.*

2° *if $t \to t_0$ and $e^{\sqrt{-1}\theta} \in \sigma(\gamma(t))$ with $|\theta - \varphi| < \delta$, then $\theta \to \varphi$ and*

$$\liminf_{t\to t_0} \frac{\theta - \varphi}{t - t_0} \geq \min\left\{-\frac{\langle B(t_0)x, x\rangle}{\langle Gx, x\rangle} \mid x \in E_\omega\right\} > 0, \tag{16}$$

$$\limsup_{t\to t_0} \frac{\theta - \varphi}{t - t_0} \leq \max\left\{-\frac{\langle B(t_0)x, x\rangle}{\langle Gx, x\rangle} \mid x \in E_\omega\right\} < +\infty. \tag{17}$$

Proof: 1° follows from Theorem 1.3.9.

Let $x_k \in \mathbf{C}^{2n}$ and $t_k \to t_0$ satisfy

$$\gamma(t_k)x_k = e^{\sqrt{-1}\theta_k}x_k, \quad \langle Gx_k, x_k\rangle = -1.$$

Thus $\gamma(t_k)^T Jx_k = e^{-\sqrt{-1}\theta_k}Jx_k$. Since the x_ks are bounded, we may assume $x_k \to x_0$ as $k \to \infty$ such that

$$\gamma(t_0)x_0 = e^{\sqrt{-1}\theta_0}x_0, \quad \langle Gx_0, x_0\rangle = -1.$$

Then we obtain $\theta_0 = \varphi$ if $\delta > 0$ is small enough and

$$\begin{aligned} \langle(\gamma(t_k) - \gamma(t_0))x_0, Jx_k\rangle &= \langle x_0, \gamma(t_k)^T Jx_k\rangle - \langle e^{\sqrt{-1}\theta_0}x_0, Jx_k\rangle \\ &= (e^{\sqrt{-1}\theta_k} - e^{\sqrt{-1}\theta_0})\langle x_0, Jx_k\rangle. \end{aligned} \tag{18}$$

This then yields

$$\begin{aligned}
\lim_{t_k\to t_0}\frac{\theta_k-\theta_0}{t_k-t_0} &= \lim_{t_k\to t_0}\left(\frac{e^{\sqrt{-1}\theta_k}-e^{\sqrt{-1}\theta_0}}{t_k-t_0}\langle x_0, Jx_k\rangle e^{-\sqrt{-1}\theta_0}\right)\\
&= \lim_{t_k\to t_0}\left(\left\langle\frac{\gamma(t_k)-\gamma(t_0)}{t_k-t_0}x_0, Jx_k\right\rangle e^{-\sqrt{-1}\theta_0}\right)\\
&= \langle\dot\gamma(t_0)x_0, Jx_0\rangle e^{-\sqrt{-1}\theta_0}\\
&= \langle\dot\gamma(t_0)x_0, Jx_0\rangle e^{-\sqrt{-1}\theta_0}\\
&= \langle JB(t_0)\gamma(t_0)x_0, Jx_0\rangle e^{-\sqrt{-1}\theta_0}\\
&= \langle B(t_0)x_0, x_0\rangle.
\end{aligned}$$

This implies (16) and (17). □

As a direct consequence of Theorem 11, we obtain

Corollary 12 *Let $f\in C([0,1], \mathrm{Sp}(2n)\cap O(2n))$ and $\omega\in\sigma(f(0))$ possess Krein type (p,q). Let $\epsilon>0$ be so small that ω is the only eigenvalue of $f(0)$ in $\mathbf{U}\cap B_\epsilon(\omega)$, where $B_\epsilon(\omega)=\{z\in\mathbf{C}\mid |z-\omega|<\epsilon\}$. Then for any sufficiently small $t>0$, the Krein-type numbers p and q are the sum of the corresponding Krein-type numbers of all eigenvalues of $f(t)$ in $\mathbf{U}\cap B_{\epsilon/3}(\omega)$ respectively.*

1.4 Normal forms for the eigenvalue 1

The results in Sections 4, 5, 6, and 7 on the normal forms of symplectic matrices are based upon recent works [LoD1] and [HaL1] by D. Dong, J. Han, and the author, which were proved in order to establish our index iteration theory, and were done when we were not aware of earlier existing related results. For other different types of normal forms for symplectic matrices and linear Hamiltonian systems we refer to works of J. Williamson [Wil1] in 1937, N. Burgoyne and R. Cushman [BuC1] in 1973, A. Laub and K. Meyer [LaM1] in 1974, H. Wimmer [Wim1] in 1991, and L. Hörmander [Hom2] in 1995. But results contained in these works are developed for different purposes and may not be convenient for our study on the topology of symplectic groups and the index theory for Hamiltonian systems.

Given any two real matrices of the square block form

$$M_1=\begin{pmatrix}A_1 & B_1\\ C_1 & D_1\end{pmatrix}_{2i\times 2i},\quad M_2=\begin{pmatrix}A_2 & B_2\\ C_2 & D_2\end{pmatrix}_{2j\times 2j},$$

we define the $\diamond$-product of M_1 and M_2 to be the $2(i+j)\times 2(i+j)$ matrix $M_1\diamond M_2$ given by

$$M_1\diamond M_2 = \begin{pmatrix} A_1 & 0 & B_1 & 0 \\ 0 & A_2 & 0 & B_2 \\ C_1 & 0 & D_1 & 0 \\ 0 & C_2 & 0 & D_2 \end{pmatrix}.$$

It is easy to verify that the multiplication defined by $\diamond$ among even order square block form matrices is associative. We denote by $M^{\diamond k}$ the k-fold $\diamond$-product $M\diamond\cdots\diamond M$. Note that the $\diamond$-product of two symplectic matrices is still symplectic. The $\diamond$-product is introduced according to the form of the standard symplectic matrix J chosen in Definition 1.1.

In this section we derive the normal forms of symplectic matrices possessing the eigenvalue 1, and prove the main Theorem 1 below. Note that in this section we consider Jordan forms and root vectors defined on the field $\mathbf{R}$. Define

$$N_1(\lambda, b) = \begin{pmatrix} \lambda & b \\ 0 & \lambda \end{pmatrix}, \qquad \forall\, b\in\mathbf{R},\ \lambda\in\mathbf{R}. \tag{1}$$

For $k\ge 2$ define the matrix $N_k(\lambda, b)$ by

$$N_k(\lambda, b) = \begin{pmatrix} A_k(\lambda) & B_k(\lambda, b) \\ 0 & C_k(\lambda) \end{pmatrix}, \tag{2}$$

where $A_k(\lambda)$ is a $k\times k$ Jordan block of the eigenvalue $\lambda\in\mathbf{C}$:

$$A_k(\lambda) = \begin{pmatrix} \lambda & 1 & 0 & \cdots & 0 & 0 \\ 0 & \lambda & 1 & \cdots & 0 & 0 \\ 0 & 0 & \lambda & \cdots & 0 & 0 \\ \cdot & \cdot & \cdot & \cdots & \cdot & \cdot \\ \cdot & \cdot & \cdot & \cdots & \cdot & \cdot \\ 0 & 0 & 0 & \cdots & \lambda & 1 \\ 0 & 0 & 0 & \cdots & 0 & \lambda \end{pmatrix}, \tag{3}$$

$C_k(\lambda)$ is a $k\times k$ lower triangle matrix of the following form:

$$C_k(\lambda) = \begin{pmatrix} \lambda^{-1} & 0 & 0 & \cdots & 0 & 0 \\ -\lambda^{-2} & \lambda^{-1} & 0 & \cdots & 0 & 0 \\ \lambda^{-3} & -\lambda^{-2} & \lambda^{-1} & \cdots & 0 & 0 \\ \cdot & \cdot & \cdot & \cdots & \cdot & \cdot \\ \cdot & \cdot & \cdot & \cdots & \cdot & \cdot \\ -(-\lambda)^{1-p} & -(-\lambda)^{2-p} & -(\lambda)^{3-p} & \cdots & \lambda^{-1} & 0 \\ -(-\lambda)^{-p} & -(\lambda)^{1-p} & -(-\lambda)^{2-p} & \cdots & -\lambda^{-2} & \lambda^{-1} \end{pmatrix}. \tag{4}$$

$B_k(\lambda, b)$ is a $k \times k$ lower triangle matrix of the following form with $b = (b_1, \dots, b_k) \in \mathbf{R}^k$:

$$B_k(\lambda, b) = \begin{pmatrix} b_1 & 0 & 0 & \cdots & 0 & 0 \\ b_2 & (-\lambda)b_2 & 0 & \cdots & 0 & 0 \\ b_3 & (-\lambda)b_3 & \lambda^2 b_3 & \cdots & 0 & 0 \\ \cdot & \cdot & \cdot & \cdots & \cdot & \cdot \\ \cdot & \cdot & \cdot & \cdots & \cdot & \cdot \\ b_{k-1} & (-\lambda)b_{k-1} & \lambda^2 b_{k-1} & \cdots & (-\lambda)^{k-2} b_{k-1} & 0 \\ b_k & (-\lambda)b_k & \lambda^2 b_k & \cdots & (-\lambda)^{k-2} b_k & (-\lambda)^{k-1} b_k \end{pmatrix}. \quad (5)$$

Since $A_k(\lambda)^T C_k(\lambda) = I$ and $B_k(\lambda, b)^T C_k(\lambda) = C_k(\lambda)^T B_k(\lambda, b)$, we obtain $N_k(\lambda, b) \in \mathrm{Sp}(2k)$.

Theorem 1 *Suppose $M \in \mathrm{Sp}(2n)$ possesses the eigenvalue 1. Then there exist a matrix $P \in \mathrm{Sp}(2n)$ and an integer $m \in \mathbf{N}$ such that*

$$P^{-1} M P = M_1 \diamond \cdots \diamond M_m \diamond M_0. \quad (6)$$

Here $M_0 \in \mathrm{Sp}(2k_0)$ with $k_0 \geq 0$ and $1 \notin \sigma(M_0)$. For $1 \leq i \leq m$, each $M_i \in \mathrm{Sp}(2k_i)$ with $k_i \geq 1$. Let $E_1(M)$ denote the invariant root vector space of M belonging to the eigenvalue 1. Then $2\sum_{i=0}^m k_i = 2k_0 + \dim E_1(M) = 2n$, $M_i = N_{k_i}(1, b)$ with $b \in \mathbf{R}^{k_i}$, and if $k_i = 1$, then $b = \pm 1$, or 0. Moreover,

$$\dim_{\mathbf{R}} \ker_{\mathbf{R}}(M_i - I) = 1, \qquad \text{if and only if } b_{i,k_i} \neq 0, \quad (7)$$

$$\dim_{\mathbf{R}} \ker_{\mathbf{R}}(M_i - I) = 2, \qquad \text{if and only if } b_{i,k_i} = 0. \quad (8)$$

We start the proof of Theorem 1 from the following lemmas.

Lemma 2 *Suppose $M \in \mathrm{Sp}(2n)$ possesses root vectors $\{\xi_i | 1 \leq i \leq p\}$ for some $p \in \mathbf{N}$, which form a Jordan block of M belonging to the eigenvalue 1 such that letting $\xi_0 = 0$,*

$$M\xi_i = \xi_i + \xi_{i-1}, \quad \text{for } 1 \leq i \leq p. \quad (9)$$

Then it holds that

$$\xi_{i-1}^T J \xi_{j-1} + \xi_{i-1}^T J \xi_j + \xi_i^T J \xi_{j-1} = 0, \qquad \text{for } 1 \leq i, j \leq p, \quad (10)$$

$$\xi_i^T J \xi_j = 0, \qquad \text{for } 1 \leq i \leq p - j, \quad 1 \leq j \leq \left[\frac{p}{2}\right], \quad (11)$$

where we denote $[a] = \max\{i \in \mathbf{Z} | i \leq a\}$ for $a \in \mathbf{R}$.

Proof: From (9) and $\xi_i^T J \xi_j = (M\xi_i)^T J(M\xi_j)$, we obtain (10) and then (11). □

Lemma 3 *Suppose $M \in \mathrm{Sp}(2n)$ possesses root vectors $\{\xi_i, \eta_j | 1 \le i \le p,\ 1 \le j \le q\}$ for some $p, q \in \mathbf{N}$ with $p \le q$, which form two different Jordan blocks of M belonging to the eigenvalue* 1 *such that, setting $\xi_0 = \eta_0 = 0$,* (9) *holds and*

$$M\eta_j = \eta_j + \eta_{j-1}, \quad for \ \ 1 \le j \le q. \tag{12}$$

Then it holds that

$$\xi_{i-1}^T J\eta_{j-1} + \xi_{i-1}^T J\eta_j + \xi_i^T J\eta_{j-1} = 0, \quad for \quad 1 \le i \le p,\ 1 \le j \le q, \tag{13}$$

$$\xi_i^T J\eta_j = 0, \qquad for \quad 1 \le j \le q - i,\ 1 \le i \le p. \tag{14}$$

Proof: Computing $\xi_i^T J\eta_j$ via (9) and (12) yields the conclusion. □

In order to study the normal forms of symplectic matrices with the eigenvalue 1, we consider the following three basic cases.

Case 1 The matrix M possesses a Jordan block belonging to the eigenvalue 1, which is given by root vectors $\{\xi_i | 1 \le i \le p\}$ satisfying (9) such that

$$\xi_1^T J\xi_p < 0. \tag{15}$$

Note that (15) implies $p > 1$. If $p = 2k + 1$ for some $k \in \mathbf{N}$, by (11) we obtain $\xi_k^T J\xi_{k+1} = 0$. By (10) and an induction argument this implies $\xi_1^T J\xi_p = 0$, a contradiction. Thus we must have $p = 2k$ for some $k \in \mathbf{N}$. Let $F = \mathrm{span}\{\xi_1, \ldots, \xi_{2k}\}$ and $G = \mathrm{span}\{\xi_1, \ldots, \xi_k\}$.

Note that G is an isotropic subspace of $\mathbf{R}^{2n}$ by (11), i.e.,

$$\xi_i^T J\xi_j = 0, \qquad \text{for } \ 1 \le i, j \le k. \tag{16}$$

By (15), if necessary multiplying all ξ_is by $1/|-\xi_1^T J\xi_{2k}|^{1/2}$, without loss of generality, we may assume $\xi_1^T J\xi_{2k} = -1$.

Subcase 1.1 $k = 1$.

Let $\delta_1 = \xi_1$ and $\delta_2 = \xi_2$. They form a symplectic base for F, and

$$M(\delta_1, \delta_2) = (\delta_1, \delta_2)M_1, \tag{17}$$

where $M_1 = N_1(1, 1) \in \mathrm{Sp}(2)$ is defined by (1). In this book, for any given matrices $A_1, \ldots, A_m$, each of which possesses the same number of rows, we denote by $(A_1, \ldots, A_m)$ the matrix formed by their column vectors. Here (δ_1, δ_2) is the $2n \times 2$ matrix formed by δ_1 and dl_2. We can extend $\{\delta_1, \delta_2\}$ to a matrix $\Delta \in \mathrm{Sp}(2n)$ such that δ_1 and δ_2 form the 1-st and the $(n + 1)$-st columns of Δ and for some $M_0 \in \mathrm{Sp}(2(n - k))$,

$$M\Delta = \Delta(M_1 \diamond M_0). \tag{18}$$

Subcase 1.2 $k > 1$.

By (10) and (11) for this subcase we obtain

$$\xi_i^T J \xi_{2k-i+1} = (-1)^i, \qquad \text{for } 1 \le i \le k. \tag{19}$$

By (9),

$$M\xi_i = \xi_i + \xi_{i-1}, \qquad \text{if} \quad 1 \le i \le k, \tag{20}$$

$$M\xi_i = \xi_i + \mathrm{com}\{\xi_{i-1}\}, \qquad \text{if} \quad k+1 \le i \le 2k. \tag{21}$$

Here and below in this chapter by $\mathrm{com}\{v_1, \dots, v_j\}$ we denote some $\mathbf{R}$-linear combination of the vectors $\{v_1, \dots, v_j\}$.

In the following we continue our study based on (11), (19), (20), and (21) only.

Define $\delta_0 = 0$ and

$$\delta_i = \xi_i, \qquad \text{for } 1 \le i \le k,$$

$$\delta_{2k} = (-1)^{k+1}\xi_{k+1},$$

$$\hat{\delta}_{k+i} = (-1)^{i+1}\xi_{2k+1-i} + \sum_{j=i+1}^{k} a_{k+i,j}\xi_{2k+1-j}, \quad \text{for } 1 \le i \le k-1, \tag{22}$$

where the constants $a_{k+i,j}$ are chosen inductively for $j = i+1, \dots, k$ by the equations

$$\xi_j^T J \left\{ (-1)^{i+1}\xi_{2k+1-i} + \sum_{h=i+1}^{j} a_{k+i,h}\xi_{2k+1-h} \right\} = 0. \tag{23}$$

By (19), the constants $a_{k+i,j}$ are uniquely determined.

Fix $i \in [1, k]$. We obtain $\delta_m^T J \hat{\delta}_{k+i} = 0$ for $1 \le m \le i-1$ by (11), and for $i+1 \le m \le k$ by (11) and (23). Thus

$$\hat{\delta}_{k+i} \perp \{\delta_1, \dots, \delta_{i-1}, \delta_{i+1}, \dots, \delta_k\}, \qquad \text{for } 1 \le i \le k. \tag{24}$$

Note that by (11) and (19),

$$\delta_i^T J \hat{\delta}_{k+i} = -1, \qquad \text{for } 1 \le i \le k. \tag{25}$$

By (22) we also have

$$M\hat{\delta}_{k+i} = \hat{\delta}_{k+i} + \mathrm{com}\{\hat{\delta}_{k+i+1}, \dots, \hat{\delta}_{2k}\}, \qquad \text{for } 1 \le i \le k.$$

Next we further modify the $\hat{\delta}_i$s to get a symplectic base of F. We define

$$\delta_i = \hat{\delta}_i + \sum_{h=i+1}^{2k} \hat{\delta}_i^T J \hat{\delta}_j \delta_{h-k}, \quad \text{for} \quad i = k+1, \dots, 2k-1. \tag{26}$$

Now for $k+1 \le i \le 2k-1$, by (11), (24), and (25), we get

$$\delta_{i-k}^T J \delta_i = -1,$$
$$\delta_i \perp \{\delta_1, \dots, \delta_{i-k-1}, \delta_{i-k+1}, \dots, \delta_k\}.$$

Especially by (11), (24), and (26), we obtain $\delta_i^T J \delta_j = 0$ for $k+1 \le i < j \le 2k$.
Then from all the above results we obtain $F = \mathrm{span}\{\delta_i | 1 \le i \le 2k\}$ and

$$\delta_i^T J \delta_{k+i} = -1, \quad \text{for} \quad 1 \le i \le k, \tag{27}$$

$$\delta_i^T J \delta_j = 0, \quad \text{for} \quad 1 \le i, j \le 2k \text{ and } |i-j| \ne k, \tag{28}$$

$$M\delta_i = \delta_i + \delta_{i-1}, \qquad \text{for} \quad 1 \le i \le k, \tag{29}$$

$$\begin{aligned} M\delta_i = \; & \delta_i + \mathrm{com}\{\delta_{i-k}, \dots, \delta_k, \delta_{i+1}, \dots, \delta_{2k}\}, \\ & \text{for} \quad k+1 \le i \le 2k. \end{aligned} \tag{30}$$

Therefore by (27) and (28), the vectors $\{\delta_1, \dots, \delta_{2k}\}$ form a symplectic base of F, and we have

$$M(\delta_1, \dots, \delta_{2k}) = (\delta_1, \dots, \delta_{2k}) M_1, \tag{31}$$

where M_1 is of the form

$$M_1 = \begin{pmatrix} A & B \\ 0 & C \end{pmatrix},$$

with $k \times k$ matrices A, B, and C. By (29) and (30) the matrix $A = A_k(1)$ is of the form (3), B and C are lower triangular matrices and the diagonal elements of C are all 1. Because $M \in \mathrm{Sp}(2n)$ and $\{\delta_1, \dots, \delta_{2k}\}$ form a symplectic base, (3) implies $M_1 \in \mathrm{Sp}(2k)$. Therefore, $A^T C = I$ and $B^T C = C^T B$. These imply that $C = C_k(1)$ is of the form (4), and that $B = B_k(1, b)$ is of the form (5) for some $b \in \mathbf{R}^k$.

Note that in this case, $b_k \ne 0$ in (5). Thus (7) holds in this case.

Now by (27) and (28) we can extend $\{\delta_1, \dots, \delta_{2k}\}$ to a matrix $\Delta \in \mathrm{Sp}(2n)$ such that $\{\delta_1, \dots, \delta_k\}$ form the 1-st to k-th columns and $\{\delta_{k+1}, \dots, \delta_{2k}\}$ form the $(n+1)$-st to $(n+k)$-th columns of Δ, and for some $M_0 \in \mathrm{Sp}(2(n-k))$,

$$M\Delta = \Delta(M_1 \diamond M_0) \tag{32}$$

Case 2 The matrix M possesses a Jordan block for the eigenvalue 1, which is given by root vectors $\{\xi_i | 1 \le i \le p\}$ satisfying (7) such that

$$\xi_1^T J \xi_p > 0. \tag{33}$$

From (33) as in the study of the case 1, $p > 1$ and p is even. Suppose $p = 2k$ for some $k \in \mathbf{N}$. Define

$$\eta_i = \xi_i, \qquad \text{for } 0 \le i \le k,$$
$$\eta_i = -\xi_i, \qquad \text{for } k+1 \le i \le 2k.$$

Then $\eta_1^T J \eta_{2k} < 0$. Without loss of generality, we can assume $\eta_1^T J \eta_{2k} = -1$.

Subcase 2.1 $k = 1$.

Let $\delta_1 = \eta_1$ and $\delta_2 = \eta_2$. They form a symplectic base of F, and (17) holds, where $M_1 = N_1(1, -1) \in \mathrm{Sp}(2)$ is defined by (1). As in the Subcase 1.1, we can also extend $\{\delta_1, \delta_2\}$ to a matrix $\Delta \in \mathrm{Sp}(2n)$ to get the corresponding (18).

Subcase 2.2 $k > 1$.

Then by (9), (10), and (11) for the ξ_is we obtain (11), (19), (20), and (21) for the η_is. Now we can carry out the proof of Case 1 to obtain a new symplectic base of $F = \mathrm{span}\{\eta_1, \dots, \eta_{2k}\}$ such that (31) and (32) hold.

Note that in this case, $b_k \neq 0$ in (5). Thus (7) holds in this case.

Case 3 The matrix M possesses two equal size Jordan blocks belonging to the eigenvalue 1, which are given by root vectors $\{\xi_i, \eta_j | 1 \le i, j \le k\}$ satisfying (9) and (12) with $p = q = k$ such that

$$\xi_1^T J \eta_k \neq 0, \quad \xi_1^T J \xi_k = 0, \quad \eta_1^T J \eta_k = 0. \tag{34}$$

If necessary multiplying all η_j by $-1/(\xi_1^T J \eta_k)$, without loss of generality, we suppose $\xi_1^T J \eta_k = -1$. Denote $F = \mathrm{span}\{\xi_i, \eta_j | 1 \le i, j \le k\}$.

Subcase 3.1 $k = 1$.

Let $\delta_1 = \xi_1$ and $\delta_2 = \eta_1$. They form a symplectic base of F, and (17) holds, where $M_1 = N_1(1, 0) = I_2 \in \mathrm{Sp}(2)$. We can also extend $\{\delta_1, \delta_2\}$ to a matrix $\Delta \in \mathrm{Sp}(2n)$ to get the corresponding (18).

Subcase 3.2 $k > 1$.

By $\xi_1^T J \eta_k = -1$ and (13) we obtain

$$\xi_i^T J \eta_{k-i+1} = (-1)^i, \qquad \forall 1 \le i \le k. \tag{35}$$

We first prove the following lemma.

Lemma 4 *Without loss of generality, we can further assume that*

$$\xi_i^T J \xi_j = 0, \qquad \forall 1 \le i, j \le k. \tag{36}$$

Proof: Assume (36) does not hold. Together with (34) we must have $k > 2$. Let $m = m(\{\xi_i\}_{1 \le i \le k})$ be the least integer in $[1, k]$ such that there exists the least integer $q \in [m+1, k]$ satisfying $\xi_m^T J \xi_q \neq 0$. By (11) and (34) we must have $1 < m < k$.

If $q < k$, by (11) we obtain

$$\xi_{m-1}^T J\xi_q + \xi_{m-1}^T J\xi_{q+1} + \xi_m^T J\xi_q = 0.$$

By the minimality of m this yields $\xi_m^T J\xi_q = 0$, a contradiction. Thus we must have $q = k$.

If $k - m = 2p$ for some $p \in \mathbf{N}$, using (10) inductively we obtain

$$0 = \xi_{m+p}^T J\xi_{m+p} = \xi_{m+p}^T J\xi_{k-p} = \pm\xi_m^T J\xi_k \neq 0.$$

This contradiction shows that $k - m$ must be odd.

Thus by (35) we obtain

$$\xi_m^T J\eta_{k-m+1} + \eta_1^T J\xi_k = 2(-1)^m. \tag{37}$$

Let $a = (-1)^{m+1}\xi_m^T J\xi_k/2$. Define $\delta_0 = 0$ and

$$\begin{aligned} \delta_i &= \xi_i, \qquad \text{if} \quad 1 \le i < m, \\ \delta_i &= \xi_i + a\eta_{i-m+1}, \qquad \text{if} \quad m \le i \le k. \end{aligned}$$

Replacing ξ_i by δ_i we obtain that the corresponding relations (9) and (12) still hold, and that $F = \text{span}\{\delta_i, \eta_j | 1 \le i, j \le k\}$. From the fact $k-m+1 \le k-1$, by (34) and (14) we obtain $\delta_1^T J\delta_k = 0$. Thus $\{\delta_i, \eta_j | 1 \le i, j \le k\}$ satisfies the corresponding (34). By (9) for the δ_is and (12), we obtain (13) and (14) for the δ_is and η_js. Thus from the fact $\delta_1^T J\eta_k = \xi_1^T J\eta_k = -1$ we obtain that $\{\delta_i, \eta_j | 1 \le i, j \le k\}$ satisfies the corresponding (35).

For any $1 \le i < m$ and $1 \le j \le k$ it holds that $j - m + 1 \le k - i$. Then by the minimality of m and (14) we obtain

$$\delta_i^T J\delta_j = \xi_i^T J\xi_j + a\xi_i^T J\eta_{j-m+1} = 0. \tag{38}$$

For any $1 \le j \le k-1$, it holds that $j-m+1 \le k-m \le k-1$. Thus by the definition of m, (14), and (11), we obtain $\delta_m^T J\delta_j = 0$. Note that by the fact $k - m + 1 \le k - 1$, (11), (37), and the definition of a, we obtain

$$\delta_m^T J\delta_k = \xi_m^T J\xi_k + a(\xi_m^T J\eta_{k-m+1} + \eta_1^T J\xi_k) + a^2\eta_1^T J\eta_{k-m+1} = 0. \tag{39}$$

If $\{\delta_i | 1 \le i \le k\}$ still do not satisfy (36), by (38) to (39) then

$$m(\{\delta_i\}_{1\le i\le k}) \ge m + 1 = m(\{\xi_i\}_{1\le i\le k}) + 1.$$

Thus by induction, we can modify the ξ_is to complete the proof of the lemma. □

Now suppose (36) holds. We define $\xi_{k+i} = \eta_i$ for $1 \le i \le k$. Using (35), (36), (9) for $\{\xi_i\}_{1\le i\le k}$, and (12) for $\{\eta_i\}_{1\le i\le k}$, we obtain (11), (19), (20), and (21) for $\{\xi_i\}_{1\le i\le 2k}$, and the fact that $F = \text{span}\{\xi_i | 1 \le i \le 2k\}$. Thus similar to the discussion of Case 1, we obtain a new symplectic base $\{\delta_i | 1 \le i \le 2k\}$ of F such that (31) and

(32) hold, where the matrix M_1 in (31) is given by (2)–(5) with $b_k = 0$. Thus (8) holds in this case.

This completes our study of the three basic cases.

Now we can give the proof of the main result of this section.

Proof of Theorem 1: Firstly by the fact $1 \in \sigma(M)$, we claim that at least one of Cases 1 to 3 must happen. If not, denote the set of all the Jordan blocks formed by linearly independent root vectors of M belonging to the eigenvalue 1 by $\mathcal{J} = \mathcal{J}_0 \cup \mathcal{J}_1$, where $\mathcal{J}_1$ contains only all largest size Jordan blocks, and $\mathcal{J}_0$ contains all other possible smaller size Jordan blocks. Since $1 \in \sigma(M)$ then $\mathcal{J}_1 \neq \emptyset$. Let $J_1 = \{\xi_i | 1 \leq i \leq q\}$ be a Jordan block in $\mathcal{J}_1$. Since Cases 1 and 2 do not happen, together with (11) we obtain

$$\xi_1^T J \xi_j = 0, \qquad \forall \xi_j \in J_1. \tag{40}$$

For any $J_0 = \{\eta_j | 1 \leq j \leq p\} \in \mathcal{J}_0$, it must hold that $p < q$ by definition. Thus by (14) we have

$$\xi_1^T J \eta_j = 0, \qquad \forall \eta_j \in J_0. \tag{41}$$

For any $J_2 = \{\gamma_j | 1 \leq j \leq q\} \in \mathcal{J}_1 \backslash \{J_1\}$, since Cases 1 and 2 do not happen, it must hold that $\gamma_1^T J \gamma_q = 0$. Since Case 3 does not happen, together with (40) it must hold that $\xi_1^T J \gamma_q = 0$. Thus together with (14) we obtain

$$\xi_1^T J \gamma_j = 0, \qquad \forall \gamma_j \in J_2. \tag{42}$$

Then (40) to (42) show that ξ_1 is an isotropic element of the invariant root vector subspace $E_1(M)$ of M belonging to the eigenvalue 1. This contradicts the fact that $E_1(M)$ is a symplectic subspace of $\mathbf{R}^{2n}$, and proves our claim.

Then by Case 1, 2, or 3, we can get (31) and (32), and carry out the induction to complete the proof of the theorem.

1.5 Normal forms for the eigenvalue -1

In this section we introduce the normal forms of symplectic matrices having eigenvalue -1. Using lemmas corresponding to Lemmas 4.2 and 4.3, we obtain the following Theorem 1. Since the proofs are similar to those of Section 1.4, we only describe the results and omit all the proofs.

Theorem 1 *Suppose $M \in \mathrm{Sp}(2n)$ possesses the eigenvalue -1. Then there exist $P \in \mathrm{Sp}(2n)$ and $m \in \mathbf{N}$ such that*

$$P^{-1} M P = M_1 \diamond \cdots \diamond M_m \diamond M_0. \tag{1}$$

Here $M_0 \in \mathrm{Sp}(2k_0)$ with $k_0 \geq 0$ and $-1 \notin \sigma(M_0)$. For $1 \leq i \leq m$, each matrix M_i is of the normal form

$$M_i = N_{k_i}(-1, b_i) \tag{2}$$

is defined by (4.1) *or* (4.2) *for some integer* $k_i \geq 1$ *and* $b_i \in \mathbf{R}^{k_i}$. *Denote by* $E_{-1}(M)$ *the invariant root vector space of M belonging to the eigenvalue* -1; *then* $2\sum_{i=0}^{m} k_i = 2k_0 + \dim E_{-1}(M) = 2n$. *Moreover, corresponding* (4.7) *and* (4.8) *also hold.*

1.6 Normal forms for eigenvalues in $\mathbf{U}\backslash\mathbf{R}$

In this section, we derive the normal forms of symplectic matrices having non-real eigenvalues on the unit circle. Throughout this section, we fix $\omega \in \mathbf{U}\backslash\mathbf{R}$ and denote by $\omega = \cos\theta + \sqrt{-1}\sin\theta$ for some $0 \leq \theta \leq 2\pi$. Note that in this case it always holds that $\sin\theta \neq 0$. For $\varphi \in \mathbf{R}$ define

$$R(\varphi) = \begin{pmatrix} \cos\varphi & -\sin\varphi \\ \sin\varphi & \cos\varphi \end{pmatrix}.$$

Lemma 1 *Suppose* $M \in \mathrm{Sp}(2n)$ *possesses root vectors* $\Omega = \{\xi_i, \eta_i \in \mathbf{R}^{2n} | 1 \leq i \leq p\}$ *for some* $p \in \mathbf{N}$ *in a Jordan block belonging to the eigenvalue* $\omega \in \mathbf{U}\backslash\mathbf{R}$ *such that for* $1 \leq i \leq p$ *and* $\xi_0 = \eta_0 = 0$,

$$M\xi_i = (\cos\theta)\xi_i - (\sin\theta)\eta_i + \xi_{i-1}, \tag{1}$$

$$M\eta_i = (\sin\theta)\xi_i + (\cos\theta)\eta_i + \eta_{i-1}. \tag{2}$$

Then for all $1 \leq i, j \leq p$,

$$\xi_i^T J\xi_j = \eta_i^T J\eta_j, \qquad \xi_i^T J\eta_j = -\eta_i^T J\xi_j, \tag{3}$$

$$\begin{aligned} &\xi_{i-1}^T J\xi_{j-1} + (\cos\theta)(\xi_{i-1}^T J\xi_j + \xi_i^T J\xi_{j-1}) \\ &\quad - (\sin\theta)(\xi_{i-1}^T J\eta_j + \eta_i^T J\xi_{j-1}) = 0, \end{aligned} \tag{4}$$

$$\begin{aligned} &\xi_{i-1}^T J\eta_{j-1} + (\cos\theta)(\xi_{i-1}^T J\eta_j + \xi_i^T J\eta_{j-1}) \\ &\quad + (\sin\theta)(\xi_{i-1}^T J\xi_j - \eta_i^T J\eta_{j-1}) = 0. \end{aligned} \tag{5}$$

Proof: Using (1) and (2) to compute $\xi_i^T J\xi_j$, $\eta_i^T J\eta_j$, $\xi_i^T J\eta_j$, and $\eta_i^T J\xi_j$, together with the facts $\sin\theta \neq 0$ and $\det R(\theta) \neq 0$, by an induction on $i + j$ we obtain (3) to (5). □

Lemma 2 *Suppose* (1) *and* (2) *of Lemma* 1 *hold for* Ω. *Let* $V_i = \mathrm{span}\{\xi_i, \eta_i\}$ *for* $1 \leq i \leq p$. *Then*

$$V_i \perp V_j, \qquad \forall 1 \leq j \leq p - i, \quad 1 \leq i \leq p. \tag{6}$$

Proof: From (3) to (5), the fact $\det R(\theta) \neq 0$, we obtain $V_1 \perp V_{j-1}$. Then by an induction argument we obtain the lemma. □

Lemma 3 *Suppose* (1) *and* (2) *of Lemma* 1 *hold for* Ω. *Then*

$$\xi_i^T J\eta_i = 0, \qquad \forall 1 \le i \le \left[\frac{p}{2}\right]. \tag{7}$$

Proof: Let $j = i + 1$ in (4). Then $i + j - 1 \le p$. Thus (7) follows from Lemma 2. □

Definition 4 Suppose (1) and (2) of Lemma 1 hold for Ω. By (3) we define

$$\begin{aligned} a_i(\Omega) &= \eta_i^T J\eta_{p-i+1} = \xi_i^T J\xi_{p-i+1}, \qquad \forall i = 1, \dots, p, \\ b_i(\Omega) &= \eta_i^T J\xi_{p-i+1} = -\xi_i^T J\eta_{p-i+1}, \qquad \forall i = 1, \dots, p. \end{aligned}$$

Lemma 5 *Suppose* (1) *and* (2) *of Lemma* 1 *hold for* Ω. *Then*

$$(b_i(\Omega), a_i(\Omega))^T = -R(2\theta)(b_{i-1}(\Omega), a_{i-1}(\Omega))^T, \qquad \forall i = 2, \dots, p. \tag{8}$$

If p is odd, then $a_{[p/2]+1}(\Omega) = 0$.

Proof: For $2 \le i \le p$ letting $j = p - i + 2$ in (4) and (5), by (6) we obtain (8). □

Using the root vectors $\{\eta_i, \xi_i\}$ here to replace the role of root vectors ξ_i in Cases 1 and 2 of Section 4, the normal forms can be derived for the symplectic matrices which possess non-real eigenvalues on the unit circle. We start with the one Jordan block case with even size number $p = 2k$.

Case 1 Suppose $M \in \mathrm{Sp}(2n)$ possesses root vectors $\Omega = \{\xi_i, \eta_i \in \mathbf{R}^{2n} | 1 \le i \le p\}$ for some $p = 2k \in 2\mathbf{N}$ in a Jordan block belonging to the eigenvalue $\omega \in \mathbf{U}\backslash\mathbf{R}$ satisfying (1) (2) and the condition

$$(b_k, a_k) \equiv (b_k(\Omega), a_k(\Omega)) \neq 0. \tag{9}$$

Denote by $(b_i, a_i) = (b_i(\Omega), a_i(\Omega))$ for $1 \le i \le k$. For column vectors $v_1, \dots, v_m \in \mathbf{R}^{2n}$ we denote by $(v_1, \dots, v_m)$ the $2n \times m$ matrix formed by them. Let $V = \mathrm{span}(\Omega)$. Based upon the root vectors Ω, we need to construct $2n \times 2$ matrices P_i for $1 \le i \le 2k$ such that the column vectors of the matrix $P = (P_1, \dots, P_{2k})$ form a symplectic base of V. We carry out the discussion in three steps.

Step 1 The definition of P_i for $1 \le i \le k$.

For $1 \le i \le k$ define

$$\delta_{2i-1} = \eta_i, \quad \delta_{2i} = \xi_i, \quad \text{and} \quad P_i = (\delta_{2i-1}, \delta_{2i}). \tag{10}$$

Step 2 The definition of $\hat{P}_{k+i}$ for $i = k, \dots, 1$.

Let $\hat{P}_{2k+1}$ be the $2n \times 2$ zero matrix. We define $\hat{P}_{k+i}$ for $i = k, \ldots, 1$ inductively in two subcases below according to the value of (b_i, a_i). Note that by (8) and (9), we have $(b_i, a_i) \neq (0, 0)$ for all these is.

Subcase 1 $b_i \neq 0$.

Define $\hat{P}_{k+i} = (\delta_{2(k+i)-1}, \delta_{2(k+i)})$, where

$$\delta_{2(k+i)-1} = -\frac{b_i}{a_i^2 + b_i^2}\left(\xi_{2k-i+1} + \frac{a_i}{b_i}\eta_{2k-i+1} + \alpha_i \xi_i\right), \tag{11}$$

$$\delta_{2(k+i)} = \frac{b_i}{a_i^2 + b_i^2}\left(\eta_{2k-i+1} - \frac{a_i}{b_i}\xi_{2k-i+1}\right), \tag{12}$$

the constant α_i is chosen so that $\delta_{2(k+i)-1} \perp \delta_{2(k+i)}$. Thus $\operatorname{span}\hat{P}_{k+i}$ becomes an isotropic subspace of $\mathbf{R}^{2n}$. By the definition (11), (12) and a direct computation, we obtain

$$P_i^T J \hat{P}_{k+i} = -I_2. \tag{13}$$

By direct computation we obtain

$$M\hat{P}_{k+i} = \hat{P}_{k+i}R(\theta) + \operatorname{com}\{\hat{P}_{k+i+1}, \max(k-i, 0)P_{i+1}, P_i, P_{i-1}\}. \tag{14}$$

Subcase 2 $b_i = 0$, and then $a_i \neq 0$.

Define $\hat{P}_{k+i} = (\delta_{2(k+i)-1}, \delta_{2(k+i)})$, where

$$\begin{aligned}\delta_{2(k+i)-1} &= -\frac{1}{a_i}\eta_{2k-i+1} + \alpha_i \xi_i,\\ \delta_{2(k+i)} &= -\frac{1}{a_i}\xi_{2k-i+1},\end{aligned}$$

the constant α_i is chosen so that $\delta_{2(k+i)-1} \perp \delta_{2(k+i)}$. Thus $\operatorname{span}\hat{P}_{k+i}$ becomes an isotropic subspace of $\mathbf{R}^{2n}$. Similar to the computation of Subcase 1 we also obtain (13) and (14).

Step 3 The definition of P_{k+i} for $i = k, \ldots, 1$.

Note that by Lemma 2 and Step 2,

$$P_j \perp \hat{P}_{k+i}, \qquad \text{for } 1 \leq j < i,\ 1 \leq i \leq k. \tag{15}$$

Define $P_{2k+1} = 0$, $P_{2k} = \tilde{P}_{2k} = \hat{P}_{2k}$, and for $i = 1, \ldots, k-1$,

$$\tilde{P}_{k+i} = \hat{P}_{k+i} + \sum_{j=i+1}^{k} \hat{P}_{k+j}A_{k+i,j}, \tag{16}$$

where the 2×2 matrices $A_{k+i,j}$ for $i=1,\dots,k-1$ and $j=i+1,\dots,k$ are defined inductively so that $P_j^T J\tilde{P}_{k+i}=0$ for $1\le i<j\le k$. By (15) we obtain

$$P_j^T J\tilde{P}_{k+i}=0, \qquad \text{for} \quad 1\le j<i\le k. \tag{17}$$

$$P_i^T J\tilde{P}_{k+i}=-I_2. \tag{18}$$

For $1\le i\le k-1$ define

$$P_{k+i}=\tilde{P}_{k+i}+\sum_{j=k+i+1}^{2k}\hat{P}_{j-k}B_{k+i,j}, \tag{19}$$

where the 2×2 matrices $B_{k+i,j}$ for $i=1,\dots,k-1$ and $j=2k,\dots,k+i+1$ are defined inductively so that $P_j^T JP_{k+i}=0$ for $k+1\le k+i<j\le 2k$.

By (14), (16), and (19) we obtain

$$\begin{aligned} MP_{k+i} &= P_{k+i}R(\theta) \\ &\quad + \mathrm{com}\{P_{k+h+1}, \max(k-h,0)P_{h+1}, P_h, P_{h-1}\,|\, i\le h\le k\}. \end{aligned} \tag{20}$$

Summarizing the results we also have

$$P_i^T JP_{k+i}=-I_2, \qquad \text{for} \quad 1\le i\le k, \tag{21}$$

$$P_i\perp P_j \quad \text{if} \quad |i-j|\ne k,0, \qquad \text{and} \quad P_i\perp P_i \quad \forall i, \tag{22}$$

$$MP_i=P_iR(\theta)+P_{i-1}, \qquad \text{for} \quad 1\le i\le k. \tag{23}$$

Therefore the column vectors of $P\equiv(P_1,\dots,P_{2k})$ form a symplectic base of $V=\mathrm{span}\Omega$, and we have

$$MP=PM_1, \tag{24}$$

where M_1 is of the following form with $2k\times 2k$ matrices $A_{2k}(\omega)$, $B_{2k}(b)$, and $C_{2k}(\omega)$:

$$M_1=N_{2k}(\omega,b)=\begin{pmatrix} A_{2k}(\omega) & B_{2k}(b) \\ 0 & C_{2k}(\omega)\end{pmatrix}. \tag{25}$$

By (23), $A_{2k}(\omega)$ is a $2k\times 2k$ Jordan block form matrix of the eigenvalue $\omega=e^{\theta\sqrt{-1}}$:

$$A_{2k}(\omega)=\begin{pmatrix} R(\theta) & I_2 & 0 & \cdots & 0 & 0 \\ 0 & R(\theta) & I_2 & \cdots & 0 & 0 \\ 0 & 0 & R(\theta) & \cdots & 0 & 0 \\ \cdot & \cdot & \cdot & \cdots & \cdot & \cdot \\ \cdot & \cdot & \cdot & \cdots & \cdot & \cdot \\ 0 & 0 & 0 & \cdots & R(\theta) & I_2 \\ 0 & 0 & 0 & \cdots & 0 & R(\theta) \end{pmatrix}. \tag{26}$$

By (20), $C_{2k}(\omega)$ is a $2k \times 2k$ matrix of 2×2 blockwise lower triangular matrices, and the diagonal elements of $C_{2k}(\omega)$ are all $R(\theta)$. Because $M \in \mathrm{Sp}(2n)$, the column vectors of P form a symplectic base of V and $M_1 \in \mathrm{Sp}(2k)$. Therefore $A_{2k}(\omega)^T C_{2k}(\omega) = I$, and then $C_{2k}(\omega)$ is of the following form:

$$C_{2k}(\omega)$$

$$= \begin{pmatrix} R(\theta) & 0 & \cdots & 0 & 0 \\ -R(2\theta) & R(\theta) & \cdots & 0 & 0 \\ \cdot & \cdot & \cdots & \cdot & \cdot \\ \cdot & \cdot & \cdots & \cdot & \cdot \\ (-1)^k R((k-1)\theta) & (-1)^{k-1} R((k-2)\theta) & \cdots & R(\theta) & 0 \\ (-1)^{k+1} R(k\theta) & (-1)^k R((k-1)\theta) & \cdots & -R(2\theta) & R(\theta) \end{pmatrix}, \quad (27)$$

and (24) implies that $B_{2k}(b)$ is a $2k \times 2k$ matrix formed by 2×2 real matrices $b_{i,j}$:

$$B_{2k}(b) = (b_{i,j})_{1 \le i,j \le k}, \qquad b_{i,j} = 0 \text{ if } j > i+1. \quad (28)$$

Since M_1 is symplectic, $B_{2k}(b)^T C_{2k}(\omega) = C_{2k}(\omega)^T B_{2k}(b)$.

Now by (21)–(22) we can extend P to a matrix $\Delta \in \mathrm{Sp}(2n)$ such that the column vectors of $(P_1, \ldots, P_k)$ form the 1-st to $2k$-th columns and the column vectors of $(P_{k+1}, \ldots, P_{2k})$ form the $(n+1)$-st to $(n+2k)$-th columns of Δ, and for some $M_0 \in \mathrm{Sp}(2(n-2k))$,

$$M\Delta = \Delta(M_1 \diamond M_0). \quad (29)$$

Now we continue the study of the normal forms of symplectic matrices having eigenvalue $\omega \in \mathbf{U}\backslash\mathbf{R}$ which possess a Jordan block with odd size number $p = 2k+1$.

Case 2 Suppose $M \in \mathrm{Sp}(2n)$ possesses root vectors $\Omega = \{\xi_i, \eta_i \in \mathbf{R}^{2n} | 1 \le i \le p\}$ for some $p = 2k+1$ with an integer $k \ge 0$ in a Jordan block belonging to the eigenvalue $\omega \in \mathbf{U}\backslash\mathbf{R}$ satisfying (1), (2), and the condition

$$b_{k+1} \equiv b_{k+1}(\Omega) \neq 0. \quad (30)$$

Let $V = \mathrm{span}(\Omega)$. Similar to the study of Case 1, we construct a symplectic base $P = (P_1, \ldots, P_{2k})$ of V through several steps. Since the discussion is very similar to that in Case 1, we are rather sketchy here. If $k = 0$, we only need to carry out the following Step 2 to get the symplectic base $\{\delta_1, \delta_2\}$ of V, (24), and (29) with

$$M_1 = N_1(\hat{\omega}, 0) = R(\hat{\theta}), \quad \hat{\omega} = e^{\hat{\theta}\sqrt{-1}}, \quad \hat{\theta} = \theta \quad \text{or} \quad -\theta, \quad (31)$$

according to $b_1 < 0$ or $b_1 > 0$ respectively. Thus we suppose $k \ge 1$ in the following.

Step 1 The definition of P_i for $1 \le i \le k$.

For $1 \le i \le k$ define

$$\delta_{2i-1} = \eta_i, \quad \delta_{2i} = \xi_i, \quad \text{and} \quad P_i = (\delta_{2i-1}, \delta_{2i}). \quad (32)$$

Step 2 The definition of P_{2k+1}.

By Lemma 5, we have $a_{k+1} = 0$. By (30) we distinguish our study in two subcases according to the sign of b_{k+1}.

If $b_{k+1} < 0$, multiplying all ξ_i and η_i by $1/|b_{k+1}|^{1/2}$, without loss of generality we may assume $b_{k+1} = -1$. We define

$$\delta_{2k+1} = \eta_{k+1}, \quad \delta_{4k+2} = \xi_{k+1}, \quad P_{2k+1} = (\delta_{2k+1}, \delta_{4k+2}). \tag{33}$$

If $b_{k+1} > 0$, multiplying all ξ_i and η_i by $1/|b_{k+1}|^{1/2}$, without loss of generality we may assume $b_{k+1} = 1$. We define

$$\delta_{2k+1} = \xi_{k+1}, \quad \delta_{4k+2} = \eta_{k+1}, \quad P_{2k+1} = (\delta_{2k+1}, \delta_{4k+2}). \tag{34}$$

Step 3 The definition of P_{k+i} for $i = k, \ldots, 1$.

Based on $(\eta_{2k-i+2}, \xi_{2k-i+2})$ for $i = k, \ldots, 1$, similar to the study in Steps 2 and 3 of Case 1, we can define $P_{k+i} = (\delta_{2(k+i)}, \delta_{2(k+i)+1})$ so that

$$\begin{aligned}
P_i^T J P_{k+i} &= -I_2, \qquad \text{for} \quad 1 \le i \le k, \\
P_i \perp P_j \quad & \text{if} \quad |i-j| \neq k, 0 \qquad \text{and} \quad P_i \perp P_i \quad \forall i \neq 2k+1, \\
M P_i &= P_i R(\theta) + P_{i-1}, \qquad \text{for} \quad 1 \le i \le k. \\
M P_{k+i} &= P_{k+i} R(\theta) \\
&\quad + \mathrm{com}\{P_{k+h+1}, \max(k-h, 0) P_{h+1}, P_h, P_{h-1} \,|\, i \le h \le k\}.
\end{aligned}$$

Thus we obtain that the column vectors of

$$P \equiv (P_1, \ldots, P_k, \delta_{2k+1}, P_{k+1}, \ldots, P_{2k}, \delta_{4k+2})$$

form a symplectic base of $V = \mathrm{span}\Omega$, and (24) holds. Here M_1 is of the following form with $2k \times 2k$ matrices A, B, C, and $2k \times 1$ matrices D, E, F, G:

$$M_1 = N_{2k+1}(\hat{\omega}, b) \equiv \begin{pmatrix} A & D & B & E \\ 0 & \cos\hat{\theta} & F^T & -\sin\hat{\theta} \\ 0 & 0 & C & 0 \\ 0 & \sin\hat{\theta} & G^T & \cos\hat{\theta} \end{pmatrix} \tag{35}$$

where A is given by (26), C is given by (27), and

$$\hat{\theta} = \theta, \qquad D = (0, \ldots, 0, 1, 0)^T, \quad E = (0, \ldots, 0, 0, 1)^T, \quad \text{if} \quad b_{k+1} = -1, \tag{36}$$

$$\hat{\theta} = -\theta, \qquad D = (0, \ldots, 0, 0, 1)^T, \quad E = (0, \ldots, 0, 1, 0)^T, \quad \text{if} \quad b_{k+1} = 1. \tag{37}$$

Note that in this case, the last two rows of B can not be formed by only 2×2 zero matrices. Now we can extend P to a matrix $\Delta \in \mathrm{Sp}(2n)$ such that (29) holds for some $M_0 \in \mathrm{Sp}(2(n-2k-1))$.

Note that $\dim_{\mathbf{C}} \ker_{\mathbf{C}}(M_1 - \omega I) = 1$ in Cases 1 and 2, because M_1 comes from a single Jordan block of M.

To continue our study for the case of more than one Jordan block, we need the following Lemmas 6, 7, and 9, whose proofs are similar to those of Lemmas 1 to 5, and are omitted.

Lemma 6 *Suppose $M \in \mathrm{Sp}(2n)$ possesses root vectors $\Omega = \{\xi_i, \eta_i \in \mathbf{R}^{2n} | 1 \le i \le p\}$ and $\hat{\Omega} = \{\hat{\xi}_i, \hat{\eta}_i \in \mathbf{R}^{2n} | 1 \le i \le q\}$ for some $p, q \in \mathbf{N}$ in two Jordan blocks belonging to the eigenvalue $\omega \in \mathbf{U}\backslash\mathbf{R}$ such that* (1) *and* (2) *hold for Ω, and that for $1 \le j \le q$ and $\hat{\xi}_0 = \hat{\eta}_0 = 0$,*

$$\begin{aligned} M\hat{\xi}_j &= (\cos\theta)\hat{\xi}_j - (\sin\theta)\hat{\eta}_j + \hat{\xi}_{j-1}, \\ M\hat{\eta}_j &= (\sin\theta)\hat{\xi}_j + (\cos\theta)\hat{\eta}_j + \hat{\eta}_{j-1}. \end{aligned}$$

Then for all $1 \le i \le p$ and $1 \le j \le q$,

$$\xi_i^T J\hat{\xi}_j = \eta_i^T J\hat{\eta}_j, \qquad \xi_i^T J\hat{\eta}_j = -\eta_i^T J\hat{\xi}_j,$$

$$\begin{aligned} &\xi_{i-1}^T J\hat{\xi}_{j-1} + (\cos\theta)(\xi_{i-1}^T J\hat{\xi}_j + \xi_i^T J\hat{\xi}_{j-1}) \\ &\quad - (\sin\theta)(\xi_{i-1}^T J\hat{\eta}_j + \eta_i^T J\hat{\xi}_{j-1}) = 0, \end{aligned}$$

$$\begin{aligned} &\xi_{i-1}^T J\hat{\eta}_{j-1} + (\cos\theta)(\xi_{i-1}^T J\hat{\eta}_j + \xi_i^T J\hat{\eta}_{j-1}) \\ &\quad + (\sin\theta)(\xi_{i-1}^T J\hat{\xi}_j - \eta_i^T J\hat{\eta}_{j-1}) = 0. \end{aligned}$$

Lemma 7 *Under the assumption of Lemma* 6, *we suppose $q \ge p$. Let $\hat{V}_i = \mathrm{span}\{\hat{\xi}_i, \hat{\eta}_i\}$ for $1 \le i \le q$. Then*

$$V_i \perp \hat{V}_j, \qquad \forall 1 \le j \le q - i, 1 \le i \le p.$$

Definition 8 Under the assumption of Lemma 6, suppose $q = p$. We define

$$\begin{aligned} A_i \equiv A_i(\Omega, \hat{\Omega}) &= \eta_i^T J\hat{\eta}_{p-i+1} \\ &= \xi_i^T J\hat{\xi}_{p-i+1}, \qquad \forall i = 1, \ldots, p, \\ B_i \equiv B_i(\Omega, \hat{\Omega}) &= \eta_i^T J\hat{\xi}_{p-i+1} \\ &= -\xi_i^T J\hat{\eta}_{p-i+1}, \qquad \forall i = 1, \ldots, p. \end{aligned}$$

Lemma 9 *Under the assumption of Lemma* 6, *suppose $q = p$. Then*

$$(A_i, B_i)^T = -R(2\theta)(A_{i-1}, B_{i-1})^T, \qquad \forall i = 2, \ldots, p.$$

Now we continue our study on the normal forms.

Case 3 Suppose $M \in \mathrm{Sp}(2n)$ satisfies the assumptions of Lemma 6 with $q = p = k$ for some $p \in \mathbf{N}$ and the conditions

$$(b_1, a_1) \equiv (b_1(\Omega), a_1(\Omega)) = (0, 0), \tag{38}$$

$$(\hat{b}_1, \hat{a}_1) \equiv (b_1(\hat{\Omega}), a_1(\hat{\Omega})) = (0, 0), \tag{39}$$

$$(B_k, A_k) \equiv (B_k(\Omega, \hat{\Omega}), A_k(\Omega, \hat{\Omega})) \neq (0, 0). \tag{40}$$

Note that by (38)–(40) and Lemmas 5 and 9,

$$b_i = a_i = \hat{b}_i = \hat{a}_i = 0, \quad (B_i, A_i) \neq (0, 0), \qquad \forall 1 \le i \le k. \tag{41}$$

Next we reduce our study of Case 3 to the study of Case 1.

Denote by $V_i = \mathrm{span}\{\xi_i, \eta_i\}$ for $1 \le i \le k$. We need the following lemma.

Lemma 10 *Under the assumption of Case* 3, *without loss of generality, we may assume that*

$$V_i \perp V_j, \qquad \forall 1 \le i, j \le k. \tag{42}$$

Proof: We follow the idea of the proof of Lemma 4.2. Assume (42) does not hold. Let $m = m(\Omega)$ be the least integer in $[1, k]$ such that there exists the least integer $q \in [m, k]$ satisfying $V_m \not\perp V_q$. By the minimality of m, we must have $q = k$. By (38), we must have $m > 1$.

Denote $P_i = (\eta_i, \xi_i)$, $Q_i = (\hat{\eta}_i, \hat{\xi}_i)$, and $R_i = (\hat{\xi}_i, -\hat{\eta}_i)$ for $0 \le i \le k$. Define

$$\begin{aligned} X_i &= P_i, \qquad \text{for } 1 \le i < m, \\ X_i &= P_i + Q_{i-m+1}A + R_{i-m+1}B, \qquad \text{for } m \le i \le k, \end{aligned}$$

where A and B are 2×2 matrices defined by $A = aR(\alpha)$ and $B = bR(\beta)$ with real numbers a, b, α, β to be determined later. Let $\Theta = (X_1, \ldots, X_k)$.

Then we have

$$MX_i = X_i R(\theta) + X_{i-1}, \qquad \forall 1 \le i \le k.$$

By the fact $m > 1$, it holds that $\mathrm{span}(\Theta, \hat{\Omega}) = \mathrm{span}(\Omega, \hat{\Omega})$ and

$$\begin{aligned} (b_1(\Theta), a_1(\Theta)) &= (b_1, a_1) = (0, 0), \\ (B_k(\Theta, \hat{\Omega}), A_k(\Theta, \hat{\Omega})) &= (B_k, A_k) \neq (0, 0). \end{aligned}$$

Thus the conditions of Case 3 still hold for Θ and $\hat{\Omega}$.

On the other hand, for $1 \le i < m$ and $1 \le j \le k$ we obtain $X_i^T J X_j = 0$ by Lemma 7 and the fact $j - m + 1 \le k - (m + 1) \le k - i$. For $1 \le j < m$ we obtain

$X_m^TJX_j = 0$ by Lemma 7. For $m \le j < k$ we obtain $X_m^TJX_j = 0$ by Lemma 7. Finally by Lemma 7 again and $m > 1$ we have

$$\begin{aligned} X_m^TJX_k = P_m^TJP_k &+ (P_m^TJQ_{k-m+1}A + A^TQ_1^TJP_k) \\ &+ (P_m^TJR_{k-m+1}B + B^TR_1^TJP_k). \end{aligned} \tag{43}$$

By (3) we obtain

$$P_m^TJP_k = \begin{pmatrix} x & y \\ -y & x \end{pmatrix}.$$

Denote $A_\pm = A_m \pm A_k$, $B_\pm = B_m \pm B_k$, $c_1 = \cos\alpha$, $s_1 = \sin\alpha$, $c_2 = \cos\beta$, and $s_2 = \sin\beta$. By direct computation we obtain

$$P_m^TJQ_{k-m+1}A + A^TQ_1^TJP_k = a\begin{pmatrix} c_1A_- + s_1B_+ & -s_1A_+ + c_1B_- \\ s_1A_+ - c_1B_- & c_1A_- + s_1B_+ \end{pmatrix},$$

and

$$P_m^TJR_{k-m+1}B + B^TR_1^TJP_k = b\begin{pmatrix} -s_2A_+ + c_2B_- & -c_2A_- - s_2B_+ \\ c_2A_- + s_2B_+ & -s_2A_+ + c_2B_- \end{pmatrix}.$$

Thus to make the right-hand side of (43) be zero, we need to solve the following equation systems for a, b, α, and β,

$$(c_1A_- + s_1B_+)a + (c_2B_- - s_2A_+)b = -x, \tag{44}$$

$$(c_1B_- - s_1A_+)a + (-c_2A_- - s_2B_+)b = -y. \tag{45}$$

The coefficient determinant of this linear system on a and b is given by

$$D = -c_1c_2(A_-^2 + B_-^2) - s_1s_2(A_+^2 + B_+^2) - (c_1s_2 + s_1c_2)(A_-B_+ - A_+B_-).$$

Since by (40) it holds that $(A_k, B_k) \ne (0,0)$, we must have

$$(A_-, A_+) \ne (0,0) \quad \text{and} \quad (B_-, B_+) \ne (0,0).$$

Thus we can choose α and $\beta \in \mathbf{R}$ suitably such that $D \ne 0$. Then we define a and $b \in \mathbf{R}$ by (44) and (45). By this choice we obtain $X_m^TJX_k = 0$ when $m > 1$.

Now if (42) still does not hold for Θ, by the above discussion we obtain

$$m(\Theta) \ge m + 1 = m(\Omega) + 1.$$

Thus by induction, we can modify Ω to complete the proof of the lemma. □

Now suppose (42) holds. As in Section 2, we can define $\xi_{k+i} = \hat{\xi}_i$ and $\eta_{k+i} = \hat{\eta}_i$ for $1 \le i \le k$, and then carry out the study of Case 1 to obtain the normal form given by (25)–(28). Note that $\dim_{\mathbf{C}} \ker_{\mathbf{C}}(M_1 - \omega I) = 2$ in Case 3, because M_1 comes from two Jordan blocks of M for the eigenvalue ω.

We have completed our study of the three basic cases.

Now the following is the main result in this section.

Theorem 11 *Suppose* $M \in \mathrm{Sp}(2n)$ *possesses the eigenvalue* $\omega \in \mathbf{U}\backslash\mathbf{R}$. *Then there exist* $P \in \mathrm{Sp}(2n)$ *and* $m \in \mathbf{N}$ *such that*

$$P^{-1}MP = M_1\diamond\cdots\diamond M_m\diamond M_0, \tag{46}$$

where $M_0 \in \mathrm{Sp}(2k_0)$ *with* $k_0 \geq 0$ *and* $\omega \notin \sigma(M_0)$, $k_i \geq 1$ *and* $M_i \in \mathrm{Sp}(2k_i)$ *is of the normal form* $N_{k_i}(\omega_i, b_i)$ *with* $\omega_i = \omega$ *or* $\overline{\omega}$ *defined by* (25)–(28) *with* $\dim_{\mathbf{C}} \ker_{\mathbf{C}}(M_i - \omega I) = 1$ *or* 2, (31) *or* (35)–(37) *with* $\dim_{\mathbf{C}} \ker_{\mathbf{C}}(M_i - \omega I) = 1$ *for* $1 \leq i \leq m$. *We denote by* $E_\omega(M)$ *the invariant root vector space of* M *belonging to the eigenvalue* ω; *then* $2\sum_{i=0}^m k_i = 2k_0 + \dim E_\omega(M) = 2n$.

Proof: Based upon the three basic cases we studied earlier, the proof of this theorem follows from that of Theorem 4.1, and therefore is omitted. □

1.7 Normal forms for eigenvalues outside U

In this section we derive the normal forms of symplectic matrices having the eigenvalue pair $\Lambda = \{\lambda, \lambda^{-1}\} \subset \mathbf{R}\backslash\{0, \pm 1\}$ or the eigenvalue quadruple $\Lambda = \{\rho\omega, \rho\bar{\omega}, \rho^{-1}\omega, \rho^{-1}\bar{\omega}\} \subset \mathbf{C}\backslash(\mathbf{U} \cup \mathbf{R})$. The following are two suggestive examples which point out the difference between the eigenvalue ± 1 and $\beta \in \mathbf{R}\backslash\{\pm 1, 0\}$.

$$\begin{pmatrix} 0 & -1 \\ 1 & \alpha/(\beta - \beta^{-1}) \end{pmatrix}\begin{pmatrix} \beta & \alpha \\ 0 & \beta^{-1} \end{pmatrix}\begin{pmatrix} \alpha/(\beta - \beta^{-1}) & 1 \\ -1 & 0 \end{pmatrix} = \begin{pmatrix} \beta^{-1} & 0 \\ 0 & \beta \end{pmatrix},$$

and

$$\begin{pmatrix} -1 & \alpha/(\beta^{-1} - \beta) \\ 0 & -1 \end{pmatrix}\begin{pmatrix} \beta & \alpha \\ 0 & \beta^{-1} \end{pmatrix}\begin{pmatrix} -1 & \alpha/(\beta - \beta^{-1}) \\ 0 & -1 \end{pmatrix} = \begin{pmatrix} \beta & 0 \\ 0 & \beta^{-1} \end{pmatrix}.$$

The main results of this section are the following theorems. Since the discussion is similar to those in Sections 4 and 6, we only list the results here. The details of the proofs in this section can be found in [HaL1].

For the case of an eigenvalue pair, we have

Theorem 1 *Suppose* $M \in \mathrm{Sp}(2n)$ *possesses the eigenvalue pair* $\Lambda = \{\lambda, \lambda^{-1}\} \subset \mathbf{R}\backslash\{0, \pm 1\}$. *Then there exist* $P \in \mathrm{Sp}(2n)$ *and* $m \in \mathbf{N}$ *such that*

$$P^{-1}MP = M_1\diamond\cdots\diamond M_m\diamond M_0, \tag{1}$$

where $M_0 \in \mathrm{Sp}(2k_0)$ *with* $k_0 \geq 0$ *and* $\Lambda \cap \sigma(M_0) = \emptyset$, $M_i \in \mathrm{Sp}(2k_i)$ *for* $k_i \geq 1$ *has the form*

$$M_i = \begin{pmatrix} A_{k_i}(\lambda) & 0 \\ 0 & C_{k_i}(\lambda) \end{pmatrix} \tag{2}$$

with $A_{k_i}(\lambda)$ and $C_{k_i}(\lambda)$ defined by (4.3) *and* (4.4) *respectively for* $1 \le i \le m$, *and* $2\sum_{i=0}^{m} k_i = 2k_0 + \dim E_\Lambda(M) = 2n$ *for* $E_\Lambda(M)$ *defined to be the invariant root vector space of* M *belonging to the eigenvalue pair* Λ.

Note that by the proof of Theorem 4.1, considering the simplest case of $P^{-1}MP = M_1$ in (1), the M in the above Theorem 1 can be reduced to

$$M_1 = \begin{pmatrix} A_n(\lambda) & B_n(\lambda, b) \\ 0 & C_n(\lambda) \end{pmatrix}, \tag{3}$$

where $A_n(\lambda)$, $C_n(\lambda)$, and $B_n(\lambda, b)$ with some $b = (b_1, \ldots, b_n) \in \mathbf{R}^n$ are given by (4.3)–(4.5) in Section 1.4. Thus the proof of Theorem 1 is reduced to the proof of the following claim:

Claim: $B_n(\lambda, b) = 0$ *in* (3).

In fact, let $j \in [1, n]$ be the greatest integer such that $b_j \neq 0$. Then the vectors

$$\{\delta_1, \ldots, \delta_{n+j-1}, \delta_{n+j} + \alpha\delta_j, \delta_{n+j+1}, \ldots, \delta_{2n}\},$$

where $\alpha = -(-\lambda)^{j-1} b_j / (\lambda - \lambda^{-1})$, also form a symplectic basis of the space $\mathbf{R}^{2n}$. In terms of this basis, the above matrix M_1 still has the form (3), and furthermore the $(j-1, j)$-th entry of M_1 satisfies $d_{j-1,j} = \alpha$. Since in this case, the corresponding $B_n(\lambda, b)$ block of M_1 must be still lower triangular, this proves $\alpha = 0$, i.e. $b_j = 0$. The claim is proved.

For the case of eigenvalue quadruple $\Lambda = \{\rho\omega, \rho\bar{\omega}, \rho^{-1}\omega, \rho^{-1}\bar{\omega}\} \subset \mathbf{C}\backslash(\mathbf{U} \cup \mathbf{R})$, fix $\omega \in \mathbf{U}\backslash\mathbf{R}$ and denote $\omega = \cos\theta + \sin\theta\sqrt{-1}$ for some $0 \le \theta \le 2\pi$. Note that in this case there always holds $\sin\theta \neq 0$. Define

$$N_{2k}(\rho, \theta) = \begin{pmatrix} A_{2k}(\rho, \theta) & 0 \\ 0 & C_{2k}(\rho, \theta) \end{pmatrix}, \tag{4}$$

where $A_{2k}(\rho, \theta)$ is a $2k \times 2k$ Jordan block form matrix of the eigenvalue group $\Lambda_+ = \{\rho\omega, \rho\overline{\omega}\}$:

$$A_{2k}(\rho, \theta) = \begin{pmatrix} \rho R(\theta) & I_2 & 0 & \cdots & 0 & 0 \\ 0 & \rho R(\theta) & I_2 & \cdots & 0 & 0 \\ 0 & 0 & \rho R(\theta) & \cdots & 0 & 0 \\ \cdot & \cdot & \cdot & \cdots & \cdot & \cdot \\ \cdot & \cdot & \cdot & \cdots & \cdot & \cdot \\ 0 & 0 & 0 & \cdots & \rho R(\theta) & I_2 \\ 0 & 0 & 0 & \cdots & 0 & \rho R(\theta) \end{pmatrix}. \tag{5}$$

$C_{2k}(\rho, \theta)$ is a 2×2 blockwise lower triangular $2k \times 2k$ matrix, the diagonal elements of $C_{2k}(\rho, \theta)$ are all $\rho^{-1}R(\theta)$, and $A_{2k}(\rho, \theta)^T C_{2k}(\rho, \theta) = I$. Thus we have

$$C_{2k}(\rho, \theta) = \begin{pmatrix} \rho^{-1}R(\theta) & 0 & \cdots & 0 \\ -\rho^{-2}R(2\theta) & \rho^{-1}R(\theta) & \cdots & 0 \\ \cdot & \cdot & \cdots & \cdot \\ \cdot & \cdot & \cdots & \cdot \\ -(-\rho^{-1})^{k-1}R((k-1)\theta) & -(-\rho^{-1})^{k-2}R((k-2)\theta) & \cdots & 0 \\ -(-\rho^{-1})^{k}R(k\theta) & -(-\rho^{-1})^{k-1}R((k-1)\theta) & \cdots & \rho^{-1}R(\theta) \end{pmatrix}. \quad (6)$$

Theorem 2 *Suppose $M \in \mathrm{Sp}(2n)$ possesses an eigenvalue group $\Lambda = \{\rho\omega, \rho\bar{\omega}, \rho^{-1}\omega, \rho^{-1}\bar{\omega}\} \subset \mathbf{C}\backslash(\mathbf{U} \cup \mathbf{R})$ with $\rho \in \mathbf{R}\backslash\{0, \pm 1\}$ and $\omega = e^{\theta\sqrt{-1}} \in \mathbf{U}\backslash\mathbf{R}$. Then there exist $P \in \mathrm{Sp}(2n)$ and $m \in \mathbf{N}$ such that*

$$P^{-1}MP = M_1 \diamond \cdots \diamond M_m \diamond M_0, \quad (7)$$

where $M_0 \in \mathrm{Sp}(2k_0)$ with $k_0 \geq 0$ and $\Lambda \cap \sigma(M_0) = \emptyset$. For $1 \leq i \leq m$, it holds that $k_i \geq 1$, $M_i \in \mathrm{Sp}(4k_i)$ is of the form $M_i = N_{k_i}(\rho, \theta)$ given by (4), and $2k_0 + 4\sum_{i=0}^{m} k_i = 2k_0 + \dim E_\Lambda(M) = 2n$ for $E_\Lambda(M)$ being the root vector space of M belonging to the eigenvalue group Λ.

Summarizing our discussion on normal forms of symplectic matrices in this chapter, we obtain the following result.

Theorem 3 *For any $M \in \mathrm{Sp}(2n)$, there exist $P \in \mathrm{Sp}(2n)$, an integer $p \in [0, n]$, normal forms $M_i \in \mathrm{Sp}(2k_i)$ of eigenvalues λ_i for $1 \leq i \leq p$ defined in the Sections 1.4 to 1.7, such that $\sum_{i=1}^{p} k_i = n$ and it holds that*

$$PMP^{-1} = M_1 \diamond \cdots \diamond M_p. \quad (8)$$

1.8 Basic normal forms

Note that for any $M \in \mathrm{Sp}(2n)$,

$$\dim_{\mathbf{C}} \ker_{\mathbf{C}}(M \pm I) = \dim_{\mathbf{R}} \ker_{\mathbf{R}}(M \pm I). \quad (1)$$

Definition 1 For any $M \in \mathrm{Sp}(2n)$ and $\omega \in \mathbf{U}$, we define the **ω-nullity** $\nu_\omega(M)$ of M by

$$\nu_\omega(M) = \dim_{\mathbf{C}} \ker_{\mathbf{C}}(M - \omega I). \quad (2)$$

We define the **hyperbolic index** $\alpha(M)$ of M by the mod 2 number of the total algebraic multiplicity of negative eigenvalues of M which are strictly less than -1, and the **elliptic height** $e(M)$ of M by the total algebraic multiplicity of all eigenvalues of M on $\mathbf{U}$. An $M \in \mathrm{Sp}(2n)$ is

truly hyperbolic,	if $e(M) = 0$,
hyperbolic,	if $1 \in \sigma(M)$ *and* $e(M) = 2$,
elliptic,	if $e(M) = 2n$,
strongly elliptic,	if $\sigma(M) \subset \mathbf{U}\backslash\{1\}$.

We denote by $\mathrm{Sp}^{th}(2n)$, $\mathrm{Sp}^{h}(2n)$, $\mathrm{Sp}^{e}(2n)$, and $\mathrm{Sp}^{se}(2n)$ the subsets of all truly hyperbolic, hyperbolic, elliptic, strongly elliptic matrices in $\mathrm{Sp}(2n)$ respectively.

By Theorem 3.1, for any $M \in \mathrm{Sp}(2n)$, it holds that $e(M) \in 2\mathbf{Z}$ and $0 \le e(M) \le 2n$. Note that the concept of hyperbolicity comes from the study of periodic solutions of autonomous Hamiltonian systems. In this case, 1 is always a Floquet multiplier with multiplicity at least two of the given periodic solution. We shall further study the structures of these subsets of $\mathrm{Sp}(2n)$ in the next chapter.

Another important function we need is the following modification of the characteristic function of M defined by (3.2): for $\omega \in \mathbf{U}$ and $M \in \mathrm{Sp}(2n)$, we define

$$D_\omega(M) = (-1)^{n-1}\omega^{-n}\det(M - \omega I). \tag{3}$$

Lemma 2 *D is a real and real smooth function on $\mathbf{U} \times \mathrm{Sp}(2n)$.*

Proof: From (3.3) we obtain $\overline{D_\omega(M)} = D_\omega(M)$. Thus $D_\omega(M)$ is real. It is clearly smooth. □

For $\omega \in \mathbf{U}$, we define the **ω-singular set** $\mathrm{Sp}(2n)^0_\omega$ of $\mathrm{Sp}(2n)$ and its subsets $\mathcal{M}^k_\omega(2n)$ with $0 \le k \le 2n$ by

$$\mathrm{Sp}(2n)^0_\omega = \{M \in \mathrm{Sp}(2n) \mid D_\omega(M) = 0\}, \tag{4}$$

$$\mathcal{M}^k_\omega(2n) = \{M \in \mathrm{Sp}(2n) \mid \nu_\omega(M) = k\}. \tag{5}$$

We define the **ω-regular sets** of $\mathrm{Sp}(2n)$ by

$$\mathrm{Sp}(2n)^\pm_\omega = \{M \in \mathrm{Sp}(2n) \mid \pm D_\omega(M) < 0\}, \tag{6}$$

$$\mathrm{Sp}(2n)^*_\omega = \mathrm{Sp}(2n)^+_\omega \cup \mathrm{Sp}(2n)^-_\omega. \tag{7}$$

By Theorem 3.1 we obtain

$$\mathrm{Sp}(2n)^0_\omega = \mathrm{Sp}(2n)^0_{\overline{\omega}}, \qquad \mathrm{Sp}(2n)^*_\omega = \mathrm{Sp}(2n)^*_{\overline{\omega}}.$$

Note that when $\omega = 1$, we have

$$\mathcal{M}^k_1(2n) = \{M \in \mathrm{Sp}(2n)^0_1 \mid \dim_{\mathbf{R}} \ker_{\mathbf{R}}(M - I) = k\}, \qquad \forall 0 \le k \le 2n.$$

For simplicity, we omit the subindex ω for $\omega = 1$ in these notations, when there is no confusion. The set $\mathrm{Sp}(2n)^*$ is called the **regular part** of $\mathrm{Sp}(2n)$. The codimension 1 hypersurface $\mathrm{Sp}(2n)^0$ is called the **singular part** of $\mathrm{Sp}(2n)$. We write $\mathcal{M}_\omega(2n)$ for $\mathcal{M}^1_\omega(2n)$, which is the regular part of the hypersurface $\mathrm{Sp}(2n)^0_\omega$.

In order to study the topological structures of subsets of $\mathrm{Sp}(2n)$ with $n \ge 2$ as well as the index functions, we study perturbations on normal form matrices defined in Sections 4 to 7 to connect the given matrix to an easier one. To carry out the concrete computations, we need the following lemma.

Lemma 3 *Suppose* $M = \begin{pmatrix} A & B \\ C & D \end{pmatrix} \in \mathcal{L}(\mathbf{R}^{2n})$ *is of the square block form, A is invertible, and* $AC = CA$. *Then*

$$\det M = \det(AD - CB).$$

Proof: Since A is invertible, we have the matrix identity

$$\begin{pmatrix} I & 0 \\ -CA^{-1} & I \end{pmatrix} \begin{pmatrix} A & B \\ C & D \end{pmatrix} \begin{pmatrix} I & -A^{-1}B \\ 0 & I \end{pmatrix} = \begin{pmatrix} A & 0 \\ 0 & D - CA^{-1}B \end{pmatrix}.$$

Thus

$$\det M = (\det A)(\det(D - CA^{-1}B)) = \det(AD - ACA^{-1}B).$$

From $AC = CA$, the lemma follows. □

Lemma 4 *For any* $M \in \mathrm{Sp}(2n)$ *and a path* $\gamma \in C([0,1], \mathrm{Sp}(2n))$, *suppose* $\lambda \in \sigma(M)$. *Then the algebraic and geometric multiplicities and the Krein type of* $\lambda \in \sigma(\gamma(t)^{-1}M\gamma(t))$ *are the same as those of M for any* $t \in [0,1]$.

Proof: By direct computations. □

Definition 5 For any $M \in \mathrm{Sp}(2n)$ and $\omega \in \mathbf{U}$, we define the **ω-homotopy set** $\Omega_\omega(M)$ of M in $\mathrm{Sp}(2n)$ by

$$\Omega_\omega(M) = \{N \in \mathrm{Sp}(2n) \mid \nu_\omega(N) = \nu_\omega(M)\}, \tag{8}$$

and the **homotopy set** $\Omega(M)$ of M in $\mathrm{Sp}(2n)$ by

$$\begin{aligned} \Omega(M) = \{N \in \mathrm{Sp}(2n) \mid \sigma(N) \cap \mathbf{U} = \sigma(M) \cap \mathbf{U}, \text{ and} \\ \nu_\lambda(N) = \nu_\lambda(M)\ \forall \lambda \in \sigma(M) \cap \mathbf{U}\}. \end{aligned} \tag{9}$$

We denote by $\Omega^0(M)$ (or $\Omega^0_\omega(M)$) the path connected component of $\Omega(M)$ ($\Omega_\omega(M)$) which contains M, and call it the **homotopy component** (or **ω-homotopy component**) of M in $\mathrm{Sp}(2n)$.

For any $M \in \mathrm{Sp}(2n)$, we define the **conjugacy class** of M in $\mathrm{Sp}(2n)$ by

$$[M] = \{N \in \mathrm{Sp}(2n) \mid N = P^{-1}MP \text{ for some } P \in \mathrm{Sp}(2n)\}.$$

Since $\mathrm{Sp}(2n)$ is path connected, $[M] \subset \Omega^0(M) \subset \Omega^0_\omega(M)$ for all $\omega \in \mathbf{U}$.

For any $M \in \mathrm{Sp}(2n)$, to find much simpler elements in $\Omega^0(M)$ or $\Omega^0_\omega(M)$ than the $\diamond$-product of the normal forms in (7.8), we consider perturbations on the normal forms. For $\alpha \in \mathbf{R}$, we define

$$R_k(\alpha) = I_{2k-2} \diamond R(\alpha) \diamond I_{2n-2k}, \quad \forall k = 1, \ldots, n.$$

Here I_0 means that this component does not appear at all.

Definition 6 A normal form $N_n(\omega, B) \in \mathrm{Sp}(2n)$ belonging to the eigenvalue $\omega \in \mathbf{U}$ is of the **first type** and denoted by $t(N_n(\omega, B)) = 1$, if $N_n(\omega, B)R_1(\alpha) \in \mathrm{Sp}(2n)^*_\omega$ for all sufficiently small $|\alpha| > 0$, and is of the **second type** and denoted by $t(N_n(\omega, B)) = 2$, if $N_n(\omega, B)R_1(\alpha) \in \mathrm{Sp}(2n)^0_\omega$ for all sufficiently small $|\alpha| > 0$.

Remark 7 By studies below, when $t(N_n(\omega, B)) = 2$, it must hold that $M \equiv N_n(\omega, B) \in \mathcal{M}^2_\omega(2n)$, $n \geq 2$ if $\omega = \pm 1$, $n \geq 3$ if $\omega \in \mathbf{U}\backslash\mathbf{R}$, and

$$MR_1(\alpha) \in \mathcal{M}^1_\omega(2n), \tag{10}$$

$$MR_{n+1}(\alpha) \in \mathcal{M}^1_\omega(2n), \tag{11}$$

$$\nu_\omega(MR_1(\beta)) - \nu_\omega(MR_1(\beta)R_{n+1}(\alpha)) = 1, \tag{12}$$

if $|\alpha| > 0$ and $|\beta| \geq 0$ are small enough.

The next example shows that the property (12) for normal forms need not hold if we replace $N_2(1, b)$ by some M which is not a normal form. This example is due to Nancy Hingston.

Example 8 (N. Hingston). Define $M \in \mathrm{Sp}(4)$ by

$$M = \begin{pmatrix} 1 & 0 & 0 & 0 \\ 0 & 1 & 0 & 0 \\ 1 & -1 & 1 & 0 \\ -1 & 1 & 0 & 1 \end{pmatrix}.$$

Note that there exists $P \in \mathrm{Sp}(4)$ such that $P^{-1}MP = I_2 \diamond N_1(1, -1)$ and $\nu_1(M) = 3$. Then by direct computation we get

$$\begin{aligned} \nu_1(MR_1(\theta)) &= 1, && \text{if and only if } \theta \neq 0 \bmod 2\pi, \\ \nu_1(MR_2(\theta)) &= 1, && \text{if and only if } \theta \neq 0 \bmod 2\pi, \\ \nu_1(MR_1(\theta_1)R_2(\theta_2)) &= 0, && \text{if and only if both } \theta_1 \text{ and } \theta_2 \neq 0 \bmod 2\pi. \end{aligned}$$

Thus for $\theta_1 \in [-1, 1]$ and $\theta_2 \in [-1, 1]\backslash\{0\}$ we obtain

$$\nu_1(MR_1(\theta_1)) - \nu_1(MR_1(\theta_1)R_2(\theta_2)) = \begin{cases} 1, & \text{if } \theta_1 \neq 0, \\ 2, & \text{if } \theta_1 = 0. \end{cases}$$

So this matrix M does not satisfy (12).

Next using normal forms, we study the representations of $M \in \mathrm{Sp}(2n)$ in its homotopy components, i.e., for any $M \in \mathrm{Sp}(2n)$, to find the simplest matrix in $\Omega^0(M)$ or $\Omega^0_\omega(M)$. Let $D(\lambda) = \mathrm{diag}(\lambda, \lambda^{-1})$ for $\lambda \in \mathbf{R}\backslash\{0\}$.

Case 1 $M = N_n(1, B) \in \mathcal{M}^1_1(2n)$ with $n \geq 2$. For $\epsilon > 0$ small enough, we define

$$f(t) = M[I_2 \diamond D(1 + t\epsilon)^{\diamond(n-1)}], \qquad \forall t \in [0, 1].$$

Applying Theorem 7.3 to $f(1)$, we obtain $P \in \mathrm{Sp}(2n)$ such that $Pf(1)P^{-1} = N_1(\lambda, c)\diamond M_0$ with $\sigma(M_0) \cap \mathbf{U} = \emptyset$. By the path connectedness of $\mathrm{Sp}(2n)$, M can be connected to $N_1(1, c)\diamond M_0$ within $\Omega^0(M)$.

Case 2 $M = N_n(1, B) \in \mathcal{M}_1^2(2n)$ with $n \geq 2$. We define $f(t) = N_n(1, (1-t)B)$ for $t \in [0, 1]$ and

$$g(t) = f(1)[I_2\diamond D(1+t)^{\diamond(n-2)}\diamond I_2], \qquad \forall t \in [0, 1].$$

Applying Theorem 7.3 to $g(1)$, we obtain a path h in $\Omega_1(g(1))$ which connects $g(1)$ to $N_2(1, c)\diamond M_0$ with $\sigma(M_0) \cap \mathbf{U} = \emptyset$. For $t \in [0, 1]$ we define 4×4 matrix paths $\phi(t) = N_2(1, (1-t)c)$ and

$$\psi(t) = \begin{pmatrix} 1 & 1-t^2 & 0 & 0 \\ 0 & 1+t & 0 & 0 \\ 0 & 0 & 1 & 0 \\ 0 & 0 & t-1 & 1/(1+t) \end{pmatrix}. \tag{13}$$

Then the path $[(\psi * \phi)\diamond M_0] * h * g * f$ connects M to $I_2\diamond D(2)\diamond M_0$ in $\Omega^0(M)$.

Case 3 $M = N_n(\omega, B) \in \mathcal{M}_\omega^1(2n)$ is a normal form of the eigenvalue $\omega \in \mathbf{U}\backslash\mathbf{R}$ with $n \geq 2$. For $\epsilon > 0$ small enough, we define

$$f(t) = N_n(\omega, B)[D(1+t\epsilon)^{\diamond(n-k)}\diamond I_{2k}], \qquad \forall t \in [0, 1],$$

where $k = 2$ if M is given by (6.25), and $k = 1$ if M is given by (6.35). Then $f(t) \in \Omega^0(M)$ for all $t \in [0, 1]$. Thus together with Theorem 7.3, M can be connected to $N_k(\hat{\omega}, c)\diamond M_0$ within $\Omega^0(M)$ with $\hat{\omega} = \omega$ or $\overline{\omega}$, $\sigma(M_0) \cap \mathbf{U} = \emptyset$, and some c.

Case 4 $M = N_n(\omega, B) \in \mathcal{M}_\omega^2(2n)$ is a normal form of the eigenvalue $\omega \in \mathbf{U}\backslash\mathbf{R}$ with $n \geq 2$ given by (6.25). We define $f(t) = N_n(\omega, (1-t)B)$ and

$$g(t) = f(1)[I_4\diamond D(1+t)^{\diamond(n-2)}], \qquad \forall t \in [0, 1].$$

Then there exists a path h in $\mathcal{M}_\omega^2(2n)$ which connects $g(1)$ to $N_2(\hat{\omega}, 0)\diamond M_0$ with $\hat{\theta} = \theta$ or $2\pi - \theta$ and $M_0 \in \mathrm{Sp}(2n-4)$ satisfies $\sigma(M_0) \cap \mathbf{U} = \emptyset$. For any $t \in [0, 1]$ we define

$$h(t) = \begin{pmatrix} c & -s(1-t) & -st & s\sqrt{2t-2t^2} \\ s(1-t) & c & s\sqrt{2t-2t^2} & st \\ st & -s\sqrt{2t-2t^2} & c & -s(1-t) \\ -s\sqrt{2t-2t^2} & -st & s(1-t) & c \end{pmatrix}, \tag{14}$$

where $c = \cos\hat{\theta}$ and $s = \sin\hat{\theta}$. By direct computation we obtain $h(t) \in \mathcal{M}_\omega^2(4)$ for all $t \in [0, 1]$. Note that $N_2(\hat{\omega}, 0)$ is connected to $R(\hat{\theta})\diamond R(-\hat{\theta})$ within $\mathcal{M}_\omega^2(4)$ by the path h. Thus $R(\hat{\theta})\diamond R(-\hat{\theta})\diamond M_0$ is in $\Omega^0(M)$.

Note that $N_n(\omega, B)$ for $\omega \in \mathbf{U}$ in other cases can also be connected to the $\diamond$-product of basic normal forms defined below and matrices possessing no eigenvalues on $\mathbf{U}$. Since the proofs are similar, they are omitted.

Definition 9 A normal form matrix $M = N_n(\omega, B) \in \mathrm{Sp}(2n)$ defined above is a **basic normal form** of eigenvalue $\omega \in \mathbf{U}$ if $n = 1$ and $\omega \in \mathbf{U}$, or $M = N_2(\omega, B) \in \mathcal{M}_\omega^1(4)$ and $\omega \in \mathbf{U}\backslash\mathbf{R}$. $D(2)$ and $D(-2)$ are **basic normal forms** of eigenvalues outside $\mathbf{U}$.

More explicitly, the following are the basic normal forms for eigenvalues in $\mathbf{U}$:

$$N_1(\lambda, b) = \begin{pmatrix} \lambda & b \\ 0 & \lambda \end{pmatrix}, \qquad \lambda = \pm 1, b = \pm 1, 0, \tag{15}$$

$$R(\theta) = \begin{pmatrix} \cos\theta & -\sin\theta \\ \sin\theta & \cos\theta \end{pmatrix}, \qquad \theta \in (0, \pi) \cup (\pi, 2\pi), \tag{16}$$

$$\begin{aligned} N_2(\omega, b) &= \begin{pmatrix} R(\theta) & b \\ 0 & R(\theta) \end{pmatrix}, \quad \theta \in (0, \pi) \cup (\pi, 2\pi), \\ b &= \begin{pmatrix} b_1 & b_2 \\ b_3 & b_4 \end{pmatrix}, \quad b_i \in \mathbf{R}, b_2 \neq b_3. \end{aligned} \tag{17}$$

We also denote $N_1(e^{\theta\sqrt{-1}}, 0) = R(\theta)$.

Together with Theorem 7.3 and our above discussion, we obtain the following result:

Theorem 10 *For any $M \in \mathrm{Sp}(2n)$, there is a path $f : [0, 1] \to \Omega^0(M)$ such that $f(0) = M$ and*

$$f(1) = M_1(\omega_1)\diamond \cdots \diamond M_k(\omega_k)\diamond M_0, \tag{18}$$

where each $M_i(\omega_i)$ is a basic normal form of some eigenvalue $\omega_i \in \mathbf{U}$ for $1 \le i \le k$, and the symplectic matrix M_0 satisfies $\sigma(M_0) \cap \mathbf{U} = \emptyset$.

Note that in Corollary 2.3.8 below, we shall further prove that M_0 can be connected within $\Omega^0(M_0)$ to a $\diamond$-product of basic normal forms $D(2)$ and $D(-2)$.

Definition 11 A basic normal form matrix $M \in \mathrm{Sp}(2n)$ is **trivial**, if for sufficiently small $\alpha > 0$, $MR((t-1)\alpha)^{\diamond n}$ possesses no eigenvalues on $\mathbf{U}$ for $t \in [0, 1)$, and is **non-trivial** otherwise.

Note that among all the basic normal forms, the matrices $N_1(1, -1)$, $N_1(-1, 1)$, $D(2)$, $D(-2)$, $N_2(\omega, b)$ and $N_2(\overline{\omega}, b)$ with $\omega = \exp(\theta\sqrt{-1}) \in \mathbf{U}\backslash\mathbf{R}$ and $(b_2 - b_3)\sin\theta > 0$ are trivial, and any other basic normal form matrix $M = N_n(\omega, B) \in \mathrm{Sp}(2n)$ with $n = 1$ or 2 and $\omega \in \mathbf{U}$ is non-trivial by our study in Section 9 below.

For our later study of the splitting numbers in the index theory, we introduce the following concept of ultimate type of any $\omega \in \mathbf{U}$ as an eigenvalue of $M \in \mathrm{Sp}(2n)$.

Definition 12 For any basic normal form $M \in \mathrm{Sp}(2n)$ of eigenvalue $\omega \in \mathbf{U} \cap \sigma(M)$, we define the **ultimate type** (p, q) of ω for M to be its usual Krein type if M is non-trivial, and to be $(0, 0)$ if M is trivial. For any $M \in \mathrm{Sp}(2n)$, we define the ultimate type of ω for M to be $(0, 0)$ if $\omega \in \mathbf{U}\backslash\sigma(M)$. For any $M \in \mathrm{Sp}(2n)$, by Theorem 10 there exists a $\diamond$-product expansion (18) in the homotopy component $\Omega^0(M)$ of M where each M_i is a basic normal form of eigenvalue in $\mathbf{U}$ for $1 \le i \le k$ and $\sigma(M_0) \cap \mathbf{U} = \emptyset$. Denote the ultimate type of ω for M_i by (p_i, q_i) for $0 \le i \le k$. Let $p = \sum_{i=0}^k p_i$ and $q = \sum_{i=0}^k q_i$. We define the **ultimate type** of ω for M by (p, q).

Proposition 13 *The ultimate type of $\omega \in \mathbf{U}$ for $M \in \mathrm{Sp}(2n)$ is uniquely determined by ω and M, therefore is well defined. It is constant on $\Omega^0(M)$ for fixed $\omega \in \mathbf{U}$.*

Proof: It suffices to notice that the $\diamond$-product (18) is uniquely determined by M independent of the arranging order of $\{M_1, \ldots, M_k, M_0\}$ and the choice of the symplectic matrix M_0 satisfying $\omega \notin \sigma(M_0)$, and (18) holds for any $N \in \Omega^0(M)$. But these two factors make no change in the ultimate type defined above. □

Lemma 14 *For $\omega \in \mathbf{U}$ and $M \in \mathrm{Sp}(2n)$, denote the Krein type and the ultimate type of ω for M by (P, Q) and (p, q). Then*

$$P - p = Q - q \ge 0. \tag{19}$$

Proof: Note that (19) holds for all basic normal forms of eigenvalues in $\mathbf{U}$. By Theorems 3.9 and 3.11, whenever the eigenvalue ω leaves $\mathbf{U}$ by a perturbation on M, both the Krein positive and negative type numbers of ω must decrease by the same integer. Therefore by the proof of Theorem 10, we obtain (19). □

Remark 15 The homotopy component, basic normal forms, and the ultimate type of symplectic matrices were introduced in [Lon16] of 1999 in order to give an algebraic characterization of splitting numbers of the index theory. These concepts will be further studied in Chapters 8 and 9. The results contained in this section are due to [Lon16].

1.9 Perturbations of basic normal forms

Next we study perturbations on the basic normal forms.

Part 1 Perturbations on the eigenvalue $\omega = 1$.

We study first the perturbations of the basic normal form matrix $N_1(1, b)$ of the eigenvalue 1 with $b = \pm 1$ or 0 defined by (4.1). Here to simplify the discussion, we use the $\mathbf{R}^3$-cylindrical coordinate decomposition of Sp(2) introduced in Section 2.1. Denote $\mathbf{R}^+ = \{r \in \mathbf{R} \,|\, r > 0\}$ and $\mathbf{R}^- = \{r \in \mathbf{R} \,|\, r < 0\}$. By direct computation we obtain

$$N_1(1, b) \in \begin{cases} \mathrm{Sp}(2)^0_{1,+}, & \text{for } b < 0, \\ \mathrm{Sp}(2)^0_{1,-}, & \text{for } b > 0. \end{cases} \tag{1}$$

Since $N_1(1,1) \in \mathrm{Sp}(2)^0_{1,-}$, there exists $\theta_0 > 0$ such that

$$\sigma(N_1(1,1)R(\theta)) \subset \begin{cases} \mathbf{R}^+, & \text{if } 0 < \theta \le \theta_0, \\ \mathbf{U}\backslash\mathbf{R}, & \text{if } 0 < -\theta \le \theta_0. \end{cases} \tag{2}$$

Since $N_1(1,-1) \in \mathrm{Sp}(2)^0_{1,+}$, there exists $\theta_0 > 0$ such that

$$\sigma(N_1(1,-1)R(\theta)) \subset \begin{cases} \mathbf{U}\backslash\mathbf{R}, & \text{if } 0 < \theta \le \theta_0, \\ \mathbf{R}^+, & \text{if } 0 < -\theta \le \theta_0. \end{cases} \tag{3}$$

Note that $N_1(1,0) = I_2$ and

$$\sigma(N_1(1,0)R(\theta)) \subset \mathbf{U}\backslash\{1\}, \qquad \text{if and only if } \theta \ne 0 \bmod 2\pi. \tag{4}$$

Part 2 Perturbations on the eigenvalue $\omega \in \mathbf{U}\backslash\{1\}$.

Using the map from M to $-M$, the case of $\omega = -1$ can be reduced to that of $\omega = 1$; the details are left to the readers.

When $n = 1$, the perturbations on the basic normal forms of the eigenvalue $\omega \in \mathbf{U}$ can be read out directly from Figure 2.1.2. Next we consider the case of $\omega \in \mathbf{U}\backslash\mathbf{R}$ for general $n \ge 1$.

Lemma 1 *For A and $B \in \mathbf{R}$ suppose the polynomial equation,*

$$\lambda^4 - 4A\lambda^3 + B\lambda^2 - 4A\lambda + 1 = 0, \tag{5}$$

possesses no real roots. Then

1° (5) *possesses two pairs of conjugate simple roots on* $\mathbf{U}$ *if and only if* $4A^2 + 2 > B$.

2° (5) *possesses one pair of conjugate double roots on* $\mathbf{U}$ *if and only if* $4A^2 + 2 = B$.

3° (5) *possesses four simple roots away from* $\mathbf{U} \cup \mathbf{R}$ *if and only if* $4A^2 + 2 < B$.

Proof: By direct computation. □

Lemma 2 *For $\omega = e^{\theta\sqrt{-1}}$ with $\theta \in (0, 2\pi)\backslash\{\pi\}$ and $b = (b_1, b_2, b_3, b_4) \in \mathbf{R}^4$, consider the matrix*

$$M = N_2(\omega, b) = \begin{pmatrix} \cos\theta & -\sin\theta & b_1 & b_2 \\ \sin\theta & \cos\theta & b_3 & b_4 \\ 0 & 0 & \cos\theta & -\sin\theta \\ 0 & 0 & \sin\theta & \cos\theta \end{pmatrix}. \tag{6}$$

Then

1° $M \in \mathrm{Sp}(4)$ *if and only if* $(b_2 - b_3)\cos\theta + (b_1 + b_4)\sin\theta = 0$. *In the following we always suppose* $M \in \mathrm{Sp}(4)$.

2° $\omega = \cos\theta + \sqrt{-1}\sin\theta$ *and* $\overline{\omega}$ *are double eigenvalues of* M.

3° $\nu_\omega(M) = 1$ *if and only if* $b_2 - b_3 \neq 0$.
4° $\nu_\omega(M) = 2$ *if and only if* $b_2 - b_3 = 0$.
5° *It holds that*

$$\begin{aligned}\det(M[R(\theta_1)\diamond R(\theta_2)] - \lambda I) &= \lambda^4 - 4A(\theta_1,\theta_2)\lambda^3 \\ &+ B(\theta_1,\theta_2)\lambda^2 - 4A(\theta_1,\theta_2)\lambda + 1,\end{aligned} \tag{7}$$

where denoting by $c = \cos\theta$, $s = \sin\theta$, $c_i = \cos\theta_i$, *and* $s_i = \sin\theta_i$ *for* $i = 1$ *and* 2, *we have*

$$4A(\theta_1,\theta_2) = 2c(c_1 + c_2) + b_1 s_1 + b_4 s_2, \tag{8}$$

$$\begin{aligned}B(\theta_1,\theta_2) \;=\; & 2(c^2 + c^2 c_1 c_2 + c_1 c_2 - s^2 s_1 s_2) \\ & - s(b_2 - b_3)(s_1 c_2 + s_2 c_1) \\ & + 2c(b_1 s_1 c_2 + b_4 s_2 c_1) + (b_1 b_4 - b_2 b_3) s_1 s_2.\end{aligned} \tag{9}$$

Furthermore,

$$\begin{aligned}&4A^2(\theta_1,\theta_2) + 2 - B(\theta_1,\theta_2) \\ &\quad = s^2(s_1 + s_2)^2 - s_1^2 - s_2^2 + 2(1 - c_1 c_2) \\ &\quad\quad + s_1[s(b_2 - b_3)c_2 + c(c_1 - c_2)b_1] \\ &\quad\quad + s_2[s(b_2 - b_3)c_1 + c(c_2 - c_1)b_4] \\ &\quad\quad + \frac{1}{4}(b_1 s_1 + b_4 s_2)^2 - (b_1 b_4 - b_2 b_3) s_1 s_2.\end{aligned} \tag{10}$$

6° *Using the notation of* 5°,

$$\begin{aligned}&D_\omega(M[R(\theta_1)\diamond R(\theta_2)]) \\ &\quad = 2(s^2 - c^2) + 8cA(\theta_1,\theta_2) - B(\theta_1,\theta_2) \\ &\quad = s_1[s(b_2 - b_3)c_2 + 2b_1 c(1 - c_2)] + 2s^2(1 - c_1) \\ &\quad\quad + s_2[s(b_2 - b_3)c_1 + 2b_4 c(1 - c_1)] + 2s^2(1 - c_2) \\ &\quad\quad + s_1 s_2[2s^2 - (b_1 b_4 - b_2 b_3)] - 2(1 + c^2)(1 - c_1)(1 - c_2).\end{aligned} \tag{11}$$

7° *Suppose* $\dim_{\mathbf{C}} \ker_{\mathbf{C}}(M - \omega I) = 1$; *then there exists* $\alpha_0 > 0$ *small enough such that* $D_\omega(M[R(\alpha)\diamond I_2]) \neq 0$ *if* $0 < |\alpha| \leq \alpha_0$. *Using the notation of* 5°, *let* $\psi(\alpha) = 4A^2(\alpha,\alpha) + 2 - B(\alpha,\alpha)$ *for* $\alpha \in \mathbf{R}$. *Then* $\psi(0) = 0$ *and* $\psi'(0) = s(b_2 - b_3)$.

8° *For* $M = N_2(\omega, b)$ *with* $b = (b_1, b_2, b_3, b_4) \in \mathbf{R}^4$ *satisfying* $M \in \mathcal{M}_\omega(4)$, M *can be connected within* $\mathcal{M}_\omega(4)$ *to* $N_2(\omega, d)$ *with* $d = (d_1, d_2, d_3, d_4) \in \mathbf{R}^4$ *such that for any* $\epsilon > 0$ *there exists a perturbation path* $\gamma : [0, 1] \to \mathcal{M}_\omega(4)$ *such that* $\gamma(0) = N_2(\omega, d)$, $\sigma(\gamma(t))$ *possesses two pairs of simple eigenvalues* $\{\omega, \overline{\omega}\}$ *and* $\lambda \in C([0, 1], \mathbf{U})$ *satisfies* $\lambda(0) = \omega$, $\lambda(t) \neq \omega$ *and* $|\lambda(t) - \omega| \leq \epsilon$ *for* $0 < t \leq 1$.

9° *For $M = N_2(\overline{\omega}, b)$ with $b = (b_1, b_2, b_3, b_4) \in \mathbf{R}^4$ satisfying $M \in \mathcal{M}_\omega(4)$, M can be connected within $\mathcal{M}_\omega(4)$ to $N_2(\overline{\omega}, d)$ with $d = (d_1, d_2, d_3, d_4) \in \mathbf{R}^4$ such that the conclusion of* 8° *holds with $\gamma(0) = N_2(\overline{\omega}, d)$.*

10° *suppose $\nu_\omega(M) = 2$, then $D_\omega(M[R(\alpha)\diamond I_2]) = 0$ if and only if $\alpha = 0$* mod 2π.

Proof: 1° to 6° follow from direct computations. Note that to get (10) and (11) we have used 1°. Setting $s_2 = 0$ and $c_2 = 1$ in (11), we obtain

$$D_\omega(M[R(\theta_1)\diamond I_2]) = s_1 s(b_2 - b_3) + 2s^2(1 - c_1).$$

Thus 7° and 10° follow.

According to the signs of $(b_2 - b_3)\sin\theta$ and $\cos\theta$, there are four cases in the proof of 8°. Next we only prove the case for $(b_2 - b_3)\sin\theta > 0$ and $\cos\theta > 0$. The other three cases are left to the readers.

Using the notation in 5°, for $|\theta_1| + |\theta_2|$ sufficiently small, define

$$f(s_1, s_2) = D_\omega(M[R(\theta_1)\diamond R(\theta_2)]), \tag{12}$$

$$g(s_1, s_2) = 4A^2(\theta_1, \theta_2) + 2 - B(\theta_1, \theta_2). \tag{13}$$

Then for $s_1 > 0$ sufficiently small and $0 \le -s_2 \le s_1$,

$$f(s_1, 0) = s_1 s(b_2 - b_3) + 2s^2(1 - c_1) > 0, \tag{14}$$

$$f(s_1, -s_1) = [2s^2 + (b_1 b_4 - b_2 b_3)]s_1^2 + o(|s_1|^2) < 0, \tag{15}$$

$$f'_{s_2}(s_1, s_2) = s(b_2 - b_3)c_1 + o(|s_1|) \ne 0, \tag{16}$$

provided

$$2s^2 + (b_1 b_4 - b_2 b_3) < 0. \tag{17}$$

Suppose (17) holds. By the implicit function theorem there is a $\theta_0 > 0$ sufficiently small and a unique continuous function $\alpha : [0, \theta_0] \to [-\theta_0, 0]$ such that $\alpha(0) = 0$, $-t\theta_0 < \alpha(t) < 0$ for $0 < t \le 1$, and $f(\sin(t\theta_0), \sin\alpha(t\theta_0)) = 0$ for $0 \le t \le 1$. Define

$$\gamma(t) = M[R(t\theta_0)\diamond R(\alpha(t\theta_0))], \qquad \forall\, t \in [0, 1]. \tag{18}$$

Then from $D_\omega(\gamma(t)) = f(\sin(t\theta_0), \sin(\alpha(t\theta_0)) = 0$ we obtain $\{\omega, \overline{\omega}\} \subset \sigma(\gamma(t))$ for all $t \in [0, 1]$.

On the other hand, by further requiring the above $\theta_0 > 0$ to be much smaller we obtain that for $0 \leq -s_2 \leq s_1 \leq \sin\theta_0$,

$$g(s_1, 0) = s_1 s(b_2 - b_3) + o(|s_1|) > 0, \tag{19}$$

$$g(s_1, -s_1) = s_1^2 \left[\frac{1}{4}(b_1 - b_4) + (b_1 b_4 - b_2 b_3)\right] > 0, \tag{20}$$

$$g'_{s_2}(s_1, s_2) = s(b_2 - b_3)c_1 + o(1) \neq 0, \tag{21}$$

provided

$$\frac{1}{4}(b_1 - b_4) + (b_1 b_4 - b_2 b_3) > 0. \tag{22}$$

Thus $g(s_1, s_2)$ is strictly monotone for $\theta_1 = t\theta_0$ and $\theta_2 = \alpha(t\theta_0)$ with $t \in [0, 1]$, and then is always positive there. By 1° of Lemma 1, we obtain $\{\omega, \overline{\omega}\} \neq \{\lambda(t), \overline{\lambda(t)}\} \subset \sigma(\gamma(t))$ for all $t \in (0, 1]$.

By 1° we also have

$$(b_2 - b_3)\sin\theta = -\frac{\sin^2\theta}{\cos\theta}(b_1 + b_4). \tag{23}$$

Thus we need to find b such that (17), (22), (23), and $(b_2 - b_3)\sin\theta > 0$ hold.

Define $d = (d_1, d_2, d_3, d_4)$ by

$$d_1 = 1/8, \quad d_4 = -48, \quad d_3 = 0, \quad d_2 = \frac{\sin\theta}{\cos\theta}(48 - 1/8).$$

Then (17), (22), (23), and $(d_2 - d_3)\sin\theta > 0$ hold for $b = d$. Thus $N_2(\omega, b)$ can be connected to $N_2(\omega, d)$ within $\mathcal{M}_\omega(4)$. By 1° of Lemma 1 we obtain that the path γ and $N_2(\omega, d) \in \mathrm{Sp}(4)$ satisfy the requirements of 8°.

The proof of 8° is complete. Then 9° follows by the same argument. □

By our above discussion, we obtain

Theorem 3 *For any $M \in \mathrm{Sp}(2n)$ and $\omega \in \mathbf{U}$, there exists a path $f : [0, 1] \to \Omega_\omega^0(M)$ satisfying $f(0) = M$ and*

$$f(1) = M_1 \diamond \cdots \diamond M_k \diamond M_0, \tag{24}$$

where each $M_i \in \mathrm{Sp}(2)$ is a normal form of the eigenvalues ω or $\overline{\omega}$ for $1 \leq i \leq k$ and the symplectic matrix M_0 satisfies $\omega \notin \sigma(M_0)$.

Proof: This theorem follows from Theorem 8.10, Lemma 2, and the path connectedness of $\mathrm{Sp}(2n)$. □

Similarly to our above discussion, we obtain

Corollary 4 *For any $M \in \mathrm{Sp}(2n)$, there is a continuous path $f : [0,1] \to \mathrm{Sp}(2n)$ such that $f(0) = M$ and $f(t)$ possesses only simple eigenvalues for $0 < t \le 1$.*

Next we consider perturbations on more general matrices.

Theorem 5 *For $M \in \mathrm{Sp}(2n)$, suppose $\omega \in \sigma(M) \cap (\mathbf{U} \backslash \mathbf{R})$ has Krein-type number (p, q) with $p \ge 1$ and $q \ge 1$. Then there exists a path $f : [0,1] \to \mathrm{Sp}(2n)$ such that $f(0) = M$, for any $t \in (0,1]$, ω is an eigenvalue of $f(t)$ with Krein-type number $(p-1, q-1)$ if $p+q > 2$, $\{d^t\omega, d^t\overline{\omega}, d^{-t}\omega, d^{-t}\overline{\omega}\}$ is a simple eigenvalue quadruple of $f(t)$ for some $d > 1$, and all the eigenvalues of $f(t)$ other than $\{\omega, \overline{\omega}\}$ are the same as those eigenvalues of M.*

Proof: By the normal form Theorem 7.3, there exists $P \in \mathrm{Sp}(2n)$ such that $P^{-1}MP = M_1 \diamond \cdots \diamond M_k$ is a $\diamond$-product of normal forms. Then using a perturbation path $g(t)$ on the corresponding normal forms with $g(0) = I$ as in our discussion in Section 8, we obtain the required path $f(t) = MPg(t)P^{-1}$. □

Remark 6 The results contained in this section are due to [Lon15] and [Lon16].

Chapter 2

Topological aspects

For any fixed ω belonging to the unit circle $\mathbf{U}$ in the complex plane $\mathbf{C}$, all the symplectic matrices having ω as an eigenvalue form a codimension 1 hypersurface in the symplectic group $\mathrm{Sp}(2n) = \mathrm{Sp}(2n, \mathbf{R})$. Chapter 2 is devoted to the study of the topological structures of $\mathrm{Sp}(2n)$, these hypersurfaces and their complements in $\mathrm{Sp}(2n)$.

In this chapter we first introduce the $\mathbf{R}^3$-cylindrical coordinate representations of $\mathrm{Sp}(2)$. This representation gives a complete and intuitive picture for $\mathrm{Sp}(2)$, and is very suggestive for the higher dimensional case of $\mathrm{Sp}(2n)$. Then we study the topologies of $\mathrm{Sp}(2n)$, its regular subset $\mathrm{Sp}(2n)_\omega^*$ and singular hypersurface $\mathrm{Sp}(2n)_\omega^0$ for $\omega \in \mathbf{U}$. In this study, we need the following set of paths in $\mathrm{Sp}(2n)$.

$$\mathcal{P}_\tau(2n) = \{\gamma \in C([0, \tau], \mathrm{Sp}(2n)) \mid \gamma(0) = I\}, \qquad \forall \tau > 0. \tag{1}$$

For notational simplicity, when we deal with real vector spaces, we shall omit the subindex $\mathbf{R}$.

2.1 Structures of Sp(2) and its subsets

In this section, we introduce a geometric representation of $\mathrm{Sp}(2) = \mathrm{Sp}(2, \mathbf{R})$ in $\mathbf{R}^3$ given in [Lon6] of 1991. It provides an intuitive and suggestive model for our further study on the general theory of the topological structures of the symplectic group $\mathrm{Sp}(2n, \mathbf{R})$ and its subsets. To the author's knowledge, the only known different model of $\mathrm{Sp}(2)$ was given by I. Gelfand and V. Lidsikii in [GeL1] of 1955, where $\mathrm{Sp}(2)$ is mapped to an open solid torus via the toral coordinates and hyperbolic functions.

For any matrix $M \in \mathrm{Sp}(2)$, by the polar decomposition, M can be written in the form

$$M = \begin{pmatrix} r & z \\ z & (1+z^2)/r \end{pmatrix} \begin{pmatrix} \cos\theta & -\sin\theta \\ \sin\theta & \cos\theta \end{pmatrix}, \tag{1}$$

where $(r, \theta, z) \in \mathbf{R}^+ \times S^1 \times \mathbf{R}$, $\mathbf{R}^+ = \{r \in \mathbf{R} \mid r > 0\}$, $S^1 = \mathbf{R}/(2\pi\mathbf{R} - \pi)$, and (r, θ, z) is uniquely determined by M. Viewing (r, θ, z) as the cylindrical coordinates in $\mathbf{R}^3$, we obtain a representation of $\mathrm{Sp}(2)$ in $\mathbf{R}^3$. In fact, this defines a C^∞ global diffeomorphism Φ from $\mathrm{Sp}(2)$ to $\mathbf{R}^3\backslash\{z\text{–axis}\}$. In the following, we identify elements in $\mathrm{Sp}(2)$ and its Φ-image in $\mathbf{R}^3$.

Under this representation, it is easy to see that the two eigenvalues of M are

$$\lambda = \frac{1}{2r}\{(r^2 + z^2 + 1)\cos\theta \pm \sqrt{(1 + r^2 + z^2)^2\cos^2\theta - 4r^2}\},$$

which are either two reciprocal real numbers or two conjugate complex numbers on the unit circle **U** in the complex plane **C**. For $\omega = \cos\varphi + \sqrt{-1}\sin\varphi \in \mathbf{U}$ and M in the form (1), we obtain

$$D_\omega(M) = 2\cos\varphi - \left(r + \frac{1+z^2}{r}\right)\cos\theta,$$

and

$$\begin{aligned} \mathrm{Sp}(2)^{\pm}_{\omega} &= \{(r,\theta,z) \in \mathbf{R}^+ \times S^1 \times \mathbf{R} \mid \pm(r^2+z^2+1)\cos\theta > 2r\cos\varphi\}, \\ \mathrm{Sp}(2)^{0}_{\omega} &= \{(r,\theta,z) \in \mathbf{R}^+ \times S^1 \times \mathbf{R} \mid (r^2+z^2+1)\cos\theta = 2r\cos\varphi\}. \end{aligned}$$

Let $\mathrm{Sp}(2)^0_{\omega,\pm} = \{(r,\theta,z) \in \mathrm{Sp}(2)^0_\omega \mid \pm\sin\theta > 0\}$. We then obtain

$$\begin{aligned} \mathrm{Sp}(2)^0_{\pm 1} &= \mathrm{Sp}(2)^0_{\pm 1,+} \cup \{I\} \cup \mathrm{Sp}(2)^0_{\pm 1,-}, \\ \mathrm{Sp}(2)^0_{\omega} &= \mathrm{Sp}(2)^0_{\omega,+} \cup \mathrm{Sp}(2)^0_{\omega,-}. \end{aligned}$$

Here we especially interested in the cylindrical coordinate representation of the singular hypersurfaces $\mathrm{Sp}(2)^0_\omega$ for $\omega \in \mathbf{U}$. In Figure 1, we give this $\mathbf{R}^3$-cylindrical coordinate representation of $\mathrm{Sp}(2)^0$ with the Descartes coordinates $(x, y, z) = (r\cos\theta, r\sin\theta, z)$.

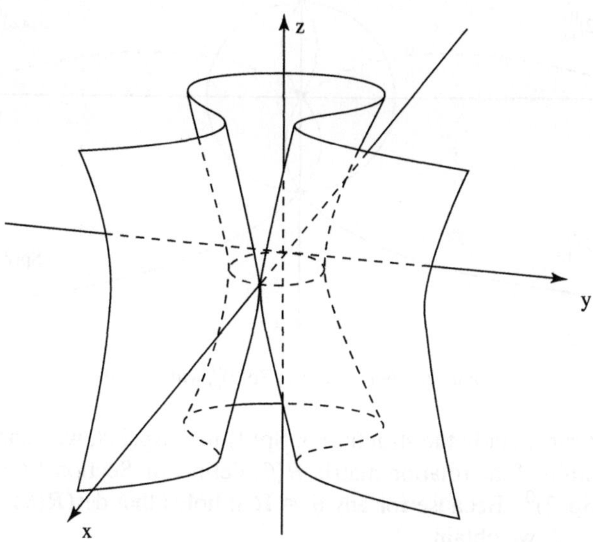

Figure 1: *The* $\mathbf{R}^3$*-cylindrical coordinate representation of* $\mathrm{Sp}(2)^0_1$

In Figure 2, the intersection of the plane $\{z = 0\}$ with the $\mathbf{R}^3$-cylindrical coordinate representation of $\mathrm{Sp}(2)^0_1$, $\mathrm{Sp}(2)^0_{-1}$, and $\mathrm{Sp}(2)^0_\omega$ for some $\omega \in \mathbf{U}\setminus\mathbf{R}$ are given. As shown in Figure 2, for any $\omega \in \mathbf{U}$ the set $\mathrm{Sp}(2)^*_\omega$ possesses precisely two path connected components $\mathrm{Sp}(2)^+_\omega$ and $\mathrm{Sp}(2)^-_\omega$, and it holds that

$$D(2) \equiv \mathrm{diag}\left(2, \frac{1}{2}\right) \in \mathrm{Sp}(2)^+_\omega \quad \text{and} \quad D(-2) \equiv \mathrm{diag}\left(-2, -\frac{1}{2}\right) \in \mathrm{Sp}(2)^-_\omega.$$

Fix an $\omega \in \mathbf{U}\backslash\mathbf{R}$. It is easy to see that $\mathrm{Sp}(2)^-_1$, $\mathrm{Sp}(2)^+_{-1}$ and $\mathrm{Sp}(2)^\pm_\omega$ are homeomorphic to $\mathbf{R}^3$. $\mathrm{Sp}(2)^+_1$ and $\mathrm{Sp}(2)^-_{-1}$ are homeomorphic to $\mathbf{R}^3\backslash\{(x, y, z) \in \mathbf{R}^3 \mid x^2 + z^2 \le y^2\}$. Thus for any $\omega \in \mathbf{U}$, the set $\mathrm{Sp}(2)^*_\omega$ is simply connected in $\mathrm{Sp}(2)$, i.e., any closed curve inside $\mathrm{Sp}(2)^+_\omega$ or $\mathrm{Sp}(2)^-_\omega$ can be continuously contracted inside $\mathrm{Sp}(2)$ to a point.

When $\omega = 1$, in the singular hypersurface $\mathrm{Sp}(2)^0 \equiv \mathrm{Sp}(2)^0_1$, the identity matrix I_2 is the only element which satisfies $\dim\ker(M - I) = 2$. The regular part $\mathcal{M}(2)$ of $\mathrm{Sp}(2)^0_1$ possesses precisely two path connected components $\mathrm{Sp}(2)^0_+ = \mathrm{Sp}(2)^0_{1,+}$ and $\mathrm{Sp}(2)^0_- = \mathrm{Sp}(2)^0_{1,-}$, both of which are smooth hypersurfaces diffeomorphic to $\mathbf{R}^2\backslash\{0\}$.

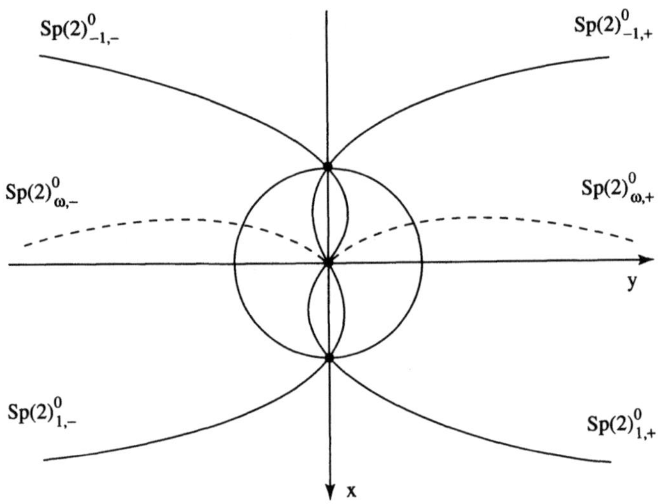

Figure 2: *Intersection of* $\mathrm{Sp}(2)^0_\omega$ *with* $\{z = 0\}$

In order to further study the structure of $\mathrm{Sp}(2)$ near $\mathrm{Sp}(2)^0$, we consider the effect of the perturbation of the rotation matrix $R(\theta)$ defined in Section 1.6 on the singular hypersurface $\mathrm{Sp}(2)^0$. Because for any $\theta \in \mathbf{R}$ it holds that $\det(R(\theta) - I) = 2(1 - \cos\theta)$, for $M = I$, we obtain

$$\nu_1(M) - \nu_1(MR(\theta)) = 2, \quad \text{if and only if} \quad \theta \neq 0 \bmod 2\pi,$$

and

$$\det(MR(\theta) - I) = 0 \ \text{if and only if} \ \theta = 0 \ \bmod\ 2\pi. \tag{2}$$

This shows that rotation perturbations starting from I_2 always go into the same path connected component Sp$(2)^-$ of Sp$(2)^*$. For any element in the regular part $\mathcal{M}(2)$ of Sp$(2)^0$, we have the following results.

Lemma 1 *For any* $M \in \mathcal{M}(2)$*, there exists* $\theta_0 > 0$ *such that for* $0 < |\theta| \le \theta_0$,

$$\nu_1(M) - \nu_1(MR(\theta)) = 1, \tag{3}$$

$$\det(MR(\theta) - I)\det(MR(-\theta) - I) < 0. \tag{4}$$

Proof: Let $M = \begin{pmatrix} a & b \\ c & d \end{pmatrix}$. Then from $\det(M - I) = 0$ and $\det M = 1$, we obtain $a + d = 2$ and $bc = -(a-1)^2$. Thus from $M \neq I$ we obtain $b \neq c$. Then for any $\theta \in \mathbf{R}$,

$$\det(MR(\theta) - I) = 2(1 - \cos\theta) + (c - b)\sin\theta.$$

Thus

$$\det(MR(\theta) - I) = (\sin\theta)\left[(c - b) + \frac{2(1 - \cos\theta)}{\sin\theta}\right], \qquad \forall \theta \neq 0 \bmod 2\pi.$$

Then from $\lim_{\theta\to 0}(1 - \cos\theta)/\sin\theta = 0$, we obtain the lemma. □

Here (4) in Lemma 1 points out that the rotation perturbations of opposite direction starting from an element in $\mathcal{M}(2)$ go into different path connected components. This can also be seen from Figures 1 and 2.

Note that from the $\mathbf{R}^3$-cylindrical coordinate representation introduced above, $\mathcal{M}(2)$ is orientable, and so is Sp$(2)^0$. In particular we can define the positive direction of Sp(2) at any of its elements M to be the direction of $\frac{d}{dt}MR(t\theta)|_{t=0}$ for small enough $\theta > 0$. We call this orientation the **natural orientation** of Sp$(2)^0$.

The properties of Sp$(2)^0_{-1}$ and Sp$(2)^0_\omega$ for some $\omega \in \mathbf{U}\backslash\mathbf{R}$ can be obtained similarly.

In the following sections, we shall prove that the higher dimensional symplectic group Sp$(2n)$ and its corresponding subsets possess similar topological structures.

2.2 The global structure of Sp$(2n, \mathbf{R})$

In this section, we study the global structure of the symplectic group Sp$(2n)$ established by I. Gelfand and V. Lidskii in [GeL1] and J. Moser [Mos1] in the 1950s.

Using the polar decomposition of symplectic matrices and the homeomorphism

$$\mathrm{Sp}(2n, \mathbf{R}) \cap O(2n, \mathbf{R}) \cong U(n, \mathbf{C}) \equiv U(n),$$

established in Section 1.1, we study the global topological structure of Sp$(2n)$ = Sp$(2n, \mathbf{R})$ in this section. The main result is the following theorem.

Theorem 1 (I. Gelfand and V. Lidskii [GeL1], J. Moser [Mos1]). *The symplectic group* $\mathrm{Sp}(2n)$ *is homeomorphic to the topological product of the unit circle* $\mathbf{U}$ *in the complex plane* $\mathbf{C}$ *and a simply connected topological space.*

The proof of this theorem is carried out by the following lemmas. Denote the set of all $2n \times 2n$ positive definite symmetric symplectic matrices by $\mathrm{PSS}(2n)$.

Lemma 2 *The symplectic group* $\mathrm{Sp}(2n)$ *is homeomorphic to the topological product of* $\mathrm{PSS}(2n)$ *and* $\mathrm{Sp}(2n) \cap O(2n)$, *where* $O(2n) = O(2n, \mathbf{R})$.

Proof: The polar decomposition proved in Lemma 1.1.3 gives this homeomorphism. □

Lemma 3 $\mathrm{PSS}(2n)$ *is homeomorphic to the Euclidean space* $\mathbf{R}^{n(n+1)}$.

Proof: Denote by $S(n)$ the subgroup of $\mathcal{L}(\mathbf{R}^n)$ formed by all $n \times n$ symmetric matrices. In the proof of Lemma 1.1.4, we have established a one to one correspondence between elements in $\mathrm{PSS}(2n)$ and matrices in $\mathcal{L}(\mathbf{R}^{2n})$ with the form $M = \begin{pmatrix} A & B \\ B & -A \end{pmatrix}$ with A and B belonging to $S(n)$. This yields a homeomorphism between $\mathrm{PSS}(2n)$ and $S(n) \times S(n)$. Since $S(n)$ is homeomorphic to the Euclidean space $\mathbf{R}^{n(n+1)/2}$, we obtain the lemma. □

Lemma 4 *The orthogonal symplectic group* $\mathrm{Sp}(2n) \cap O(2n)$ *is homeomorphic to the unitary group* $U(n) = U(n, \mathbf{C})$.

Proof: In the proof of Lemma 1.1.5, a one to one correspondence between any orthogonal symplectic matrix $U = \begin{pmatrix} A & B \\ -B & A \end{pmatrix}$ and the unitary matrix $A + \sqrt{-1}B$ is established. This correspondence gives the homeomorphism from $\mathrm{Sp}(2n) \cap O(2n)$ to $U(n)$. □

Remark 5 The homeomorphism obtained in Lemma 4 is also a group isomorphism.

Lemma 6 *The unitary group* $U(n)$ *is homeomorphic to the topological product of the unit circle* $\mathbf{U}$ *in the complex plane* $\mathbf{C}$ *and the unitary unimodule group* $\mathrm{SU}(n)$. *Recall that* $\mathrm{SU}(n)$ *is the subgroup of* $U(n)$ *formed by those matrices with determinant* 1.

Proof: Let A be any unitary matrix. Denote $e^{\sqrt{-1}a} = \det A$ for some $a \in \mathbf{R}$. Define an $n \times n$ matrix $B = \mathrm{diag}(e^{-\sqrt{-1}a}, 1, \ldots, 1)A$. Then B is unitary and unimodule, i.e., $\det B = 1$. Define an $n \times n$ matrix $D = \mathrm{diag}(e^{\sqrt{-1}a}, 1, \ldots, 1)$. Then

$$A = DB. \tag{1}$$

Denote the set of all matrices in the form of D by $\mathcal{D}(n)$. Then it is easy to see that the decomposition (1) of any unitary matrix A in $\mathcal{D}(n) \times \mathrm{SU}(n)$ is unique. This decomposition yields the homeomorphism from $U(n)$ to $\mathcal{D}(n) \times \mathrm{SU}(n)$. Note that $\mathcal{D}(n)$ is homeomorphic to the unit circle $\mathbf{U}$ in $\mathbf{C}$. The lemma is proved. □

Lemma 7 (cf. H. Weyl [Weyl1]). *The unitary unimodule group* SU(n) *is simply connected.*

Proof: Since SU(1) contains only one element 1, it is simply connected. Suppose $n > 1$. Then SU$(n-1)$ can be viewed as a subgroup of SU(n) consisting of all matrices of the form

$$M = \begin{pmatrix} 1 & 0 & \cdots & 0 \\ 0 & * & \cdots & * \\ \cdot & \cdot & \cdots & \cdot \\ 0 & * & \cdots & * \end{pmatrix}.$$

Two elements A and A' in SU(n) are left equivalent mod this subgroup, i.e. they belong to the same coset of SU$(n-1)$, if and only if they possess the same first row. But their first row must have the form

$$a = (a_1, \ldots, a_n) \in \mathbf{C}^n, \tag{2}$$

satisfying

$$\sum_{i=1}^{n} |a_i|^2 = 1. \tag{3}$$

On the other hand, given any vector $a \in \mathbf{C}^n$ satisfying (3), it can be completed to a unitary unimodule basis $\{a, e_2, \ldots, e_n\}$ of $\mathbf{C}^n$. Thus any such a vector can be viewed as the first row of some unitary unimodule matrix. Therefore the set $\mathcal{M}$ formed by the above mentioned cosets is homeomorphic to the unit sphere S^{2n-1} in the real Euclidean space $\mathbf{R}^{2n}$ with $\{\mathrm{Re}(a_i), \mathrm{Im}(a_i) \mid i = 1, \ldots, n\}$ as real coordinates. When $n > 1$, the sphere S^{2n-1} is simply connected. Given any closed curve C in SU(n), it is also a closed curve in $\mathcal{M}$. Thus it can be contracted to the coset corresponding to the point $(1, 0, \ldots, 0)$. Therefore closed curves in SU(n) can be contracted into SU$(n-1)$. By induction, we complete the proof. □

Now we can give

The proof of Theorem 1: By Lemmas 2, 3, 4, and 6, we obtain homeomorphisms:

$$\begin{aligned} \mathrm{Sp}(2n) &\cong \mathrm{PSS}(2n) \times (\mathrm{Sp}(2n) \cap O(2n)) \\ &\cong \mathbf{R}^{n(n+1)} \times U(n) \\ &\cong \mathbf{R}^{n(n+1)} \times \mathbf{U} \times \mathrm{SU}(n). \end{aligned} \tag{4}$$

By Lemma 7, SU(n) is simply connected. This proves the theorem.

Note that in our proof of Theorem 1, we have used results from [MoZ1] and [Weyl1] too.

From Theorem 1 we obtain immediately

Corollary 8 *The symplectic group* Sp$(2n)$ *is path connected.*

Theorem 9 *The symplectic group* Sp$(2n)$ *form a* $2n^2+n$ *real dimensional manifold.*

Proof: Note that the homeomorphisms in (4) are locally smooth. Thus Sp$(2n)$ form a real manifold. We claim

$$\dim_{\mathbf{R}} \mathrm{SU}(n) = n^2 - 1. \tag{5}$$

We prove this claim by induction. When $n = 1$, (5) holds trivially. Suppose the claim holds for SU$(n-1)$. Then by the proof of Lemma 7, there exists a homeomorphism from SU(n) to SU$(n-1) \times S^{2n-1}$. Thus we obtain the claim:

$$\dim_{\mathbf{R}} \mathrm{SU}(n) = (n-1)^2 - 1 + 2n - 1 = n^2 - 1.$$

By (4) we obtain

$$\dim_{\mathbf{R}} \mathrm{Sp}(2n) = n(n+1) + 1 + (n^2 - 1) = 2n^2 + n.$$

The proof is complete. □

Theorem 10 *The symplectic group* Sp$(2n)$ *possesses a natural orientation induced from the Euclidean spaces via the representation* (4).

Proof: By the proof of Lemma 7, SU(n) possesses a natural orientation induced from unit spheres in the Euclidean spaces. Thus (4) induces a natural orientation from the Euclidean spaces to Sp$(2n)$. □

2.3 Hyperbolic symplectic matrix set

In this section, we study the topologies of the sets of truly hyperbolic and hyperbolic symplectic matrices in Sp$(2n)$ defined in Section 1.8. Here we need the hyperbolic index $\alpha : \cup_{n\geq 1}\mathrm{Sp}(2n) \to \{0, 1\}$ of Definition 1.8.1. Results in this section were proved in [Lon18] of 2000, and by T. An and the author in [LoA1] of 1998. Some of this discussion also uses ideas from [CoZ2] and [DoL1].

Theorem 1 *The set* $\mathrm{Sp}^{th}(2)$ *possesses precisely two path connected components defined by*

$$\mathrm{Sp}^{th}_i(2) = \{M \in \mathrm{Sp}^{th}(2) \,|\, \alpha(M) = i\}, \qquad \text{for } i = 0, 1. \tag{1}$$

It holds that

$$\mathrm{Sp}^{th}_0(2) = \mathrm{Sp}(2)^+, \tag{2}$$
$$\mathrm{Sp}^{th}_1(2) = -\mathrm{Sp}(2)^+ \equiv \{-M \,|\, M \in \mathrm{Sp}(2)^+\}.$$

They are simply connected in Sp(2). *The hyperbolic set* $\mathrm{Sp}^h(2)$ *is path connected, simply connected, and*

$$\mathrm{Sp}^h(2) = \mathrm{Sp}(2)^0. \tag{3}$$

Proof: This theorem follows from the $\mathbf{R}^3$-cylindrical coordinate representation of Sp(2) introduced in Section 2.1. □

Since $\mathrm{Sp}_0^{th}(2) = \mathrm{Sp}(2)_1^+$ and $\mathrm{Sp}_1^{th}(2) = \mathrm{Sp}(2)_{-1}^-$, we got an intuitive picture of them in Figure 1.2. For the topological structures of the higher dimensional $\mathrm{Sp}^{th}(2n)$ and $\mathrm{Sp}^h(2n)$, we have the following main results.

Theorem 2 *The set* $\mathrm{Sp}^{th}(2n)$ *with* $n \geq 1$ *possesses precisely two path connected components defined by*

$$\mathrm{Sp}_i^{th}(2n) = \{M \in \mathrm{Sp}^{th}(2n) \mid \alpha(M) = i\}, \qquad \textit{for } i = 0, 1. \tag{4}$$

Both $\mathrm{Sp}_0^{th}(2n)$ *and* $\mathrm{Sp}_1^{th}(2n)$ *are simply connected in* $\mathrm{Sp}(2n)$.

Theorem 3 *The set* $\mathrm{Sp}^h(2n)$ *with* $n \geq 2$ *possesses precisely two path connected components defined by*

$$\mathrm{Sp}_i^h(2n) = \{M \in \mathrm{Sp}^h(2n) \mid \alpha(M) = i\}, \qquad \textit{for } i = 0, 1. \tag{5}$$

Both $\mathrm{Sp}_0^h(2n)$ *and* $\mathrm{Sp}_1^h(2n)$ *are simply connected in* $\mathrm{Sp}(2n)$.

For any two curves ψ and $\phi : [0, \tau] \to \mathrm{Sp}(2n)$ with $\tau > 0$ satisfying $\psi(\tau) = \phi(0)$, we define their joint curve by

$$\phi * \psi(t) = \begin{cases} \psi(2t), & \text{if } \ 0 \leq t \leq \tau/2, \\ \phi(2t - \tau), & \text{if } \ \tau/2 \leq t \leq \tau. \end{cases} \tag{6}$$

The proofs of Theorems 2 and 3 are completed by the following lemmas.

Lemma 4 *The hyperbolic index function* $\alpha : \mathrm{Sp}(2n) \to \{0, 1\}$ *is locally constant.*

Proof: For any $M \in \mathrm{Sp}(2n)$, the only way that, under a small perturbation, any of its real eigenvalue that are not on $\mathbf{U}$ can leave $\mathbf{R}$ is to leave as a pair of reciprocal eigenvalues of an even multiplicity and to yield an eigenvalue quadruple. Thus the lemma holds. □

Lemma 5 *For* $n \geq 1$ *and any* $M \in \mathrm{Sp}^{th}(2n)$*, there exists a path* $f \in C([0, 1], \mathrm{Sp}^{th}(2n))$ *such that* $f(0) = M$ *and*

$$f(1) = M_n^+ \equiv D(2)^{\diamond n}, \qquad \textit{if} \quad \alpha(M) = 0, \tag{7}$$

$$f(1) = M_n^- \equiv D(-2)\diamond D(2)^{\diamond(n-1)}, \qquad \textit{if} \quad \alpha(M) = 1, \tag{8}$$

where we recall $D(a) = \mathrm{diag}\{a, a^{-1}\}$ *for* $a \in \mathbf{R}\setminus\{0\}$.

Proof: Fix $M \in \mathrm{Sp}^{th}(2n)$. The case of $n = 1$ has been studied in Theorem 1. We prove the lemma for $n \geq 2$ in six steps.

Step 1 Since $\mathrm{Sp}^{th}(2n)$ is open in $\mathrm{Sp}(2n)$, by Corollary 1.9.4, we can connect M to a nearby matrix A_1 by a path ϕ_1 within $\mathrm{Sp}^{th}(2n)$ such that A_1 possesses only simple eigenvalues.

By Theorems 1.7.1 and 1.7.2, we obtain a matrix $P_1 \in \mathrm{Sp}(2n)$ such that $A_2 \equiv P_1^{-1}A_1P_1 = M_1\diamond\cdots\diamond M_m$ for some integer $m \in [1, n]$, where either $N_i \in \mathrm{Sp}(2)$ is of the normal form (1.7.2), or $M_i \in \mathrm{Sp}(4)$ is of the normal form (1.7.4). Let $\zeta \in \mathcal{P}_1(2n)$ such that $\zeta(1) = P_1$ and let $\phi_2(t) = \zeta(t)^{-1}A_1\zeta(t)$ for all $t \in [0, 1]$. Then ϕ_2 is a path in $\mathrm{Sp}^{th}(2n)$ which connects A_1 to A_2.

Next we study the M_is with $1 \le i \le m$ according to different types of normal forms.

Step 2 If $M_i \in \mathrm{Sp}(2)$ is of the normal form (1.7.2) with parameters $\lambda > 0$ and $\lambda \neq 1$, then $M_i \in \mathrm{Sp}_0^{th}(2) = \mathrm{Sp}(2)^+$. By Theorem 1, i.e. the path-connectedness of $\mathrm{Sp}(2)^+$, there is a path ξ_i in $\mathrm{Sp}(2)^+$ which connects N_i to $D(2)$.

Step 3 If $M_i \in \mathrm{Sp}(2)$ is of the form (1.7.2) with the parameter $\lambda < 0$ and $\lambda \neq -1$, then $N_i \in \mathrm{Sp}_1^{th}(2) = -\mathrm{Sp}(2)^+$. By Theorem 1, i.e., the path-connectedness of $-\mathrm{Sp}(2)^+$, there is a path ξ_i in $-\mathrm{Sp}(2)^+$ which connects N_i to $D(-2)$.

Step 4 If $M_i = N_2(\rho_i, \theta_i, b) \in \mathrm{Sp}(4)$ is of the form (1.7.4) with parameters $\rho_i \in \mathbf{R}\setminus \{0, \pm 1\}$, $\theta_i \in (0, \pi)\cup(\pi, 2\pi)$, and $b \in \mathbf{R}^4$. Define $\eta_1(t) = N_2(\rho_i, (1-t)\theta_i, (1-t)b)$ for $0 \le t \le 1$ via (1.7.4). Then this path η_1 connects M_i to $D(\rho_i)\diamond D(\rho_i)$ within $\mathrm{Sp}^{th}(4)$.

If $\rho_i > 0$, by Step 2 there is a path η_2 that connects $D(\rho_i)$ to $D(2)$ within $\mathrm{Sp}(2)^+$. Then the path $\xi_i \equiv (\eta_2\diamond\eta_2) * \eta_1$ connects N_i to $D(2)\diamond D(2)$ within $\mathrm{Sp}^{th}(4)$.

If $\rho_i < 0$, by Step 3 there is a path η_3 that connects $D(\rho_i)$ to $D(-2)$ within $\mathrm{Sp}(2)^+$. Then the path $\xi_i \equiv (\eta_3\diamond\eta_3) * \eta_1$ connects N_i to $D(-2)\diamond D(-2)$ within $\mathrm{Sp}^{th}(4)$.

Step 5 Define $\phi_3 = \xi_1\diamond\cdots\diamond\xi_m$. Then the path ϕ_3 connects A_2 to a $\diamond$-product A_3 of $D(-2)$s and $D(2)$s within $\mathrm{Sp}^{th}(2n)$. By changing the symplectic basis of $\mathbf{R}^{2n}$, we obtain a path ϕ_4 within $\mathrm{Sp}^{th}(2n)$ that connects A_3 to

$$A_4 = D(-2)\diamond\cdots\diamond D(-2)\diamond D(2)\diamond\cdots\diamond D(2).$$

Step 6 Define

$$\gamma(t) = \mathrm{diag}\left(2R((1-t)\pi), \frac{1}{2}R((1-t)\pi)\right), \qquad \forall t \in [0, 1].$$

Then γ is a path connecting $D(-2)\diamond D(-2)$ to $D(2)\diamond D(2)$ within $\mathrm{Sp}^{th}(4)$. Now we apply the path γ inductively to as many as possible $D(-2)\diamond D(-2)$ components of A_4. In such a way we obtain a path $\phi_5 : [0, 1] \to \mathrm{Sp}^{th}(2n)$ such that $\phi_5(0) = A_4$ and

$$\phi_5(1) = \begin{cases} M_n^+, & \text{if} \quad \alpha(A_4) = 0, \\ M_n^-, & \text{if} \quad \alpha(A_4) = 1. \end{cases}$$

Define $\phi = \phi_5 * \cdots * \phi_1$. Then ϕ is a continuous path connecting M to M_n^+ or M_n^- within $\mathrm{Sp}^{th}(2n)$ according to the value of $\alpha(A_4)$. By Lemma 4, the function α is constant along the path ϕ. Thus $\alpha(M) = \alpha(A_4)$. This completes the proof of the lemma. □

Now we can give

The proof of Theorem 2: When $n = 1$, this is Theorem 1. When $n \geq 2$, by Lemma 5 the set $\mathrm{Sp}^{th}(2n)$ possesses at most two path connected components. On the other hand, we have

$$\alpha(M_n^+) = 0 \qquad \text{and} \qquad \alpha(M_n^-) = 1.$$

Thus by Lemma 4, the set $\mathrm{Sp}^{th}(2n)$ possesses at least two path connected components, and then precisely two .

For $i = 0$ or 1, let $\gamma : S^1 \to \mathrm{Sp}_i^{th}(2n)$ be continuous. Now we consider the function $\rho_n : \mathrm{Sp}(2n) \to \mathbf{U}$ defined by Lemma 4.3 below. Note that for $i = 0$ and 1,

$$\rho_n(M) = (-1)^i, \qquad \forall\, M \in \mathrm{Sp}_i^{th}(2n).$$

Then $\rho_n \circ \gamma$ is contractible in S^1. Thus γ is contractible in $\mathrm{Sp}(2n)$ by Lemma 4.7 below, i.e., $\mathrm{Sp}_i^{th}(2n)$ is simply connected in $\mathrm{Sp}(2n)$. The proof is complete. □

Lemma 6 *For $n \geq 2$ and any $M \in \mathrm{Sp}^h(2n)$, there exist an $N \in \mathrm{Sp}(2)^0$ and a path $f \in C([0,1], \mathrm{Sp}^h(2n) \cap \Omega^0(M))$ such that $f(0) = M$ and for some $b = \pm 1$ or 0, which is uniquely determined by M,*

$$\nu_1(f(t)) = \nu_1(M), \qquad \forall t \in [0,1], \tag{9}$$

$$f(1) = N_1(1,b)\diamond M_{n-1}^+, \qquad \text{if} \quad \alpha(M) = 0, \tag{10}$$

$$f(1) = N_1(1,b)\diamond M_{n-1}^-, \qquad \text{if} \quad \alpha(M) = 1. \tag{11}$$

Proof: Since $n \geq 2$ and $M \in \mathrm{Sp}^h(2n)$, by Theorem 1.4.1 there exists a matrix $P \in \mathrm{Sp}(2n)$ such that $P^{-1}MP = N\diamond M_0$ for some $N \in \mathrm{Sp}(2)^0$ and some $M_0 \in \mathrm{Sp}^{th}(2n-2)$. Let $\xi \in \mathcal{P}_1(2n)$ such that $\xi(1) = P$. Define $\phi_1(t) = \xi(t)^{-1}M\xi(t)$ for all $t \in [0,1]$. Then ϕ_1 connects M to $N\diamond M_0$ within $\mathrm{Sp}^h(2n)$. By Theorem 2, there is a path $\phi_2 : [0,1] \to \mathrm{Sp}^{th}(2n-2)$ connecting M_0 to M_{n-1}^+ or M_{n-1}^- according to the value of $\alpha(M_0) = \alpha(M)$ being 0 or 1. Define $f = (I_2\diamond\phi_2) * \phi_1$. Then f is the path claimed by the lemma. □

By the above proof, the integer $b = b(M)$ in (10) and (11) is uniquely determined by $M \in \mathrm{Sp}^h(2n)$. Thus the following definition is meaningful and will be used in our study of iterations.

Definition 7 For any $M \in \mathrm{Sp}^h(2n)$, the integer pair

$$(\alpha(M), b(M)) \in \{0,1\} \times \{-1,0,1\} \tag{12}$$

is called the **hyperbolic character** of M.

Now we can give

Proof of Theorem 3: By Lemma 6 and the path connectedness of $\mathrm{Sp}(2)^0$, the set $\mathrm{Sp}^h(2n)$ possesses at most two path connected components. Since $n \geq 2$, by Lemma 4 and the facts

$$\alpha(N\diamond M_{n-1}^+) = 0 \quad \text{and} \quad \alpha(N\diamond M_{n-1}^-) = 1,$$

we obtain that $\mathrm{Sp}^h(2n)$ possesses at least two and thus precisely two path connected components. Using the function $\rho_n : \mathrm{Sp}(2n) \to \mathbf{U}$ defined by Lemma 4.3 below, we obtain

$$\rho_n(M) = (-1)^i, \qquad \forall\, M \in \mathrm{Sp}_i^h(2n).$$

Thus similar to the proof of Theorem 2, $\mathrm{Sp}_i^{th}(2n)$ is simply connected in $\mathrm{Sp}(2n)$ for $i = 0$ and 1 by Lemma 4.7 below. The proof is complete. □

Note that the simply connectedness in Theorems 2 and 3 are not used in the proofs of the following Section 4.

The following is a direct consequence of Theorem 1.8.10 and Lemma 5.

Corollary 8 *For any $M \in \mathrm{Sp}(2n)$, there is a path $f : [0, 1] \to \Omega^0(M)$ such that $f(0) = M$ and*

$$f(1) = M_1(\omega_1)\diamond\cdots\diamond M_k(\omega_k)\diamond M_0, \tag{13}$$

where each $M_i(\omega_i)$ is a basic normal form of the eigenvalue $\omega_i \in \mathbf{U}$ for $1 \leq i \leq k$, and the symplectic matrix $M_0 = M_j^+$ or M_j^- for some non-negative integer j.

2.4 Structure of regular sets

In this section, we study the topological structure of the ω-regular set $\mathrm{Sp}(2n)_\omega^0$ for any $\omega \in \mathbf{U}$. Note that in Section 1 we have seen that $\mathrm{Sp}(2)_\omega^*$ with $\omega \in \mathbf{U}$ contains precisely two path connected components, both of which are simply connected in $\mathrm{Sp}(2)$. Here we prove that this is also true for the regular set $\mathrm{Sp}(2n)_\omega^*$ of the symplectic group $\mathrm{Sp}(2n)$ when $n \geq 2$. We still omit the subindex $\omega = 1$ when there is no confusion.

Theorem 1 (C. Conley and E. Zehnder [CoZ2], Y. Long [Lon15]). *For $\omega \in \mathbf{U}$, the set $\mathrm{Sp}(2n)_\omega^*$ possesses precisely two path connected components $\mathrm{Sp}(2n)_\omega^+$ and $\mathrm{Sp}(2n)_\omega^-$ and $M_n^+ \in \mathrm{Sp}(2n)_\omega^+$ and $M_n^- \in \mathrm{Sp}(2n)_\omega^-$, where $M_n^\pm$ are defined by* (3.7) *and* (3.8).

Proof: Fix $\omega = 1$. Note that $\mathrm{Sp}(2n)^*$ is open in $\mathrm{Sp}(2n)$. By direct computation we obtain $M_n^+ \in \mathrm{Sp}(2n)^+$ and $M_n^- \in \mathrm{Sp}(2n)^-$. Thus $\mathrm{Sp}(2n)^*$ possesses at least two path connected components.

Now fix any $M \in \mathrm{Sp}(2n)^*$. Next we construct a path f in $\mathrm{Sp}(2n)^*$ which connects M to M_n^+ or M_n^-.

Since $\mathrm{Sp}(2n)^*$ is open in $\mathrm{Sp}(2n)$, by Corollary 1.9.4, there is a path $f_1 \in C([0, 1], \mathrm{Sp}(2n)^*)$ such that $f_1(0) = M$ and $f_1(t)$ possesses only simple eigenvalues for all

$t \in (0, 1]$. By Theorem 1.7.3 and the path connectedness of $\mathrm{Sp}(2n)$ we obtain a path $f_2 : [0, 1] \to \mathrm{Sp}(2n)^*$ such that $f_2(0) = f_1(1)$ and $f_2(1)$ is a $\diamond$-product of normal forms of simple eigenvalues according to $\sigma(f_2(1))$.

For $A = R(\alpha)$ with $\alpha \in (0, \pi) \cup (\pi, 2\pi)$, or $A = D(-a)$ with $a > 0$, since $\mathrm{Sp}(2)^-$ is path connected, there is a path connecting A to $D(-2)$ within $\mathrm{Sp}(2)^-$. For $A = D(a)$ with $a \in \mathbf{R}^+\backslash\{1\}$, since $\mathrm{Sp}(2)^+$ is path connected, we also can connect $D(a)$ to $D(2)$ within $\mathrm{Sp}(2)^+$. Thus we can connect $f_2(1)$ by a path $f_3 : [0, 1] \to \mathrm{Sp}(2n)^*$ to a matrix $f_3(1) \in \mathrm{Sp}^{th}(2n)$. Then by Theorem 3.3, we obtain a path f_4 in $\mathrm{Sp}^{th}(2n)$ which connects $f_3(1)$ to M_n^+ or M_n^-. This proves that $\mathrm{Sp}(2n)^*$ possesses at most two and thus precisely two path connected components.

The cases of $\omega = -1$ and $\omega \in \mathbf{U}\backslash\mathbf{R}$ can be studied similarly, and the details are left to the readers. The proof is complete. □

Theorem 2 (C. Conley and E. Zehnder [CoZ2], D. Salamon and E. Zehnder [SaZ2], Y. Long [Lon15]). *For any $\omega \in \mathbf{U}$, the set $\mathrm{Sp}(2n)^*_\omega$ is simply connected in* $\mathrm{Sp}(2n)$.

We start the proof of this theorem from the following lemmas.

Lemma 3 *For every $n \in \mathbf{N}$, there exists a function $\rho_n \in C(\mathrm{Sp}(2n), \mathbf{U})$ such that*

$$\rho_n(B^{-1}AB) = \rho_n(A), \qquad \forall A, B \in \mathrm{Sp}(2n), \tag{1}$$

$$\begin{aligned} \rho_n(A\diamond B) = \rho_k(A)\rho_h(B),& \\ \forall A \in \mathrm{Sp}(2k),\ B \in \mathrm{Sp}(2h),\ \text{and } k+h = n,& \end{aligned} \tag{2}$$

$$\rho_n(M) = (-1)^{\alpha(M)}, \quad \text{if} \quad \sigma(M)\cap\mathbf{U} = \emptyset,\ M \in \mathrm{Sp}(2n), \tag{3}$$

$$\rho_1(R(\theta)) = e^{\sqrt{-1}\theta}, \qquad \forall\theta \in \mathbf{R}. \tag{4}$$

Proof: We carry out the proof in several steps.

Step 1 The Definition of $\{\rho_n\}$.

Suppose $M \in \mathrm{Sp}(2n)$ possesses only simple eigenvalues. Then its eigenvalues can be arranged in a sequence

$$\lambda_1, \ldots, \lambda_n, \lambda_1^{-1}, \ldots, \lambda_n^{-1},$$

where $\lambda_1, \ldots, \lambda_n$ are either Krein-negative eigenvalues of M on $\mathbf{U}$ or eigenvalues inside the open unit disc in $\mathbf{C}$ of M. Denote by $\arg\lambda$ the argument angle in $[0, 2\pi)$ of the polar decomposition of the complex number λ. Define $\mathrm{Arg}(M)$ in $\mathbf{R}/(2\pi\mathbf{Z})$ by

$$\mathrm{Arg}(M) = \sum_{j=1}^{n} \arg(\lambda_j) \qquad \mathrm{mod} \quad 2\pi. \tag{5}$$

Define

$$\rho_n(M) = \exp(\sqrt{-1}\mathrm{Arg}(M)) = \prod_{j=1}^{n} \frac{\lambda_j}{|\lambda_j|}. \tag{6}$$

Since in fact Arg(M) only depends on Krein-negative eigenvalues on $\mathbf{U}\backslash\mathbf{R}$ and negative eigenvalues of M, the map ρ_n is continuous on symplectic matrices possessing simple eigenvalues.

Next we generalize the definition of ρ_n to the whole Sp$(2n)$. For any $M \in \mathrm{Sp}(2n)$, suppose $\lambda \in \sigma(M) \cap \mathbf{U}\backslash\{\pm 1\}$ has Krein type $(p_\lambda(M), q_\lambda(M))$. Denote by $2m^0(M)$ the total multiplicity of the eigenvalue -1 of M. We define

$$\rho_n(M) = \begin{cases} (-1)^{m_0(M)+\alpha(M)} \prod_{\lambda\in\sigma(M)\cap(\mathbf{U}\backslash\{\pm 1\})} \lambda^{q_\lambda(M)}, \\ \qquad\qquad \text{if} \quad \sigma(M) \cap (\mathbf{R}^- \cup (\mathbf{U}\backslash\{1\})) \neq \emptyset, \\ 1, \qquad\quad \text{if} \quad \sigma(M) \cap (\mathbf{R}^- \cup (\mathbf{U}\backslash\{1\})) = \emptyset, \end{cases} \tag{7}$$

where $\mathbf{R}^-$ is the set of all negative real numbers, and $\alpha(M)$ is the hyperbolic index of M defined in Section 1.8. This definition coincides with (6) if M possesses only simple eigenvalues.

Step 2 The continuity of $\rho_n : \mathrm{Sp}(2n) \to \mathbf{U}$.

In fact, fix any $M \in \mathrm{Sp}(2n)$. Let $f : [0, 1] \to \mathrm{Sp}(2n)$ be any path satisfying $f(0) = M$. Then for $t > 0$ sufficiently small, we consider the possible changes of eigenvalues along the path f which may contribute to any change of the value of ρ_n defined in (7).

If the eigenvalue -1 and nearby negative eigenvalues of M have some exchanges when $t > 0$, then

$$m_0(f(t)) + \alpha(f(t)) = \text{constant mod } 2, \qquad \text{for } t > 0 \text{ sufficiently small.} \tag{8}$$

Thus they make no contribution to any change of (7) for small $t > 0$.

If the eigenvalue -1 and the nearby quadruple eigenvalues in $\mathbf{C}\backslash(\mathbf{U} \cup \mathbf{R})$ of M have some exchanges when $t > 0$, then $m_0(M)$ must change by an even integer. Thus (8) still holds, and they make no contribution to any change of (7) for small $t > 0$.

If the eigenvalue -1 and nearby complex conjugate eigenvalues on $\mathbf{U}\backslash\mathbf{R}$ of M have some exchanges, by our discussion in Section 1.3, then the total multiplicity of the changed eigenvalues is even. Thus $m_0(M)$ must change by an integer $k(t)$, and at the same time there are Krein-negative eigenvalues on $\mathbf{U}\backslash\mathbf{R}$ sufficiently close to -1 of $f(t)$ that appear or disappear with total multiplicity $k(t)$ whenever $t > 0$ is small enough. Thus

$$(-1)^{m_0(M)} \prod_{\lambda\in\sigma(M)\cap(\mathbf{U}\backslash\{\pm 1\})} \lambda^{m_\lambda^+(M)} \tag{9}$$

changes as little as we want if $t > 0$ is small enough.

If the eigenvalues on $\mathbf{U}\setminus\mathbf{R}$ and quadruple eigenvalues in $\mathbf{C}\setminus(\mathbf{U}\cup\mathbf{R})$ of M have some exchange, by Theorem 1.3.9 this can happen only for multiples of a quadruple of eigenvalues $\{\omega_1, \omega_2, \omega_3, \omega_4\}$ with $\omega_1 = \omega_2 = \omega$ and $\omega_3 = \omega_4 = \overline{\omega}$ in $\mathbf{U}\setminus\mathbf{R}$; ω_1 and ω_3 are Krein positive, ω_2 and ω_4 are Krein negative. Thus $\omega_1\omega_3 = \omega\overline{\omega} = 1$ and these exchanges make no contribution to any change of (7).

If the eigenvalues on $\mathbf{U}\setminus\mathbf{R}$ and nearby eigenvalues on $\mathbf{U}$ have some exchanges, then the corresponding changes of (7) can be required to be as small as we want if $t > 0$ is sufficiently small.

If the eigenvalue 1 and nearby complex conjugate eigenvalues on $\mathbf{U}\setminus\mathbf{R}$ of M have some exchanges, noting that the eigenvalue 1 has also appeared in the product of (7), then the contribution of these exchanges to the change of (7) can be required to be as small as we want if $t > 0$ is sufficiently small.

If the eigenvalue 1 and the nearby quadruple eigenvalues in $\mathbf{C}\setminus(\mathbf{U}\cup\mathbf{R})$ of M have some exchanges, or quadruple eigenvalues in $\mathbf{C}\setminus(\mathbf{U}\cup\mathbf{R})$ of M have some changes, these changes make no contribution to any change of (7).

Thus by the above discussion, for any $\epsilon > 0$, there exists $\delta > 0$ sufficiently small such that whenever $N \in \mathrm{Sp}(2n)$ and $\|N - M\|_{\mathcal{L}(\mathbf{R}^{2n})} \le \delta$, we always have

$$|\rho_n(N) - \rho_n(M)| \le \epsilon. \tag{10}$$

This proves the continuity of ρ_n for every $n \in \mathbf{N}$.

Step 3 Basic properties of the maps $\rho_n : \mathrm{Sp}(2n) \to \mathbf{U}$ for $n \in \mathbf{N}$.

In fact, for any A and $B \in \mathrm{Sp}(2n)$, since $B^{-1}AB$ and A possess the same eigenvalues with the same algebraic and geometric multiplicities and the same Krein type numbers, we obtain (1).

By the definition of the $\{\rho_n\}$, we obtain (2) and (3). When $e^{\sqrt{-1}\theta}$ and $e^{-\sqrt{-1}\theta} \in \mathbf{U}\setminus\mathbf{R}$, by direct computation we obtain that $e^{\sqrt{-1}\theta}$ is a Krein-negative and $e^{-\sqrt{-1}\theta}$ is a Krein-positive eigenvalue of $R(\theta)$ respectively. Thus by (7) we obtain (4) for this case. By the continuity of ρ_1 the case of $e^{\sqrt{-1}\theta} = e^{\sqrt{-1}\theta} = \pm 1$ follows.

The proof of Lemma 3 is complete. □

Lemma 4 *The map* $\{\rho_n\}$ *with* $\rho_n \in C(\mathrm{Sp}(2n), \mathbf{U})$ *satisfying* (1)–(4) *of Lemma* 3 *is unique.*

Proof: Suppose $\zeta_n \in C(\mathrm{Sp}(2n), \mathbf{U})$ are defined for every $n \in \mathbf{N}$ and also satisfy (1)–(4). We need to prove

$$\zeta_n(M) = \rho_n(M), \qquad \forall M \in \mathrm{Sp}(2n). \tag{11}$$

In fact, fix an $M \in \mathrm{Sp}(2n)$ and any $\epsilon > 0$. By Corollary 1.9.4, there exists a sufficiently short path ϕ in $\mathrm{Sp}(2n)$ connecting M to a nearby matrix $A \in \mathrm{Sp}(2n)$ such that A possesses only simple eigenvalues and

$$|\zeta_n(M) - \zeta_n(A)| \le \epsilon, \qquad |\rho_n(M) - \rho_n(A)| \le \epsilon. \tag{12}$$

Here we have used the continuities of both ρ_n and ζ_n. Since A possesses only simple eigenvalues, $\sigma(A)$ contains only simple complex conjugate eigenvalues on $\mathbf{U}\backslash\mathbf{R}$, real reciprocal eigenvalues on $\mathbf{R}\backslash\{0, \pm 1\}$, and simple quadruple eigenvalues in $\mathbf{C}\backslash(\mathbf{U} \cup \mathbf{R})$.

By Theorem 1.7.3, there exists a matrix $P \in \mathrm{Sp}(2n)$ such that $N \equiv P^{-1}AP = N_1 \diamond \cdots \diamond N_k$ is a $\diamond$-product of normal form matrices for some $1 \le k \le n$, where each $N_i \in \mathrm{Sp}(2m_i)$ is one of the forms $R(\theta)$, $D(\lambda)$, and $\mathrm{diag}(\lambda R(\theta), \lambda^{-1}R(\theta))$ with $\theta \in \mathbf{U}\backslash\mathbf{R}$ and $\lambda \in \mathbf{R}\backslash\{0, \pm 1\}$. Thus by (1) and (2) we obtain

$$\zeta_n(A) = \zeta_n(N) = \prod_{1 \le i \le k} \zeta_{m_i}(N_i). \tag{13}$$

Similarly we have

$$\rho_n(A) = \rho_n(N) = \prod_{1 \le i \le k} \rho_{m_i}(N_i). \tag{14}$$

By (3) and (4), we obtain

$$\zeta_{m_i}(N_i) = \rho_{m_i}(N_i), \qquad \forall 1 \le i \le k. \tag{15}$$

Thus by (12)–(15), we obtain

$$|\zeta_n(M) - \rho_n(M)| \le 2\epsilon.$$

Since $\epsilon > 0$ is arbitrary, this proves (11) and the lemma. □

Note that in the above uniqueness proof, we only need (3) to hold for $n = 1$ and 2.

By Lemma 1.1.3, every $M \in \mathrm{Sp}(2n)$ has its unique polar decomposition $M = A(M)U(M)$, where $A(M) = (MM^T)^{1/2}$ is symmetric positive definite and symplectic, $U(M)$ is orthogonal and symplectic. Therefore by Lemma 1.1.5, $U(M)$ has the form

$$U(M) = \begin{pmatrix} u_1(M) & -u_2(M) \\ u_2(M) & u_1(M) \end{pmatrix}, \tag{16}$$

and $u(M) \equiv u_1(M) + \sqrt{-1}u_2(M) \in U(n, \mathbf{C})$.

Lemma 5 *The maps $\{\rho_n\}$ defined by Lemma 3 further satisfy*

$$\rho_n(M) = \det u(M), \qquad \forall M \in \mathrm{Sp}(2n) \cap O(2n), \tag{17}$$

$$\rho_n(M^{-1}) = \rho_n(M^T) = \overline{\rho_n(M)}, \qquad \forall M \in \mathrm{Sp}(2n). \tag{18}$$

Proof: Note that $\det u(M)$ is the product of all the eigenvalues of $u(M)$ for any $M \in \mathrm{Sp}(2n) \cap O(2n)$. Therefore (16) follows from Theorem 1.3.10 immediately.

Note that for $M \in \mathrm{Sp}(2n)$, by $J^{-1}M^T J = M^{-1}$ and (1), we obtain the first equality in (18). It also holds that

(i) $\sigma(M) = \sigma(M^T)$ and $\alpha(M) = \alpha(M^T)$;

(ii) $\lambda \in \sigma(M) \cap (\mathbf{U}\backslash\mathbf{R})$ possesses Krein type (p, q) if and only if $\overline{\lambda} \in \sigma(M^T) \cap (\mathbf{U}\backslash\mathbf{R})$ possesses Krein type (p, q);

(iii) if $\{\lambda, \overline{\lambda}, \lambda^{-1}, (\overline{\lambda})^{-1}\} \subset \mathbf{C}\backslash(\mathbf{U}\backslash\mathbf{R})$ form an eigenvalue quadruple for M, they are eigenvalues of M^T with the same multiplicities.

From these three facts, we obtain the second equality in (18). The proof is complete. □

Remark 6 It holds that

$$A \equiv \begin{pmatrix} \frac{1}{\sqrt{2}} & \frac{3}{\sqrt{2}} \\ 0 & \sqrt{2} \end{pmatrix} = \begin{pmatrix} 2 & 1 \\ 1 & 1 \end{pmatrix} \begin{pmatrix} \frac{1}{\sqrt{2}} & \frac{1}{\sqrt{2}} \\ -\frac{1}{\sqrt{2}} & \frac{1}{\sqrt{2}} \end{pmatrix}.$$

Therefore together with the first two identities in Section 1.7, we obtain that the function $\det u(A)$ does not satisfy (2) for this matrix A. Consequently (17) does not hold for every symplectic matrix in general.

Lemma 7 *The map* $\rho_n : \mathrm{Sp}(2n) \to \mathbf{U}$ *defined by Lemma* 3 *induces an isomorphism on the fundamental groups*

$$(\rho_n)_* : \pi_1(\mathrm{Sp}(2n)) \xrightarrow{\cong} \pi_1(\mathbf{U}). \tag{19}$$

Proof: Note that $\mathrm{Sp}(2n) \cap O(2n)$ is homeomorphic to the unitary group $U(n, \mathbf{C})$ by Lemma 1.1.5 and the remarks next to it via the map $u : \mathrm{Sp}(2n) \cap O(2n) \to U(n, \mathbf{C})$ defined by (16). Note also that the determinant map $\det : U(n, \mathbf{C}) \to \mathbf{C}$ induces an isomorphism $\det_* : \pi_1(U(n, \mathbf{C})) \to \pi_1(\mathbf{U})$ by the exact homotopy sequence (cf. [Spa 1])

$$0 = \pi_1(SU(n, \mathbf{C})) \to \pi_1(U(n, \mathbf{C})) \xrightarrow{\det_*} \pi_1(\mathbf{U}) \to \pi_0(SU(n, \mathbf{C})) = 0.$$

Therefore by Lemma 5 the map ρ_n induces an isomorphism

$$(\rho_n)_* : \pi_1(\mathrm{Sp}(2n)) \cap O(2n)) \to \pi_1(\mathbf{U}). \tag{20}$$

Since by (2.4), the quotient space $\mathrm{Sp}(2n)/(\mathrm{Sp}(2n)) \cap O(2n))$ is contractible, (19) holds by properties of ρ_n. □

Now we can give

The proof of Theorem 2: We distinguish three cases according to the value of ω.

Case 1 $\omega = 1$.

The proof is carried out in two steps. We still omit the subindex $\omega = 1$.

Step 1 An alternative definition of ρ_n on $\mathrm{Sp}(2n)^*$.

In this step, we shall define functions $b_i \in C(\mathrm{Sp}(2n)^*, [0, 2\pi])$ for $1 \le i \le n$ such that for any $M \in \mathrm{Sp}(2n)^*$,

$$\rho_n(M) = \exp\left\{\sqrt{-1}\sum_{i=1}^{n} b_i(M)\right\}. \tag{21}$$

Let $\mathbf{R}^+$ be the set of all positive real numbers. Fix an $M \in \mathrm{Sp}(2n)^*$. We pick up n eigenvalues of M by

$$\Lambda = \{\lambda_1, \ldots, \lambda_k, a_1, \ldots, a_{n-k}\}, \tag{22}$$

where a_i for $1 \le i \le n-k$ denote all the eigenvalues of M in $(0, 1)$, and λ_i for $1 \le i \le k$ are chosen in the following way:

If $\lambda \in \mathbf{U}$ has Krein type (p, q), we define $\lambda_i = \lambda$ for q times.

If $\lambda \in \mathbf{C}\backslash\mathbf{R}^+$ and $|\lambda| < 1$ has multiplicity m, we define $\lambda_i = \lambda$ for m times.

By Theorem 1.3.1 and Lemma 1.3.8, the total number of λ_is we have defined must be k.

We construct the $b_i(M)$s as follows.

(A) For each λ_i in Λ, we choose a real number $b_i(M)$ for $1 \le i \le k$ such that

$$0 \le b_i(M) \le 2\pi, \tag{23}$$

$$\exp\{\sqrt{-1}b_i(M)\} = \frac{\lambda_i}{|\lambda_i|}. \tag{24}$$

Note that these $b_i(M)$s are uniquely determined in such a way.

Then we define $b_i(M)$ according to a_i.

(B) If $\det(M - I) > 0$, by Theorem 1.3.1 there must hold $n - k = 2m_1$ for some integer $m_1 \ge 0$. We define $b_i(M) = 0$ and $b_i(M) = 2\pi$ for m_1 times respectively.

(C) If $\det(M - I) < 0$, by Theorem 1.3.1 there must hold $n - k = 2m_1 + 1$ for some integer $m_1 \ge 0$. We define $b_i(M) = 0$ for m_1+1 times and $b_i(M) = 2\pi$ for m_1 times.

Note that by Theorems 1.3.1, 1.3.9 and 1.9.5, we have defined n of the $b_i(M)$s and they are continuous on $\mathrm{Sp}(2n)^*$.

Note that any λ and $\bar{\lambda} \in \sigma(M) \cap (\mathbf{C}\backslash\mathbf{R})$ with $|\lambda| < 1$ only contribute 1 on the right-hand side of (21). Thus by Theorem 1.3.1 and the definition of $b_i(M)$, the function defined on $\mathrm{Sp}(2n)^*$ by the right-hand side of (21) in the above way coincides with the function ρ_n defined by (7). This coincidence can also be verified by the uniqueness in Lemma 4.

Step 2 For $\tau > 0$ let $\gamma : S_\tau \equiv \mathbf{R}/(\tau\mathbf{Z}) \to \mathrm{Sp}(2n)^*$ be a continuous closed curve in $\mathrm{Sp}(2n)^*$. Then $\rho_n \circ \gamma : S_\tau \to \mathbf{U}$ is a continuous cycle on $\mathbf{U}$. Using the functions $b_i(\gamma(t))$ for $1 \le i \le n$ we define

$$\hat{\Delta}_\gamma(t) = \sum_{i=1}^{n} b_i(\gamma(t)), \qquad \forall 0 \le t \le \tau, \tag{25}$$

$$\hat{\Delta}_\tau(\gamma) = \hat{\Delta}_\gamma(\tau) - \hat{\Delta}_\gamma(0). \tag{26}$$

Note that each $b_i \circ \gamma : S_\tau \to [0, 2\pi]$ is contractible, and then by

$$\rho_n \circ \gamma(t) = \exp(\sqrt{-1}\hat{\Delta}_\gamma(t)),$$

$\rho_n \circ \gamma : S_\tau \to \mathbf{U}$ is contractible too. Then by Lemma 7, the loop γ is contractible in $\mathrm{Sp}(2n)$.

Thus $\mathrm{Sp}(2n)^*$ is simply connected in $\mathrm{Sp}(2n)$.

Case 2 $\omega = -1$

In this case, we have $D_{-1}(M) = D_1(-M)$. Thus the map $M \mapsto -M$ gives a homeomorphism from $\mathrm{Sp}(2n)_{-1}^{\pm}$ to $\mathrm{Sp}(2n)_1^{\pm}$. Then the case of $\omega = -1$ follows from that of $\omega = 1$.

Case 3 $\omega \in \mathbf{U}\backslash\mathbf{R}$.

Let $\theta \in (0, \pi) \cup (\pi, 2\pi)$ such that $\omega = e^{\sqrt{-1}\theta}$. Similar to Case 1, we can construct functions $b_i \in C(\mathrm{Sp}(2n)_\omega^*, \mathbf{R})$ for $1 \le i \le n$ such that for any $M \in \mathrm{Sp}(2n)_\omega^*$,

$$b_i(M) \in [\theta, 2\pi + \theta], \tag{27}$$

$$\exp\left(\sqrt{-1}\sum_{i=1}^{n} b_i(M)\right) = \rho_n(M). \tag{28}$$

The details of this definition are left to the readers. Now similarly to Case 1, for any loop $\gamma : S^1 \to \mathrm{Sp}(2n)_\omega^*$, by (28) and properties of $b_i(M)$, we obtain that $\rho_n \circ \gamma : S^1 \to \mathbf{U}$ is contractible in $\mathbf{U}$. Then by Lemma 7, the loop γ is contractible in $\mathrm{Sp}(2n)$.

The proof is complete.

Corollary 8 (D. Salamon and E. Zehnder [SaZ2]). *For any $\tau > 0$, $\omega \in \mathbf{U}$, and $M \in \mathrm{Sp}(2n)_\omega^*$, let $\gamma : [0, \tau] \to \mathrm{Sp}(2n)_\omega^*$ be a continuous path satisfying $\gamma(0) = M$ and $\gamma(\tau) = M_n^+$ or M_n^-. Then*

$$|\hat{\Delta}_\tau(\gamma)| < n\pi. \tag{29}$$

Proof: We only prove the case of $\omega = 1$. Other cases are similar.

When $n = 1$, the estimate (29) follows from Figures 1.1 and 1.2 immediately. When $n \geq 2$, by our above definitions for $b_i(M)$ in the proof of Theorem 2, we obtain

$$\sum_{i=1}^{n} b_i(\gamma(\tau)) = \sum_{i=1}^{n} b_i(M_n^+) = \sum_{i=1}^{n} b_i(M_n^-) = n\pi,$$

and

$$0 < \sum_{i=1}^{n} b_i(\gamma(0)) < 2n\pi.$$

Thus (29) holds. □

Next we study a subset $\mathrm{Sp}(2n)^{**}$ of the regular set of $\mathrm{Sp}(2n)$ defined by:

$$\mathrm{Sp}(2n)^{**} = \{M \in \mathrm{Sp}(2n)_1^* \mid \lambda \in \sigma(M) \cap \mathbf{U} \ \textit{implies} \ \lambda = -1\}. \tag{30}$$

Proposition 9 1° $\mathrm{Sp}(2n)^*$ *can be deformed into* $\mathrm{Sp}(2n)^{**}$ *for* $n \geq 1$.

2° $\mathrm{Sp}(2)^{**}$ *is a strong deformation retract of* $\mathrm{Sp}(2)^*$.

3° $\mathrm{Sp}(2n)^{**}$ *is not a deformation retract of* $\mathrm{Sp}(2n)^*$ *if* $n \geq 2$.

Proof: 2° follows from the study of Sp(2) in Section 2.1. Assume 3° does not hold, we prove it indirectly. By definition there exists $\delta \in C(\mathrm{Sp}(2n)^* \times [0,1], \mathrm{Sp}(2n)^*)$ such that

$$\delta(M,0) = M, \quad \delta(M,1) \in \mathrm{Sp}(2n)^{**}, \qquad \forall M \in \mathrm{Sp}(2n)^*, \tag{31}$$

$$\delta(M,1) = M, \qquad \forall M \in \mathrm{Sp}(2n)^{**}. \tag{32}$$

Since $n \geq 2$, for $\rho \in [1,2]$ we can define

$$M_\rho = \begin{pmatrix} \rho R(\frac{\pi}{2}) & 0 \\ 0 & \rho^{-1} R(\frac{\pi}{2}) \end{pmatrix} \diamond (-I_{2n-4}). \tag{33}$$

Then we have

$$\sigma(M_\rho) = \{-1, \rho\sqrt{-1}, \rho^{-1}\sqrt{-1}, -\rho\sqrt{-1}, -\rho^{-1}\sqrt{-1}\},$$

where -1 has its multiplicity $(2n-4)$. For $1 < \rho \leq 2$, we have $M_\rho \in \mathrm{Sp}(2n)^{**}$, and then by (33) we obtain $\delta(M_\rho, 1) = M_\rho$ and $\sigma(\delta(M_\rho,1)) = \sigma(M_\rho)$. Thus by the continuities of δ and $\sigma(M_\rho)$ in $\rho \in [1,2]$,

$$\begin{aligned} \sigma(\delta(M_1,1)) &= \lim_{\rho\to 1, \rho>1} \sigma(\delta(M_\rho,1)) \\ &= \lim_{\rho\to 1, \rho>1} \sigma(M_\rho) \\ &= \{-1, \sqrt{-1}, \sqrt{-1}, -\sqrt{-1}, -\sqrt{-1}\}. \end{aligned}$$

This contradicts (31) and (30) and completes the proof of 3°.

We leave the proof of 1° to the readers. □

Remark 10 Theorem 1 on the path connected property of $\mathrm{Sp}(2n)_1^*$ was proved by C. Conley and E. Zehnder in [CoZ2] of 1984 for the case of $n \geq 2$. The case of $n = 1$ follows from the $\mathbf{R}^3$-cylindrical coordinate representation immediately.

Theorem 2 on the simply connectedness of $\mathrm{Sp}(2n)_1^*$ in $\mathrm{Sp}(2n)$ was formulated by C. Conley and E. Zehnder in [CoZ2] of 1984 for the case of $n \geq 2$. Proposition 9 was proved by the author in 1988 during his visit to ETH-Zürich, and was put into [Lon15] which was published later in 1999. It shows that Lemma 1.4 of [CoZ2] is incorrect and therefore the proof of Theorem 2 for the simply connectedness of $\mathrm{Sp}(2n)_1^*$ in $\mathrm{Sp}(2n)$ given in [CoZ2] is not complete. The complete proof of Theorem 2 for $\mathrm{Sp}(2n)_1^*$ was given by D. Salamon and E. Zehnder in [SaZ1] and [SaZ2] in 1990–1992, in which the function $\rho_n(M)$ of (7) was defined. Corollary 8 was proved in [SaZ2].

Theorems 1 and 2 for $\omega \in \mathbf{U}\backslash\{1\}$ were proved in [Lon15] of 1999.

2.5 Structure of singular sets

In this section we study the topological structure of the singular set $\mathrm{Sp}(2n)_\omega^0$ in $\mathrm{Sp}(2n)$ for $\omega \in \mathbf{U}$. When $n = 1$, this is completely clear by our $\mathbf{R}^3$-cylindrical coordinate representation of $\mathrm{Sp}(2)_\omega^0$. But when $n \geq 2$, the structure becomes rather complicated. Our study is realized via normal forms introduced in Chapter 1. Note that in this section we are working in the field $\mathbf{R}$.

Theorem 1 *$\mathrm{Sp}(2n)_1^0$ and $\mathrm{Sp}(2n)_{-1}^0$ are path connected respectively. For $\omega \in \mathbf{U}\backslash\mathbf{R}$, the set $\mathrm{Sp}(2)_\omega^0$ has precisely two path connected components, and the set $\mathrm{Sp}(2n)_\omega^0$ is path connected if $n \geq 2$.*

Proof: Let $M \in \mathrm{Sp}(2n)_\omega^0$. By Theorem 1.9.3, there is a path $f : [0, 1] \to \mathrm{Sp}(2n)_1^0$ and an integer $k \in [1, n]$ such that $f(0) = M$ and

$$f(1) = M_1 \diamond \cdots \diamond M_k \diamond M_0 \equiv N, \tag{1}$$

where $M_0 \in \mathrm{Sp}(2k_0)$ for some $k_0 \geq 0$ and $\omega \notin \sigma(M_0)$, and each M_i for $1 \leq i \leq k$ is a basic normal form matrix for the eigenvalue ω.

When $\omega = 1$, each $M_i \in \mathrm{Sp}(2)$ in (1) has the form $N_1(1, b_i)$ with $b_i = \pm 1$ or 0 for $1 \leq i \leq k$. Connecting each M_i to I_2 within $\mathrm{Sp}(2)_1^0$ and M_0 to I_{2k_0} in $\mathrm{Sp}(2k_0)$, we obtain the path connectedness of $M \in \mathrm{Sp}(2n)_1^0$.

When $\omega = -1$, since $M \mapsto -M$ yields a homeomorphism between $\mathrm{Sp}(2n)_1^0$ and $\mathrm{Sp}(2n)_{-1}^0$, the later one is also path connected.

When $\omega \in \mathbf{U}\backslash\mathbf{R}$ and $n = 1$, the conclusion follows from the $\mathbf{R}^3$-cylindrical coordinate representation of $\mathrm{Sp}(2)$.

When $\omega \in \mathbf{U}\backslash\mathbf{R}$ and $n \geq 2$, by our study in Case 4 before Definition 1.8.9, we can connect the basic normal form $N_2(\omega, b)$ or $N_2(\bar{\omega}, b)$ to $N_2(\omega, 0)$ or $N_2(\bar{\omega}, 0)$ within $\mathrm{Sp}(2n)_\omega^0$. Then we connect $N_2(\omega, 0)$ or $N_2(\bar{\omega}, 0)$ to $R(\theta)\diamond R(-\theta)$ within $\mathrm{Sp}(2n)_\omega^0$ where $\omega = e^{\sqrt{-1}\theta}$. Therefore we may assume every M_i in (1) is of the form $R(\theta)$

or $R(-\theta)$. Keep one of the $R(\theta)$s and $R(-\theta)$s fixed, and connect all the other M_j and M_0 to the identity matrices respectively. Then M has been connected to $R(\theta)\diamond I$ or $R(-\theta)\diamond I$ within $\mathrm{Sp}(2n)^0_\omega$. Then within $\mathrm{Sp}(2n)^0_\omega$ we can connect $R(-\theta)\diamond I$ to $R(-\theta)\diamond R(\theta)\diamond I_{2n-4}$, then to $R(\theta)\diamond R(\theta)\diamond I_{2n-4}$, and finally to $R(\theta)\diamond I$. Thus $\mathrm{Sp}(2n)^0_\omega$ is path connected if $n \geq 2$. □

By the definition of $\mathcal{M}^k_\omega(2n)$ in Section 1.8, we have $\mathcal{M}^0_\omega(2n) = \mathrm{Sp}(2n)^*_\omega$, which is simply connected in $\mathrm{Sp}(2n)$ by Theorem 4.2. Note that $\mathcal{M}^{2n}_1(2n) = \{I_{2n}\}$ is certainly simply connected.

By our discussion in Section 1, the set $\mathcal{M}_\omega(2)$ possesses precisely two path connected components $\mathrm{Sp}(2)^0_{\omega,+}$ and $\mathrm{Sp}(2)^0_{\omega,-}$, each of which is simply connected in $\mathrm{Sp}(2)$ for $\omega \in \mathbf{U}$, but is not simply connected in itself if $\omega = 1$ or -1.

Proposition 2 *When $n \geq 2$ and $\omega \in \mathbf{U}$, the singular hypersurface $\mathrm{Sp}(2n)^0_\omega$ is not simply connected in $\mathrm{Sp}(2n)$. Thus it is also not simply connected in itself.*

Proof: Let $e^{\sqrt{-1}\theta} = \omega \in \mathbf{U}$. Define

$$f(t) = I_{2n-4}\diamond R(\theta)\diamond R(2t\pi), \qquad \forall 0 \leq t \leq 1.$$

Then the function $\hat{\Delta}_1$ defined by (4.25) satisfies $\hat{\Delta}_1(f) = 2\pi$ and $f(t) \in \mathrm{Sp}(2n)^0_\omega$ for all $t \in [0, 1]$. Thus $\mathrm{Sp}(2n)^0_\omega$ with $n \geq 2$ is not simply connected in $\mathrm{Sp}(2n)$, and is not simply connected in itself. □

Proposition 3 1° *The conjugacy class $[M]$ in $\mathrm{Sp}(2n)$ is path connected in itself for any $M \in \mathrm{Sp}(2n)$.*

2° *The homotopy component $\Omega^0(M)$ and thus the conjugacy class $[M]$ is simply connected in $\mathrm{Sp}(2n)$ for any $M \in \mathrm{Sp}(2n)$.*

Proof: 1° The path connectedness of $[M]$ follows from the normal form results in Chapter 1.

2° Fix $M \in \mathrm{Sp}(2n)$. Let $\gamma : [0, 1]/\{0, 1\} \to \Omega^0(M)$ be a continuous loop. By the definition of $\Omega^0(M)$, any $N \in \Omega^0(M)$ satisfies $\sigma(N) \cap \mathbf{U} = \sigma(M) \cap \mathbf{U}$. Thus by the definition of the function $\rho_n : \mathrm{Sp}(2n) \to \mathbf{U}$ via (4.7), it must hold that

$$k(\gamma) \equiv {}^{\#}\{\rho_n(\gamma(t)) \,|\, t \in [0, 1]\} < +\infty.$$

By Lemma 4.3, the function ρ_n is continuous. Thus $k(\gamma) = 1$, i.e., γ is contractible in $\mathrm{Sp}(2n)$. □

Note that in Proposition 3, if $M \in \mathrm{Sp}(2n)^*_\omega$, the conclusion is a consequence of Theorem 4.2.

Note that the structure of $\mathrm{Sp}(2n)^0_{-1}$ is similar to that of $\mathrm{Sp}(2n)^0_1$. The details are left to the readers. When $\omega \in \mathbf{U}\backslash\mathbf{R}$, the set $\mathrm{Sp}(2)^0_\omega$ possesses two path connected components and $\mathrm{Sp}(2n)^0_\omega$ is path connected when $n \geq 2$.

Remark 4 The above results in this section are based on the ideas and results in [Lon6] of 1991, [Lon13] of 1997, and [Lon15] of 1999.

In [LZu2], the following further results on the differentiable structures of $\mathcal{M}_\omega^k(2n)$ for $1 \le k \le 2n$ defined in Section 1.6 were studied. The following results were proved based upon the ideas of V.I. Arnold in [Arn1] and of our study on the normal forms. These results form the basis for the intersection number definition of the index theory for symplectic paths. We shall not use them in this book, and refer the interested readers to [LZu2] of C. Zhu and the author for detailed proofs.

Theorem 5 (cf. Theorem 1.1 of [LZu2]). *For every $0 \le k \le 2n$, $\mathcal{M}_{\pm1}^k(2n)$ is a codimension $\frac{k(k+1)}{2}$ smooth submanifold of* Sp$(2n)$, *and $\frac{d}{dt}|_{t=0}$ (Me^{Jt}), with $M \in \mathcal{M}_{\pm1}(2n)$, forms a transverse structure of $\mathcal{M}_{\pm1}(2n)$ in* Sp$(2n)$. *Moreover,*

$$\overline{\mathcal{M}_{\pm1}^k(2n)} = \bigcup_{h \ge k} \mathcal{M}_{\pm1}^h(2n).$$

Theorem 6 (cf. Theorem 1.2 of [LZu2]). *For any $\omega \in \mathbf{U}\setminus\mathbf{R}$, the set $\mathcal{M}_\omega(2n)$ is a codimension* 1 *smooth submanifold of* Sp$(2n)$, *and $\frac{d}{dt}|_{t=0}$ (Me^{Jt}), with $M \in \mathcal{M}_\omega(2n)$, forms a transverse structure of $\mathcal{M}_\omega(2n)$ in* Sp$(2n)$. *Moreover, $\cup_{l \ge k}\mathcal{M}_\omega^l(2n)$ is a codimension k^2 real algebraic subvariety of* Sp$(2n)$ *for every $0 \le k \le n$, and is an empty set for $n < k \le 2n$.*

2.6 Transversality of rotation paths

For $\theta \in \mathbf{R}$ and $1 \le k \le n$, we define the rotation matrix $R_k(\theta)$ by

$$R_k(\theta) = I_{2k-2} \diamond R(\theta) \diamond I_{2n-2k}. \tag{1}$$

Here I_0 means that this component in (1) does not appear. Recall $R(\theta) = \begin{pmatrix} \cos\theta & -\sin\theta \\ \sin\theta & \cos\theta \end{pmatrix}$. It is easy to verify the following properties of these rotation matrices. For any $\alpha, \beta \in \mathbf{R}$ and $1 \le i, j \le n$,

$$R_i(\alpha)R_i(\beta) = R_i(\alpha+\beta), \tag{2}$$

$$R_i(\alpha)R_j(\beta) = R_j(\beta)R_i(\alpha). \tag{3}$$

In this section we study the topological structure of Sp$(2n)$ near Sp$(2n)_\omega^0$ and $\mathcal{M}_\omega(2n)$ for $\omega \in \mathbf{U}$ defined in Section 1.8, especially the transversality of rotation paths with respect to Sp$(2n)_\omega^0$. When $n = 1$, this follows from our discussion in Section 1 via the $\mathbf{R}^3$-cylindrical coordinate representation of Sp$(2)_\omega^0$. But when $n \ge 2$, the structure becomes rather complicated. We use the normal forms introduced in Chapter 1, together with the discussion of the changes of dimension of the null space and the function $D_\omega(M)$ for any $M \in \mathrm{Sp}(2n)^0_\omega$ under the continuous actions of rotation matrices $R_k(\theta)$. Our main results in this section are the following Theorems 1 to 6. Note that in this section we are working in the field $\mathbf{C}$.

Theorem 1 *For $\omega \in \mathbf{U}$ and $M \in \mathcal{M}_\omega(2n)$, there exist an integer k, $1 \le k \le n$, and a small real number $\theta_0 > 0$ such that*

$$\nu_\omega(MR_k(\theta)) = 0, \qquad \text{if } 0 < |\theta| \le \theta_0. \tag{4}$$

Theorem 2 *For $n \ge 2$ and $\omega \in \mathbf{U}$, suppose $M \in \mathcal{M}_\omega(2n)$, and there exist integers k_1, k_2, $1 \le k_1 < k_2 \le n$, and a real number $\theta_0 > 0$ such that*

$$\nu_\omega(MR_{k_1}(\theta)) = \nu_\omega(MR_{k_2}(\theta)) = 0, \quad \text{for } 0 < |\theta| \le \theta_0. \tag{5}$$

Then

$$D_\omega(MR_{k_i}(\theta))D_\omega(MR_{k_i}(-\theta)) < 0, \qquad \text{for } 0 < |\theta| \le \theta_0, \tag{6}$$

$$D_\omega(MR_{k_1}(\theta))D_\omega(MR_{k_2}(\theta)) > 0, \qquad \text{for } 0 < |\theta| \le \theta_0. \tag{7}$$

Theorem 3 *Suppose $\omega \in \mathbf{U}$, $M \in \mathrm{Sp}(2n)^0_\omega$, and there exists $P \in \mathrm{Sp}(2n)$ such that*

$$PMP^{-1} = H_0 \diamond M_0 \equiv N, \tag{8}$$

with $H_0 \in \mathrm{Sp}(2h)^0_\omega$ being a normal form of the eigenvalue ω for some integer $h \in [1, n]$ and $M_0 \in \mathrm{Sp}(2(n-h))^*_\omega$. Suppose

$$\nu_\omega(M) - \nu_\omega(MP^{-1}R_1(\theta)P) = 2, \qquad for \ 0 < |\theta| \le \theta_0, \tag{9}$$

with some small $\theta_0 > 0$. Then $h = 1$ and $H_0 = \pm I_2$ if $\omega = \pm 1$, or $h = 2$ and $H_0 = N_2(\varphi, b) \in \mathcal{M}^2_\omega(4)$ with $\varphi = \omega$ or $\overline{\omega}$ if $\omega \in \mathbf{U}\backslash\mathbf{R}$. In particular

$$D_1(MP^{-1}R_1(\theta)P) = 0 \qquad \text{if and only if} \qquad \theta = 0 \mod 2\pi. \tag{10}$$

Note that $\mathrm{Sp}(2n)^*_\omega$ contains precisely two path connected components $\mathrm{Sp}(2n)^+_\omega$ and $\mathrm{Sp}(2n)^-_\omega$. The conclusion (6) of Theorem 2 points out that under the condition there, $MR_{k_i}(\theta)$ and $MR_{k_i}(-\theta)$ belong to different path connected components of $\mathrm{Sp}(2n)^*_\omega$ for $0 < \theta \le \theta_0$. The conclusion (7) of Theorem 2 points out that under the condition there, $MR_{k_1}(\theta)$ and $MR_{k_2}(\theta)$ belong to the same path connected component of $\mathrm{Sp}(2n)^*_\omega$ for $0 < |\theta| \le \theta_0$. Theorem 3 points out that under conditions there $MP^{-1}R_k(\theta)P$ belongs to the same path connected component of $\mathrm{Sp}(2n)^*_\omega$ for $0 < \theta < 2\pi$.

The above results describe the topological structures near the singular hypersurface $\mathrm{Sp}(2n)^0$, as we shall see in the next section and in later chapters. We start the proofs of these theorems from the following study on the rotation matrices.

Proposition 4 *For any $\omega \in \mathbf{U}$, $M \in \mathrm{Sp}(2n)^0_\omega$, integer k with $1 \le k \le n$ and $\theta \in \mathbf{R}$,*

$$\nu_\omega(M) - \nu_\omega(MR_k(\theta)) \le 2. \tag{11}$$

Proof: We only consider the case $\omega = 1$ and work in the field $\mathbf{R}$. Other cases are left to the readers. Let $m = \dim\ker(M - I)$. If $m \le 2$, then (11) already holds. So we suppose $m \ge 3$ and that $\{a_1, \dots, a_m\}$ is a base of $\ker(M - I)$. Fix $k \in \{1, \dots, n\}$. Without loss of generality we may suppose that the k-th and $(n+k)$-th components of a_i are all zero for $3 \le i \le m$. Therefore for $3 \le i \le m$ we have $(MR_k(\theta) - I)a_i = (M - I)a_i = 0$, hence $\dim\ker(MR_k(\theta) - I) \ge m - 2$. This completes the proof. □

Note that $\{M \in \mathrm{Sp}(2n)^0_\omega \mid \nu_\omega(M) \ge m\}$ is a closed subset in $\mathrm{Sp}(2n)^0_\omega$ for any $m \ge 0$. Therefore when $|\theta|$ is sufficiently small, the left-hand side of (11) is always non-negative for any integer $k \in [1, n]$.

In the following, we only prove Theorems 1 and 2 for $\omega = 1$ and work in the field $\mathbf{R}$. Proofs for other cases are similar and are left to the readers.

Proof of Theorem 1: Fix $M \in \mathcal{M}(2n)$. It suffices to prove the existence of an integer $k \in [1, n]$ and $\theta_0 > 0$ such that

$$\nu_1(M) - \nu_1(MR_k(\theta)) \ge 1, \quad \text{for } 0 < |\theta| \le \theta_0. \tag{12}$$

Let $\{a_1, \dots, a_m\} \subset \mathbf{R}^{2n}$ be a cyclic base of a maximal cyclic space V of M belonging to the eigenvalue 1, i.e., they form a base of V and satisfy $(M - I)a_k = a_{k-1}$ for $1 \le k \le m$ and $a_0 = 0$. Here $m \in [1, n]$ and without loss of generality we assume

$$a_1^T J a_m = -1. \tag{13}$$

We choose an invariant base $\{d_1, \dots, d_p\}$ of the root vector subspace W of M belonging to eigenvalues other than 1 with $p = 2n - 2m$, so that $e_i = (M - I)d_i \in W$ for $1 \le i \le p$. Then $\{a_1, \dots, a_m, d_1, \dots, d_p\}$ form a base of $\mathbf{R}^{2n}$. Denote by $A = (a_1, \dots, a_m)$, $A_1 = (a_2, \dots, a_m)$, $A_m = (a_1, \dots, a_{m-1})$, $D = (d_1, \dots, d_p)$, and $E = (e_1, \dots, e_p)$ the matrices formed by the corresponding column vectors. Here we further assume

$$\det(A, D) = 1. \tag{14}$$

Then

$$\mathrm{rank}(A_m, E) = 2n - 1. \tag{15}$$

In fact, if column vectors of (A_m, E) were linearly dependent, a non-trivial linear combination of column vectors of (A_1, D) would belong to $\ker(M - I)$. Then this contradicts $M \in \mathcal{M}(2n)$. The column vectors of (A_m, E) are also symplectically orthogonal to a_1, i.e.,

$$a_1^T J x = 0, \qquad \forall\, x \in X = \mathrm{span}\{\text{column vectors of } (A_m, E)\}. \tag{16}$$

This follows from the symplectic orthogonality between V and W and the fact that $M \in \mathrm{Sp}(2n)$.

Write $a_1 = (a_{1,1}, \ldots, a_{2n,1})^T \in \mathbf{R}^{2n}$ and other vectors similarly. Without loss of generality by (13) we assume that

$$a_{1,1} \neq 0. \tag{17}$$

Note that in this notation from (16) for any $x = (x_1, \ldots, x_{2n})^T \in X$,

$$\sum_{i=1}^{n} a_{n+i,1} x_i - \sum_{i=1}^{n} a_{i,1} x_{n+i} = 0. \tag{18}$$

By (17) and (18), the $(n+1)$-st row vector of the $2n \times (2n-1)$ matrix (A_m, E) is a linear combination of the other row vectors of (A_m, E). Let H be the $(2n-1) \times (2n-1)$ matrix obtained from (A_m, E) by cancelling its $(n+1)$-st row. Then by (17) we obtain

$$\det H \neq 0. \tag{19}$$

From $(MR_1(\alpha) - I)R_1(-\alpha) = (M - I) + (I - R_1(-\alpha))$ we obtain

$$\begin{aligned} &(MR_1(\alpha) - I)R_1(-\alpha)(a_1, A_1, D) \\ &= ((I - R_1(-\alpha))a_1, A_m, E) + (I - R_1(-\alpha))(0, A_1, D). \end{aligned} \tag{20}$$

Thus for sufficiently small $|\alpha|$ we obtain

$$\begin{aligned} &D_1(MR_1(\alpha)) \\ &= (-1)^{n-1} \det[(MR_1(\alpha) - I)R_1(-\alpha)(a_1, A_1, D)] \\ &= (-1)^{n-1}\{\det[((I - R_1(-\alpha))a_1, A_m, E)] + o(|\alpha|)\} \\ &= (-1)^{n-1}[(-1)^n (\sin\alpha)(|a_{1,1}|^2 + |a_{n+1,1}|^2) \det H + o(|\alpha|)] \\ &= (\sin\alpha)[(|a_{1,1}|^2 + |a_{n+1,1}|^2)G + o(1)], \end{aligned} \tag{21}$$

where $f(\alpha) = o(|\alpha|^k)$ if $\lim_{\alpha\to 0} |f(\alpha)|/|\alpha|^k = 0$ for $k = 0$ or 1, and $G = -(\det H)/a_{1,1}$. To get the third equality in (21), we have added $-a_{n+i,1}/a_{1,1}$ times the i-th row and $a_{i,1}/a_{1,1}$ times the $(n+i)$-th row of the determinant in (21) to its $(n+1)$-st row, and used the above definiton of H. From (17), (19), and (21), Theorem 1 follows.

Proof of Theorem 2: By the assumption of Theorem 2,

$$|a_{k_1,1}|^2 + |a_{n+k_1,1}|^2 \neq 0, \qquad |a_{k_2,1}|^2 + |a_{n+k_2,1}|^2 \neq 0.$$

Without loss of generality, we assume $a_{k_1,1} \neq 0$. Similar to (21) we obtain

$$\begin{aligned} &D_1(MR_{k_1}(\alpha))D_1(MR_{k_1}(-\alpha)) \\ &= -(\sin\alpha)^2[(|a_{k_1,1}|^2 + |a_{n+k_1,1}|^2)^2 G^2 + o(1)] \\ &< 0, \end{aligned}$$

for $|\alpha| > 0$ sufficiently small and some $G \neq 0$. This proves (6). Similarly we also obtain

$$\begin{aligned} &D_\omega(MR_{k_1}(\alpha))D_\omega(MR_{k_2}(\alpha)) \\ &\quad = (\sin\alpha)^2[(|a_{k_1,1}|^2 + |a_{n+k_1,1}|^2)(|a_{k_2,1}|^2 + |a_{n+k_2,1}|^2)G^2 + o(1)] \\ &\quad > 0, \end{aligned}$$

for $|\alpha| > 0$ sufficiently small and the same $G \neq 0$. This proves (7).

The proof of Theorem 3: Fix $M \in \mathrm{Sp}(2n)_1^0$ (or $\mathrm{Sp}(2n)_{-1}^0$). By the definition of the normal forms in Section 1.4 (or 1.5), the condition (9) implies $H_0 = I_2$ (or $-I_2$). Thus by a direct computation we obtain (10). When $M \in \mathrm{Sp}(2n)_\omega^0$ for $\omega \in \mathbf{U}\backslash\mathbf{R}$, the result follows from the normal forms in Section 1.6 and Lemma 1.9.2.

Remark 5 Results in this section were proved in [Lon13] and [Lon15].

2.7 Orientability of $\mathcal{M}_\omega(2n)$ in Sp$(2n)$

In this section, we study the structure near $\mathcal{M}_\omega(2n)$ for $\omega \in \mathbf{U}$ defined in Section 1.8 by reducing the study near any $M \in \mathcal{M}_\omega(2n)$ to that near a matrix in the $\diamond$-product $\mathrm{Sp}(2)_\omega^0 \diamond \mathrm{Sp}(2n-2)_\omega^*$.

Theorem 1 *For any $\omega = e^{\theta\sqrt{-1}} \in \mathbf{U}$ and $M \in \mathcal{M}_\omega(2n)$, by Theorem 1.7.3 there exists $P \in \mathrm{Sp}(2n)$ such that*

$$PMP^{-1} = N_k(\hat{\omega}, b)\diamond M_1 \equiv N, \tag{1}$$

where $N_k(\hat{\omega}, b) \in \mathcal{M}_\omega(2k)$ is a normal form of the eigenvalue $\hat{\omega}$ with $\hat{\omega} = \omega$ or $\overline{\omega}$, and $M_1 \in \mathrm{Sp}(2n-2k)_\omega^*$. Then there exist $M_0 \in \mathrm{Sp}(2n-2)_\omega^*$ and a continuous path $\sigma : [0, 1] \to \mathcal{M}_\omega(2n)$ such that

$$\sigma(0) = M, \qquad \sigma(1) = N_1(\tilde{\omega}, b)\diamond M_0, \tag{2}$$

where $\tilde{\omega} = \pm 1$ and $b \in \{1, -1\}$ if $\omega = \pm 1$, $\tilde{\omega} = \omega$ or $\overline{\omega}$ and $b = 0$ if $\omega \in \mathbf{U}\backslash\mathbf{R}$. Moreover for any sufficiently small $\alpha > 0$, there exist continuous paths $\sigma^\pm : [0, 1] \to \mathrm{Sp}(2n)_\omega^*$ such that

$$\sigma^-(0) = MP^{-1}R_1(-\alpha)P, \qquad \sigma^-(1) = [N_1(\tilde{\omega}, b)R(-\alpha)]\diamond M_0, \tag{3}$$

$$\sigma^+(0) = MP^{-1}R_1(\alpha)P, \qquad \sigma^+(1) = [N_1(\tilde{\omega}, b)R(\alpha)]\diamond M_0, \tag{4}$$

and $(N_1(\tilde{\omega}, b)\phi_{\theta_0}(t))\diamond M_0 \in \mathrm{Sp}(2n)^*$ for all $t \in [-1, 1]\backslash\{0\}$, where $\phi_{\theta_0}(t)$ is defined by

$$\phi_{\theta_0}(t) = R(t\theta_0), \qquad \forall\, t \in [0, 1]. \tag{5}$$

Furthermore, the distance between paths $\sigma^\pm$ and σ is not greater than twice that of N and $NR_1(\alpha)$.

From now on, we fix $\omega = 1$ and only carry out the proof for this case. We still omit the subindex $\omega = 1$ in notations. Other cases can be proved similarly and therefore are left to the readers. To prove Theorem 1, we need the following basic lemma.

Lemma 2 *Let $q : [0, 1] \to \mathcal{M}(2n)$ be a continuous curve. Fix small $\theta_0 > 0$ sufficiently close to 0 so that for all $s \in [0, 1]$ and $|\theta| \le \theta_0$ the matrix*

$$Q_\theta(s) \equiv q(s) R_{m(s)}(\theta) \tag{6}$$

satisfies $Q_\theta(s) \in \mathrm{Sp}(2n)^*$ for $0 < |\theta| \le \theta_0$ and $s \in [0, 1]$, and $m(s) \in [1, n]$ is the least positive integer which has this property. The existence of $m(s)$ is given by Theorem 6.1. Then there exist two path connected sets Ω^+ and Ω^- such that

$$\{Q_\theta(s) \,|\, 0 \le s \le 1,\ 0 < \pm\theta \le \theta_0\} \subset \Omega^\pm \subset \mathrm{Sp}(2n)^*. \tag{7}$$

Furthermore, Ω^+ and Ω^- belong to different path connected components of $\mathrm{Sp}(2n)^*$, and we can choose two continuous curves in $f^\pm : [0, 1] \to \Omega^\pm$ so that $f^\pm(s)$ is as close to $q(s)$ as we want.

Proof: We first construct Ω^+ in several steps.

Step 1 Note that for fixed $s \in [0, 1]$, the set

$$G(s) \equiv \{Q_\theta(s) \,|\, 0 < \theta \le \theta_0\}$$

defines a continuous curve in $\mathrm{Sp}(2n)^*$.

Next we fix a $\theta \in (0, \theta_0]$ and define an auxiliary set $\Omega(\theta)$ in the following Steps 2 to 5.

Step 2 In general the map $Q_\theta : [0, 1] \to \mathrm{Sp}(2n)^*$ may not be continuous. Suppose there are k integers $A(\theta) \equiv \{m_1, \ldots, m_k\}$ appearing in (6) as $m(s)$ for all $s \in [0, 1]$ satisfying $1 \le m_1 < \cdots < m_k \le n$ for some $k \in [1, n]$.

If $k = 1$, the map Q_θ is continuous and we define $\Omega(\theta) = Q_\theta([0, 1])$, which is path connected.

Suppose $k > 1$. Then the map Q_θ is not continuous. We construct $\Omega(\theta)$ as follows.

Step 3 Suppose $\nu_1(q(s_0)R_{m_i}(\alpha)) = 0$ for some $s_0 \in [0, 1]$ and any $0 < \alpha \le \theta$. Without loss of generality we assume $m_i = 1$ here. For fixed $\alpha \in (0, \theta]$, by the effect of the rotation matrix $R_1(\alpha)$, there exists an open subinterval neighborhood $N(\alpha)$ of s_0 in $[0, 1]$ depending on α such that

$$\det(q(s)R_1(\alpha) - I) \ne 0, \qquad \forall s \in N(\alpha).$$

By the computation formula (6.21) for the effect of $R_1(\alpha)$,

$$\det(q(s)R_1(\alpha) - I) = (-1)^n \sin\alpha[(1 + a_{n+1,1}^2(s))b(s) + o(1)], \tag{8}$$

where by the continuity of q, both $a_{n+1,1}(s)$ and $b(s)$ are real continuous functions in s, independent of α, and $b(s_0) \neq 0$. The term $o(1)$ is defined as $\lim_{\alpha\to 0} o(1) = 0$. Thus we can further require $\theta \in (0, \theta_0]$ to be sufficiently small so that there exists an open subinterval neighborhood N of s_0 in $[0, 1]$ satisfying $N \subset \cap\{N(\alpha) \mid 0 < \alpha \le \theta\}$. Thus we have proved the existence of an open subinterval neighborhood N of s_0 in $[0, 1]$ such that

$$\nu_1(q(s)R_{m_i}(\theta)) = 0, \qquad \forall s \in N,\ 0 < \alpha \le \theta. \tag{9}$$

Thus for each $m_i \in A(\theta)$ the set $\{s \in [0, 1] \mid Q_\theta(s) = q(s)R_{m_i}(\theta)\}$ is a union of subintervals $\{N_{i,j}\}_{j\in E_i}$ in $[0, 1]$ with some subindex set E_i. Denote by $N^\circ_{i,j}$ the interior of $N_{i,j}$ in $[0, 1]$. Note that some $N_{i,j}$ may contain only one point. The unit interval $[0, 1]$ is a disjoint union of all these subintervals $N_{i,j}$ for $j \in E_i$ and $i = 1, \ldots, k$.

Step 4 Let s_0 be a discontinuous point of Q_θ. By the definition (6), we suppose $Q_\theta(s_0) = q(s_0)R_{m_i}(\theta)$ for some $m_i \in A(\theta)$. By the continuous effect of $R_{m_i}(\alpha)$ for $0 < \alpha \le \theta$ on $q(s)$ with s near s_0 given by (9) and the openness of Sp(2n)* in Sp(2n), we obtain a small open interval neighborhood $B(s_0)$ of s_0 in $[0, 1]$ such that

$$\nu_1(q(s)R_{m_i}(\alpha)) = 0, \qquad \forall s \in B(s_0),\ 0 < \alpha \le \theta.$$

For any $s \in B(s_0)$, by definition $Q_\theta(s) = q(s)R_{m(s)}(\theta)$ for some $m(s) \in A(\theta)$. If $m(s) \neq m_i$, by Theorem 6.2, there exists a continuous curve $\varphi_{\theta,s} : [0, 1] \to \mathrm{Sp}(2n)^*$ such that $\varphi_{\theta,s}(0) = q(s)R_{m_i}(\theta)$ and $\varphi_{\theta,s}(1) = q(s)R_{m(s)}(\theta)$. If $m(s) = m_i$, define $\varphi_{\theta,s}([0, 1]) = Q_\theta(s)$. Define

$$F_\theta(B(s_0)) = \cup\{\varphi_{\theta,s}([0, 1]) \mid s \in B(s_0)\}.$$

Then $F_\theta(B(s_0))$ is a path connected subset of Sp(2n)* containing the non-path connected set $Q_\theta(B(s_0))$ and the path connected set $\{q(s)R_{m_i}(\theta) \mid s \in B(s_0)\}$.

The set $F_\theta(B(s_0))$ is shown in Figure 3 by thick black curves with $m(s) < m_i = m(s_0)$ and $s \in B(s_0)$.

Step 5 Define $\mathcal{F}$ to be the family of open subintervals of $[0, 1]$ which consists of $B(s)$ for all discontinuity point s of Q_θ and $N^\circ_{i,j}$ for all $j \in E_i$ and $i = 1, \ldots, k$. Then $\mathcal{F}$ is an open covering of $[0, 1]$. Thus $\mathcal{F}$ possesses a finite subcovering $\mathcal{F}_1$ of $[0, 1]$. Define the set $\Omega(\theta)$ to be the union of the sets

$$\cup\{F_\theta(B(s)) \mid B(s) \in \mathcal{F}_1, s \in [0, 1]\}$$

and

$$\cup\{Q_\theta(N_{i,j}) \mid N^\circ_{i,j} \in \mathcal{F}_1, j \in E_i, 1 \le i \le k\}.$$

Then by our construction in Step 4, the set $\Omega(\theta)$ is path connected.

Step 6 Fix a $\theta \in (0, \theta_0]$ such that (9) holds. Using the set $\Omega(\theta)$ constructed in Step 2 or 3–5, and the set $G(s)$ defined in (A), we define

$$\Omega^+ = (\cup\{G(s) \mid s \in [0, 1]\}) \cup \Omega(\theta).$$

Then Ω^+ is path connected and satisfies (7).

By a similar proof we obtain the path connected set Ω^- satisfying (7).

That Ω^+ and Ω^- belong to different path connected components follows from Theorem 6.2. The last conclusion of Lemma 2 follows from our construction of $\Omega^{\pm}$. □

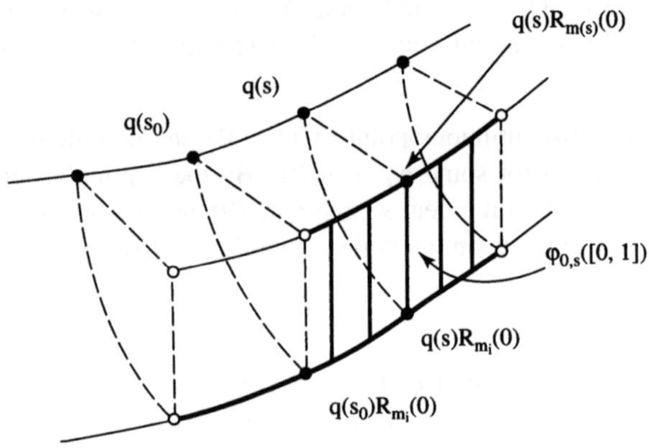

Figure 3: *The set* $F_\theta(B(s_0))$

With the aid of Lemma 2, we now give

Proof of Theorem 1 for $\omega = 1$**:** By our discussion in Section 1, it suffices to study the case for $n \geq 2$. Fix $M \in \mathcal{M}(2n)$ and assume that (1) holds. By our study in Section 3, we obtain a path $f : [0, 1] \to \mathcal{M}(2n)$ such that $f(0) = N$, $f(1) = N_1(1, b)\diamond M_0$ for some $N_1(1, b) \in \mathrm{Sp}(2)^0$ with $b = 1$ or -1 and a matrix $M_0 \in \mathrm{Sp}(2n-2)^*$. By Lemma 2, we obtain two paths $f^{\pm} : [0, 1] \to \mathrm{Sp}(2n)^*$ such that

$$f^{\pm}(0) = NR_1(\pm\theta), \qquad f^{\pm}(1) = [N_1(1, b)R(\pm\theta)]\diamond M_0,$$

and the distance between $f^{\pm}$ and f is not greater than twice the distance between N and $NR_1(\theta)$. For P defined in (1), choose $p \in C([0, 1], \mathrm{Sp}(2n))$ such that $p(0) = P$ and $p(1) = I$. Define

$$\begin{aligned} \sigma(t) &= p(t)^{-1}f(t)p(t), \qquad \forall\, t \in [0, 1], \\ \sigma^{\pm}(t) &= p(t)^{-1}f^{\pm}(t)p(t), \qquad \forall\, t \in [0, 1]. \end{aligned}$$

Then these three paths satisfy the requirements of the theorem and complete the proof.

□

Theorem 3 *For any $\omega \in \mathbf{U}$, the regular part $\mathcal{M}_\omega(2n)$ of the singular hypersurface* $\mathrm{Sp}(2n)^0_\omega$ *in* $\mathrm{Sp}(2n)$ *possesses a natural orientation.*

Proof: Since $\mathcal{M}_\omega(2n)$ is a smooth codimension 1 open submanifold in $\mathrm{Sp}(2n)$, it is two-sided locally near any $M \in \mathcal{M}_\omega(2n)$. Fix $M \in \mathcal{M}_\omega(2n)$. By Theorem 6.1, there exists an integer $k \in [1, n]$ such that (6.4) holds. We can define the positive normal direction to be the direction of the unit normal vector $\nu(\mathcal{M}_\omega(2n), M)$ of $\mathcal{M}_\omega(2n)$ at M such that $\epsilon\nu(\mathcal{M}_\omega(2n), M)$ with sufficiently small $\epsilon > 0$ points to the path connected component of $\mathrm{Sp}(2n)^*_\omega$ which contains $MR_k(\theta)$ for sufficiently small $\theta > 0$. By Theorems 1 and 6.2, this orientation is globally well defined. □

Remark 4 Results in this section were proved in [Lon13] and [Lon15].

Part II
The Variational Method

In his pioneering work [Rab2] in 1978, P. Rabinowitz proved the existence of periodic solutions of nonlinear Hamiltonian systems via variational methods. Since then, a large number of research papers have appeared in this area. Our goal in Part II is mainly to introduce the variational structure of Hamiltonian systems and two most popular finite dimensional reduction methods in the study of periodic solutions for Hamiltonian systems: the saddle point reduction method and the Galerkin approximation method.

Chapter 3

Hamiltonian systems and canonical transformations

In Sections 1 and 2 of this chapter we introduce the classical theory of Hamiltonian systems, study the relation between such systems and the canonical transformations, and then study the generating functions of canonical transformations.

On the studies of the N-body problems, Hamiltonian analysis, canonical transformations, and generating functions there exist a great amount of research articles and books. Historically, many famous mathematicians and physicists made great contributions to this field. We refer the readers to classical books [AbM1] of R. Abraham and J. Marsdan, [Arn1] of V. Arnold, [AKN1] of V. Arnold, V. Kozlov, and A. Neishtadt, [Mos4] of J. Moser, [Poi1] of H. Poincaré, [SiM1] fo C.L. Siegel and J. Moser, [Win1] of A. Wintner, and the references therein. Chinese readers can also find introductory materials in Chinese books [Wan1] of J.-H. Wang, [Yi1] of Zh.-H. Yi, and [YiS1] of Zh.-H. Yi and Y.-S. Sun.

In recent years, researches in these and related fields have being developed greatly. For these new achievements we refer the readers to the excellent books and articles by A. Ambrosetti and Coti Zelati [AmC2], K.C. Chang [Cha3], C.-Q. Cheng and Y.-S. Sun [ChS1] (in Chinese), I. Ekeland [Eke3], Y. Eliashberg [Eli1], H. Hofer and E. Zehnder [HoZ1], J. Mather and G. Forni [MFo1], J. Mawhin and Willem [MaW1], D. McDuff and D. Salamon [McS1, 2], K. Meyer and G. Hall [MeH1], P. Rabinowitz [Rab7, 8], Y. Ruan [Rua1, 2], M. Struwe [Str1], G. Tian [Tia2, 3], A. Weinstein [Wei1], Z.-H. Xia [Xia1, 2], V. Yakubovich and V. Starzhinski [YaS1], and the references therein.

Results in Sections 1 and 2 are rather classical and elementary, and can be found in the above mentioned books.

3.1 Canonical transformations

The most basic problem in celestial mechanics is to study the motion of many planets under Newton's law of universal attraction. The three body problem on the motion of sun, earth and moon is a typical example. These problems usually reduced to the study of the following type of differential equation systems:

$$\ddot{q} + V'(t, q) = 0, \qquad \text{for} \quad q \in \mathbf{R}^n, \tag{1}$$

where n is a given natural number, $q : \mathbf{R} \to \mathbf{R}^n$ describes the positions of these particles, $\dot{q}$ denotes the derivative of q with respect to the time variable t. $V : \mathbf{R} \times \Omega \to \mathbf{R}$ is the potential function of the motion defined on an open subset Ω of $\mathbf{R}^n$. $V'(t, q)$ denotes its gradient with respect to the variable q. Here we only consider the case $\Omega = \mathbf{R}^n$.

Let $p = \dot{q}$ and $x = (p^T, q^T)^T$. To simplify the notation, in the remainder of this book, when there is no confusion, we shall simply write $x = (p, q) \in \mathbf{R}^{2n}$ in this case. Let $H(t, x)$ be the energy function of the system (1) defined by

$$H(t, x) = \frac{1}{2}|p|^2 + V(t, q).$$

Then the system (1) reduces to a special case of the differential equation system

$$\dot{x} = JH'(t, x), \tag{2}$$

where $x \in \mathbf{R}^{2n}$, $H'(t, x)$ denotes the gradient of H with respect to the variable x, J is the standard symplectic matrix defined in Section 1.1. In general we let $H : \mathbf{R} \times \mathbf{R}^{2n} \to \mathbf{R}$ be a smooth function. In this book usually we require H to be at least C^2. Any ordinary differential equation system which is of the form (2) is called a **Hamiltonian system**. H is called the energy function and also the Hamiltonian function of (2). When the Hamiltonian function H is independent of the time variable t, the system (2) becomes an autonomous system of the form

$$\dot{x} = JH'(x). \tag{3}$$

If we apply an arbitrary coordinate transformation to the system (3), in general, this will change the form of the differential equation system. We want to find a class of transformations such that they transform Hamiltonian systems into other Hamiltonian systems, i.e., the so called canonical transformations.

Rewrite the system (2) in the form

$$\dot{x} = JH_x(x). \tag{4}$$

Here we denote by H_x the gradient of H with respect to the variable x. Let $x = u(w)$ be a coordinate transformation. Define

$$K(w) = H(u(w)).$$

Then $\dot{x} = u_w\dot{w}$, where u_w denotes the gradient matrix of u with respect to the variable w, and

$$K_w(w) = (u_w)^T H_x(u(w)).$$

Thus (4) becomes

$$-u_w^T J u_w \dot{w} = K_w(w).$$

If we require

$$u_w^T J u_w = J, \tag{5}$$

then via this transformation, the system (4) still becomes a Hamiltonian system,

$$\dot{w} = JK_w(w). \tag{6}$$

Clearly, the condition (5) requires the derivative of the transformation $x = u(w)$ to be a symplectic matrix at every point.

Definition 1 If a diffeomorphism $x = u(w)$ satisfies (5), it is called a **canonical transformation**, or a **symplectomorphism.**

Clearly, canonical transformations form a group. By the condition (5), linear canonical transformations have the form $x = Mw$, where $M \in \mathrm{Sp}(2n)$.

In order to further study the properties of the canonical transformations, denote by $x = (p, q) \in \mathbf{R}^{2n}$, i.e., $x_i = p_i$ and $x_{n+i} = q_i$ for $i = 1, \dots, n$. We consider the standard symplectic space $(\mathbf{R}^{2n}, \omega)$, where ω is the standard symplectic form on $\mathbf{R}^{2n}$. By the notation of the exterior differentiations, ω has the representation

$$\begin{aligned}\omega &= \frac{1}{2} J dx \wedge dx \\ &= \frac{1}{2}\sum_{i=1}^{n}(dp_i \wedge dq_i - dq_i \wedge dp_i) \\ &= \sum_{i=1}^{n} dp_i \wedge dq_i .\end{aligned}$$

Given a differentiable map $u : w \mapsto x = u(w)$, the pull-back map u^* induced by u yields a new differentiable 2-form,

$$\begin{aligned}u^*\omega(w) &= \frac{1}{2} J du(w) \wedge du(w) \\ &= \frac{1}{2} J u_w dw \wedge u_w dw \\ &= \frac{1}{2}(u_w)^T J u_w dw \wedge dw.\end{aligned}$$

From this we immediately obtain that $x = u(w)$ is a canonical transformation if and only if

$$u^*\omega = \omega. \tag{7}$$

Consider the ordinary differential equation system defined on the space $\mathbf{R}^{2n}$,

$$\dot{x} = f(x), \tag{8}$$

where $f : \mathbf{R}^{2n} \to \mathbf{R}^{2n}$ is a smooth map. Denote by $\phi^t(x)$ the flow on $\mathbf{R}^{2n}$ defined by (8), i.e., the solution map of (8) satisfying the initial value condition $\phi^0(x) = x$. The following theorem further discusses the relation between Hamiltonian systems and canonical transformations.

Theorem 2 *The differential equation system* (8) *defined on* $\mathbf{R}^{2n}$ *is a Hamiltonian system, if and only if the flow* $\phi^t(x)$ *defined by* (8) *is canonical at every time* t.

Proof: Denote the Jacobian matrix of $\phi^t(x)$ by $\Phi(t,x) = \phi^t_x(x)$. By (8), it satisfies

$$\frac{d}{dt}\Phi(t,x) = (f(\phi^t(x)))_x = F(t,x)\Phi(t,x),$$

where $F(t,x) = f_x(\phi^t(x))$. We then obtain

$$\frac{d}{dt}(\Phi^T J\Phi) = \Phi^T(-(JF)^T + JF)\Phi. \tag{9}$$

Necessity. If (8) is a Hamiltonian system, then there exists a Hamiltonian function $H : \mathbf{R}^{2n} \to \mathbf{R}$ such that $f = JH_x$. Thus $F = JH_{xx}$. By the symmetry of the Hessian matrix H_{xx}, the right-hand side of (9) is zero, i.e., $\Phi^T J\Phi$ is independent of t. When $t = 0$ it equals J. Thus Φ is symplectic. This proves that ϕ^t is canonical.

Sufficiency. Suppose ϕ^t is canonical. Thus Φ is symplectic. Let $g(x) = -Jf(x)$. Then

$$g_x(\phi^t(x)) = -Jf_x(\phi^t(x)) = -JF(t,x).$$

Since Φ is symplectic, (9) yields

$$0 = \frac{d}{dt}(\Phi^T J\Phi) = \Phi^T(-(JF)^T + JF)\Phi.$$

Therefore $g_x(\phi^t(x))$ is a symmetric matrix. Especially at the time $t = 0$ we obtain

$$g_{i,x_j}(x) = g_{j,x_i}(x), \qquad \forall x = (x_1, \ldots, x_{2n})^T \in \mathbf{R}^{2n}.$$

Since $\mathbf{R}^{2n}$ itself is open and convex, by the calculus of several variables there exists a function $H : \mathbf{R}^{2n} \to \mathbf{R}$ such that $g(x) = H_x(x)$. Thus $f = Jg = JH_x$, i.e., (8) is a Hamiltonian system.

The proof is complete. □

Remark 3 If we discuss this problem on a subdomain of $\mathbf{R}^{2n}$ or a manifold, in general the sufficiency part of the above theorem holds only locally. In fact, in the proof of Theorem 2 we have proved that the fundamental solution of a linear symmetric Hamiltonian system is symplectic at any time.

Corollary 4 *Let $\gamma \in C^1([0,\tau], \mathcal{L}(\mathbf{R}^{2n}))$ for some $\tau > 0$ satisfy the initial value problem of the linear Hamiltonian system,*

$$\begin{cases} \dot{\gamma}(t) = JA(t)\gamma(t), \\ \gamma(0) = I, \end{cases} \tag{10}$$

where $A \in C([0,\tau], \mathcal{L}(\mathbf{R}^{2n}))$. Then $A^T(t) = A(t)$ if and only if $\gamma(t) \in \mathrm{Sp}(2n)$ for all $t \in [0,\tau]$. γ is called the **associated symplectic path** *of A.*

Proof: The conclusion follows from (9). □

Remark 5 Theorem 2 and Corollary 4 explain why the topological structures of the symplectic group Sp($2n$) and its subsets must get into the study of Hamiltonian systems.

Similarly we also have

Proposition 6 *Let $a < b$ be two real numbers. If $x \in C^1((a,b), \mathbf{R}^{2n})$ is a solution of the autonomous Hamiltonian system*

$$\dot{x} = JH'(x),$$

where $H \in C^1(\mathbf{R}^{2n}, \mathbf{R})$, then $H(x(t))$ is constant for all $t \in (a,b)$.

Proof: In fact, for $a < t < b$,

$$\frac{d}{dt}H(x(t)) = H'(x(t)) \cdot \dot{x}(t) = H'(x(t)) \cdot JH'(x(t)) = 0.$$

Thus the proposition follows. □

3.2 Generating functions

In this section we continue our study on canonical transformations. By (1.7) a diffeomorphism $u : w \mapsto x = u(w)$ on $(\mathbf{R}^{2n}, \omega)$ is canonical if and only if $u^*\omega = \omega$, where $x = (\xi, \eta)$, $w = (q, p)$, $\xi = \xi(q, p)$, $\eta = \eta(q, p)$,

$$\omega = \sum_{i=1}^{n} dp_i \wedge dq_i \equiv dp \wedge dq.$$

If Ω is any oriented bounded surface of dimension 2,we denote its boundary by $\partial\Omega$. Then by the Stokes formula we have that $u^*\omega = \omega$ is equivalent to the condition that u is area preserving, i.e.,

$$\int_{\Omega} \omega = \int_{u(\Omega)} \omega, \quad \forall\, \Omega.$$

Since $\int_{\partial\Omega} pdq = \int_{u(\partial\Omega)} pdq$, it is further equivalent to

$$\int_{\partial\Omega} pdq = \int_{u(\partial\Omega)} pdq, \quad \forall\, \Omega.$$

Thus we obtain

$$\int_{\partial\Omega} pdq - \eta d\xi = 0,$$

i.e. when (q_0, p_0) is fixed, the integral

$$S(q,p) \equiv \int_{(q_0,p_0)}^{(q,p)} pdq - \eta d\xi$$

depends only on (q, p) and is independent of the choice of the integral paths. Therefore

$$dS(q, p) = pdq - \eta d\xi.$$

Suppose now we can choose (ξ, q) as independent variables in a neighborhood of some point (q_0, p_0), i.e., we suppose on the point (q_0, p_0) that

$$\det \frac{\partial(q, \xi)}{\partial(q, p)} = \det \frac{\partial \xi}{\partial p} \neq 0. \tag{1}$$

By the implicit function theorem, p can be represented as a function of (q, ξ). Thus near this point the function S can be represented by

$$S(q, p(q, \xi)) \equiv V(q, \xi). \tag{2}$$

Definition 1 The function $V(q, \xi)$ is called the **generating function** of the canonical transformation $x = u(w)$.

Lemma 2 *Suppose $V(q, \xi)$ is a function defined in a neighborhood of a point (q_0, ξ_0), and satisfies*

$$\det \frac{\partial^2 V}{\partial q \partial \xi}|_{(q_0, \xi_0)} \neq 0. \tag{3}$$

Then V is the generating function of some canonical transformation.

Proof: We consider the following equation on ξ:

$$\frac{\partial V(q, \xi)}{\partial q} = p.$$

Let $p_0 = (\frac{\partial V(q,\xi)}{\partial q})|_{(q_0, \xi_0)}$. By the implicit function theorem, in a neighborhood of the point (q_0, ξ_0) this equation can be solved in terms of ξ to obtain a function

$$\xi = \xi(q, p).$$

Let $g(q, \xi) = -\frac{\partial V(q,\xi)}{\partial \xi}$. We define

$$\eta = g(q, \xi(q, p)).$$

Then in the above neighborhood, the map $u : (q, p) \mapsto (\xi, \eta) = u(q, p)$ is a canonical transformation in $\mathbf{R}^{2n}$. Since

$$pdq - \eta d\xi = \frac{\partial V(q, \xi)}{\partial q} dq + \frac{\partial V(q, \xi)}{\partial \xi} d\xi,$$

and u satisfies (1) by (3), i.e.,

$$\det \frac{\partial \xi}{\partial p} = \det \left(\frac{\partial^2 V(q, \xi)}{\partial q \partial \xi} \right)^{-1} \neq 0,$$

then V is the generating function of the canonical transformation $(\xi, \eta) = u(q, p)$. $\square$

By the uniqueness of the function determined by the implicit function theorem, for a given canonical transformation $x = u(w)$, its generating function $V(q, \xi)$ in this lemma generates precisely $x = u(w)$ itself.

The above discussion establishes a relation between canonical transformations and generating functions. But it guarantees only that those canonical transformations which satisfy the condition (1) possess generating functions. Note that the simplest canonical transformation, i.e., the identity map $\xi = q$ and $\eta = p$, does not satisfy this condition. Next we show that for any given canonical transformation, we always can choose the coordinate function system to construct its generating function.

Denote the variables in $\mathbf{R}^{2n}$ by $q = (q_1, \ldots, q_n)$ and $p = (p_1, \ldots, p_n)$. Let

$$\sigma(i, j) = \begin{pmatrix} 1 & \cdots & k & k+1 & \cdots & n \\ i_1 & \cdots & i_{n-k} & j_1 & \cdots & j_k \end{pmatrix}$$

be an n-permutation. There are 2^n such permutations in total. For a given n-permutation $\sigma(i, j)$, we denote

$$q(\sigma, i) = (q_{i_1}, \ldots, q_{i_{n-k}}), \quad p(\sigma, j) = (p_{j_1}, \ldots, p_{j_k}).$$

We consider the n-dimensional isotropic subspace defined by any n variables of $(q_1, \ldots, q_n, p_1, \ldots, p_n)$ in the symplectic space $(\mathbf{R}^{2n}, \omega)$. This is a Lagrange subspace. For every n-permutation of $(1, \ldots, n)$, we associate an isotropic coordinate plane in $\mathbf{R}^{2n}$ whose n corresponding coordinates $(q(\sigma, i), p(\sigma, j))$ are all identically zero. We simply denote this isotropic coordinate plane by $(q(\sigma, i), p(\sigma, j))$. There are 2^n such isotropic coordinate planes in total.

Remark 3 Two subspaces L_1 and L_2 of a given vector space L are **transverse**, if $L_1 + L_2 = L$. By this definition, two n-dimensional subspaces of $\mathbf{R}^{2n}$ are transverse if and only if their intersection contains only the origin.

Lemma 4 *Any Lagrange subspace A of $(\mathbf{R}^{2n}, \omega)$ must be transverse to at least one of these 2^n isotropic coordinate planes.*

Proof: Let P denote the isotropic coordinate plane $(q_1, \ldots, q_n)$, (cf. Figure 1). Let $T = A \cap P$, and $k = \dim T$ with $0 \leq k \leq n$. Then T as a subspace of P must be transverse to at least an $(n-k)$ dimensional coordinate plane, say $S = (q_{i_1}, \ldots, q_{i_{n-k}})$. Thus

$$T + S = P, \qquad T \cap S = \{0\}.$$

Let $\sigma(i, j)$ be the n-permutation $(i_1, \ldots, i_{n-k}, j_1, \ldots, j_k)$ obtained from extending the $(n-k)$ integers $(i_1, \ldots, i_{n-k})$. Denote by M the n-dimensional isotropic plane according to $\sigma(i, j)$. Then $S = M \cap P$. By $T \subset A$ and $A \perp A$ (symplectic orthogonal), we obtain $T \perp A$. From $S \subset M$ and $M \perp M$, there holds $S \perp M$. Thus

$(T+S) \perp (A \cap M)$. This yields $P \perp (A \cap M)$. But $P = P^{\perp}$, therefore $A \cap M \subset P$. Then

$$A \cap M = (A \cap P) \cap (M \cap P) = T \cap S = \{0\}.$$

This proves the lemma. □

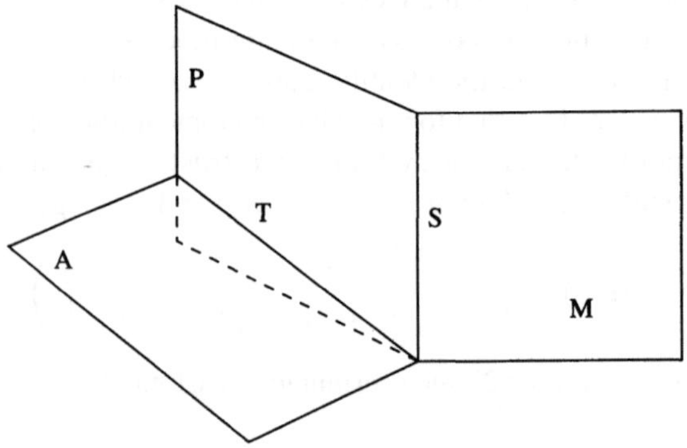

Figure 1: *Subspaces A, P, and M*

Given any canonical transformation, next we prove the existence of at least one n-permutation $\sigma(i, j)$ of $(1, \ldots, n)$ such that the corresponding functions $(\xi(\sigma, i), \eta(\sigma, j))$ are independent of q, from which therefore the generating function can be constructed.

Theorem 5 (cf. V.I. Arnold, p. 267 of [Arn2]). *Let $u : \mathbf{R}^{2n} \to \mathbf{R}^{2n}$ be a canonical transformation given by $\xi = \xi(q, p)$ and $\eta = \eta(q, p)$. In a neighborhood of any point $w_0 = (q_0, p_0)$, among the 2^n set functions $(q, \xi(\sigma, i), \eta(\sigma, j))$ corresponding to all n-permutations $\{\sigma(i, j)\}$, there exists at least one which gives independent coordinates of $\mathbf{R}^{2n}$, i.e., for this $\sigma(i, j)$,*

$$\det \frac{\partial(q, \xi(\sigma, i), \eta(\sigma, j))}{\partial(q, p)} = \det \frac{\partial(\xi(\sigma, i), \eta(\sigma, j))}{\partial(p_1, \ldots, p_n)} \neq 0. \tag{4}$$

In a neighborhood of this point w_0, the canonical transformation u has generating function

$$V(q, \xi(\sigma, i), \eta(\sigma, j)) = S(q, p(\xi(\sigma, i), \eta(\sigma, j))),$$

where

$$S(q, p) \equiv \int_{w_0}^{(q,p)} p dq - \eta d\xi,$$

and $p = p(\xi(\sigma, i), \eta(\sigma, j))$ *is the function relation uniquely determined by the implicit function theorem and the condition (4).* u *can be redefined implicitly by the equation:*

$$p = \frac{\partial V}{\partial q}, \quad \xi_j = \frac{\partial V}{\partial \eta_j}, \quad \eta_i = -\frac{\partial V}{\partial \xi_i}. \tag{5}$$

Vice versa, suppose a given function $V(q, \xi(\sigma, i), \eta(\sigma, j))$ *satisfies*

$$\det\left(\frac{\partial^2 V}{\partial q \partial \xi(\sigma, i) \partial \eta(\sigma, j)}\right)|_{(q_0, \xi(\sigma,i)_0, \eta(\sigma,j)_0)} \neq 0.$$

Then (5) gives a canonical transformation in a neighborhood of the point $(q_0, \xi(\sigma, i)_0, \eta(\sigma, j)_0)$.

Proof: This proof is almost the same as the special case of $\det \frac{\partial \xi}{\partial p} \neq 0$ discussed in Lemma 2. It suffices to prove that among all 2^n function groups there is at least one $(q, \xi(\sigma, i), \eta(\sigma, j))$ which makes (4) hold.

Thus, viewing $\mathbf{R}^{2n}$ to be the tangent space at the point (q_0, p_0), we obtain that $u_w(w_0) : \mathbf{R}^{2n} \to \mathbf{R}^{2n}$ is a symplectic linear map, and denote it by S.

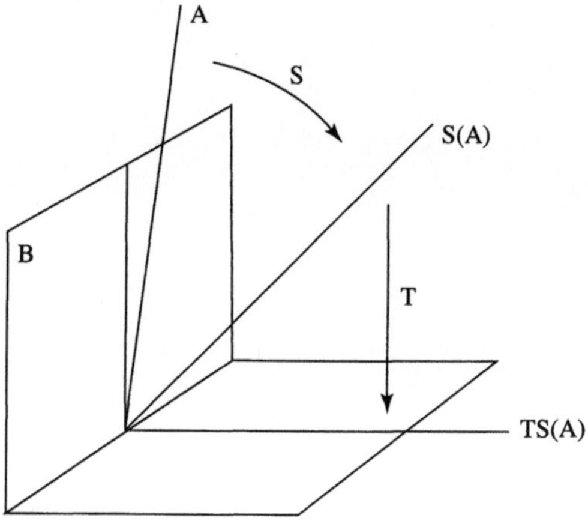

Figure 2: *The maps S and T*

Let A be the n-dimensional isotropic plane $(p_1, \ldots, p_n)$ in $\mathbf{R}^{2n}$ (cf. Figure 2). This is a Lagrange subspace, whose image SA under S is also a Lagrange subspace. By Lemma 4, SA must be transverse to one of the 2^n isotropic coordinate plane of dimension n defined there. Denote this isotropic plane by $B = (q(\sigma, i), p(\sigma, j))$.

Define an n-dimensional isotropic plane $M = (q(\sigma, j), p(\sigma, i))$, and denote by T the projection from SA to M. The condition (4) is equivalent to the composition map $TS : A \to M$ being non-singular. Since S is non-singular, and T is non-singular too by the transversality, TS must be non-singular. □

Remark 6 Theorem 5 and its proof is taken from Section 48 in [Arn2] of V.I. Arnold.

Chapter 4
The variational functional

In this chapter, we study finite dimensional approximation methods for Hamiltonian systems. In Section 1, we study the Galerkin approximation method for Hamiltonian systems. Then in Section 2 we define the functional corresponding to the general nonlinear Hamiltonian systems on the space L of square integrable periodic functions and study its basic properties. In Section 3 we introduce the saddle point reduction method. In Section 4 we study the dimension of the kernal of the reduced functional corresponding to the given Hamiltonian system. In Section 5, we derive certain useful estimates on the reduced functional. Results in Sections 2 to 5 form the most basic part of the saddle point reduction method.

For the theory of Sobolev spaces we refer readers to the books of R. Adams [Ada1] and D. Gilbarg and N. Trudinger [GiT1]. For discussions on nonlinear functional analysis and critical point theory, we refer to the books of J. Aubin and I. Ekeland [AuE1], K.C. Chang [Cha6], W.Y. Chen [Che1] (in Chinese), K. Deimling [Dei1], D.J. Guo [Guo1] (in Chinese), M. Krasnosel'skii [Kra1], L. Nirenberg [Nir1], P.H. Rabinowitz [Rab7], J. Schwartz [Stz1], and E. Zeidler [Zei1]. The spectrum decomposition theory of the unbounded linear selfadjoint operators can be found in many text books of functional analysis, for example those of F. Riesz and B.Sz. Nagy [RiN1] and K.C. Chang, M.Z. Guo, and Y.Q. Lin [CGL1] (in Chinese).

4.1 The Galerkin approximation

We give an introduction to the Galerkin approximation method for Hamiltonian systems in this section. We consider the Hamiltonian system

$$\dot{x} = JH'(t, x), \qquad x \in \mathbf{R}^{2n}. \tag{1}$$

In the following, we always suppose that the Hamiltonian function H satisfies the following conditions:

(H1) $H \in C^2(\mathbf{R} \times \mathbf{R}^{2n}, \mathbf{R})$, *and for some* $\tau > 0$ *it is* τ*-periodic in the time variable* t*, i.e.,*

$$H(t + \tau, x) = H(t, x), \qquad \forall (t, x) \in \mathbf{R} \times \mathbf{R}^{2n}.$$

(H2) *There exist constants* $a > 0$ *and* $p > 1$ *such that*

$$|H''(t, x)| \le a(1 + |x|^p), \qquad \forall (t, x) \in \mathbf{R} \times \mathbf{R}^{2n}.$$

In the study of its τ-periodic solutions, a rather important property of the Hamiltonian system is that it possesses a variational structure, i.e., it is the Euler equation of a certain functional defined on a suitable function space. In order to study the existence and other properties of τ-periodic solutions of the Hamiltonian system (1), in the following we introduce a direct variational method, and define a functional corresponding to (1) whose critical points correspond to τ-periodic solutions of (1) precisely. Then our problem can be solved by using the critical point theory which has been developed greatly in recent years.

Definition 1 Let X and Y be two Banach spaces. U is an open subset of X. $f : U \to Y$ is a given map. f is **Gateaux differentiable** at $x \in U$, if for any $h \in X$ there exists $y \in Y$ such that

$$f(x+th) - f(x) - ty = o(t), \quad \text{as } t \to 0^+, x + th \in U,$$

where $o(t)$ as $t \to 0^+$ is defined by $\lim_{t\to 0, t>0} o(t) = 0$. In this case y is called the **Gateaux derivative** of f at x along the direction h, and is denoted by $df(x)h$.

f is **Fréchet differentiable** (or simply **differentiable**) at $x \in U$, if there exists a continuous linear map $A : X \to Y$ such that for any $h \in X$ and $x + h \in U$,

$$\|f(x+h) - f(x) - Ah\|_Y = o(\|h\|_X), \quad \text{as } \|h\|_X \to 0.$$

In this case A is called the **Fréchet derivative** of f at x, and is denoted it by $f'(x)$. If f is differentiable everywhere on U and f' is continuous on U, we denote it by $f \in C^1(U, Y)$. Similarly the higher order derivatives of f can be defined.

Let $f \in C^1(X, \mathbf{R})$. If $f'(x) = 0$ for some $x \in X$, the point x is a **critical point** of f and $c = f(x)$ is the **critical value** of f at x. Define

$$\mathcal{K}_c(f) = \{x \in X \mid f'(x) = 0, f(x) = c\} \quad \text{and} \quad \mathcal{K}(f) = \{x \in X \mid f'(x) = 0\}.$$

Recall that a bounded linear operator $T : X \to Y$ is **Fredholm**, if the range $R(T)$ is closed, and both dim ker (K) and codim$R(T)$ are finite.

Let W be a separable Hilbert space with inner product $\langle \cdot, \cdot \rangle$ and norm $\|\cdot\|$. Suppose that $A : W \to W$ is a bounded self-adjoint operator with finite dimensional kernel N, and the restriction $A|_{N^\perp}$ is invertible. Denote by $P : W \to N$ the corresponding orthogonal projection. Let $g \in C^2(W, \mathbf{R})$ possess a compact derivative g'. Our aim is to solve the operator equation

$$Ax = g'(x). \tag{2}$$

Define

$$f(x) = \frac{1}{2}\langle Ax, x \rangle - g(x), \qquad \forall x \in W. \tag{3}$$

Then $f \in C^2(W, \mathbf{R})$, and the critical point of f corresponds to solutions of the equation (1). To transform this infinite dimensional problem into a finite dimensional one, we consider the following reduction:

Definition 2 Let $\Gamma = \{P_m | m \in \mathbf{N}\}$ be a sequence of orthogonal projections. Γ is a Galerkin approximation scheme with respect to A if the following three conditions hold:

1° $W_m \equiv P_m W$ is finite dimensional for all $m \in \mathbf{N}$,

2° $P_m \to I$ strongly as $m \to \infty$,

3° $[P_m, A] \equiv P_m A - A P_m \to 0$ in the operator norm.

Definition 3 f satisfies the **(P.S)* condition** on W with respect to $\{P_m\}$, if any sequence $\{x_m\} \subset W$ satisfying $x_m \in W_m$ and for which there exists a constant $b > 0$ such that

$$\begin{aligned} f_m'(x_m) &\to 0 \quad \text{in} \quad W \quad \text{as} \quad m \to \infty, \\ |f_m(x_m)| &\le b, \qquad \forall m \ge 0, \end{aligned}$$

possesses a convergent subsequence in W.

The following result follows from the above definition immediately.

Theorem 4 *Suppose the functional f satisfies the (P.S)* condition on W with respect to $\{P_m\}$. If there exist two constants $b_1 < b_2$ in $\mathbf{R}$ such that f_m possesses a critical point x_m with $b_1 \le f_m(x_m) \le b_2$ for any $m \in \mathbf{N}$, the functional f possesses a critical point x in W with $f(x) \in [b_1, b_2]$.*

To apply the Galerkin approximation method to the Hamiltonian system (1), recall that $W = W^{1/2,2}(S_\tau, \mathbf{R}^{2n})$ with $\tau > 0$ and $S_\tau = \mathbf{R}/(\tau\mathbf{Z})$ is defined to be the subspace of $L^2(S_\tau, \mathbf{R}^{2n})$ which consists of all elements

$$z(t) = \sum_{k \in \mathbf{Z}} \exp\left(\frac{2k\pi t}{\tau} J\right) a_k, \qquad a_k \in \mathbf{R}^{2n},$$

satisfying

$$\|z\|^2 \equiv \sum_{k \in \mathbf{Z}} (1 + |k|)\tau |a_k|^2 < +\infty. \tag{4}$$

Then W becomes a Hilbert space. We denote its inner product and norm by $\langle \cdot, \cdot \rangle$ and $\|\cdot\|$. By the well-known Sobolev embedding theorem (cf. [Ada1], [GiT1]), for any $s \in [1, +\infty)$, there exists a constant $C_s > 0$ such that

$$\|z\|_{L^s} \le C_s \|z\|, \quad \forall z \in W. \tag{5}$$

Define the operator $A : W \to W$ by

$$\langle Ax, y \rangle = \int_0^\tau (-J\dot{x} \cdot y) dt, \qquad \forall x, y \in W. \tag{6}$$

Then A is bounded linear and self-adjoint on W. The range of A is closed. Under the norm of the space W, the spectrum of the operator A is

$$\sigma(A) = \left\{ \frac{2k\pi}{(1 + |k|)\tau} \mid k \in \mathbf{Z} \right\}.$$

It is a point spectrum, i.e., it contains only eigenvalues, the multiplicity of each eigenvalue $(2k\pi)/((1+|k|)\tau)$ is $2n$. Note that $\ker A = \mathbf{R}^{2n} \equiv F_0$ and other eigenvector spaces are given by $F_k \equiv \exp((2k\pi t/\tau)J)\mathbf{R}^{2n}$ for all $k \in \mathbf{Z}\backslash\{0\}$.

Suppose the conditions (H1) and (H2) hold. We can define a functional on the space W by

$$g(x) = \int_0^\tau H(t, x(t))dt, \qquad \forall x \in W. \tag{7}$$

Then $g \in C^2(E, \mathbf{R})$ has compact derivative, and

$$g'(x) = H'(\cdot, x(\cdot)), \tag{8}$$

$$g''(x)y = H''(\cdot, x(\cdot))y(\cdot). \tag{9}$$

Define a functional $f : W \to \mathbf{R}$ by (3) with the above g. Then $f \in C^2(W, \mathbf{R})$, and critical points of f correspond to solutions of the abstract equation (3), and then correspond to 1-periodic solutions of the Hamiltonian system (1).

In order to find critical points of f, for any $B \in C(S_\tau, \mathcal{L}_s(\mathbf{R}^{2n}))$, define an operator $B : W \to W$ by

$$\langle Bx, y\rangle = \int_0^\tau B(t)x(t)\cdot y(t)dt, \qquad \forall x, y \in W. \tag{10}$$

Then B is a compact linear self-adjoint operator.

Lemma 5 *For any $B \in C(S_\tau, \mathcal{L}_s(\mathbf{R}^{2n}))$, the value 0 is isolated in $\{0\} \cup \sigma(A-B)$.*

Proof: Let $\mathrm{sign}(k) = k/|k|$ when $k \in \mathbf{R}\backslash\{0\}$. Let $\delta_{k,j} = 0$ if $k \neq j$, and $\delta_{k,j} = 1$ otherwise. Since the operator B is compact and self-adjoint on the Hilbert space W, there exists a sequence $\lambda_k \to 0$ and a basis $\{e_k\}$ of $\oplus_{k\neq 0} F_k$ such that

$$\langle Ae_k, e_j\rangle = \mathrm{sign}(k)\delta_{k,j}, \tag{11}$$

$$\langle Be_k, z\rangle = \langle A\lambda_k e_k, z\rangle, \quad \forall z \in W. \tag{12}$$

Thus for any $z \in W$ given by $z = \sum_{k\neq 0} b_k e_k$ we obtain

$$\begin{aligned}\langle (A-B)z, z\rangle &= \sum_{k\neq 0}\mathrm{sign}(k)b_k^2 - \sum_{k\neq 0}\lambda_k b_k^2 \\ &= \sum_{k\neq 0}(\mathrm{sign}(k) - \lambda_k)b_k^2.\end{aligned}$$

Thus 0 is isolated in $\{0\} \cup \sigma(A-B)$. □

Fix $B_\infty \in C(S_\tau, \mathcal{L}_s(\mathbf{R}^{2n}))$. By Lemma 5, we can list the set $\{0\} \cup \sigma(A - B_\infty)$ by

$$\ldots \le \lambda_{-2} \le \lambda_{-1} < 0 < \lambda_1 \le \lambda_2 \le \ldots, \tag{13}$$

where an eigenvalue λ_i of $A - B_\infty$ should be repeated k times in (11) if its multiplicity is k, and the eigenvector of $A - B_\infty$ belonging to λ_i by e_i, and $W_0 = \ker(A - B_\infty)$. Define

$$W_m = W_0 \cup \operatorname{span}\{e_i \mid 0 < |i| \le m\}, \qquad \forall m \in \mathbf{N}. \tag{14}$$

Let $P_m : W \to W_m$ be the corresponding orthogonal projection. Then $\{P_m\}_{m\in\mathbf{N}}$ form a Galerkin approximation scheme with respect to $A - B_\infty$. We shall use such schemes to solve the system (1) in various cases.

The next corollary studies the truncated kernel of the operator $A - B$.

Corollary 6 (G. Fei and Q. Qiu [FeQ1]). *Let $\{P_m : W \to W_m\}$ be the Galerkin approximation scheme with respect to $A - B_\infty$. For any $B \in C(S_\tau, \mathcal{L}_s(\mathbf{R}^{2n}))$, there exists $m_0 > 0$ such that*

$$\dim\ker(P_m(A - B)P_m) \le \dim\ker(A - B), \qquad \forall m \ge m_0. \tag{15}$$

Proof: (15) follows from Lemma 5 and 2° of Definition 2. □

Remark 7 Results in this section on Galerkin method for Hamiltonian systems can be found in K.C. Chang, J.Q. Liu, and M.J. Liu's paper [CLL1] of 1997, and G. Fei and Q. Qiu's paper [FeQ1] of 1996. Corollary 6 was proved by G. Fei and Q. Qiu in [FeQ1] of 1996. The proof given here is based on Lemma 5 and is simpler than the original one in [FeQ1].

4.2 The L^2-variational structure

We consider the Hamiltonian system

$$\dot{x} = JH'(t, x), \qquad x \in \mathbf{R}^{2n}. \tag{1}$$

In this chapter we always suppose H satisfies the condition (H1) in Section 1 and

(H3) *There exists a constant $C(H) > 0$ such that*

$$|H''(t, x)| \le C(H), \qquad \forall (t, x) \in \mathbf{R} \times \mathbf{R}^{2n}.$$

Our aim is to find τ-periodic solutions of the system (1), i.e., the solutions of (1) which satisfy the periodic boundary value condition

$$x(\tau) = x(0). \tag{2}$$

Let $S_\tau = \mathbf{R}/(\tau\mathbf{Z})$. Define $L = L^2(S_\tau, \mathbf{R}^{2n})$ equipped with the usual norm

$$\|x\|_L = \left(\int_0^\tau |x(t)|^2 dt\right)^{1/2}, \qquad \forall x \in L.$$

Then L becomes a Hilbert space. Denote the corresponding inner product of L by $\langle \cdot, \cdot \rangle_L$. Let $W = W^{1,2}(S_\tau, \mathbf{R}^{2n})$ be the space of all vector functions from S_τ to $\mathbf{R}^{2n}$ possessing square integrable first order derivatives. On the space we define the norm

$$\|x\|_W = \left(\int_0^\tau (|\dot{x}(t)|^2 + |x(t)|^2) dt \right)^{1/2}, \qquad \forall x \in W.$$

Under this norm, W becomes a Hilbert space. This is the Sobolev space of τ-periodic functions. W is a dense subspace of L. One of the most important properties of this space is that the following famous Sobolev and Rellich embedding theorem holds in it.

Theorem 1 *Suppose $\tau > 0$. Then there exists a constant $K > 0$ such that*

$$\|x\|_C \leq K \|x\|_W, \qquad \forall x \in W,$$

where $\|x\|_C = \max_{0 \leq t \leq \tau} |x(t)|$, and the embedding map from W to the continuous function space $C = C(S_\tau, \mathbf{R}^{2n})$, equipped with the norm $\| \cdot \|_C$, is compact.

$x \in W$ is a **weak τ-periodic solution** of the Hamiltonian system (1), if

$$\int_0^\tau \left\{ \frac{1}{2}(\dot{x} \cdot Jy + \dot{y} \cdot Jx) - H'(t, x) \cdot y \right\} dt = 0, \qquad \forall y \in W. \tag{3}$$

Theorem 2 *Under the conditions* (H1) *and* (H3), *weak solutions of* (1)–(2) *are classical solutions.*

Proof: Suppose $x \in W$ is a weak solution of the periodic boundary value problem (1)–(2). By the Sobolev imbedding theorem, we have $x \in C(S_\tau, \mathbf{R}^{2n})$. Since the fact that x satisfies (3) implies that it satisfies (1) almost everywhere in S_τ, by (1) the derivative $\dot{x}$ can be extended continuously to all of S_τ. This proves $x \in C^1(S_\tau, \mathbf{R}^{2n})$, i.e., x is a classical solution of (1)–(2). □

In the Hilbert space L we define an operator $Ax = -J\dot{x}$. Then the domain of A is $\mathrm{dom} A = W$. The range of A is closed and the resolution of A is compact. Under the norm of the space L, the spectrum of the operator A is $\sigma(A) = \frac{2\pi}{\tau}\mathbf{Z}$. It is a point spectrum, i.e., it contains only eigenvalues, and the multiplicity of every eigenvalue is $2n$. The eigensubspace of A belonging to the eigenvalue $\frac{2k\pi}{\tau}$ is

$$\begin{aligned} E_k &= \exp\left(\frac{2k\pi t}{\tau} J \right) \mathbf{R}^{2n} \\ &= \left(\left(\cos \frac{2k\pi t}{\tau} \right) I + \left(\sin \frac{2k\pi t}{\tau} \right) J \right) \mathbf{R}^{2n}. \end{aligned}$$

Especially, $\ker A = E_0 = \mathbf{R}^{2n}$.

Define a functional on the space L by

$$g(x) = \int_0^\tau H(t, x(t))dt. \tag{4}$$

By the conditions (H1) and (H3), we have $g \in C^1(L, \mathbf{R})$, and

$$g'(x) = H'(t, x(t)). \tag{5}$$

$g'(x)$ is Gadeaux differentiable, its Gateaux derivative is

$$dg'(x)y = H''(t, x(t))y, \tag{6}$$

and there exists a constant $c(H) > 0$ such that

$$\|dg'(x)\|_{\mathcal{L}(L)} \leq c(H). \tag{7}$$

Define a functional

$$f(x) = \frac{1}{2}\langle Ax, x\rangle_L - g(x), \qquad \forall x \in \operatorname{dom} A = W. \tag{8}$$

Under the conditions (H1) and (H3), $f \in C^1(W, \mathbf{R})$, f' is Gadeaux differentiable. Define an equation in the space L by

$$Ax = g'(x), \qquad \text{for} \quad x \in W. \tag{9}$$

Then (9) is precisely the Euler equation of the functional f on the space L, i.e.,

$$\langle f'(x), y\rangle_L = \langle Ax - g'(x), y\rangle_L, \qquad \forall y \in W. \tag{10}$$

Therefore critical points of f are solutions of the equation (9). Note that the equation (9) is equivalent to the τ-periodic boundary value problem of the Hamiltonian system (1). Thus the problem of finding τ-periodic solutions of the Hamiltonian system (1) is equivalent to that of finding critical points of the functional f in the space W.

4.3 The saddle point reduction

In the previous section, we have defined by (2.8) the functional f corresponding to the periodic boundary value problem of the Hamiltonian system (2.1)–(2.2). Note that f is strongly indefinite, and the Morse indices of f at its critical points are always infinite. These cause substantial difficulties in finding critical points of f and in applying the Morse theory to related problems. In this section, a Lyapunov–Schmidt reduction method, the saddle point reduction method, is introduced to reduce problem in infinite dimensional spaces to discussions on finite dimensional spaces, and at the same time the finiteness of the Morse indices is achieved.

To the author's knowledge, the saddle point reduction method was introduced by H. Amann [Ama1]. The saddle point reduction Theorem 1 below was proved by H. Amann and E. Zehnder in their celebrated paper [AmZ1] of 1980 using the monotone operator theory. Then K.C. Chang gave a simpler and more direct proof in [Cha1] of 1981. Our discussion in this section is a combination of their ideas in the case of Hamiltonian systems. For the saddle point reduction method in a more general setting, we refer to [AmZ1] and [Cha6].

As in Sections 1 and 2, we suppose the Hamiltonian function H satisfies the conditions (H1) and (H3). For the case of given period $\tau > 0$, we have already defined the Hilbert spaces L and W, the operator A on W, the functional g, and the functional f corresponding to (2.1)–(2.2). Let $P_0 : L \to E_0 = \mathbf{R}^{2n}$ be the projection map. Since zero is an eigenvalue of the operator A, in order to improve the property of A we define an invertible operator A_0 on the space W by

$$A_0 x = Ax + P_0 x, \qquad \forall x \in W. \tag{1}$$

Then $\mathrm{dom} A_0 = W$, it is self-adjoint and invertible, its range is closed, and its resolution is compact as is that of A. The spectrum of A_0 is $\sigma(A_0) = \{1\} \cup \frac{2\pi}{\tau}(\mathbf{Z}\setminus\{0\})$, and is a point spectrum. If $1 \notin \frac{2\pi}{\tau}(\mathbf{Z}\setminus\{0\})$, then the multiplicity of every eigenvalue is $2n$. The eigensubspace of the operator A_0 belonging to the eigenvalue 1 is $E_0 = \mathbf{R}^{2n} = \ker A_0$. For $k \in \mathbf{Z}\setminus\{0\}$, the eigensubspaces of A_0 and A belonging to the eigenvalue $2k\pi/\tau$ are the same E_k defined in Section 2. If $1 \in \frac{2\pi}{\tau}(\mathbf{Z}\setminus\{0\})$, then the multiplicity of 1 is $4n$, and the eigensubspace belonging to 1 is $E_0 \oplus E_k$ where the integer k satisfies $1 = 2k\pi/\tau$.

Since the Hamiltonian function H satisfies the conditions (H1) and (H3), the functional g satisfies (2.7). Without loss of generality, we suppose the constant $c(H)$ there satisfies $c(H) \notin \sigma(A_0)$ and $c(H) > 1$. Denote by $\{E_\lambda\}$ the spectral resolution of the selfadjoint operator A_0; we define the projections on the space L by

$$P = \int_{-c(H)}^{c(H)} dE_\lambda, \qquad P^+ = \int_{c(H)}^{\infty} dE_\lambda, \quad P^- = \int_{-\infty}^{-c(H)} dE_\lambda, \tag{2}$$

$$Q^+ = \int_0^{\infty} dE_\lambda, \qquad Q^- = \int_{-\infty}^{0} dE_\lambda. \tag{3}$$

Then the Hilbert space L possesses an orthogonal decomposition

$$L = L^+ \oplus L^- \oplus Z, \tag{4}$$

where $Z = PL$, $L^\pm = P^\pm L$. Next we define projections $S^\pm \in \mathcal{L}(L, L^\pm)$ and $R \in \mathcal{L}(L, Z)$ on the space L according to the inverse of the square root of the operator $|A_0|$ by

$$S^+ = \int_{c(H)}^{+\infty} \lambda^{-\frac{1}{2}} dE_\lambda, \quad S^- = \int_{-\infty}^{-c(H)} (-\lambda)^{-\frac{1}{2}} dE_\lambda, \tag{5}$$

$$R = \int_{-c(H)}^{+c(H)} |\lambda|^{-\frac{1}{2}} dE_\lambda. \tag{6}$$

Then $S^{\pm}|_{L^{\pm}}$ and $R|_Z$ are injections, and

$$A_0(S^+)^2 = P^+, \qquad -A_0(S^-)^2 = P^-, \tag{7}$$

$$A_0(RQ^+)^2 = PQ^+, \qquad -A_0(RQ^-)^2 = PQ^-, \tag{8}$$

$$(S^+)^2 - (S^-)^2 + (RQ^+)^2 - (RQ^-)^2 = A_0^{-1}. \tag{9}$$

Since A_0 is self-adjoint, all the above operators are self-adjoint. Define a subspace V of L by

$$V = V^+ \oplus V^- \oplus V_0, \tag{10}$$

where

$$V^{\pm} = S^{\pm}L^{\pm}, \quad V^0 = RZ. \tag{11}$$

If the norm of the space V is defined by

$$\|v\|_V = (\|(S^+)^{-1}v^+\|_L^2 + \|(S^-)^{-1}v^-\|_L^2 + \|R^{-1}v^0\|_L^2 + \|v\|_L^2)^{\frac{1}{2}}, \tag{12}$$

where $v = v^+ + v^- + v^0 \in V^+ \oplus V^- \oplus V^0 = V$, then V is isometric to the Sobolev fraction exponent space $W^{1/2,2}(S_\tau, \mathbf{R}^{2n})$.

For any $x = x^+ + x^- + z \in L^+ \oplus L^- \oplus Z = L$, we define two functionals

$$g_0(x) = g(x) + \frac{1}{2}\langle P_0 x, x\rangle_L, \tag{13}$$

and

$$f_0(x) = \frac{1}{2}(\|x^+\|_L^2 + \|Q^+z\|_L^2 - \|Q^-z\|_L^2 - \|x^-\|_L^2) - g_0(v), \tag{14}$$

where $v = S^+x^+ + S^-x^- + Rz$ and the functional g is defined by (2.4). Note that $g \in C^1(L, \mathbf{R})$ and $g'(v)$ is Gadeaux differentiable; thus we obtain that f_0 is C^1 on L, and $f_0'(x)$ is Gadeaux differentiable. For any $x = x^+ + x^- + z \in L$, we have

$$\begin{aligned} \|x^+\|_L^2 &= \langle (S^+)^{-1}S^+x^+, (S^+)^{-1}S^+x^+\rangle_L \\ &= \langle A_0 S^+x^+, S^+x^+\rangle_L. \end{aligned} \tag{15}$$

Similar relations also hold for $\|x^-\|_L^2$, $\|Q^+z\|_L^2$, and $\|Q^-z\|_L^2$. From these, we obtain

$$\begin{aligned} f_0(x) &= \frac{1}{2}\langle A_0 v, v\rangle_L - g_0(v) \\ &= \frac{1}{2}\langle A v, v\rangle_L - g(v) \\ &= f(v), \end{aligned} \tag{16}$$

The Euler equation of the functional f_0 in the sense of the space L is given by

$$x^{\pm} = \pm S^{\pm} g_0'(v), \tag{17}$$

$$Q^{\pm} z = \pm Q^{\pm} R g_0'(v), \tag{18}$$

where $v = S^+x^+ + S^-x^- + Rz$. Thus the solutions of (17) and (18) in L are in one to one correspondence to critical points of f_0. By (16) they are also critical points of f, and therefore are in one to one correspondence to periodic solutions of the Hamiltonian system (2.1).

Now we fix $z \in Z$ and solve the equation (17) for $x^{\pm}$. We shall prove that the right-hand side of the equation (17) defines a contraction mapping on the space L, and then by the implicit function theorem the solution $x^{\pm}$ can be obtained.

By the condition (H3), we obtain

$$\|g_0'(v) - g_0'(\bar{v})\|_L \leq c(H)\|v - \bar{v}\|_L \qquad \forall\, v, \bar{v} \in L. \tag{19}$$

From the definitions of A_0 and $P^{\pm}$, as well as $c(H) \notin \sigma(A_0)$, there exists a constant $\gamma > c(H)$ such that

$$\| |A_0|^{-1/2}|_{L^+ \oplus L^-} \|_{\mathcal{L}(L)} \leq \frac{1}{\sqrt{\gamma}}. \tag{20}$$

Thus for any $x = x^+ + x^- + z$ and $\bar{x} = \bar{x}^+ + \bar{x}^- + z \in L$, defining $v = S^+x^+ + S^-x^- + Rz$ and $\bar{v} = S^+\bar{x}^+ + S^-\bar{x}^- + Rz$, we obtain

$$\begin{aligned} &\|(S^+ - S^-)[g_0'(v) - g_0'(\bar{v})]\|_L \\ &\leq \|(S^+)^2 - (S^-)^2\|_{\mathcal{L}(L)} \|(S^+ - S^-)^{-1}(g_0'(v) - g_0'(\bar{v}))\|_L \\ &\leq \frac{c(H)}{\gamma} \|(S^+ + S^-)^{-1}\|_{\mathcal{L}(L)} \|S^+(x^+ - \bar{x}^+) + S^-(x^- - \bar{x}^-)\|_L \\ &\leq \frac{c(H)}{\gamma} \|x - \bar{x}\|_L. \end{aligned} \tag{21}$$

Here in the proof of the second inequality, we have used (9), (19), and (20). This proves that the right-hand side of (17) defines a contraction mapping. Thus for any $z \in Z$, (17) possesses a unique solution $x^{\pm}(z)$. Since g_0' is symmetric and Gadeaux differentiable, the equation (17) also yields $x^{\pm} \in C^1(Z, V)$.

Noting that the projection R can be viewed as an invertible map from the space Z to itself, we define a map on the space Z by

$$\begin{aligned} u(z) &= w(z) + z, \\ w(z) &= S^+x^+(R^{-1}z) + S^-x^-(R^{-1}z). \end{aligned} \tag{22}$$

By the definitions of the operators $S^{\pm}$ and the space V, u maps the space Z into $W = \mathrm{dom} A$. Since the solution obtained from (17) satisfies $x^{\pm} \in C^1(Z, V)$, we have $u \in C^1(Z, W)$.

Now we define a functional on the space Z by

$$\begin{aligned} a(z) &= f_0(x^+(R^{-1}z)+x^-(R^{-1}z)+R^{-1}z) \\ &= \frac{1}{2}(\|x^+(R^{-1}z)\|_L^2+\|Q^+R^{-1}z\|_L^2 \\ &\quad -\|Q^-R^{-1}z\|_L^2-\|x^-(R^{-1}z)\|_L^2)-g_0(u(z)). \end{aligned} \tag{23}$$

For any $z \in Z$, by (15) we obtain

$$\begin{aligned} a(z) &= \frac{1}{2}\langle A_0u(z), u(z)\rangle_L - g_0(u(z)) \\ &= \frac{1}{2}\langle Au(z), u(z)\rangle_L - g(u(z)) \\ &= f(u(z)). \end{aligned} \tag{24}$$

From the definition (22) for any $z \in Z$,

$$u'(z)^* = R^{-1}\eta^+S^+ + R^{-1}\eta^-S^- + P, \tag{25}$$

where $\eta^\pm = (x^\pm)'(R^{-1}z)^*$. By (7)–(9) we obtain

$$\begin{aligned} u'(z)^*A_0u(z) &= R^{-1}\eta^+S^+A_0S^+x_+(R^{-1}z) \\ &\quad +R^{-1}\eta^-S^-A_0S^-x^-(R^{-1}z)+A_0z \\ &= R^{-1}\eta^+x^+(R^{-1}z)-R^{-1}\eta^-x^-(R^{-1}z)+A_0z. \end{aligned} \tag{26}$$

Therefore by (24),

$$\begin{aligned} a'(z) &= u'(z)^*[A_0u(z)-g_0'(u(z))] \\ &= A_0z - Pg_0'(u(z)) \\ &= A_0u(z)-g_0'(u(z)) \\ &= Az - Pg'(u(z)) \\ &= Au(z)-g'(u(z)). \end{aligned} \tag{27}$$

Here in the proof of (27) we have used (17) and (26).

By (19) and (27), we obtain that a' is globally Lipschitz continuous on Z. Since g' is Gateaux differentiable and $u = u(z)$ is C^1, we obtain $a \in C^2(Z, \mathbf{R})$ and

$$\begin{aligned} a''(z) &= A_0P - Pdg_0'(u(z))u'(z) \\ &= [A_0 - dg_0'(u(z))]u'(z) \\ &= AP - Pdg'(u(z))u'(z) \\ &= [A - dg'(u(z))]u'(z). \end{aligned} \tag{28}$$

Thus our above discussion on the τ-periodic boundary value problem of the Hamiltonian system (2.1) produces the following reduction theorem.

Theorem 1 (H. Amann and E. Zehnder [AmZ1], K.C. Chang [Cha6]). *Suppose the Hamiltonian function H satisfies the condition* (H1) *and* (H3). *Then there exist $a \in C^2(Z, \mathbf{R})$ and an injection map $u \in C^1(Z, L)$ such that $u : Z \to W = \mathrm{dom}A$ satisfies the following conditions:*

1° *The map u has the form $u = w(z) + z$, where $Pw(z) = 0$.*

2° *The functional a satisfies*

$$\begin{aligned} a(z) &= f(u(z)) \\ &= \frac{1}{2}\langle Au(z), u(z)\rangle_L - g(u(z)) \\ a'(z) &= Az - Pg'(u(z)) \\ &= Au(z) - g'(u(z)), \\ a''(z) &= AP - Pdg'(u(z))u'(z) \\ &= [A - dg'(u(z))]u'(z). \end{aligned}$$

And a' is globally Lipschitz continuous.

3° *$z \in Z$ is a critical point of a, i.e., $a'(z) = 0$, if and only if $u(z)$ is a solution of $Au = g'(u)$, i.e., a critical point of f, if and only if $u(z)$ is a τ-periodic solution of the Hamiltonian system* (2.1).

4° *If $g'(u) = Bu$ for all $u \in L$, where B is the induced linear operator on the space L from a constant symmetric matrix $B(t) \equiv B$ defined on $\mathbf{R}^{2n}$, then $a(z) = \frac{1}{2}\langle (A - B)z, z\rangle_L$.*

5° *If Σ is a topological space, for any $\sigma \in \Sigma$ the functional $g : \Sigma \times L \to \mathbf{R}$ satisfies $g(\sigma, \cdot) \in C^1(L, \mathbf{R})$, $g' \in C(\Sigma \times L, \mathbf{R})$, and the inequality* (18) *with the constant β being independent from $\sigma \in \Sigma$. Then the corrsponding map $u = u(\sigma, z)$ and its derivative $u'_z(\sigma, z)$ with respect to z are all continuous.*

Proof: 1°–3° have been proved. 4° follows from the fact $BP = PB$. We leave the proof of 5° to the readers. □

4.4 The dimension theorem on kernels

For the Hamiltonian system (2.1), in the previous two sections we have defined the space $L = L^2(S_\tau, \mathbf{R}^{2n})$ and the corresponding functional f. By using the saddle point reduction method the map $u : Z \to L$ from the finite dimensional space Z to the space L and the functional $a(z) = f(u(z))$ on Z are defined. The one to one correspondence relation between the critical point z of a and the critical point $u(z)$ of f is established. In this section we shall establish the relation on the dimensions of the kernels of $a''(z)$ and $df'(u(z))$.

Fix a constant $\tau > 0$. Suppose the Hamiltonian function H satisfies (H1) and (H3). Let $z \in Z$ be a critical point of a, let $u(z) = w(z) + z$ given by (3.22). Then $Pw(z) = 0$ and $u(z) \in L$ is a critical point of f. By the definition of f and the equation (2.5), $\zeta \in \ker df(u(z))$ if and only if ζ is a τ-periodic solution of the linear Hamiltonian system

$$\dot{y} = JH''(t, u(z)(t))y. \tag{1}$$

Denote by $\gamma(t)$ the fundamental solution of (1). Then by the classical Floquet theory, the dimension of the τ-periodic solution space of the linear system (1) is equal to $\dim\ker(\gamma(\tau) - I) \equiv \nu_1(\gamma(\tau))$. The following theorem yields the relation between this dimension and the dimension of the kernel of $a''(z)$.

Theorem 1 (E. Zehnder [Zeh2]). *Under the above conditions,*

$$\dim\ker a''(z) = \nu_1(\gamma(\tau)). \tag{2}$$

Proof: Using the notation of the previous section, let $P^{\perp} = P^{+} + P^{-}$. By Theorem 3.1, especially the equation (3.28), for any $\xi \in Z$ we obtain

$$a''(z)\xi = A_0 P\xi - P dg_0'(u(z))(\xi + dw(z)\xi). \tag{3}$$

On the other hand, by (3.17) we obtain

$$S^{\pm}x^{\pm}(R^{-1}z) = \pm(S^{\pm})^2 g_0'(u(z)).$$

Thus for any $\xi \in Z$,

$$\begin{aligned} dw(z)\xi &= d(S^{+}x^{+}(R^{-1}z) + S^{-}x^{-}(R^{-1}z))\xi \\ &= [(S^{+})^2 dg_0'(u(z)) - (S^{-})^2 dg_0'(u(z))]u'(z)\xi \\ &= A_0^{-1}P^{\perp}dg_0'(u(z))(\xi + dw(z)\xi). \end{aligned} \tag{4}$$

This yields

$$A_0 P^{\perp} dw(z)\xi = P^{\perp} dg_0'(u(z))(\xi + dw(z)\xi), \qquad \forall\, \xi \in Z. \tag{5}$$

Suppose $\xi \in \ker a''(z)$. Let $\eta = dw(z)\xi$ and $\zeta = \xi + \eta$. Then from (3) and (5) we obtain

$$A_0 P\xi = P dg_0'(u(z))\zeta, \tag{6}$$

$$A_0 P^{\perp}\eta = P^{\perp} dg_0'(u(z))\zeta. \tag{7}$$

That is

$$A_0\zeta = dg_0'(u(z))\zeta.$$

By the definitions of A_0 and g_0, this yields

$$A\zeta = dg'(u(z))\zeta,$$

i.e., ζ is a solution of the linear system (1).

On the other hand, suppose ζ is a solution of the linear system (1). Let $\xi = P\zeta$ and $\eta = P^{\perp}\zeta$. Then $\zeta = \xi + \eta$, They satisfy (6) and (7) respectively. Since the operator A_0 is invertible, (7) yields

$$\eta = A_0^{-1}P^{\perp}dg_0'(u(z))(\xi + \eta). \tag{8}$$

Since (3.19) and (3.20) hold, for any η_1 and $\eta_2 \in P^{\perp}L$, we have

$$\|A_0^{-1}P^{\perp}dg_0'(u(z))(\xi + \eta_1) - A_0^{-1}P^{\perp}dg_0'(u(z))(\xi + \eta_2)\|_L \leq \frac{\beta}{\gamma}\|\eta_1 - \eta_2\|_L.$$

Since $\beta\gamma < 1$, when $\xi \in Z$ is fixed, the right-hand side of (8) is a contraction mapping on $P^{\perp}L$. Therefore η is the unique solution of (8). But by (5), for $\xi = P\zeta$, $dw(z)\xi$ is also a solution of (8). Therefore,

$$\eta = dw(z)\xi.$$

Then (3) and (6) yield $\xi \in \ker a''(z)$.

By the relation $\zeta = \xi + dw(z)\xi$ in our above argument, linearly independent ξs in $\ker a''(z)$ correspond to linearly independent solutions ζ of (1). Thus the theorem is proved. □

Remark 2 The dimension Theorem 4.1 on kernels is due to E. Zehnder [Zeh2]. The proof given here is a modification of his original one.

4.5 Certain estimates

Fix a constant $\tau > 0$. In this section we prove inequalities satisfied by the functional $a(z) = f(u(z))$ and the injection $u(z)$ on Z defined by Theorem 3.1 with notations there.

Part 1 Estimates on the reduced functional

Theorem 1 (H. Amann and E. Zehnder [AmZ1]). *Suppose the Hamilton function H satisfies* (H1) *and* (H3). *Then for any $z \in Z$, functional $a(z)$ defined by* (3.23) *satisfies*

$$a(z) \leq \frac{1}{2}\langle A(P^{-}w(z) + z), (P^{-}w(z) + z\rangle_L - g(P^{-}w(z) + z), \tag{1}$$

where P^{-} is defined by (3.2), $w(z) = u(z) - z$ is defined by (3.22).

Proof: Using notations in Section 2, we first consider the function f_0 defined by (3.14). For any $x = x^{+} + x^{-} + z \in L^{+} \oplus L^{-} \oplus Z = L$, we obtain

$$(f_0)'_{x^{+}}(x) = x^{+} - S^{+}g_0(S^{+}x^{+} + S^{-}x^{-} + z). \tag{2}$$

Therefore, for the $x^+ = x^+(z)$ determined by (3.17),

$$(f_0)'_{x^+}(x^+(z) + x^-(z) + z) = 0, \quad \forall z \in Z. \tag{3}$$

For any $h + z \in L^+ \oplus Z$ and $t \in \mathbf{R}$, let

$$\begin{aligned} \phi(t) &= \frac{1}{2}(\|x^+(z) + th\|_L^2 + \|Q^+z\|_L^2 - \|Q^-z\|_L^2 - \|x^-(z)\|_L^2) \\ &\quad - g_0(S^+(x^+(z) + th) + S^-x^-(z) + z). \end{aligned} \tag{4}$$

Then by (3) and the fact that g_0' is Gadeaux differentiable, we obtain

$$\begin{aligned} \phi'(t) &= \langle x^+(z), h\rangle_L + t\|h\|_L^2 \\ &\quad - \langle g_0'(S^+(x^+(z) + th) + S^-x^-(z) + z), S^+h\rangle_L, \end{aligned} \tag{5}$$

$$\phi''(t) = \|h\|_L^2 - \langle dg_0'(S^+(x^+(z) + th) + S^-x^- + z)S^+h, S^+h\rangle_L. \tag{6}$$

By (3.9) we obtain $(S^+)^2 = A_0^{-1}|_{L^+}$. Since $c(H) \notin \sigma(A_0)$, by the definition (3.5) of S^+, there exists a constant $\mu > c(H)$ such that

$$\|A_0|_{L^+}\|^{-1} \le \frac{1}{\mu}. \tag{7}$$

Since S^+ is selfadjoint, by (2.6), from (6) and (7) we obtain

$$\phi''(t) \ge \left(1 - \frac{c(H)}{\mu}\right)\|h\|_L^2 > 0, \quad \forall h \in L^+\backslash\{0\}, t \in \mathbf{R}. \tag{8}$$

Thus for any $z \in Z$, from (3) and (8) as a function of $x^+ \in L^+$, the functional $f_0(x^+ + x^-(z) + z)$ is convex along any line passing through $x^+(z)$, and thus takes a global minimal value at $x^+(z)$, i.e.,

$$f_0(x^+(z) + x^-(z) + z) \le f_0(x^+ + x^-(z) + z), \quad \forall x^+ + z \in L^+ \oplus Z. \tag{9}$$

Therefore for any $z \in Z$ by the definition (3.23) of $a(z)$ we have

$$\begin{aligned} a(z) &= f_0(x^+(R^{-1}z) + x^-(R^{-1}z) + R^{-1}z) \\ &\le f_0(x^-(R^{-1}z) + R^{-1}z). \end{aligned} \tag{10}$$

By (3.16) and the definition of v below (3.14) this implies

$$\begin{aligned} a(z) &\le \frac{1}{2}\langle A(S^-x^-(R^{-1}z) + z, S^-x^-(R^{-1}z) + z\rangle_L \\ &\quad - g(S^-x^-(R^{-1}z) + z). \end{aligned} \tag{11}$$

By the definition of P^- and the definition of $w(z) = u(z) - z$ in (3.22), we have $P^-w(z) = S^-x^-(R^{-1}z)$. Thus (11) implies (1). This proves the theorem. □

The proof of the following proposition is similar to that of Theorem 1, and thus is left to the readers.

Proposition 2 (H. Amann and E. Zehnder [AmZ1]). *Suppose H satisfies* (H1), (H3), *and*

(H4) $H(t,x) = o(x)$ *at* $x = 0$.

Then

$$a(z) \geq \frac{1}{2}\langle A(P^+w(z)+z), (P^+w(z)+z)\rangle_L - o(\|z\|^2), \quad \text{as} \quad z \to 0 \ \text{in} \ Z. \tag{12}$$

Part 2 Estimates on the injection map

Fix a Hamiltonian function H satisfying conditions (H1) and (H3). Based on objects defined and notation introduced in Section 2, for $s \geq c(H)$ and $s \notin \frac{2\pi}{\tau}\mathbf{Z}$, we define new projection operators P_s, $P_s^{\pm}$, $S_s^{\pm}$, and R_s using formulae (3.2), (3.5) and (3.6) with the integration limit $c(H)$ there being replaced by s. Then we get the corresponding spaces decomposition $L = L_s^+ \oplus L_s^- \oplus Z_s$ and $V = V_s^+ \oplus V_s^- \oplus V_s^0$.

By our study in Section 2, the system (3.17)–(3.18) becomes

$$x^{\pm} = \pm S_s^{\pm} g_0'(x), \tag{13}$$

$$Q^{\pm}z = \pm Q^{\pm}R_s g_0'(v_s), \tag{14}$$

where $v_s = S_s^+x^+ + S_s^-x^- + R_s z$. Then for every $z \in Z_s$ the system (13) possesses a unique solution pair $x_s^{\pm}(z)$ satisfying $x_s^{\pm} \in C^2(Z_s, V_s^{\pm}) \cap C^2(Z_s, W)$. Define $u_s(z) = S_s^+x_s^+(z) + S_s^-x_s^-(z) + R_s z$ for all $z \in Z_s$. Then the functional $a_s : Z_s \to \mathbf{R}$ is given by Theorem 3.1:

$$a_s(z) = f(u_s(z)) = \frac{1}{2}\langle Au_s(z), u_s(z)\rangle_L - g(u_s(z)), \quad \forall z \in Z_s. \tag{15}$$

By Theorem 3.1, $a_s \in C^2(Z_s, \mathbf{R})$, $u_s \in C^2(Z_s, W)$ is an injection.

Proposition 3 *For any $s \geq c(H)$, $s \notin \sigma(A)$, $x \in W$,*

$$\|u_s(P_s x) - x\|_W \to 0, \quad \text{as } s \to +\infty. \tag{16}$$

Proof: For $x \in W$, by Theorem 3.1 we have $u_s(P_s x) \in W$. Let $P_s^{\perp} = P_s^+ + P_s^-$. Then by (13) similarly to (3.22) we obtain

$$\begin{aligned} u_s(P_s x) - x &= S_s^+ x_s^+(R_s^{-1}P_s x) + S_s^- x_s^-(R_s^{-1}P_s x) - P_s^{\perp}x \\ &= [(S_s^+)^2 - (S_s^-)^2]g_0'(u_s(P_s x)) - P_s^{\perp}x. \end{aligned} \tag{17}$$

As in (3.20), there exists a constant $c_s > c(H)$ such that

$$\|A_0^{-1}|_{L_s^+ \oplus L_s^-}\| \le 1/c_s. \tag{18}$$

Then as in (3.21) by (3.9) we obtain

$$\begin{aligned}\|u_s(P_s x) - x\|_W &\le \|(S_s^+)^2 - (S_s^-)^2\|_{\mathcal{L}(L_s^+ \oplus L_s^-)} \|dg_0'\|_{\mathcal{L}(L)} \|u_s(P_s x) - x\|_L \\ &\quad + \|(S_s^+)^2 - (S_s^-)^2\|_{\mathcal{L}(L_s^+ \oplus L_s^-)} \|g_0'(x)\|_L + \|P_s^\perp x\|_W \\ &\le \frac{c(H)}{c_s} \|u_s(P_s x) - x\|_W + \frac{1}{c_s} \|g_0'(x)\|_L + \|P_s^\perp x\|_W,\end{aligned}$$

where the first inequality is obtained by using the mean value inequality for some element y_s, and the third inequality is obtained by (2.7), (3.9), and (18). Thus we obtain

$$\|u_s(P_s x) - x\|_W \le \frac{1}{c_s - c(H)} \|g_0'(x)\|_L + \frac{c_s}{c_s - c(H)} \|P_s^\perp x\|_W.$$

Since $c_s \to +\infty$ and $\|P_s^\perp x\|_W \to 0$ as $s \to +\infty$, (16) holds. □

Remark 4 The estimate Theorem 1 and Proposition 2 were proved by H. Amann and E. Zehnder in [AmZ1]. Here we modified their proof to our case. Proposition 3 was proved by the author in his 1994 IFM of ETH–Zürich preprint quoted in [Lon17], and can also be found in [AnL1] of T. An and the author of 1998.

Part III
Index Theory

In recent years the classical Morse index theory has been applied to the case of functionals on infinite dimensional spaces, and plays an important role in the study of many nonlinear problems. Unfortunately, the Morse indices of the functional corresponding to a Hamiltonian system at its critical point are always infinite. This produces a great difficulty for applications of the Morse index theory on Hamiltonian systems. The main goal of Part III is to introduce an index function theory parameterized by the unit circle in the complex plane for paths in symplectic groups. Especially the index theory corresponding to the parameter 1 yields a finite representation of the above mentioned infinite Morse indices for Hamiltonian systems.

This index function theory for paths in symplectic groups parameterized by $\omega \in \mathbf{U}$ is established in Chapter 5. In Chapter 6 we study further properties of this index function theory, especially we establish its relation theorem with the Morse index theory of the reduced functional obtained via the saddle point reduction method. We also give an axiom characterization of it. Finally in Chapter 7 we study its relationship with other Morse index theories. Such studies are based on the algebraic and topological structures of $\mathrm{Sp}(2n)$ and its subsets, variational structures of Hamiltonian systems obtained in the previous two parts.

For historical remarks on this index function theory, the readers are referred to the Introduction of this book.

Chapter 5

Index functions for symplectic paths

In this chapter, we introduce an index function theory for paths in the symplectic groups started from the identity, i.e., elements in $\mathcal{P}_\tau(2n)$ with $\tau > 0$ defined by (2.0.1):

$$\mathcal{P}_\tau(2n) = \{\gamma \in C([0,\tau], \mathrm{Sp}(2n)) \mid \gamma(0) = I\}.$$

For $\tau > 0$ and $\omega \in \mathbf{U}$, we further define the set of **ω-non-degenerate paths** by

$$\mathcal{P}^*_{\tau,\omega}(2n) = \{\gamma \in \mathcal{P}_\tau(2n)) \mid \gamma(\tau) \in \mathrm{Sp}(2n)^*_\omega\}, \tag{1}$$

and the set of **ω-degenerate paths** by

$$\mathcal{P}^0_{\tau,\omega}(2n) = \mathcal{P}_\tau(2n) \backslash \mathcal{P}^*_{\tau,\omega}(2n). \tag{2}$$

Based upon the topological structures of the symplectic group $\mathrm{Sp}(2n)$ and its subsets, we can give a complete homotopy classification of all paths in $\mathcal{P}^*_{\tau,\omega}(2n)$ for any $\omega \in \mathbf{U}$. Based on this homotopy classification, the index function theory is defined for all such non-degenerate paths. Then we define the index function of any path $\gamma \in \mathcal{P}^0_{\tau,\omega}(2n)$ to be the index of a suitable perturbation path $\beta \in \mathcal{P}^*_{\tau,\omega}(2n)$ of γ. In the next chapter, we shall prove that this index of γ is actually the minimum of those of the nearby ω-non-degenerate paths. Specially this index function theory assigns two integers $(i_\omega(\gamma), \nu_\omega(\gamma))$ to any path $\gamma \in \mathcal{P}_\tau(2n)$ for each $\omega \in \mathbf{U}$.

To make the concept of this homotopy classification for these paths become precise, we need the following definitions introduced in [Lon5] and [Lon16].

Definition 1 For any $\tau > 0$, $\gamma \in \mathcal{P}_\tau(2n)$, and $\omega \in \mathbf{U}$, using Definition 1.8.1 we define

$$\nu_\omega(\gamma) = \nu_\omega(\gamma(\tau)) = \dim_{\mathbf{C}} \ker_{\mathbf{C}}(\gamma(\tau) - \omega I). \tag{3}$$

A path $\gamma \in \mathcal{P}_\tau(2n)$ is called **ω-degenerate** if $\nu_\omega(\gamma) > 0$, otherwise it is called **ω-non-degenerate**.

Definition 2 Given two curves ψ_0 and $\psi_1 \in C([a,b], \mathrm{Sp}(2n))$ with $\psi_0(a) = \psi_1(a)$ and $\psi_0(b) = \psi_1(b)$, if there is a map $\delta \in C([0,1] \times [a,b], \mathrm{Sp}(2n))$ such that $\delta(0,\cdot) = \psi_0(\cdot)$, $\delta(1,\cdot) = \psi_1(\cdot)$, $\delta(s,a) = \psi_0(a)$, and $\delta(s,b) = \psi_0(b)$, then ψ_0 and ψ_1 are **homotopic on** $[a,b]$ **with fixed end points** and we write $\psi_0 \sim \psi_1$ on $[a,b]$ with fixed end points.

Definition 3 For $\tau > 0$ and $\omega \in \mathbf{U}$, given two paths γ_0 and $\gamma_1 \in \mathcal{P}_\tau(2n)$, if there exists a map $\delta \in C([0,1] \times [0,\tau], \mathrm{Sp}(2n))$ such that $\delta(0,\cdot) = \gamma_0(\cdot)$, $\delta(1,\cdot) = \gamma_1(\cdot)$, $\delta(s,0) = I$, and $\nu_\omega(\delta(s,.))$ is constant for $0 \le s \le 1$, then γ_0 and γ_1 are **ω-homotopic on** $[0,\tau]$ **along** $\delta(\cdot,\tau)$ and we write $\gamma_0 \sim_\omega \gamma_1$. If $\gamma_0 \sim_\omega \gamma_1$ on $[0,\tau]$ along $\delta(\cdot,\tau)$

for all $\omega \in \mathbf{U}$, then γ_0 and γ_1 are **U-homotopic on** $[0, \tau]$ **along** $\delta(\cdot, \tau)$ and we write $\gamma_0 \approx \gamma_1$.

In the following Section 1, we establish the index function theory for paths in Sp(2) based on the $\mathbf{R}^3$-cylindrical coordinate representation of Sp(2) introduced in Section 2.1. This study yields a beautiful intuitive model for the higher dimensional case. Then we establish the index function theory for any non-degenerate path in Sp(2n) and introduce two families of standard paths in Section 2 based on the topological properties of Sp(2n) and its subsets studied in Chapter 2. In Section 3, we study basic properties of the index function for non-degenerate paths. In Section 4, we study perturbations of degenerate paths and define the index function theory for degenerate symplectic paths.

5.1 Paths in Sp(2)

Fix $\tau > 0$. In this section, we give two methods to define an index function theory parametrized by $\omega \in \mathbf{U}$ for elements in $\mathcal{P}^*_{\tau,\omega}(2)$. The first method is completely elementary based on the homotopy classification of $\mathcal{P}^*_\tau(2)$. In the second method we use the concept of intersection numbers in the algebraic topology. These two methods view the index function theory from different angles.

We also have two ways to define the index function for degenerate paths. In the first method, we perturb degenerate paths to non-degenerate paths, and then use them to define the index functions. In the second method, we use a max-min argument on non-degenerate paths in any neighborhood of a given degenerate path to define the index functions for elements in $\mathcal{P}^0_{\tau,\omega}(2)$.

All these methods are strongly based on our $\mathbf{R}^3$-cylindrical coordinate representation of Sp(2) introduced in Section 2.1. This coordinate transformation yields a C^∞-diffeomorphism from Sp(2) to $\mathbf{R}^3\setminus\{z\text{−axis}\}$. All the considerations in this section should be understood under this diffeomorphism without further explanations.

Non-degenerate Method 1. The homotopy classification method.

For any $\tau > 0$ and $k \in \mathbf{Z}$ we define a sequence of **zigzag standard paths** $\hat{\alpha}_k$ in $\mathcal{P}^*_\tau(2)$ as follows. Set

$$\phi_{\tau,\theta}(t) = R\left(\theta\frac{t}{\tau}\right) \qquad \forall\, t \in [0, \tau],\ \theta \in \mathbf{R}, \tag{1}$$

$$\hat{\alpha}_0(t) = D(1 + t/\tau) \equiv \xi_+(t), \qquad \forall\, t \in [0, \tau]. \tag{2}$$

For $0 \le t \le 1$ we define

$$\hat{\alpha}_k(t) = [D(2)\phi_{\tau,k\pi}] * \xi_+(t), \qquad \text{if } k \in \mathbf{Z}\setminus\{0\}. \tag{3}$$

Then $\hat{\alpha}_k \in \mathcal{P}_\tau(2)$ and satisfies

$$\nu_\omega(\hat{\alpha}_k) = 0, \qquad \forall\, \omega \in \mathbf{U},\ k \in \mathbf{Z}. \tag{4}$$

For any $\gamma \in \mathcal{P}_\tau(2)$ and $t \in [0, \tau]$, we denote by $\theta = \theta(t)$ a real function satisfying (2.1.1) in the polar decomposition of $\gamma(t)$. Because this decomposition is uniquely determined by $\gamma(t)$ as proved by Lemma 1.1.3, we can choose this function $\theta(t)$ to be continuous. We define

$$\Delta_\tau(\gamma) = \theta(\tau) - \theta(0). \tag{5}$$

Note that here the function $\theta(t)$ can be replaced by $\theta_k(t) = \theta(t) + 2k\pi$ for any $k \in \mathbf{Z}$, and this exhausts all continuous real functions satisfying (2.1.1) with $M = \gamma(t)$ for all $t \in [0, \tau]$. The important point here is that the function $\Delta_\tau : \mathcal{P}_\tau(2) \to \mathbf{R}$ is continuous and independent of the choice of $k \in \mathbf{Z}$.

For any $\omega \in \mathbf{U}$ and every $\gamma \in \mathcal{P}^*_{\tau,\omega}(2)$, by our study in Section 2.1, we can connect $\gamma(\tau)$ to $M_1^+ = D(2)$ or $M_1^- = D(-2)$ by a path β within $\mathrm{Sp}(2)^*_\omega$. Let

$$k = \frac{\Delta_\tau(\beta * \gamma)}{\pi}.$$

Then we have $k \in \mathbf{Z}$ and $\gamma \sim_\omega \hat{\alpha}_k$ in the sense of Definition 0.3. Note that this k is uniquely determined by γ. This yields a homotopy classification of elements in $\mathcal{P}^*_{\tau,\omega}(2)$ and every such homotopy class contains one and only one path $\hat{\alpha}_k$. An illustration of this study is given in Figure 1.

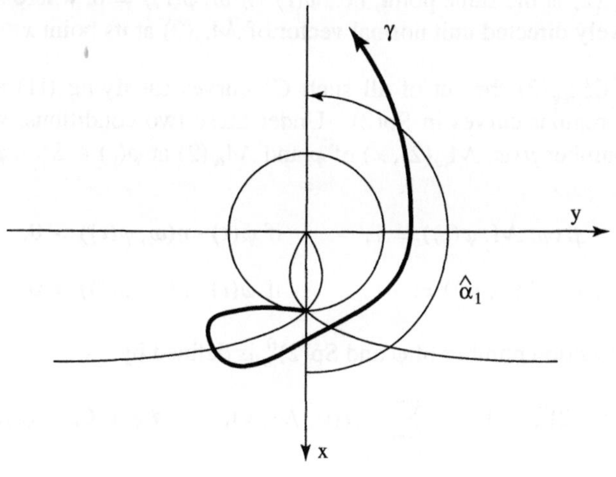

Figure 1: *The case of* $\gamma \sim_1 \hat{\alpha}_1$

Definition 1 For any $\omega \in \mathbf{U}$, $\tau > 0$, and $\gamma \in \mathcal{P}^*_{\tau,\omega}(2)$, we define

$$i_\omega(\gamma) = k, \qquad \text{if} \quad \gamma \sim_\omega \hat{\alpha}_k. \tag{6}$$

Non-degenerate Method 2. The intersection number method.

This definition needs certain terminology from algebraic topology which we refer to classical books on the topics without further explanations.

Fix $\omega \in \mathbf{U}$. In Section 2.1, an orientation of $\mathrm{Sp}(2)^0_\omega$ is defined. Therefore $\mathrm{Sp}(2)^0_\omega$ form a locally finite 2-dimensional singular homological cycle (cf. Lefshetz [Lef1]). For any $\tau > 0$ and path $\beta \in \mathcal{P}_\tau(2)$ denote $\beta^{-1}(t) = \beta(\tau - t)$ for $t \in [0, \tau]$.

Definition 2 We define

$$i_\omega(\gamma) = [\mathrm{Sp}(2)^0_\omega : \gamma * \xi_+^{-1}], \quad \forall\, \gamma \in \mathcal{P}^*_{\tau,\omega}(2). \tag{7}$$

For any path $\gamma \in \mathcal{P}^*_{\tau,\omega}(2)$ the two end points of the joint path $\gamma * \xi_+^{-1}$ are not located on $\mathrm{Sp}(2)^0_\omega$. Thus the algebraic homological intersection number in (7) is well defined.

To further explain this definition, fixing an $\omega \in \mathbf{U}$, we consider the smooth paths first. Let $\varphi \in C^1([0,\tau], \mathrm{Sp}(2))$ such that $\varphi(0) = D(2)$ and $\varphi(\tau) \in \mathrm{Sp}(2)^*_\omega$. Then the direction of φ at the point $\varphi(t)$ is defined to be the tangent direction $\dot\varphi(t)$ of φ at that point. Now we assume the following conditions on φ.

(I1) It holds that

$$\varphi([0,\tau]) \cap \mathrm{Sp}(2)^0_\omega \subset \varphi([0,\tau]) \cap \mathcal{M}_\omega(2) \equiv S(\omega, \varphi). \tag{8}$$

(I2) φ intersects $\mathcal{M}_\omega(2)$ transversally, i.e., at any intersection point $\varphi(t) \in S(\omega,\varphi)$, the tangent vector of φ at the point $\varphi(t)$ is not contained in the tangent plane of $\mathcal{M}_\omega(2)$ at the same point, i.e., $\dot\varphi(t) \cdot \eta(\omega, \varphi(t)) \neq 0$, where $\eta(\omega, x)$ is the positively directed unit normal vector of $\mathcal{M}_\omega(2)$ at its point x.

Denote by $C^1_{\tau,\mathrm{reg}}(2)$ the set of all such C^1 curves satisfying (I1) and (I2), and call them the regular curves in $\mathrm{Sp}(2)$. Under these two conditions, we define the intersection number $\mu(\varphi, \mathcal{M}_\omega(2), x)$ of φ and $\mathcal{M}_\omega(2)$ at $\varphi(t) \in S(\omega, \varphi)$ by

$$\mu(\varphi, \mathcal{M}, \varphi(t)) = 1, \qquad \text{if } \dot\varphi(t) \cdot \eta(\omega, \varphi(t)) > 0, \tag{9}$$

$$\mu(\varphi, \mathcal{M}, \varphi(t)) = -1, \qquad \text{if } \dot\varphi(t) \cdot \eta(\omega, \varphi(t)) < 0. \tag{10}$$

Then the intersection number of φ and $\mathrm{Sp}(2)^0_\omega$ is defined by

$$[\mathrm{Sp}(2)^0_\omega : \varphi] = \sum_{x \in S(\omega,\varphi)} \mu(\varphi, \mathcal{M}, x), \qquad \forall\, \varphi \in C^1_{\tau,\mathrm{reg}}(2). \tag{11}$$

This definition is also illustrated in Figure 1.

Define by $C^1_\tau(2)$ the set of paths in $C^1([0,\tau], \mathrm{Sp}(2))$ started from $D(2)$. Let $C^1_{\tau,\mathrm{reg}}(2, D(2)) = C^1_\tau(2) \cap C^1_{\tau,\mathrm{reg}}(2)$. Now fix a path $\gamma \in \mathcal{P}_\tau(2)$. We consider the C^0-approximations $\varphi \in C^1_{\tau,\mathrm{reg}}(2, D(2))$ of $\gamma * \xi_+^{-1}$ satisfying $\varphi(\tau) = \gamma(\tau)$. It is easy to see that if φ is sufficiently C^0-close to $\gamma * \xi_+^{-1}$, the intersection number $[\mathrm{Sp}(2)^0_\omega : \varphi]$ is independent of the particular choices of φ. Therefore we obtain

$$[\mathrm{Sp}(2)^0_\omega : \gamma * \xi_+^{-1}] = [\mathrm{Sp}(2)^0_\omega : \varphi], \tag{12}$$

for all such smooth C^0-approximations of $\gamma * \xi_+^{-1}$. Thus Definition 2 is well defined.

Note that the indices defined by the above two methods are homotopy invariant, i.e., for any $\omega \in \mathbf{U}$, γ_0 and $\gamma_1 \in \mathcal{P}_\tau(2)$, $\gamma_0 \sim_\omega \gamma_1$ implies

$$i_\omega(\gamma_0) = i_\omega(\gamma_1), \quad \nu_\omega(\gamma_0) = \nu_\omega(\gamma_1),$$

and they take the same values on $\hat{\alpha}_k$. Therefore the indices defined by the above two methods coincide completely on every element in $\mathcal{P}^*_{\tau,\omega}(2)$.

Next we give two methods to define the index functions for degenerate symplectic paths.

Degenerate Method 1. The perturbation method.

Fix $\tau > 0$ and $\omega \in \mathbf{U}$. Let $\gamma \in \mathcal{P}^0_{\tau,\omega}(2)$. We define a family of nearby non-degenerate paths for γ by perturbations:

$$\gamma_s(t) = \gamma(t)R(s\rho(t)\theta), \qquad \forall t \in [0, \tau],\ s \in [-1, 1], \tag{13}$$

where $\rho \in C^1([0, \tau], \mathbf{R})$ satisfies $\rho(t) = 0$ for 0_0 and $0 < \rho(t) < 1$ for $t_0 < t \le \tau$ with some $t_0 \in (0, \tau)$, and $\rho(\tau) = 1$. We choose $\theta > 0$ to be so small and $t_0 > 0$ to be so close to τ such that $\gamma_s \in \mathcal{P}^*_{\tau,\omega}(2)$ for all $s \in [-1, 1]\backslash\{0\}$. Then from our discussion in Section 2.1, we obtain

$$i_\omega(\gamma_s) = i_\omega(\gamma_{s'}), \qquad \forall ss' > 0,\ s,\ s' \in [-1, 1], \tag{14}$$

$$i_\omega(\gamma_1) = i_\omega(\gamma_{-1}) = \nu_\omega(\gamma), \tag{15}$$

$$i_\omega(\gamma_{-1}) = i_\omega(\beta) \le i_\omega(\gamma_1), \tag{16}$$

for all $\beta \in \mathcal{P}^*_{\tau,\omega}(2)$ sufficiently close to γ in $\mathcal{P}_\tau(2n)$.

Definition 3 For any $\gamma \in \mathcal{P}^0_{\tau,\omega}(2)$, we define

$$i_\omega(\gamma) = i_\omega(\gamma_{-1}). \tag{17}$$

Degenerate Method 2. The minimizing method.

We define the index functions for any degenerate path by those of its nearby non-degenerate paths.

Definition 4 For any $\tau > 0$, $\omega \in \mathbf{U}$, and $\gamma \in \mathcal{P}^0_{\tau,\omega}(2)$, we define

$$i_\omega(\gamma) = \sup_{U \in \mathcal{N}(\gamma)} \inf\{i_\omega(\beta) \mid \beta \in U \cap \mathcal{P}^*_{\tau,\omega}(2)\}, \tag{18}$$

where $\mathcal{N}(\gamma)$ is the set of all open neighborhoods of γ in $\mathcal{P}_\tau(2n)$.

Note that by the definition (13) of γ_{-s}, for any $U \in \mathcal{N}(\gamma)$ there is at least an $s \in (0, 1]$ such that $\gamma_{-s} \in U$. Therefore the right-hand side of (18) is not greater than that of (17). By (14)–(16), it is clear that there is a $U \in \mathcal{N}(\gamma)$ such that

$$\inf\{i_\omega(\beta) \mid \beta \in U \cap \mathcal{P}^*_{\tau,\omega}(2)\}$$

is not less than the right-hand side of (17). Therefore Definitions 3 and 4 coincide.

In such a way, for any $\tau > 0$, $\omega \in \mathbf{U}$ and $\gamma \in \mathcal{P}_\tau(2)$, a pair of integers

$$(i_\omega(\gamma), \nu_\omega(\gamma)) \in \mathbf{Z} \times \{0, 1, 2\}$$

is defined and is called the **index function** of γ.

Note that the topological structures of Sp(2), $\mathrm{Sp}(2)^*_\omega$, and $\mathrm{Sp}(2)^0_\omega$ used in this definition are those contained in Sections 2.1 to 2.4. This observation is very suggestive for our study on paths in the higher dimensional symplectic groups. In the following sections, we shall extend the above Non-degenerate Method 1 and Degenerate Method 1 to define the index function theory for paths in the symplectic group Sp(2n) for any $n \in \mathbf{N}$.

5.2 Non-degenerate paths in Sp(2n)

In this section we define the index function theory parametrized by $\omega \in \mathbf{U}$ for any non-degenerate path in the symplectic group Sp(2n) with $n \geq 1$.

By Lemmas 1.1.3, 1.1.5, and our discussion before Lemma 2.4.5, for any $M \in$ Sp(2n), we denote by

$$U(M) = \begin{pmatrix} u_1(M) & -u_2(M) \\ u_2(M) & u_1(M) \end{pmatrix}$$

the orthogonal and symplectic part of its unique polar decomposition. Then $u(M) = u_1(M) + \sqrt{-1}u_2(M) \in U(n, \mathbf{C})$. So in such a way, for every path $\gamma \in \mathcal{P}_\tau(2n)$ we can uniquely associate to it a path,

$$u_\gamma(t) = u(\gamma(t)), \qquad \forall t \in [0, \tau],$$

in the unitary group $U(n, \mathbf{C})$.

Lemma 1 1° *For any $\gamma \in C([0, \tau], \mathrm{Sp}(2n))$, let $\Delta : [0, \tau] \to \mathbf{R}$ be any continuous real function satisfying*

$$\det u_\gamma(t) = \exp(\sqrt{-1}\Delta(t)), \qquad \forall t \in [0, \tau]. \tag{1}$$

We define the rotation number of γ *by*

$$\Delta_\tau(\gamma) = \Delta(\tau) - \Delta(0). \tag{2}$$

For $0 \le a < b \le \tau$, *we also define*

$$\Delta_{[a,b]}(\gamma) = \Delta(b) - \Delta(a). \tag{3}$$

Then $\Delta_\tau(\gamma)$ *and* $\Delta_{[a,b]}(\gamma)$ *depend only on* γ *but not on the choice of the function* Δ *satisfying* (1).

2° *For* $\gamma \in C([0,\tau], \mathrm{Sp}(2n) \cap O(2n)) \cap \mathcal{P}_\tau(2n)$, *the function* ρ_n *defined by* (2.4.7) *satisfies*

$$\rho_n(\gamma(t)) = \exp(\sqrt{-1}\Delta(t)), \quad \forall t \in [0,\tau]; \tag{4}$$

the function $\hat{\Delta}_\tau$ *defined by* (2.4.25) *and the above function* Δ_τ *satisfy*

$$\hat{\Delta}_\tau(\gamma) = \Delta_\tau(\gamma). \tag{5}$$

3° *For any* $\gamma_j \in C([0,\tau], \mathrm{Sp}(2n_j))$ *with* $j = 0, 1$, *and* $a \in (0,\tau)$,

$$\Delta_\tau(\gamma_0 \diamond \gamma_1) = \Delta_\tau(\gamma_0) + \Delta_\tau(\gamma_1), \tag{6}$$

$$\Delta_\tau(\gamma_0) = \Delta_{[0,a]}(\gamma_0) + \Delta_{[a,\tau]}(\gamma_0). \tag{7}$$

Proof: Let $\Delta' : [0,\tau] \to \mathbf{R}$ be another continuous real function satisfying (1). By (1), for every $t \in [0,\tau]$ there exists $k(t) \in \mathbf{Z}$ such that $\Delta(t) - \Delta'(t) = 2\pi k(t)$. Since both Δ and Δ' are continuous, so is the function k. Since k is integer valued, by the connectedness of the interval $[0,\tau]$, k must be constant. Thus the first conclusion in 1° holds. (4) and (5) follow from Lemma 2.4.5 and (1). 3° follows from the definition (1). □

The following lemma studies the relation between the usual homotopy and the homotopy which fixes the end points.

Lemma 2 *Let* $\omega \in \mathbf{U}$. *Suppose* γ_0 *and* $\gamma_1 \in \mathcal{P}_\tau(2n)$ *possess common end point* $\gamma_0(\tau) = \gamma_1(\tau)$. *Suppose* $\gamma_0 \sim_\omega \gamma_1$ *on* $[0,\tau]$ *in the sense of Definition* 0.3 *via a homotopy* $\delta : [0,1] \times [0,\tau] \to \mathrm{Sp}(2n)$ *such that* $\delta(\cdot,\tau)$ *is contractible in* $\mathrm{Sp}(2n)$. *Then the homotopy* δ *can be modified to fix the end points all the time, i.e.,* $\delta(s,\tau) = \gamma_0(\tau)$ *for all* $0 \le s \le 1$.

Proof: Without loss of generality, suppose $\tau = 1$. By the assumption of the lemma, $\xi \equiv \delta(\cdot, 1)$ is a contractible curve in $\mathrm{Sp}(2n)$, i.e., there is a homotopy map α which shrinks ξ to some $M \in \mathrm{Sp}(2n)$ within $\mathrm{Sp}(2n)$. Since $\mathrm{Sp}(2n)$ is path connected, without loss of generality, we suppose $M = \gamma_0(1) = \gamma_1(1)$. Therefore

$\alpha \in C([0,1]^2, \mathrm{Sp}(2n))$ satisfies $\alpha(0,t) = \xi(t)$, $\alpha(1,t) = \alpha(s,0) = \alpha(s,1) = M$ for $(s,t) \in [0,1]^2$. For $(s,t) \in [0,1]^2$, define (cf. Figure 2 for the definition of β)

$$\beta(s,t) = \begin{cases} \gamma_0(\frac{t}{1-2s}), & 0 \le t \le 1-2s, \quad 0 \le s \le \frac{1}{4}, \\ \delta(2s-\frac{1}{2}, 2t), & 0 \le t \le \frac{1}{2}, \quad \frac{1}{4} \le s \le \frac{3}{4}, \\ \alpha(2t-1, 2s-\frac{1}{2}), & \frac{1}{2} \le t \le 1, \quad \frac{1}{4} \le s \le \frac{3}{4}, \\ \gamma_1(\frac{t}{2s-1}), & 0 \le t \le 2s-1, \quad \frac{3}{4} \le s \le 1, \\ M, & \text{otherwise}. \end{cases} \tag{8}$$

Then $\beta \in C([0,1]^2, \mathrm{Sp}(2n))$ gives the homotopy from γ_0 to γ_1 which fixes the end points all the time. □

By the same argument we obtain the following lemma. We omit its proof here.

Lemma 3 *Suppose γ_0 and $\gamma_1 \in C([0,\tau], \mathrm{Sp}(2n) \cap O(2n))$ possess common end points $\gamma_0(0) = \gamma_1(0)$ and $\gamma_0(\tau) = \gamma_1(\tau)$. Suppose $\gamma_0 \sim \gamma_1$ on $[0,\tau]$ in the sense of Definition* 0.2 *via a homotopy $\delta : [0,1] \times [0,\tau] \to \mathrm{Sp}(2n) \cap O(2n)$ such that $\delta(\cdot, t)$ with $t = 0$ and $t = \tau$ are contractible in $\mathrm{Sp}(2n) \cap O(2n)$. Then the homotopy δ can be modified to fix the end points all the time, i.e., $\delta(s,\tau) = \gamma_0(\tau)$ for all $0 \le s \le 1$.*

Lemma 4 *If γ_0 and $\gamma_1 \in \mathcal{P}_\tau(2n)$ possess common end point $\gamma_0(\tau) = \gamma_1(\tau)$, then $\Delta_\tau(\gamma_0) = \Delta_\tau(\gamma_1)$ if and only if $\gamma_0 \sim_\omega \gamma_1$ on $[0,\tau]$ with fixed end points for some $\omega \in \mathbf{U}$ in the sense of Definition* 0.3.

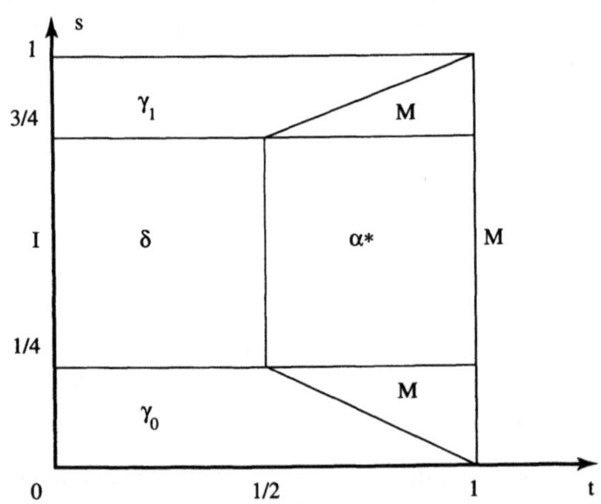

Figure 2: *The homotopy β*

Proof: Without loss of generality we suppose $\tau = 1$.

Using the polar decomposition of symplectic matrices, for any $M = AU \in \mathrm{Sp}(2n)$ with $A \in \mathrm{PSS}(2n)$ and $U \in \mathrm{Sp}(2n) \cap O(2n)$ (cf. Sections 1.1 and 2.2), there is a curve $\eta \in C([0,1], \mathrm{PSS}(2n))$ satisfying $\eta(0) = A$ and $\eta(1) = I$. Especially for any $t \in [0,1]$, all the eigenvalues of $\eta(t)$ are real. Thus by the definition (2.4.7) of the function ρ_n, we have $\rho_n(\eta(t)) = 1$ or -1. Since $\rho_n(I) = 1$, and both ρ_n and η are continuous, we get $\rho_n(\eta(t)) = 1$ for all $t \in [0,1]$. Therefore by Lemma 2.4.7 the curve η is contractible in $\mathrm{Sp}(2n)$, and so is the curve $\eta(\cdot)U$.

Similarly by the polar decomposition, each path γ_i uniquely determines a path $\beta_i : [0,1] \to \mathrm{Sp}(2n) \cap O(2n)$ by shrinking the positive definite symmetric symplectic part of the matrix $\gamma_i(t)$ for $t \in [0,1]$ and $i = 0, 1$ to the identity matrix. Note that this shrinking is continuous by Lemma 1.1.3. This yields two homotopies $\delta_i : [0,1] \to \mathcal{P}_1(2n)$ such that $\delta_i(0) = \gamma_i$, $\delta_i(1) = \beta_i$ for $i = 0, 1$. Especially by our discussion in the previous paragraph, the curve $(\delta_1(\cdot)(1))^{-1} * \delta_0(\cdot)(1) : [0,1] \to \mathrm{Sp}(2n)$ is contractible in $\mathrm{Sp}(2n)$.

We claim that $\gamma_0 \sim_\omega \gamma_1$ with fixed end points if and only if $\beta_0 \sim_\omega \beta_1$ with fixed end points in the sense of Definition 0.3.

In fact, for the necessity of the claim, suppose $\beta_0 \sim_\omega \beta_1$ with fixed end points via a homotopy η. Then $(\delta_1)^{-1} * \eta * \delta_0$ yields a homotopy ξ from γ_0 to γ_1 with $\xi(\cdot, 1)$ being contractible in $\mathrm{Sp}(2n)$ by our above discussion. Thus by Lemma 3, $\gamma_0 \sim_\omega \gamma_1$. By a similar argument, we get the sufficiency of the claim.

Note that by definition of $\Delta(t)$ in (1),

$$\Delta_1(\gamma_i) = \Delta_1(\beta_i), \qquad \text{for } i = 0, 1.$$

Thus it suffices to prove that $\Delta_1(\beta_0) = \Delta_1(\beta_1)$ if and only if $\beta_0 \sim_\omega \beta_1$ on $[0,1]$ with fixed end points in $\mathrm{Sp}(2n)$.

By Theorem 2.2.1, especially (2.2.4), $\beta_0 \sim_\omega \beta_1$ on $[0,1]$ with fixed end points if and only if $\beta \equiv \beta_1^{-1} * \beta_0$ is contractible in $\mathrm{Sp}(2n)$. By Lemma 2.4.7, the latter holds if and only if $\rho_n \circ \beta : [0,1] \to \mathbf{U}$ is contractible in $\mathbf{U}$. Fix a function $\Delta(t)$ satisfying (1). By Lemma 1 and the fact $\beta([0,1]) \subset \mathrm{Sp}(2n) \cap O(2n)$, we obtain $\Delta_1 = \hat{\Delta}_1$ defined by (2.4.26) and that $\rho_n \circ \beta$ is a loop in $\mathbf{U}$, and

$$\Delta_1(\beta_0) - \Delta_1(\beta_1) = \Delta_1(\beta) = 2k\pi,$$

for some $k \in \mathbf{Z}$. Note that by Lemma 2.4.7, $\rho_n \circ \beta$ is contractible in $\mathbf{U}$ if and only if $k = 0$, i.e., $\Delta_1(\beta_0) = \Delta_1(\beta_1)$. This proves the lemma. □

By the same argument of Lemma 4, we obtain

Lemma 5 *For any two continuous curves ψ_0 and $\psi_1 : [0,\tau] \to \mathrm{Sp}(2n)$ with $\psi_0(0) = \psi_1(0)$ and $\psi_0(\tau) = \psi_1(\tau)$, there exist continuous functions Δ^0 and $\Delta^1 : [0,\tau] \to \mathbf{R}$ satisfying* (1) *determined by the curves ψ_0 and ψ_1 respectively. Then $\psi_0 \sim \psi_1$ on $[0,\tau]$ with fixed end points in the sense of Definition* 0.2 *if and only if $\Delta^0_\tau(\psi_0) = \Delta^1_\tau(\psi_1)$, where $\Delta^i_\tau(\psi_i) = \Delta^i(\tau) - \Delta^i(0)$ for $i = 0, 1$ respectively.*

For any $\omega \in \mathbf{U}$ and $\gamma \in \mathcal{P}^*_{\tau,\omega}(2n)$, by Theorem 2.4.1 we can connect $\gamma(\tau)$ to M_n^+ or M_n^- by a path $\beta : [0,\tau] \to \mathrm{Sp}(2n)^*_\omega$.

Lemma 6 *Under the above conditions, define*

$$k \equiv \Delta_\tau(\beta * \gamma)/\pi. \tag{9}$$

Then k is an integer, is independent of the choice of the path β, and

$$\begin{cases} k \quad \text{is odd}, & \text{if} \quad \beta(\tau) = M_n^-, \\ k \quad \text{is even}, & \text{if} \quad \beta(\tau) = M_n^+. \end{cases} \tag{10}$$

We denote by $\mathcal{P}_{\tau,\omega}^k(2n)$ the set of all such paths in $\mathcal{P}_{\tau,\omega}^*(2n)$, that possess the property (9).

Proof: By the definition of $M_n^{\pm}$, k is an integer and (10) holds. Now we connect $\gamma(\tau)$ to $\beta(\tau)$ by another curve $\beta' : [0, \tau] \to \mathrm{Sp}(2n)_\omega^*$. Since $\mathrm{Sp}(2n)_\omega^*$ is simply connected in Sp(2n) by Theorem 2.4.2, we obtain $\beta \sim_\omega \beta'$ on $[0, \tau]$ with fixed points in the sense of Definition 0.3. Therefore by Lemmas 1 and 4 we obtain

$$\Delta_\tau(\beta * \gamma) = \Delta_\tau(\beta) + \Delta_\tau(\gamma) = \Delta_\tau(\beta') + \Delta_\tau(\gamma) = \Delta_\tau(\beta' * \gamma).$$

This proves the lemma. □

As in Definition 1.1, we define the index function for non-degenerate paths by

Definition 7 (C. Conley and E. Zehnder [CoZ2], Y. Long and E. Zehnder [LZe1], Y. Long [Lon16]). For any $\omega \in \mathbf{U}$ and $\tau > 0$, we define

$$i_\omega(\gamma) = k, \qquad \text{if} \;\; \gamma \in \mathcal{P}_{\tau,\omega}^k(2n). \tag{11}$$

Remark 8 Results in this section are based upon those obtained in [CoZ2] of C. Conley and E. Zehnder, [LZe1] of Y. Long and E. Zehnder, and [Lon16].

5.3 Index properties of non-degenerate paths

In this section we establish the symplectic additivity and the homotopy invariance of the index function theory for non-degenerate symplectic paths and the homotopy classification of $\mathcal{P}_{\tau,\omega}^*(2n)$.

Lemma 1 *For any $\tau > 0$, $\omega \in \mathbf{U}$, and $\gamma \in \mathcal{P}_{\tau,\omega}^*(2n)$,*

$$i_\omega(\gamma) = i_{\overline{\omega}}(\gamma), \qquad \nu_\omega(\gamma) = \nu_{\overline{\omega}}(\gamma). \tag{1}$$

Proof: By Theorem 1.3.1, the two hypersurfaces $\mathrm{Sp}(2n)_\omega^0$ and $\mathrm{Sp}(2n)_{\overline{\omega}}^0$ completely coincide. Therefore Lemma 1 follows. □

The following Theorem 2 on the symplectic additivity of the index theory is believed and used in many papers such as [CoZ2], [LZe1], [Lon5], and [DoL1] etc. A rigorous proof was given in [Lon13] which we use here.

Theorem 2 *For $\omega \in \mathbf{U}$, suppose $\gamma_j \in \mathcal{P}^*_{\tau,\omega}(2n_j)$ for $j = 0$ and 1. Then $\gamma_0\diamond\gamma_1 \in \mathcal{P}^*_{\tau,\omega}(2n_0 + 2n_1)$ and*

$$i_\omega(\gamma_0\diamond\gamma_1) = i_\omega(\gamma_0) + i_\omega(\gamma_1), \tag{2}$$

where $\gamma_0\diamond\gamma_1(t) = \gamma_0(t)\diamond\gamma_1(t)$ for all $t \in [0, \tau]$.

Proof: Without loss of generality we suppose $\tau = 1$. Let $n = n_0 + n_1$. For $i = 0$ and 1, choose paths $\beta_i : [0, \tau] \to \mathrm{Sp}(2n_i)^*_\omega$ to connect $\gamma_i(\tau)$ to $M^+_{n_i}$ or $M^-_{n_i}$ respectively. Note that

$$(\beta_0\diamond\beta_1) * (\gamma_0\diamond\gamma_1) = (\beta_0 * \gamma_0)\diamond(\beta_1 * \gamma_1).$$

Then we obtain

$$\begin{aligned} i_\omega(\gamma_0\diamond\gamma_1) &= \Delta_1(\beta_0 * \gamma_0)/\pi + \Delta_1(\beta_1 * \gamma_1)/\pi \\ &= i_\omega(\gamma_1\diamond\gamma_0). \end{aligned} \tag{3}$$

Suppose $\beta_0(1) = M^-_{n_0}$ and $\beta_1(1) = M^-_{n_1}$. We define a path

$$\eta(t) = (M^-_{n_0}\diamond M^-_{n_1})[R(t\pi)\diamond I_{2n_0-2}\diamond R(t\pi)\diamond I_{2n_1-2}], \quad \forall\, 0 \le t \le 1.$$

Then η connects $M^-_{n_0}\diamond M^-_{n_1}$ to M^+_n within $\mathrm{Sp}(2n)^*_\omega$. Note that $\Delta_1(\eta) = 0$. Thus we obtain

$$\begin{aligned} i_\omega(\gamma_0\diamond\gamma_1) &= \Delta_1(\eta * [\beta_0\diamond\beta_1] * [\gamma_0\diamond\gamma_1])/\pi \\ &= \Delta_1(\eta)/\pi + \Delta_1(\beta_0 * \gamma_0)/\pi + \Delta_1(\beta_1 * \gamma_1)/\pi \\ &= i_\omega(\gamma_0) + i_\omega(\gamma_1). \end{aligned}$$

Other cases of the values of $\beta_0(1)$ and $\beta_1(1)$ can be proved similarly. The details are left to the readers. □

Theorem 3 *Let $\omega \in \mathbf{U}$ and $\tau > 0$. If γ_0 and $\gamma_1 \in \mathcal{P}^*_{\tau,\omega}(2n)$, then $i_\omega(\gamma_0) = i_\omega(\gamma_1)$ if and only if $\gamma_0 \sim_\omega \gamma_1$ on $[0, \tau]$. Therefore $\mathcal{P}^k_{\tau,\omega}(2n)$ with $k \in \mathbf{Z}$ defined by Lemma 2.6 give a complete homotopy classification of $\mathcal{P}^*_{\tau,\omega}(2n)$.*

Proof: Connecting $\gamma_i(\tau)$ to M^+_n or M^-_n by a path $\beta_i : [0, \tau] \to \mathrm{Sp}(2n)^*$ for $i = 0$ and 1, we have

$$\gamma_i \sim_\omega \beta_i * \gamma_i \qquad \text{on} \quad [0, \tau] \quad \text{along} \quad \beta_i, \tag{4}$$

$$i_\omega(\gamma_i) = i_\omega(\beta_i * \gamma_i), \tag{5}$$

for $i = 0$ and 1. Thus $\gamma_0 \sim_\omega \gamma_1$ if and only if $\beta_0 * \gamma_0 \sim_\omega \beta_1 * \gamma_1$. Since $i_\omega(\beta_0 * \gamma_0) = i_\omega(\beta_1 * \gamma_1)$ if and only if $\Delta_\tau(\beta_0 * \gamma_0) = \Delta_\tau(\beta_1 * \gamma_1)$, which especially implies $\beta_0(\tau) = \beta_1(\tau)$. Therefore the proof is complete by Lemma 2.4. □

Next we define a family of **standard smooth paths** $\{\hat{\beta}_k\}$ in $\mathcal{P}_\tau(2n)$ which was introduced by E. Zehnder and the author in [LZe1] to simplify the computations. Note that the end point of each path $\hat{\beta}_k$ is M_n^+ or M_n^-. Each $\hat{\beta}_k$ is the fundamental solution of a linear Hamiltonian system

$$\dot{y} = JB(t)y \tag{6}$$

with $B(t)$ given by

$$\hat{B}_k(t) = -J\dot{\hat{\beta}}_k(t)\hat{\beta}_k^{-1}(t). \tag{7}$$

Here and in the rest of this section, we always assume $t \in [0, \tau]$. We define these $\hat{\beta}_k$s in two cases according to $n = 1$ and $n \geq 2$.

Case $n = 1$: Define $\omega(t) = \frac{1}{2}(1 + \cos(2\pi\frac{t}{\tau}))$, and

$$\omega_1(t) = \begin{cases} \omega(t) & 0 \leq t \leq \frac{\tau}{2}, \\ 0 & \frac{\tau}{2} \leq t \leq \tau, \end{cases}$$

$$\omega_2(t) = \begin{cases} 0 & 0 \leq t \leq \frac{\tau}{2}, \\ \omega(t) & \frac{\tau}{2} \leq t \leq \tau. \end{cases}$$

Denote $K = \begin{pmatrix} 0 & 1 \\ 1 & 0 \end{pmatrix}$. We define

$$\hat{\beta}_0(t) = D(2^{t/\tau}), \quad \hat{B}_0 = (\ln 2)K. \tag{8}$$

If $k \in 2\mathbf{Z} + 1$, we define

$$\hat{\beta}_k(t) = R\left(\frac{k\pi t}{\tau}\right), \quad \hat{B}_k = k\pi I. \tag{9}$$

If $k \in 2\mathbf{Z}\backslash\{0\}$, we define

$$\hat{\beta}_k(t) = D(2^{\omega_2(t/\tau)})R(-\omega_1(t/\tau)k\pi), \tag{10}$$

$$\hat{B}_k(t) = -\frac{k\pi}{\tau}\dot{\omega}_1(t/\tau)I + \frac{\ln 2}{\tau}\dot{\omega}_2(t/\tau)K. \tag{11}$$

We refer to Remark 6.3.14 below for further discussion of this case.

Case $n \geq 2$: Define $n \times n$ matrices

$$\begin{aligned} X_k &= \mathrm{diag}(0, (k - n + 2)\pi, \pi, \ldots, \pi), \\ Y &= \mathrm{diag}(\ln 2, 0, \ldots, 0), \\ Z_k &= \mathrm{diag}((k - n + 1)\pi, \pi, \ldots, \pi). \end{aligned}$$

Let 0 denote the zero matrix.

If $(-1)^{n+k} = -1$, i.e., n and k possess different parities, we define

$$\hat{\beta}_k(t) = \exp\left(\frac{t}{\tau} J\hat{B}_k\right), \tag{12}$$

$$\hat{B}_k = \begin{pmatrix} X_k & Y \\ Y & X_k \end{pmatrix}. \tag{13}$$

If $(-1)^{n+k} = 1$, i.e., n and k possess the same parities, we define

$$\hat{\beta}_k(t) = \exp(tJ\hat{B}_k), \tag{14}$$

$$\hat{B}_k = \begin{pmatrix} Z_k & 0 \\ 0 & Z_k \end{pmatrix}. \tag{15}$$

We also define a family of **standard zigzag paths** $\hat{\alpha}_k$ as follows. Using paths defined in (1.2) and (1.3), we define

$$\hat{\alpha}_0(t) = \xi_+^{\diamond n}(t), \tag{16}$$

$$\hat{\alpha}_k(t) = [(D(2)\phi_{1,k\pi}) * \xi_+]\diamond\xi_+^{\diamond(n-1)}(t), \quad \forall k \in \mathbf{Z}\backslash\{0\}. \tag{17}$$

Then for any $\omega \in \mathbf{U}$ and $k \in \mathbf{Z}$, we have $\hat{\alpha}_k \in \mathcal{P}_{\tau,\omega}^*(2n)$ and

$$i_\omega(\hat{\alpha}_k) = k, \tag{18}$$

$$\hat{\alpha}_k(\tau) = M_n^{\pm}, \qquad \text{if } (-1)^k = \pm 1. \tag{19}$$

The following proposition is obtained by direct computations.

Proposition 4 *For any $\omega \in \mathbf{U}$, $\tau > 0$, and $k \in \mathbf{Z}$, it holds that $\hat{\alpha}_k, \hat{\beta}_k \in \mathcal{P}_{\tau,\omega}^k(2n)$, $\hat{\beta}_k$ is C^1, and $\hat{B}_k \in C(S_\tau, \mathcal{L}_s(\mathbf{R}^{2n}))$ with $S_\tau = \mathbf{R}/(\tau\mathbf{Z})$.*

Equivalently, Theorem 3 shows that for every $\gamma \in \mathcal{P}_{\tau,\omega}^*(2n)$ there is one and only one $k \in \mathbf{Z}$ such that $\gamma \sim \hat{\alpha}_k$. This can also be used to give the definition of i_ω on $\mathcal{P}_{\tau,\omega}^*(2n)$ as in Section 2 to replace Definition 2.7.

5.4 Perturbations of degenerate paths

For $\omega \in \mathbf{U}$ and $\tau > 0$, fix a $\gamma \in \mathcal{P}_{\tau,\omega}^0(2n)$. We apply Theorem 1.7.3 and Definition 1.8.6 to $M = \gamma(\tau)$ and obtain

$$P\gamma(\tau)P^{-1} = H_1\diamond\cdots\diamond H_p\diamond K_1\diamond\cdots\diamond K_q\diamond M_0 \equiv N, \tag{1}$$

for some $P \in \mathrm{Sp}(2n)$ and non-negative integers p and q with $p+q > 0$ and $p+q \le n$, where $H_i \in \mathrm{Sp}(2h_i)_\omega^0$ for $1 \le i \le p$ is a first type normal form, $K_j \in \mathrm{Sp}(2k_j)_\omega^0$ for

$1 \le j \le q$ is a second type normal form of the eigenvalue ω, $M_0 \in \mathrm{Sp}(2r)^*_\omega$, and $\sum_{i=1}^{p} h_i + \sum_{j=1}^{q} k_j + r = n$. Setting $h_0 = h_0 = 0$, we define

$$m_i = \sum_{s=0}^{i-1} h_s + 1, \qquad \text{for } 1 \le i \le p,$$

$$m_{p+2j-1} = \sum_{s=0}^{p} h_s + \sum_{s=0}^{j} k_s + 1, \quad \text{for } 1 \le j \le q,$$

$$m_{p+2j} = \sum_{s=0}^{p} h_s + \sum_{s=0}^{j} k_s + k_j, \quad \text{for } 1 \le j \le q.$$

Let Σ_0 be the subset of $\mathrm{Sp}(2n)^0_\omega$ which contains all matrices M satisfying

$$\nu_\omega(M) > \nu_\omega(\gamma).$$

Here Σ_0 is empty if $\nu_\omega(\gamma) = 2n$. Let Σ_1 be the path connected component of $\mathrm{Sp}(2n)^0_\omega \backslash \Sigma_0$ which contains $\gamma(\tau)$. For $\epsilon > 0$ small enough, let $B_\epsilon(\gamma(\tau))$ be the open ball in $\mathrm{Sp}(2n)$ centered at $\gamma(\tau)$ with radius ϵ, here the metric on $\mathrm{Sp}(2n)$ is the one induced from that of $\mathbf{R}^{4n^2}$. Choose $\epsilon > 0$ to be sufficiently small so that $\overline{B_\epsilon(\gamma(\tau))}$ is contractible and possesses no intersection with $\mathrm{Sp}(2n)^0_\omega \backslash \Sigma_1$.

Let $\theta_0 \in (0, \frac{\pi}{8n})$. For $s_i \in [-1, 1]$ with $1 \le i \le p + 2q$, we define

$$Q(s_1, \ldots, s_{p+2q}) \equiv \gamma(\tau) P^{-1} R_{m_1}(s_1\theta_0) \cdots R_{m_{p+2q}}(s_{p+2q}\theta_0) P. \tag{2}$$

Then for all $s_i \in [-1, 1]\backslash\{0\}$ with $1 \le i \le p + 2q$,

$$Q(s_1, \ldots, s_{p+2q}) \in \mathrm{Sp}(2n)^*_\omega \cap B_\epsilon(\gamma(\tau)), \tag{3}$$

$$Q(s_1, \ldots, s_{p+2q}) P^{-1} R_{m_k}(-s_k\theta_0) P \in \mathrm{Sp}(2n)^0_\omega, \quad 1 \le k \le p + 2q, \tag{4}$$

$$\nu_\omega(Q(s_1, \ldots, s_{p+2q}) P^{-1} R_{m_k}(-s_k\theta_0) P) = c_k, \quad 1 \le k \le p + 2q, \tag{5}$$

where the constant $c_k = 1$ or 2.

For $t_0 \in (0, \tau)$, let $\rho \in C^2([0, \tau], [0, 1])$ such that $\rho(t) = 0$ for $0 \le t \le t_0$, $\dot\rho(t) \ge 0$ for $0 \le t \le \tau$, $\rho(\tau) = 1$, and $\dot\rho(\tau) = 0$. Whenever $t_0 \in (0, \tau)$ is sufficiently close to τ, for any $(s, t) \in [-1, 1] \times [0, \tau]$ the paths

$$\gamma_s(t) = \gamma(t) P^{-1} R_{m_1}(s\rho(t)\theta_0) \cdots R_{m_{p+2q}}(s\rho(t)\theta_0) P \tag{6}$$

satisfy the condition that γ_s converges to γ in $C^1([0, \tau], \mathrm{Sp}(2n))$ as $s \to 0$, and

$$\begin{cases} \gamma_0 &= \gamma, \\ \gamma_s(t) &= \gamma(t), \qquad \forall\, 0 \le t \le t_0, \quad s \in [-1, 1], \\ \gamma_s(t) &\in B_\epsilon(\gamma(\tau)), \quad \forall\, t_0 \le t \le 1, \quad s \in [-1, 1], \\ \nu_\omega(\gamma_s) &= 0 \qquad \forall s \in [-1, 1]\backslash\{0\}, \\ i_\omega(\gamma_s) &= i_\omega(\gamma_{s'}), \quad \forall s, s' \in [-1, 1] \text{ with } ss' > 0. \end{cases} \tag{7}$$

Note that the matrix P and the normal form $\diamond$-product N in (1) of $\gamma(\tau)$ need not be uniquely determined by the matrix γ. The path γ_s depends not only on $\gamma(\tau)$ but also on the choices of P and N. Nevertheless in the later sections we shall see that $i_\omega(\gamma_s)$ for any $s \neq 0$ is uniquely determined by γ.

The following is the most important property of these rotational perturbation paths.

Theorem 1 ([Lon5], [Lon13], [Lon16]). *For any $\omega \in \mathbf{U}$, $\tau > 0$, $\gamma \in \mathcal{P}^0_{\tau,\omega}(2n)$, and $0 < s \leq 1$, the rotational perturbation paths defined by* (6) *satisfy*

$$i_\omega(\gamma_s) - i_\omega(\gamma_{-s}) = \nu_\omega(\gamma). \tag{8}$$

Proof: Without loss of generality, we assume $\tau = 1$. Fix $s \in (0, 1]$. For $0 \leq k \leq p + 2q$ define

$$\begin{aligned}\alpha_k(t) &= \gamma(t)P^{-1}R_{m_1}(s\rho(t)\theta_0)\cdots R_{m_k}(s\rho(t)\theta_0)\\ &\quad R_{m_{k+1}}(-s\rho(t)\theta_0)\cdots R_{m_{p+2q}}(-s\rho(t)\theta_0)P.\end{aligned}$$

Then by definition, we have $\alpha_k \in \mathcal{P}^*_{1,\pm}(2n)$ for $0 \leq k \leq p + 2q$, and $\alpha_0 = \gamma_{-s}$, $\alpha_{p+2q} = \gamma_s$.

In Figure 1, the case of $p+2q = 2$ with three paths α_0, α_1, and α_2 are shown. The dark curves describe intuitively a part of the singular hypersurface $\mathrm{Sp}(2n)^0_\omega$.

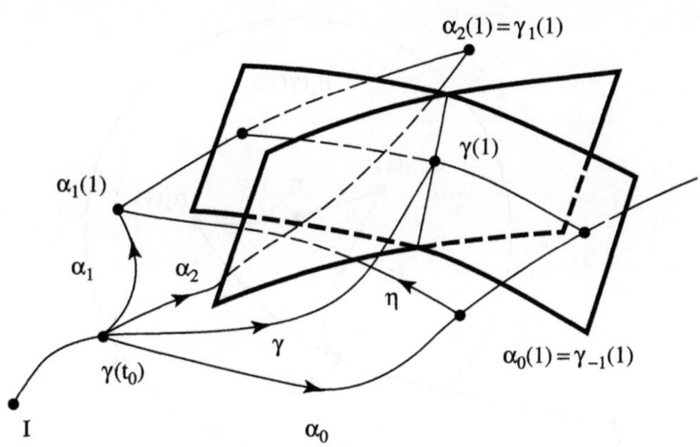

Figure 1: *The case of $p + 2q = 2$ with paths α_0, α_1, and α_2*

For $1 \leq k \leq p + 2q$, define

$$a_k = \nu_\omega(\alpha_k(1)P^{-1}R_{m_k}(-s\theta_0)P).$$

By (5), the constant a_k only takes the value 1 or 2. By the properties of normal forms, especially Remark 1.8.7, we obtain

$$\nu_\omega(\gamma) = \sum_{k=1}^{p+2q} \nu_\omega(\alpha_k(1)P^{-1}R_{m_k}(-s\theta_0)P) = \sum_{k=1}^{p+2q} a_k. \tag{9}$$

Assume the following equalities hold,

$$i_\omega(\alpha_k) - i_\omega(\alpha_{k-1}) = a_k, \quad \text{for } 1 \le k \le p+2q. \tag{10}$$

Summing (10) up from $k = 1$ to $k = p + 2q$ yields

$$i_\omega(\gamma_s) - i_\omega(\gamma_{-s}) = i_\omega(\alpha_{p+2q}) - i_\omega(\alpha_0) = \sum_{k=1}^{p+2q} a_k = \nu_\omega(\gamma).$$

Therefore the proof of (8) is reduced to that of (10).

Fix a $k \in \{1, \ldots, p+2q\}$. Let $\eta(t) = \alpha_{k-1}(1)P^{-1}R_{m_k}(2ts\theta_0)P$ for $t \in [0,1]$. Then by definition we have $\eta(0) = \alpha_{k-1}(1)$, $\eta(1) = \alpha_k(1)$, $\eta(1/2) = \alpha_k(1)P^{-1}R_{m_k}(-s\theta_0)P \in \mathrm{Sp}(2n)^0_\omega$, $\nu_\omega(\eta(1/2)) = a_k$, and $\eta(t) \in \mathrm{Sp}(2n)^*_\omega$ for $t \in [0,1]\setminus\{1/2\}$. Note that $B_\epsilon(\gamma(1))$ is contractible if $\epsilon > 0$ is small enough. By (5) and (7) we obtain that $\alpha_k|_{[t_0,1]}$ is homotopic to the joint path of $\alpha_{k-1}|_{[t_0,1]}$ and η with fixed end points in the sense of Definition 0.2. Therefore we obtain $\alpha_k \sim \eta * \alpha_{k-1}$ with fixed end points in the sense of Definition 0.3 (cf. Figure 2). Then (10) becomes

$$i_\omega(\eta * \alpha_{k-1}) - i_\omega(\alpha_{k-1}) = a_k. \tag{11}$$

The proof of (11) contains two cases according to the value of a_k.

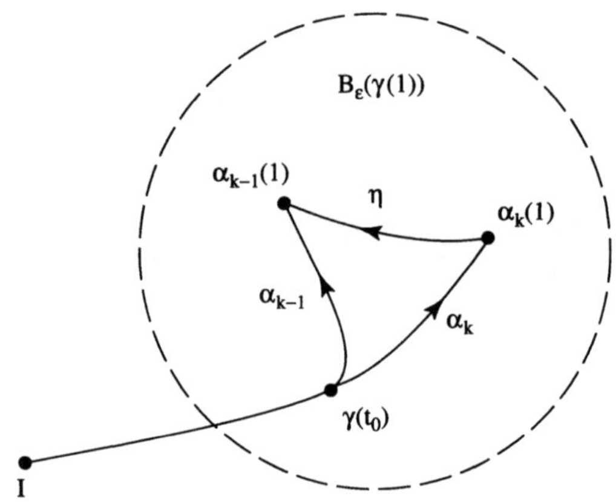

Figure 2: *The case of* $\alpha_k \sim \eta * \alpha_{k-1}$

Case 1 $a_k = 2$.

We notice that $\nu_\omega(\eta(1/2)) = a_k = 2$ and

$$\eta * \alpha_{k-1}(t) = \eta(1/2)P^{-1}R_{m_k}((t-3/4)4s\theta_0)P$$

for $1/2 \le t \le 1$. For $0 < |t - 3/4| \le 1/4$, we get

$$\nu_\omega\left(\eta\left(\frac{1}{2}\right)\right) - \nu_\omega\left(\eta\left(\frac{1}{2}\right)P^{-1}R_{m_k}\left(\left(t-\frac{3}{4}\right)4s\theta_0\right)P\right) = 2.$$

Hence from Theorem 2.6.3, when $0 < \theta < 2\pi$, $\eta(1/2)P^{-1}R_{m_k}(\theta)P$ belongs to the same path connected component of $\mathrm{Sp}(2n)^*_\omega$. For the new paths constructed below in the proof of this case, we refer to Figure 3.

Therefore we get

$$\eta * \alpha_{k-1} \sim \beta * \alpha_{k-1}, \tag{12}$$

where $\beta(t) = \alpha_{k-1}(1)P^{-1}R_{m_k}(2t\pi)P$ for $0 \le t \le 1$. By the discussion at the end of Section 3, we get $\alpha_{k-1} \sim \hat{\alpha}_j$ for some zigzag standard path $\hat{\alpha}_j$ defined there. So there exists a path $\xi \in C([0,1], \mathrm{Sp}(2n)^*_\omega)$ such that $\xi(0) = \hat{\alpha}_j(1)$ and $\xi(1) = \alpha_{k-1}(1)$ hold. Thus $\alpha_{k-1} \sim \xi * \hat{\alpha}_j$ and

$$\beta * \alpha_{k-1} \sim \beta * \xi * \hat{\alpha}_j \sim \xi^{-1} * \beta * \xi * \hat{\alpha}_j \equiv \zeta$$

hold. Here because $\beta * (\xi * \hat{\alpha}_j) \sim (\beta * \xi) * \hat{\alpha}_j$, we simply use the notation $\beta * \xi * \hat{\alpha}_j$.

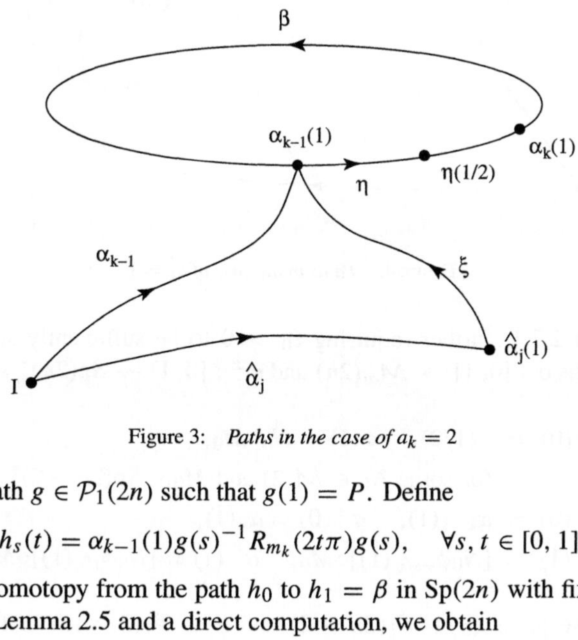

Figure 3: *Paths in the case of $a_k = 2$*

Choose a path $g \in \mathcal{P}_1(2n)$ such that $g(1) = P$. Define

$$h_s(t) = \alpha_{k-1}(1)g(s)^{-1}R_{m_k}(2t\pi)g(s), \quad \forall s, t \in [0,1].$$

This gives a homotopy from the path h_0 to $h_1 = \beta$ in $\mathrm{Sp}(2n)$ with fixed end points. Therefore by Lemma 2.5 and a direct computation, we obtain

$$\Delta_1(\beta) = \Delta_1(h_0) = 2\pi.$$

Thus by a further direct computation on smooth standard paths, we obtain $\zeta(1) = \hat{\alpha}_j(1) = \hat{\alpha}_{j+2}(1)$ and

$$\Delta_1(\zeta) = \Delta_1(\hat{\alpha}_j) + \Delta_1(\beta) = (j+2)\pi = \Delta_1(\hat{\alpha}_{j+2}).$$

Then from Lemma 2.4, we get that $\zeta \sim \hat{\alpha}_{j+2}$. Combining with (12) we get $\eta * \alpha_{k-1} \sim \hat{\alpha}_{j+2}$, and hence by Theorem 3.3,

$$i_\omega(\eta * \alpha_{k-1}) - i_\omega(\alpha_{k-1}) = i_\omega(\hat{\alpha}_{j+2}) - i_\omega(\hat{\alpha}_j) = 2.$$

Thus (11) holds in this case.

Case 2 $a_k = 1$.

If $n = 1$, the equality (11) follows from the Definitions 1.1 and 1.3, and the definition of η.

Next we study the case for $n \geq 2$. For the new paths constructed below in the proof of this case, we refer to Figure 4.

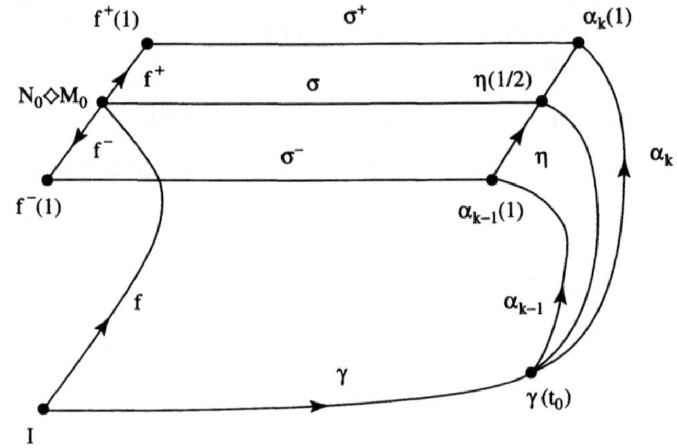

Figure 4: *Paths in the case of* $a_k = 1$

By Theorem 2.7.1, further requiring $\theta_0 > 0$ to be sufficiently small, there are continuous paths $\sigma : [0, 1] \to \mathcal{M}_\omega(2n)$ and $\sigma^\pm : [0, 1] \to \mathrm{Sp}(2n)^*_\omega$ such that

$$\begin{aligned} \sigma(0) &= \eta(1/2), \quad \sigma(1) = N_0 \diamond M_0, \\ &\quad\text{for some } N_0 \in \mathcal{M}(2) \text{ and } M_0 \in \mathrm{Sp}(2n-2)^*_\omega, \\ \sigma^-(0) &= \alpha_{k-1}(1), \quad \sigma^+(0) = \alpha_k(1), \\ \sigma^-(1) &= [N_0 \phi_{-s\theta_0}(1)] \diamond M_0, \quad \sigma^+(1) = [N_0 \phi_{s\theta_0}(1)] \diamond M_0. \end{aligned}$$

Now we choose paths $\xi_0 \in \mathcal{P}_1(2n-2)$ and $\xi \in \mathcal{P}_1(2)$ such that $\xi_0(1) = M_0$, $\xi(1) = N_0$, and $\xi(t) \in \mathrm{Sp}(2)^0_\omega$ for all $t \in [0, 1]$. Define $\eta_{1/2}(t) = \eta(t/2)$ for $0 \leq t \leq 1$. Define

$$\zeta(t) = \begin{cases} R(4m\pi t), & 0 \leq t \leq \frac{1}{2}, \\ \xi(2t-1), & \frac{1}{2} \leq t \leq 1, \end{cases}$$

where $m \in \mathbf{Z}$ is chosen so that

$$\Delta_1(\zeta) + \Delta_1(\xi_0) = \Delta_1(\sigma * \eta_{1/2} * \alpha_{k-1}).$$

Define paths $\chi^{\pm}$ and splitting paths $f^{\pm}$ and f by

$$\begin{aligned} \chi^{\pm} &= (N_0\phi_{\pm s\theta_0}) * \zeta : [0,1] \to \mathrm{Sp}(2), \\ f^{\pm} &= \chi^{\pm}\diamond\xi_0 : [0,1] \to \mathrm{Sp}(2n), \\ f &= \zeta\diamond\xi_0 : [0,1] \to \mathrm{Sp}(2n). \end{aligned}$$

Then

$$\begin{cases} f^{\pm} = [(N_0\phi_{\pm s\theta_0})\diamond I_{2n-2}] * f, \\ f^{+}(1) = \sigma^{+} * \eta * \alpha_{k-1}(1), \\ f^{-}(1) = \sigma^{-} * \alpha_{k-1}(1), \\ \Delta_1(f^{+}) = s\theta_0 + \Delta_1(f) \\ \quad = s\theta_0 + \Delta_1(\sigma * \eta_{1/2} * \alpha_{k-1}) \\ \quad = \Delta_1(\sigma^{+} * \eta * \alpha_{k-1}), \\ \Delta_1(f^{-}) = -s\theta_0 + \Delta_1(f) \\ \quad = -s\theta_0 + \Delta_1(\sigma * \eta_{1/2} * \alpha_{k-1}) \\ \quad = \Delta_1(\sigma^{-} * \alpha_{k-1}). \end{cases} \tag{13}$$

Thus by Lemma 2.4,

$$f^{+} \sim \sigma^{+} * \eta * \alpha_{k-1}, \quad f^{-} \sim \sigma^{-} * \alpha_{k-1}.$$

Therefore

$$\begin{aligned} i_\omega(\eta * \alpha_{k-1}) - i_\omega(\alpha_{k-1}) &= i_\omega(\sigma^{+} * \eta * \alpha_{k-1}) - i_\omega(\sigma^{-} * \alpha_{k-1}) \\ &= i_\omega(f^{+}) - i_1(f^{-}) \\ &= i_\omega(\chi^{+}) - i_\omega(\chi^{-}), \end{aligned} \tag{14}$$

where we have used Theorem 3.2 on $f^{\pm}$. Now by (13) and our study of the case of $n = 1$, the right-hand side of (14) must be 1. Thus (11) holds in this case.

The proof of Theorem 1 is complete. □

Based upon our rotational perturbation paths, the index function for degenerate symplectic paths can be defined as follows.

Definition 2 ([Lon5], [Lon13], [Lon16]). For any $\tau > 0$, $\omega \in \mathbf{U}$, and $\gamma \in \mathcal{P}^0_{\tau,\omega}(2n)$, define

$$i_\omega(\gamma) = i_\omega(\gamma_{-s}) \qquad \forall\, s \in (0,1], \tag{15}$$

where γ_{-s} is the rotational perturbation path of γ defined by (6).

We shall prove in the next chapter that Definition 2 is well defined, i.e. the index $i_\omega(\gamma)$ of (15) depends only on γ and ω.

Definition 3 For any $\omega \in \mathbf{U}$, $\tau > 0$, and $\gamma \in \mathcal{P}_\tau(2n)$, Definitions 0.1, 2.7, and the above Definition 2 assign a pair of integers

$$(i_\omega(\gamma), \nu_\omega(\gamma)) \in \mathbf{Z} \times \{0, 1, \ldots, 2n\}$$

to γ, which we call the **index function** of γ at ω. We also call $i_\omega(\gamma)$ the **ω-index** of γ, and $\nu_\omega(\gamma)$ the **ω-nullity** of γ.

Definition 4 For $\tau > 0$, let $B \in C(\mathbf{R}/(\tau\mathbf{Z}), \mathcal{L}_s(\mathbf{R}^{2n}))$. By Corollary 3.1.4, the fundamental solution γ_B of the linear Hamiltonian system

$$\dot{x}(t) = JB(t)x(t) \tag{16}$$

satisfies $\gamma_B \in \mathcal{P}_\tau(2n)$, and is called the **associated symplectic path** of B. For $\omega \in \mathbf{U}$, we define the index function of B via that of γ_B:

$$(i_\omega(B), \nu_\omega(B)) = (i_\omega(\gamma_B), \nu_\omega(\gamma_B)). \tag{17}$$

As usual, eigenvalues of $\gamma_B(\tau)$ are called **Floquet multipliers** of the system (16) (or B).

Let $H \in C^1(\mathbf{R}/(\tau\mathbf{Z}) \times \mathbf{R}^{2n}, \mathbf{R})$. Suppose x is a τ-periodic solution of the Hamiltonian system

$$\dot{x}(t) = JH'(t, x(t)), \tag{18}$$

such that H is C^2 along the orbit $x(\mathbf{R})$ of x. The **associated symplectic path** of x is defined to be the fundamental solution $\gamma_x \equiv \gamma_B$ of the linearized Hamiltonian system (16) with $B(t) = H''(t, x(t))$ for all t. For $\omega \in \mathbf{U}$, we define the index function of x via that of its associated symplectic path γ_x:

$$(i_\omega(x), \nu_\omega(x)) = (i_\omega(\gamma_x), \nu_\omega(\gamma_x)). \tag{19}$$

As usual, eigenvalues of $\gamma_x(\tau)$ are called **Floquet multipliers** of the solution x of the system (18).

Remark 5 In this chapter we have used the method of [CoZ2], [LZe1], [Lon5], [Lon13], and [Lon16] to define the index function theory. Such a method is completely topological. Note that in [SaZ1] and [SaZ2] such an index theory was also defined for 1-non-degenerate paths in $\mathcal{P}_\tau(2n)$. But there, instead of the function $\Delta(t)$ and $\Delta_\tau(\gamma)$ defined in Lemma 2.1, the authors used the function $\hat{\Delta}(t)$ and $\hat{\Delta}_\tau(\gamma)$ defined by (2.4.25) and (2.4.26).

Other ways of defining this index theory can be found in [ZhL1] and [LZu1], where the spectral flow method was used. We also refer to [RoS2] for other index theories with a different index for degenerate paths via the spectral method. In [LZu2], it was proved that $\mathrm{Sp}(2n)_\omega^0$ for any $\omega \in \mathbf{U}$ is a codimension 1 oriented cycle, and it induces a topological intersection number definition of this index theory. Compared with these different methods, the method described in this book is rather elementary and intuitive.

Chapter 6

Properties of index functions

In this chapter, we study basic properties of the index function theory for elements in $\mathcal{P}_\tau(2n)$ defined in Chapter 5, study its relation with the Morse indices of the reduced functional of corresponding linear Hamiltonian systems, and derive its axiom characterization. These results form the basis for our later study of the iteration theory and Morse theoretical applications to nonlinear Hamiltonian systems.

6.1 Index functions and Morse indices

Let $\omega \in \mathbf{U}$. We consider the boundary value problem of the linear Hamiltonian system

$$\dot{y}(t) = JB(t)y(t), \tag{1}$$

$$y(\tau) = \omega y(0), \tag{2}$$

with $B \in C(S_\tau, \mathcal{L}_s(\mathbf{R}^{2n}))$ and $S_\tau = \mathbf{R}/(\tau\mathbf{Z})$ with $\tau > 0$. By Corollary 3.1.4, the fundamental solution γ_B of the system (1) belongs to $\mathcal{P}_\tau(2n)$. Thus we can define the index function of the problem (1)–(2) by that of γ_B:

$$(i_\omega(B), \nu_\omega(B)) = (i_\omega(\gamma_B), \nu_\omega(\gamma_B)). \tag{3}$$

In this section we study the relationship between the index function of the boundary value problem (1)–(2) and the Morse indices of the corresponding functionals defined on the truncated loop space. For this purpose, we define

$$\hat{\mathcal{P}}_\tau(2n) = \{\gamma \in C^1([0,\tau], \mathrm{Sp}(2n)) \mid \gamma(0) = I, \dot{\gamma}(1) = \dot{\gamma}(0)\gamma(1)\}, \tag{4}$$

$$\hat{\mathcal{P}}^*_\tau(2n) = \hat{\mathcal{P}}_\tau(2n) \cap \mathcal{P}^*_\tau(2n), \quad \hat{\mathcal{P}}^0_\tau(2n) = \hat{\mathcal{P}}_\tau(2n)\backslash\hat{\mathcal{P}}^*_\tau(2n). \tag{5}$$

Note that $B = -J\dot{\gamma}(\cdot)\gamma^{-1}(\cdot) \in C(S_\tau, \mathcal{L}_s(\mathbf{R}^{2n}))$ if and only if $\gamma \in \hat{\mathcal{P}}_\tau(2n)$, i.e., the path family $\hat{\mathcal{P}}_\tau(2n)$ is formed by fundamental solutions of all the linear Hamiltonian systems (1) with continuous symmetric and τ-periodic coefficients. The topology of $\hat{\mathcal{P}}^*_\tau(2n)$ is the $C^1([0,\tau], \mathrm{Sp}(2n))$-topology induced from that of $\mathrm{Sp}(2n)$. Note that $\hat{\mathcal{P}}_\tau(2n)$ is dense in $\mathcal{P}_\tau(2n)$.

For $\tau > 0$, define L to be the Hilbert space $L^2([0,\tau], \mathbf{C}^{2n})$ with the usual L^2 inner product

$$\langle x, y\rangle_L = \int_0^\tau \langle x(t), y(t)\rangle dt, \quad \forall x, y \in L, \tag{6}$$

and $W = W^{1,2}(S_\tau, \mathbf{C}^{2n})$ as a subspace of L, where $\langle\cdot,\cdot\rangle$ is the inner product of $\mathbf{C}^{2n}$. For $\omega = \exp(\theta\sqrt{-1}) \in \mathbf{U}$, define W^ω to be the subspace $\{y \in W \mid y(\tau) = \omega y(0)\}$ of L. Any $y \in W^\omega$ has the form

$$y(t) = \sum_{k\in\mathbf{Z}} e^{\sqrt{-1}(\theta+2k\pi)t/\tau}\xi_k, \qquad \xi_k \in \mathbf{C}^{2n},$$

where

$$\sum_{k\in\mathbf{Z}} ((\theta + 2k\pi)^2 + 1)|\xi_k|^2 < +\infty.$$

Define $W_k^\omega = W_k^{\omega,+} \oplus W_k^{\omega,-}$ with

$$W_k^{\omega,\pm} = e^{\sqrt{-1}(\theta+2k\pi)t/\tau}(J \pm \sqrt{-1}I)\mathbf{R}^{2n}. \tag{7}$$

Then

$$W^\omega = \oplus W_k^\omega. \tag{8}$$

Define $A = -J\frac{d}{dt} : W^\omega \to L$. Then A is continuous and self-adjoint, i.e., it satisfies

$$\langle Ax, y\rangle_L = \langle x, Ay\rangle_L, \qquad \forall x, y \in W^\omega.$$

Viewing A as from the subspace W^ω of L to L, we have

$$\sigma(A) = \{\lambda_k^\pm \mid k \in \mathbf{Z}\}, \qquad \lambda_k^\pm = \pm\frac{\theta + 2k\pi}{\tau}, \tag{9}$$

each eigenvalue $\lambda_k^\pm$ of A has multiplicity $2n$ and the corresponding eigenspace is $W_k^{\omega,\pm}$.

Given $B \in C(S_\tau, \mathcal{L}_s(\mathbf{R}^{2n})) \subset C(S_\tau, \mathcal{L}_s(\mathbf{C}^{2n}))$, it induces a symmetric operator on L by

$$\langle Bx, y\rangle_L = \int_0^\tau \langle B(t)x(t), y(t)\rangle dt, \quad \forall x, y \in L. \tag{10}$$

Then solutions of (1)–(2) are in one to one correspondence with critical points of the functional

$$f_\omega(y) = \frac{1}{2}\langle (A - B)y, y\rangle_L, \qquad \forall y \in W^\omega \subset L. \tag{11}$$

Then $f_\omega : W^\omega \to \mathbf{R}$ is smooth in the topology of L.

Using the saddle point reduction method of Chapter 4, we obtain a subspace

$$Z^\omega = \oplus_{|k|\le k_0} W_k^\omega \tag{12}$$

with a sufficiently large $k_0 \in \mathbf{N}$, an injection map $u_\omega \in C^\infty(Z^\omega, L)$, and a smooth functional $a_\omega \in C^\infty(Z^\omega, \mathbf{R})$ defined by

$$a_\omega(z) = f_\omega(u_\omega(z)), \qquad \forall z \in Z^\omega. \tag{13}$$

Let $2d_\omega = \dim_{\mathbf{C}} Z^\omega$. Note that the origin of Z^ω as a critical point of a_ω corresponds to the origin of W^ω as a critical point of f_ω. Denote by m^* for $* = +, 0$, and $-$, the positive, null, and negative Morse indices of the functional a_ω at the origin respectively, i.e., the total multiplicities of positive, zero, and negative eigenvalues of the $2d_\omega \times 2d_\omega$ matrix $a''_\omega(0)$ respectively.

The following is the main result of this section.

Theorem 1 (C. Conley and E. Zehnder [CoZ2], Y. Long and E. Zehnder [LZe1], Y. Long [Lon5], [Lon13], [Lon16]). *For any $\omega \in \mathbf{U}$,*

$$\begin{cases} m^- &= d_\omega + i_\omega(B), \\ m^0 &= \nu_\omega(B), \\ m^+ &= d_\omega - i_\omega(B) - \nu_\omega(B). \end{cases} \tag{14}$$

The rest of this section is devoted to the proof of this theorem. *We shall only give the proof for the case of $\omega = 1$ in Theorem* 1, and leave other cases to the readers.

When $\omega = 1$, the boundary condition (2) becomes

$$y(\tau) = y(0). \tag{15}$$

We then work on real spaces $L = L^2(S_\tau, \mathbf{R}^{2n})$ and $W = W^{1,2}(S_\tau, \mathbf{R}^{2n})$. Using the inner product $a \cdot b$ in $\mathbf{R}^{2n}$ to replace $\langle\cdot,\cdot\rangle$ in $\mathbf{C}^{2n}$ in the above notation, we get the corresponding operators and functionals. Note that the Hamiltonian function corresponding to (1) is

$$H(t,x) = \frac{1}{2}B(t)x \cdot x \qquad \forall\, (t,x) \in \mathbf{R} \times \mathbf{R}^{2n}. \tag{16}$$

It satisfies the condition (H1) and (H3) in Section 4.2. For simplicity, in the following we omit all the sub and sup indices for $\omega = 1$, and simply write $f, Z, a, u : Z \to W$, and $2d = \dim Z$, etc.

The proof of this theorem is rather long, and is completed by the following sequence of lemmas. Without loss of generality, we suppose $\tau = 1$ in the remainder of this section. We start from the basic properties of the operator $A - B$.

Lemma 2 1° *The operator $A - B$ defined on* dom$A$ *is linear self-adjoint, has compact resolution, and then its spectrum is a point spectrum, i.e., $\sigma(A - B) = \sigma_p(A - B)$.*

2° *$0 \in \sigma(A - B)$ if and only if $\nu_1(B) > 0$. The multiplicity of the eigenvalue 0 of $A - B$ is precisely $\nu_1(B)$. When $\nu_1(B) = 0$, the operator $A - B$ possesses a bounded inverse.*

3° *If* $B(t) \equiv B$ *is independent of time* t*, the operator* B *defined by* (10) *is commutative with the projection map* $P : L \to Z$ *defined by* (4.3.2).

Proof: 1° and 3° follow from the definitions of A, B, and P. Suppose K is the kernal of $A - B$. Then $x \in K$ is equivalent to the fact that x is a 1-periodic solution of (1). Thus the total multiplicity of the eigenvalue 0 of the operator $A - B$ is precisely the dimension of the 1-periodic solution space of the system (1). By the Floquet theory (cf. [CoL1]) this dimension is precisely $\nu_1(B)$. This proves 2°. □

Lemma 3 *Let* $B \in C([0, 1] \times S_\tau, \mathcal{L}_s(\mathbf{R}^{2n}))$ *and denote by* $B_s(\cdot) = B(s, \cdot)$*. Denote the Morse indices of the functional* a_s *on* Z *corresponding to the system* (1) *with* $B(t) \equiv B_s$ *at* $z = 0$ *by* m_s^-, m_s^0, *and* m_s^+, *where* Z *and* $a_s : Z \to \mathbf{R}$ *are the finite dimensional space with large enough dimension* $\dim Z = 2d$ *and the functional on* Z *obtained by the saddle point reduction method for all* $s \in [0, 1]$ *with the same* Z*. Suppose*

$$m_s^0 = m_0^0, \qquad \forall s \in [0, 1]. \tag{17}$$

Then

$$m_s^- = m_0^-, \quad m_s^+ = m_0^+, \quad \forall s \in [0, 1]. \tag{18}$$

Proof: Fix any $r \in [0, 1]$. For any $s \in [0, 1]$ that is sufficiently close to r, we obtain

$$m_r^- \le m_s^-, \qquad m_r^+ \le m_s^+.$$

Thus by (17) we obtain

$$2d = m_r^- + m_r^0 + m_r^+ \le m_s^- + m_s^0 + m_s^+ = 2d.$$

This proves that m_s^- and m_s^+ are locally constants. By the connectedness of [0, 1], they are also globally constants, i.e., (18) holds. □

Remark 4 Lemma 3 and Theorem 1 give the analytical meaning of our concept of homotopy in Definition 5.0.3, i.e., the homotopy allows the Hamiltonian systems to possess the same truncated Morse indices.

Lemma 5 *Suppose* $\gamma_0, \gamma_1 \in \hat{\mathcal{P}}_1(2n)$, *and* $\gamma_0 \sim_1 \gamma_1$. *Then this homotopy can be realized in* $\hat{\mathcal{P}}_1(2n)$.

Proof: Denote the homotopy from γ_0 to γ_1 in $\mathcal{P}_1(2n)$ by $\delta : [0, 1]^2 \to \mathrm{Sp}(2n)$. Near the segments $\{0\} \times [0, 1]$ and $\{1\} \times [0, 1]$ in the square $[0, 1]^2$, pushing γ_0 and γ_1 into $[0, 1]^2$ along the direction perpendicular to these two segments, without loss of generality, we can assume that for some $\epsilon \in (0, 1/8)$,

$$\delta(s, t) = \begin{cases} \gamma_0(t), & \text{if } 0 \le s \le 2\epsilon, \\ \gamma_1(t), & \text{if } 1 - 2\epsilon \le s \le 1. \end{cases}$$

We define $\varphi \in C([0,1],[0,\epsilon])$ by

$$\varphi(s) = \begin{cases} 0, & 0 \le s \le \epsilon \ \text{ or } \ 1-\epsilon \le s \le 1, \\ s-\epsilon, & \epsilon < s < 2\epsilon, \\ \epsilon, & 2\epsilon \le s \le 1-2\epsilon, \\ 1-\epsilon-s, & 1-2\epsilon < s < 1-\epsilon, \end{cases}$$

and for any $s \in [0,1]$ a function $\xi_s : [0,1] \to [0,1]$ by

$$\xi_s(t) = \begin{cases} 0, & 0 \le t \le \varphi(s), \\ \frac{1}{2}\left(1+\cos\frac{\pi(t+1-3\varphi(s))}{1-2\varphi(s)}\right), & \varphi(s) < t < 1-\varphi(s), \\ 1, & 1-\varphi(s) \le t \le 1. \end{cases}$$

Then this function is continuous on s and t in $[0,1]$. For each $s \in [0,1]$, we have that $\xi_s \in C^1([0,1],[0,1])$ is strictly increasing from $\xi_s(\varphi(s)) = 0$ to $\xi_s(1-\varphi(s)) = 1$, and satisfies $\dot{\xi}_s(\varphi(s)) = \dot{\xi}_s(1-\varphi(s)) = 0$. Now we define a new homotopy η (cf. Figure 1) as a reparametrization of δ by

$$\eta(s,t) \doteq \begin{cases} \delta(s, (1-\frac{s}{\epsilon})t + \frac{s}{\epsilon}\xi_s(t)), & 0 \le s \le \epsilon, \\ \delta(s, \xi_s(t)), & \epsilon \le s \le 1-\epsilon, \\ \delta(s, \frac{s+\epsilon-1}{\epsilon}t + \frac{1-s}{\epsilon}\xi_s(t)), & 1-\epsilon \le s \le 1. \end{cases} \tag{19}$$

Note that the second arguments on the right-hand side of (19) as functions of t are C^1, are continuous in s, and increase from 0 to 1 as t increases from 0 to 1 for each fixed s. Then $\eta \in C([0,1]^2, \mathrm{Sp}(2n))$ gives a homotopy from γ_0 to γ_1 in $\mathcal{P}_1(2n)$, η is C^1 in the variable t in an open neighborhood of the boundary of the square $[0,1]^2$, and

$$\dot{\eta}(s,1) = \dot{\eta}(s,0)\eta(s,1), \quad \forall s \in [0,1],$$

where $\dot{\eta}$ is the derivative of η with respect to t. Thus η is not C^1 in the variable t at most on a compact subregion of $(0,1)^2$.

Now, following the standard differential topology argument (for example cf. [Hir1]), we can find an approximation ζ of η such that ζ coincides with η near the boundary of $[0,1]^2$, is C^1 in the variable t for all $s \in [0,1]$, and can be chosen to be as close to η as we want. This map ζ gives a homotopy claimed by the lemma.

Another argument to obtain the smoothness in t from η is to smooth out the generating function (cf. Section 3.2) corresponding to the symplectic matrices in the parameter t, and then obtain the map ζ claimed by the lemma. □

Next under the assumption of Theorem 1, we consider the case of $\nu_1(B) = 0$ first, i.e., the non-degenerate case.

Denote by γ the fundamental solution of the system (1). Then by definition $i_1(\gamma) = i_1(B)$ and $\nu_1(\gamma) = \nu_1(B) = 0$. By the discussion of Section 5.3, there is a smooth

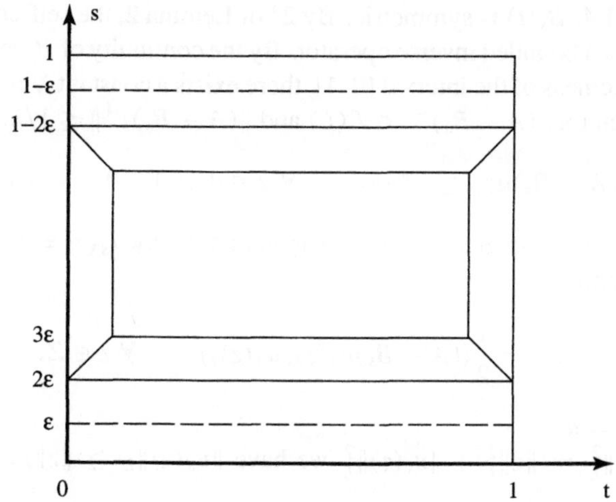

Figure 1: *The homotopy η*

standard path $\hat{\beta}_j$ which is homotopic to γ, i.e., $\hat{\beta}_j \sim \gamma$ on $[0, 1]$. By Definition 5.0.3 of the homotopy, γ can be deformed to $\hat{\beta}_j$ in the non-degenerate path set $\mathcal{P}_1^*(2n)$ continuously.

In Lemma 5, let $\gamma_0 = \gamma$ and $\gamma_1 = \hat{\beta}_j$. Then we obtain a homotopy $\zeta : [0, 1]^2 \to \mathrm{Sp}(2n)$ of γ and $\hat{\beta}_j$ in $\hat{\mathcal{P}}_1^*(2n)$. Define

$$B_s(t) = -J\dot{\zeta}(s, t)\zeta^{-1}(s, t), \qquad \forall 0 \le t \le 1. \tag{20}$$

Then by Corollary 3.1.4 we obtain $B_s \in C^1([0, 1], \mathcal{L}_s(\mathbf{R}^{2n}))$ and is continuous in $s \in [0, 1]$. Let $\gamma_s = \zeta(s, \cdot)$, which is the fundamental solution of (1) with $B = B_s$ for each $s \in [0, 1]$. By Theorem 4.3.1, for

$$H_s(t, x) = \frac{1}{2}B_s(t)x \cdot x, \quad \forall\ (t, x) \in \mathbf{R} \times \mathbf{R}^{2n}, \tag{21}$$

we obtain the corresponding space Z, functionals f_s, a_s, and the injection map u_s for $0 \le s \le 1$. Here we have used the compactness of the interval $[0, 1]$ and the continuity of $B_s(t)$ to get a uniform finite dimensional space Z with sufficiently large dimension for all $s \in [0, 1]$. Denote by m_s^-, m_s^0, and m_s^+ the Morse indices of B_s respectively.

Lemma 6 *Suppose $\nu_1(\gamma) = 0$. Then in the above discussion, $z = 0$ is an isolated critical point of a_s and*

$$m_s^+ = m^+, \quad m_s^0 = m^0 = 0, \quad m_s^- = m^-, \quad \forall\ s \in [0, 1]. \tag{22}$$

Proof: Since $\gamma_s \in \hat{\mathcal{P}}_1^*(2n)$, by Theorem 4.4.1 we obtain

$$m_s^0 = \dim\ker(\gamma_s(1) - I) = \nu_1(\gamma_s) = 0, \qquad \forall\ s \in [0, 1]. \tag{23}$$

By Corollary 3.1.4, $B_s(t)$ is symmetric. By 2° of Lemma 2, the self-adjoint operator $A - B_s$ possesses a bounded inverse operator. By the continuity of B_s in the parameter s and the compactness of the interval [0, 1], there exists a constant $\lambda > 0$ independent of $s \in [0, 1]$ such that $(A - B_s)^{-1} \in \mathcal{L}(L)$ and $\|(A - B_s)^{-1}\| \le 1/\lambda$. Therefore

$$\|(A - B_s)u\|_L \ge \lambda \|u\|_L \qquad \forall\, u \in \mathrm{dom} A, \quad 0 \le s \le 1.$$

Denote the injection map given by 5° of Theorem 4.3.1 by $u_s(z) = z + v_s(z)$, where $Pv_s(z) = 0$. Define

$$a_s(z) = \frac{1}{2}\langle (A - B_s)u_s(z), u_s(z)\rangle_L \qquad \forall\, z \in Z. \tag{24}$$

Specifically, $a_0 = a$.

From $\|u_s(z)\|_L^2 = \|z\|_L^2 + \|v_s(z)\|_L^2$ we have $\|u_s(z)\|_L \ge \|z\|_L$. Therefore we obtain

$$\|a_s'(z)\|_Z = \|(A - B_s)u_s(z)\| \ge \lambda \|z\|_Z \qquad 0 \le s \le 1.$$

Thus $z = 0$ is an isolated critical point of a_s on Z for all $s \in [0, 1]$.

Now (22) follows from (23) and Lemma 3. □

Lemma 6 reduces the study for the non-degenerate case to the study of the special smooth standard paths defined at the end of Section 5.3. A symplectic path is called an **exponential path** if it is of the form $\gamma(t) = \exp(tJB)$ for $0 \le t \le 1$ with some $B \in \mathcal{L}_s(2n)$. Such paths form a simpler subset of symplectic paths.

Lemma 7 *For any $k \in \mathbf{Z}$, denote by m_k^-, m_k^0, and m_k^+ the Morse indices of the functional a_k corresponding to the smooth standard path $\hat{\beta}_k$ defined at the end of Section 5.3. Denote $2d = \dim Z$ in the saddle point reduction. Then*

$$m_k^- = d + k, \quad m_k^0 = 0, \quad m_k^+ = d - k. \tag{25}$$

Proof: Fix $k \in \mathbf{Z}$. (23) in the proof of Lemma 6 implies $m_k^0 = 0$. Since the sum of these three Morse indices must be $2d$, it suffices to compute m_k^+. We continue our computations in two cases.

Case 1 $n = 1$ and $k \notin 2\mathbf{Z}\backslash\{0\}$ or $n \ge 2$.

In this case, the coefficient matrix function $B(t)$ of the linear system (1) corresponding to the smooth standard path $\hat{\beta}_k$ is a constant matrix $B(t) \equiv \hat{B}_k$ defined by (5.3.8), (5.3.13), or (5.3.15) respectively. In this case the Morse indices of the corresponding functional at the critical point $z = 0$ can be easily computed. Thus here we only compute the case of $n \ge 2$ and $(-1)^{n+k} = 1$, i.e., the smooth standard paths defined by (5.3.14). All the verifications of other cases are left to the readers.

By 4° of Theorem 4.3.1,

$$a_k(z) = \frac{1}{2}\langle (A - \hat{B}_k)z, z\rangle_L \qquad \forall\, z \in Z.$$

Therefore on the space Z,

$$a_k''(0) = A - \hat{B}_k. \tag{26}$$

According to the decomposition of the eigenvector subspaces of the operator A in Section 4.2, we can denote Z by

$$Z = E_0 \bigoplus \left[\bigoplus_{m=1}^{m_0} (E_m \bigoplus E_{-m}) \right]. \tag{27}$$

The space $E_0 = \mathbf{R}^{2n}$ is an invariant subspace of $A - \hat{B}_k$, on which $A - \hat{B}_k$ has the form

$$A - \hat{B}_k = -\begin{pmatrix} Z_k & 0 \\ 0 & Z_k \end{pmatrix}. \tag{28}$$

It has eigenvalues $\lambda_1 = -\pi < 0$ with multiplicity $2(n-1)$, and $\lambda_2 = -(k-n+1)\pi$ with multiplicity 2. Note that from $k \neq n-1$, we obtain $\lambda_2 < 0$ when $k > n-1$, and $\lambda_2 > 0$ when $k < n-1$.

For $m \geq 1$, the space $E_m \oplus E_{-m}$ is an invariant subspace of $A - \hat{B}_k$. Denote

$$E_m \oplus E_{-m} = \{\zeta = (\cos 2m\pi t)\xi + (\sin 2m\pi t)\eta \mid \xi, \eta \in \mathbf{R}^{2n}\}. \tag{29}$$

Then by direct computation, the operator $A - \hat{B}_k$ has the following form on this subspace,

$$\begin{aligned} 2\langle (A - \hat{B}_k)\zeta, \zeta\rangle_L &= 4m\pi J\xi \cdot \eta - \hat{B}_k\xi \cdot \xi - \hat{B}_k\eta \cdot \eta \\ &= \begin{pmatrix} -\hat{B}_k & -\mu J \\ \mu J & -\hat{B}_k \end{pmatrix} \begin{pmatrix} \xi \\ \eta \end{pmatrix} \cdot \begin{pmatrix} \xi \\ \eta \end{pmatrix}, \end{aligned} \tag{30}$$

where $\mu = 2m\pi$, ξ and $\eta \in \mathbf{R}^{2n}$. The above $4n \times 4n$ matrix on each orthogonal subspace of $E_m \oplus E_{-m}$ is a 4×4 matrix of the form

$$\begin{pmatrix} \alpha & 0 & 0 & \mu \\ 0 & \alpha & -\mu & 0 \\ 0 & -\mu & \alpha & 0 \\ \mu & 0 & 0 & \alpha \end{pmatrix}. \tag{31}$$

Thus when $\alpha = -\pi$, $\lambda_3 = (2m-1)\pi > 0$ and $\lambda_4 = -(2m+1)\pi < 0$ are eigenvalues of $A - \hat{B}_k$ with multiplicity $2(n-1)$. When $\alpha = -(k-n+1)\pi$, $\lambda_5 = (-(k-n+1) - 2m)\pi$ and $\lambda_6 = (-(k-n+1) + 2m)\pi$ are eigenvalues of $A - \hat{B}_k$ of multiplicity 2. When $k \geq n$, $\lambda_5 < 0$; if $m < (k-n+1)/2$ then $\lambda_6 < 0$; if $m > (k-n+1)/2$ then $\lambda_6 > 0$. When $k < n$, $\lambda_6 > 0$; if $m < -(k-n+1)/2$ then $\lambda_5 > 0$; if $m > -(k-n+1)/2$ then $\lambda_5 < 0$.

By these computations and the decomposition (27) of Z, when $k \geq n$ we obtain

$$m_k^+ = \sum_{m=1}^{\frac{k-n}{2}} 2(n-1) + \sum_{m=1+\frac{k-n}{2}}^{\frac{d-n}{2n}} 2n = d - k. \tag{32}$$

When $k < n$, similarly we obtain $m_k^+ = k - n$. This proves (25).

Case 2 $n = 1$ and $k \in 2\mathbf{Z}\backslash\{0\}$.

In this case $\hat{B}_k(t)$ is defined by (5.3.11), which is not a constant matrix. Note that by Lemma 2.11 below, there is no exponential paths in each subfamily $\mathcal{P}_{1,1}^k(2)$ when $k \in 2\mathbf{Z}\backslash\{0\}$. Then in such a sense the smooth standard paths defined by (5.3.10) are the simplest ones in these homotopy classes.

By (5.3.11), we define a new path $\gamma \in \mathcal{P}_1^*(4)$ by

$$\gamma(t) = F(t)G(t), \qquad \forall\, 0 \leq t \leq 1,$$

where $F(t) = D(2^{\omega_2(t)}) \diamond I_2$ and $G(t) = R(-\omega_1(t)k\pi) \diamond R(t\pi)$ for $0 \leq t \leq 1$, where $D(a)$ for $a \in \mathbf{R}\backslash\{0\}$ and $R(\theta)$ for $\theta \in \mathbf{R}$ are defined in Section 1.8. Then $\gamma(1) = M_4^+$, and for $0 \leq t \leq 1$ it holds that $\Delta(t) = -\omega_1(t)k\pi + t\pi$. This yields $\Delta_1(\gamma) = (k+1)\pi$. Therefore

$$i_1(\gamma) = k + 1. \tag{33}$$

On the other hand, the path γ is the fundamental solution of the system (1) with

$$B(t) = [-(\dot{\omega}_1(t)k\pi)I_2 + (\dot{\omega}_2(t)\ln 2)K] \diamond (\pi I_2),$$

for $t \in [0, \tau]$, where K is defined in (5.3.11). Let $x = (x_1, x_2) \equiv (x_1^T, x_2^T)^T$, $y = (y_1, y_2)$, and $z = (x_1, y_1, x_2, y_2)$. Then corresponding to $B(t)$ the quadratic form splits into

$$\langle (A - B)u(z), u(z)\rangle_L = \langle (A - \hat{B}_k)u(x), u(x)\rangle_L + \langle (A - \hat{B}_1)y, y\rangle_L, \tag{34}$$

where $\hat{B}_1$ is defined by (5.3.9). The base space corresponding to B on the left-hand side of (34) is R^4. The base spaces corresponding to $\hat{B}_k$ and $\hat{B}_1$ on the right-hand side of (34) are R^2 respectively. Denote the positive Morse index of the left-hand side by m^+. Then (34) yields

$$m^+ = m_k^+ + m_1^+. \tag{35}$$

By the assumption of the lemma, the dimension of the truncated space Z according to $\hat{B}_k$ is $2d$. Without loss of generality, we assume d has been chosen to be large enough so that it works for $\hat{B}_1$, $\hat{B}_k$, and B simultaneously. Then the truncated spaces according to $\hat{B}_1$ and B have dimensions $2d$ and $4d$ respectively. Thus applying Lemma 6 and Case 1 of this lemma, we obtain

$$2d - (k+1) = m_k^+ + (d-1).$$

This proves (25) in Case 2 and completes the proof of the lemma. □

Combining the above two lemmas, we have proved Theorem 1 for the non-degenerate case. We give next

The proof of Theorem 1 for $\omega = 1$: Without loss of generality, we suppose $\tau = 1$. By Definition 5.4.2, it suffices to prove the following for $s \in (0, 1]$ when $\nu_1(\gamma) > 0$:

$$\begin{cases} m^- &= d + i_1(\gamma_{-s}), \\ m^0 &= \nu_1(\gamma), \\ m^+ &= d - i_1(\gamma_{-s}) - \nu_1(\gamma), \end{cases} \tag{36}$$

where γ_s for $s \in [-1, 1]$ is the rotational perturbation paths of γ defined by (5.4.6).

When $\nu_1(\gamma) > 0$, $z = 0$ is a critical point of the functional a on Z corresponding to the system (1), but it need not be isolated. Then $\gamma(1) \in \mathrm{Sp}(2n)^0$. By Theorem 4.4.1,

$$m^0 = \dim\ker a''(0) = \nu_1(\gamma(1)) = \nu_1(\gamma). \tag{37}$$

Let $\epsilon > 0$ and $\gamma_s \in \hat{\mathcal{P}}_1(2n)$ with $-1 \le s \le 1$ be perturbation paths defined by (5.4.6) for γ. Note that in this case Theorem 5.4.1 holds. For $-1 \le s \le 1$ define

$$B_s(t) = -J\dot{\gamma}_s(t)\gamma_s^{-1}(t), \quad \forall t \in [0, 1], \tag{38}$$

$$H_s(t, x) = \frac{1}{2}B_s(t)x \cdot x \quad \forall (t, x) \in \mathbf{R} \times \mathbf{R}^{2n}. \tag{39}$$

Then $B_0(t) = B(t)$. When $s \neq 0$, $B_s \in C(S_1, \mathcal{L}_s(\mathbf{R}^{2n}))$, $B_s(t) = H_s''(t, x)$, and $\gamma_s \in \hat{\mathcal{P}}_1^*(2n)$. Denote the functionals defined in Section 4.2 according to B_s by f_s and a_s. Here we can also choose the dimension of the truncated space Z to be large enough so that the saddle point reduction method works for all $s \in [-1, 1]$. Denote $2d = \dim Z$.

Then a_s converges to a in C^2 as $s \to 0$. By Lemma 6, the origin $z = 0$ is an isolated critical point of a_s when $s \neq 0$. Denote the Morse indices of a_s at $z = 0$ by m_s^-, m_s^0, and m_s^+ respectively. Then by Lemmas 6 and 7 we obtain

$$m_s^- = d + i_1(\gamma_s), \quad m_s^0 = 0, \quad m_s^+ = d - i_1(\gamma_s), \tag{40}$$

if $s \in [-1, 1]\backslash\{0\}$. When $|s| > 0$ is sufficiently small, the matrix $a_s''(0)$ is a small perturbation of $a''(0)$. Thus in this case, $m^+ \le m_s^+$ and $m^- \le m_{-s}^-$. Together with Lemmas 6 and 7 we obtain that, when $0 < s \le 1$ and sufficiently close to 0,

$$m^+ \le m_s^+ = d - i_1(\gamma_s), \tag{41}$$

$$m^- \le m_{-s}^- = d + i_1(\gamma_{-s}). \tag{42}$$

By Theorem 5.4.1, the above (41) can be rewritten into

$$m^+ \le d - i_1(\gamma_s) = d - i_1(\gamma_{-s}) - \nu_1(\gamma). \tag{43}$$

Note that $\dim Z = 2d$. Together with (37) we then obtain

$$d - i_1(\gamma_{-s}) \le 2d - m^- = m^+ + m^0 = m^+ + \nu_1(\gamma) \le d - i_1(\gamma_{-s}). \tag{44}$$

Thus in (44), equalities must hold and they yield the first and last equalities in (36) by Definition 5.4.2.

The proof is complete.

Theorem 8 *For $\omega \in \mathbf{U}$ and $\tau > 0$, let $\gamma \in \mathcal{P}^0_{\tau,\omega}(2n)$. Then for any paths α and $\beta \in \mathcal{P}^*_{\tau,\omega}(2n)$ which are sufficiently C^0-close to γ, and any $s \in (0,1]$,*

$$i_\omega(\gamma_{-s}) \le i_\omega(\alpha) \le i_\omega(\gamma_s) = i_\omega(\gamma_{-s}) + \nu_\omega(\gamma), \tag{45}$$

$$|i_\omega(\beta) - i_\omega(\alpha)| \le \nu_\omega(\gamma). \tag{46}$$

Proof: Without loss of generality we suppose $\tau = 1$. We still only study the case of $\omega = 1$ and leave other cases to the readers.

Note that (46) follows from (45) and Theorem 5.4.1. We prove (45) in two steps.

Step 1 $\gamma \in \hat{\mathcal{P}}^0_1(2n)$. We claim that there exists $\epsilon = \epsilon(\gamma)$ such that (45) holds for all $\alpha \in \hat{\mathcal{P}}^*_1(2n) \cap B^1_\epsilon(\gamma)$, where $B^1_\epsilon(\gamma)$ is the open ball centered at γ with radius ϵ in $\hat{\mathcal{P}}_1(2n)$.

Fix $s \in (0,1]$. We prove the left inequality in (45) indirectly by assuming that there exist $\alpha_k \in \hat{\mathcal{P}}^*_1(2n)$ for $k \in \mathbf{N}$ such that α_k converges to γ in C^1 as $k \to \infty$, and

$$i_1(\gamma_{-s}) > i_1(\alpha_k). \tag{47}$$

Denote $B_k(t) = -J\dot{\alpha}_k(t)\alpha_k^{-1}(t)$. Then $B_k \to B$ in C^0. Therefore we can choose the truncated space Z to be large enough for all the B and B_k to carry out the saddle point reduction. Let $2d = \dim Z$. Denote by m_k^-, m_k^0, and m_k^+ the Morse indices of the functional a_k on Z corresponding to α_k, and by m^-, m^0, and m^+ the Morse indices of the functional a on Z corresponding to γ. Then whenever k is large enough, by Lemmas 6 and 7 we obtain

$$m^- \le m_k^- = d + i_1(\alpha_k) < d + i_1(\gamma_{-s}).$$

Together with Theorem 1, we get the contradiction

$$d - i_1(\gamma_{-s}) = m^+ + m^0 = 2d - m^- > d - i_1(\gamma_{-s}).$$

This proves the left inequality in (45). The right inequality in (45) follows by a similar proof.

Step 2 Fix $\gamma \in \mathcal{P}^0_1(2n)$. We choose a path $\beta \in C^1([0,1], \mathrm{Sp}(2n)) \cap \mathcal{P}_\tau(2n)$ such that $\beta(1) = \gamma(1)$ and $\beta \sim_1 \gamma$. Using the function $\rho(t) = 3t^2 - 2t^3$ for $0 \le t \le 1$, define

$$\hat{\gamma}(t) = \beta(\rho(t)), \quad \forall t \in [0,1]. \tag{48}$$

Then we obtain $\hat{\gamma} \in \hat{\mathcal{P}}_1(2n)$ and $\hat{\gamma} \sim_1 \gamma$ with fixed end points. Specifically this implies $\Delta_1(\hat{\gamma}) = \Delta_1(\gamma)$. Thus by the construction in the proof of Theorem 5.4.1, we obtain $\hat{\gamma}_s(1) = \gamma_s(1)$ and $\hat{\gamma}_s \sim \gamma_s$ for all $s \in [-1, 1]$ provided $\hat{\gamma}$ is sufficiently C^0 close to γ. By Theorem 5.3.3, we obtain

$$i_1(\hat{\gamma}_s) = i_1(\gamma_s), \quad \nu_1(\hat{\gamma}_s) = \nu_1(\gamma_s) = 0, \quad \forall s \in [-1, 1]\setminus\{0\}.$$

For the constant $\epsilon = \epsilon(\hat{\gamma}) > 0$ given by Step 1, let $B^0_{\epsilon/2}(\gamma)$ be the open ball centered at γ with radius $\epsilon/2$ in $\mathcal{P}_1(2n)$ under the $C([0,1], \mathrm{Sp}(2n))$ topology. By further requiring this $\epsilon > 0$ to be much smaller, we can assume

$$|\Delta_1(\hat{\gamma}) - \Delta_1(\hat{\alpha})| \le 1/2, \quad \forall \hat{\alpha} \in B^1_\epsilon(\hat{\gamma}),$$

$$|\Delta_1(\gamma) - \Delta_1(\alpha)| \le 1/2, \quad \forall \alpha \in B^0_{\epsilon/2}(\gamma).$$

Fix a path $\alpha \in \mathcal{P}^*_1(2n) \cap B^0_{\epsilon/2}(\gamma)$. Choose $\hat{\alpha} \in \hat{P}^*_1(2n) \cap B^1_\epsilon(\hat{\gamma})$ such that $\hat{\alpha}(1) = \alpha(1)$. Then we obtain

$$|\Delta_1(\hat{\alpha}) - \Delta_1(\alpha)| \le 1.$$

This implies $\Delta_1(\hat{\alpha}) = \Delta_1(\alpha)$. By Lemma 5.2.4, this implies $\hat{\alpha} \sim \alpha$. Then by Theorem 5.3.3, we obtain $i_1(\hat{\alpha}) = i_1(\alpha)$. Thus by (45), proved in Step 1 for $\hat{\gamma}$, we obtain (45), for γ, γ_{-s}, γ_s, and α.

The proof is complete. □

As a direct consequence of Theorem 8, we obtain

Corollary 9 *For any $\omega \in \mathbf{U}$, $\tau > 0$, and $\gamma \in \mathcal{P}^0_{\tau,\omega}(2n)$, denote by γ_s for $s \in [-1, 1]$ the rotation perturbation paths defined by* (5.4.6) *for γ. Then for any $s \in (0, 1]$,*

$$i_\omega(\gamma_{-s}) = \sup_{U\in\mathcal{N}(\gamma)} \inf\{i_\omega(\beta) \mid \beta \in U \cap \mathcal{P}^*_{\tau,\omega}(2n)\}, \tag{49}$$

$$i_\omega(\gamma_s) = \inf_{U\in\mathcal{N}(\gamma)} \sup\{i_\omega(\beta) \mid \beta \in U \cap \mathcal{P}^*_{\tau,\omega}(2n)\}, \tag{50}$$

where $\mathcal{N}(\gamma)$ is the set of all open neighborhoods of γ in $\mathcal{P}_\tau(2n)$.

Proof: Suppose $\gamma \in \mathcal{P}^0_{\tau,\omega}(2n)$. For any $U \in \mathcal{N}(\gamma)$, by the definition (5.4.6) of γ_s, there exists an $s > 0$ sufficiently small such that both γ_{-s} and γ_s are in U. So the right-hand side of (49) (or (50)) is not greater (or less) than $i_\omega(\gamma_{-s})$ (or $i_\omega(\gamma_s)$). When $U \in \mathcal{N}(\gamma)$ is sufficiently small, by Theorem 8 there exists an $s > 0$ sufficiently small such that the infimum (or supremum) on $U \cap \mathcal{P}^*_{\tau,\omega}(2n)$ is realized by $i_\omega(\gamma_{-s})$ (or $i_\omega(\gamma_s)$). Therefore the right-hand side of (49) (or (50)) is not less (or greater) than $i_\omega(\gamma_{-s})$ (or $i_\omega(\gamma_s)$). This proves the corollary. □

Now we can also give an alternative definition of the index function for the degenerate paths:

Definition 10 For any $\tau > 0$, $\omega \in \mathbf{U}$, and $\gamma \in \mathcal{P}^0_{\tau,\omega}(2n)$, we define the index function $i_\omega(\gamma)$ by

$$i_\omega(\gamma) = \sup_{U\in\mathcal{N}(\gamma)} \inf\{i_\omega(\beta) \mid \beta \in U \cap \mathcal{P}^*_{\tau,\omega}(2n)\}, \tag{51}$$

where $\mathcal{N}(\gamma)$ is the set of all open neighborhoods of γ in $\mathcal{P}_\tau(2n)$.

Lemma 11 *Definition* 10 *is well defined, and coincides with Definition* 5.4.2.

Proof: By (49) we obtain that (51) coincides with Definition 5.4.2. Specifically (49) implies the finiteness of (51), i.e., it is well defined. □

Note that by Theorem 8, Corollary 9 and Definition 10 are equivalent to the following more intuitive but less elegant results.

Corollary 12 *For any* $\omega \in \mathbf{U}$, $\tau > 0$, *and* $\gamma \in \mathcal{P}^0_{\tau,\omega}(2n)$, *denote by* γ_s *for* $s \in [-1, 1]$ *the rotation perturbation paths defined by* (5.4.6) *for* γ. *Then for any* $s \in (0, 1]$,

$$\begin{aligned} i_\omega(\gamma_{-s}) &= \inf\{i_\omega(\beta) \mid \beta \in \mathcal{P}^*_{\tau,\omega}(2n), \\ &\qquad \beta \textit{ is sufficiently } C^0\textit{-close to } \gamma \textit{ in } \mathcal{P}_\tau(2n)\}, \end{aligned} \tag{52}$$

$$\begin{aligned} i_\omega(\gamma_s) &= \sup\{i_\tau(\beta) \mid \beta \in \mathcal{P}^*_\tau(2n) \\ &\qquad \beta \textit{ is sufficiently } C^0\textit{-close to } \gamma \textit{ in } \mathcal{P}_\tau(2n)\}, \end{aligned} \tag{53}$$

where by saying that β *is sufficiently* C^0*-close to* γ *in* $\mathcal{P}_\tau(2n)$ *we mean that* β *is close enough to* γ *in the topology of* $\mathcal{P}_\tau(2n)$ *so that there exists a homotopy* $\delta : [0, 1] \to \mathcal{P}_\tau(2n)$ *with the properties* $\delta(0) = \beta$, $\delta(1) = \gamma$, *and* $\delta(s) \in \mathcal{P}^*_{\tau,\omega}(2n)$ *for all* $0 \le s < 1$.

Definition 13 For any $\tau > 0$, $\omega \in \mathbf{U}$, and $\gamma \in \mathcal{P}^0_{\tau,\omega}(2n)$, we define:

$$\begin{aligned} i_\omega(\gamma) = \inf\{&i_\omega(\beta) \mid \beta \in \mathcal{P}^*_{\tau,\omega}(2n), \\ &\beta \text{ is sufficiently } C^0\text{-close to } \gamma \text{ in } \mathcal{P}_\tau(2n) \\ &\text{in the sense described in Corollary 12}\}. \end{aligned} \tag{54}$$

Remark 14 The relation theorem on the Morse index and the index function is established in this section. Here Theorem 1 for i_1 and ν_1 with $\nu_1(\gamma) = 0$, i.e., the $\omega = 1$ non-degenerate case Lemmas 6 and 7, was proved by C. Conley and E. Zehnder in [CoZ2] of 1984 for the case $n \ge 2$, and by E. Zehnder and the author in [LZe1] of 1990 for the case $n = 1$. Theorem 1 for i_1 and ν_1 with $\nu_1(\gamma) > 0$, i.e. the $\omega = 1$ degenerate case was proved in [Lon5] of 1990 and [Lon13] of 1997. The description and the proof of Theorem 1 for general $\omega \in \mathbf{U}$ appeared in [Lon16]. The original proofs of Theorem 1 in [CoZ2], [LZe1], and [Lon5] use the Conley homotopy index theory. Here in this book we have replaced the Conley homotopy index argument by a rather simple variational proof based on Lemma 3 following [Lon13]. Theorem 8 was proved in [Lon5] and [Lon13] for $\omega = 1$ and in [Lon16] for general $\omega \in \mathbf{U}$. That (45) of Theorem 8 implies Definition 10 was pointed out to the author by a referee of [LZu3] from Annals of Mathematics in November 2000. Here we have used it to replace Definition 13 originally used in [Lon16].

6.2 An axiom approach and further properties

In this section we study further properties of and give an axiom description to the index function theory defined on $\mathcal{P}_\tau(2n)$ by Definition 5.4.3.

Proposition 1 *Let $\omega \in \mathbf{U}$. For any two paths γ_0 and $\gamma_1 \in \mathcal{P}_\tau(2n)$ with $\gamma_0(\tau) = \gamma_1(\tau)$, it holds that $\Delta_\tau(\gamma_0) = \Delta_\tau(\gamma_1)$ if and only if $i_\omega(\gamma_0) = i_\omega(\gamma_1)$.*

Proof: Without loss of generality, suppose $\tau = 1$. If $\nu_\omega(\gamma_0) = 0$, this result follows from Lemma 5.2.4 and Theorem 5.3.3.

Suppose $\nu_\omega(\gamma_0) > 0$. Define $\zeta(t) = Q(-t, \ldots, -t)$ for $t \in [0, 1]$ by (5.4.2) with γ replaced by γ_0. Then $\zeta(1) = (\gamma_j)_{-1}(1)$ where $(\gamma_j)_{-1}$ is defined by (5.4.6) for $j = 0, 1$. Let $\psi : [0, 1] \to \mathrm{Sp}(2n)^*_\omega$ be a path connecting $\zeta(1)$ to M_n^+ or M_n^-. Then the path $(\gamma_j)_{-1}$ is homotopic to the joint path $\zeta * \gamma_j$. Thus these two paths possess the same rotation numbers. Then for $j = 0$ and 1 by Definition 5.4.2, we obtain

$$\begin{aligned} i_\omega(\gamma_j) &= \Delta_1(\psi * \zeta * \gamma_j)/\pi \\ &= \Delta_1(\psi)/\pi + \Delta_1(\zeta)/\pi + \Delta_1(\gamma_j)/\pi \end{aligned} \tag{1}$$

where ζ depends on the normal form $\diamond$-product of $\gamma_0(1) = \gamma_1(1)$. From (1) the proposition follows. □

Corollary 2 *Let $\omega \in \mathbf{U}$. For any two paths γ_0 and $\gamma_1 \in \mathcal{P}_\tau(2n)$ with $\gamma_0(\tau) = \gamma_1(\tau)$, it holds that $i_\omega(\gamma_0) = i_\omega(\gamma_1)$ if and only if $\gamma_0 \sim_\omega \gamma_1$ on $[0, \tau]$ with fixed end points.*

Proof: This is a consequence of Lemma 5.2.4 and Proposition 1. □

Theorem 3 (Homotopy invariant). *Let $\omega \in \mathbf{U}$. For any two paths γ_0 and $\gamma_1 \in \mathcal{P}_\tau(2n)$, if $\gamma_0 \sim_\omega \gamma_1$ on $[0, \tau]$,*

$$i_\omega(\gamma_0) = i_\omega(\gamma_1), \qquad \nu_\omega(\gamma_0) = \nu_\omega(\gamma_1). \tag{2}$$

Proof: Without loss of generality, we suppose $\tau = 1$. By Corollary 2, we only need to consider the case when the end points are not fixed. If $\nu_\omega(\gamma_0) = 0$, the claim follows from Lemma 5.2.4 and Definition 5.2.7.

Suppose $\nu_\omega(\gamma_0) > 0$. By Definition 5.0.3 and $\gamma_0 \sim_\omega \gamma_1$ on $[0, 1]$, we obtain $\nu_\omega(\gamma_0) = \nu_\omega(\gamma_1)$.

To prove the first equality in (2), by Step 2 in the proof of Theorem 1.8, there are smooth paths $\xi_j \in \hat{\mathcal{P}}_1(2n)$ sufficiently close to γ_j for $j = 0$ and 1 respectively such that $\xi_j(1) = \gamma_j(1)$ and $\Delta_1(\xi_j) = \Delta_1(\gamma_j)$. Then $i_\omega(\xi_j) = i_\omega(\gamma_j)$ holds by Proposition 1, and $\xi_i \sim_\omega \gamma_i$ with fixed end points by Corollary 2 for $i = 0$ and 1.

Combining the homotopies from ξ_0 to γ_0, from γ_0 to γ_1, and from γ_1 to ξ_1, we obtain a homotopy $\delta \in C([0, 1]^2, \mathrm{Sp}(2n))$ from ξ_0 to ξ_1 in $\mathcal{P}_1(2n)$ such that $\nu_\omega(\delta_s(1)) =$ constant for all $0 \le s \le 1$. By Lemma 1.5, we can further assume

$$\delta_s \equiv \delta(s, \cdot) \in \hat{\mathcal{P}}_1(2n), \quad \forall\, s \in [0, 1]. \tag{3}$$

Consider the linear Hamiltonian system (1.1) with $B = B_s(t)$ whose fundamental solution is δ_s. By Theorem 1.1 we obtain

$$\begin{cases} m_s^+ &= d_\omega - i_\omega(\delta_s) - \nu_\omega(\delta_s), \\ m_s^0 &= \nu_\omega(\delta_s) = \nu_\omega(\gamma_0), \\ m_s^- &= d_\omega + i_\omega(\delta_s), \end{cases} \tag{4}$$

for all $s \in [0, 1]$. Then by Lemma 1.3, these Morse indices are constants for all $s \in [0, 1]$. Thus we obtain $i_1(\xi_0) = i_1(\xi_1)$. This proves the first equality of (2). □

Theorem 4 (Inverse homotopy invariant). *Let $\omega \in \mathbf{U}$. For any two paths γ_0 and $\gamma_1 \in \mathcal{P}_\tau(2n)$ with $i_\omega(\gamma_0) = i_\omega(\gamma_1)$, suppose that there exists a continuous path $h : [0, 1] \to \mathrm{Sp}(2n)$ such that $h(0) = \gamma_0(\tau)$, $h(1) = \gamma_1(\tau)$, and $\nu_\omega(h(s)) = \nu_\omega(\gamma_0)$ for all $s \in [0, 1]$. Then $\gamma_0 \sim_\omega \gamma_1$ on $[0, \tau]$ along h.*

Proof: Note that γ_0 and the joint path $h * \gamma_0 : [0, \tau] \to \mathrm{Sp}(2n)$ are homotopic on $[0, \tau]$ along h. We denote this homotopy by $\alpha : [0, 1] \times [0, \tau] \to \mathrm{Sp}(2n)$. By Theorem 3, $i_\omega(h|_{[0,s]} * \gamma_0) = i_\omega(\gamma_0)$ for all $s \in [0, 1]$, and then $i_\omega(h * \gamma_0) = i_\omega(\gamma_1)$. So from $h * \gamma_0(\tau) = \gamma_1(\tau)$ and Corollary 2 we obtain that $h * \gamma_0$ and γ_1 are homotopic on $[0, \tau]$ with fixed end points. Denote this homotopy by $\beta : [0, 1] \times [0, \tau] \to \mathrm{Sp}(2n)$.

We define a new map $\delta : [0, 1] \times [0, \tau] \to \mathrm{Sp}(2n)$ (cf. Figure 2) by

$$\delta(s, t) = \begin{cases} \alpha(2\frac{\tau s}{\tau + t}, t), & \text{if } 0 \le s \le \frac{t}{2\tau} + \frac{1}{2}, \\ \beta(2\frac{\tau s - t}{\tau - t} - 1, t), & \text{if } \frac{t}{2\tau} + \frac{1}{2} < s \le 1. \end{cases} \tag{5}$$

Then this δ gives the ω-homotopy of $\gamma_0 \sim_\omega \gamma_1$ on $[0, \tau]$ along h. □

A direct consequence of Theorem 3 is that the index function is invariant under conjugation in $\mathrm{Sp}(2n)$.

Corollary 5 *Given a path $\gamma \in \mathcal{P}_\tau(2n)$ and a matrix $M \in \mathrm{Sp}(2n)$, define $\beta(t) = M^{-1}\gamma(t)M$ for $0 \le t \le \tau$. Then*

$$i_\omega(\beta) = i_\omega(\gamma), \quad \nu_\omega(\beta) = \nu_\omega(\gamma). \tag{6}$$

Proof: Choose $\alpha \in \mathcal{P}_\tau(2n)$ so that $\alpha(\tau) = M$. Let $\delta(s, t) = \alpha^{-1}(s)\gamma(t)\alpha(s)$ for all $(s, t) \in [0, 1] \times [0, \tau]$. Then we obtain $\gamma \sim_\omega \beta$ on $[0, \tau]$ along $\delta(\cdot, \tau)$ via the homotopy δ. Thus (6) follows from Theorem 3. □

Theorem 6 (Symplectic additivity). *Suppose $\gamma_0 \in \mathcal{P}_\tau(2n_0)$ and $\gamma_1 \in \mathcal{P}_\tau(2n_1)$. Then $\gamma_0 \diamond \gamma_1 \in \mathcal{P}_\tau(2n_0 + 2n_1)$ and*

$$i_\omega(\gamma_0 \diamond \gamma_1) = i_\omega(\gamma_0) + i_\omega(\gamma_1). \tag{7}$$

Proof: Set $\tau = 1$. By Theorem 5.3.2, the arguments in Step 2 of the proof for Theorem 1.8 and Theorem 3, we only need to prove (7) when $\gamma_j \in \hat{\mathcal{P}}_1^0(2n_j)$ for

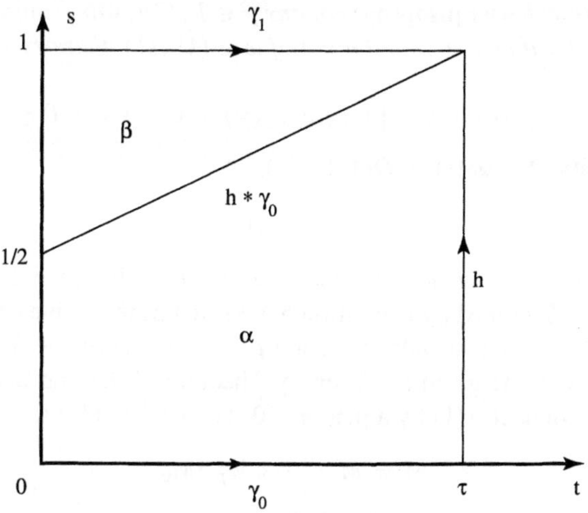

Figure 2: *The homotopy δ*

$j = 0$ and 1. Let $[\gamma_j]_{-1}$ be the rotational perturbation paths of γ_j defined by (5.4.6) respectively. Then by Definition 5.4.2 and Theorem 5.3.2 we obtain

$$\begin{aligned} i_\omega(\gamma_0 \diamond \gamma_1) &= i_\omega([\gamma_0]_{-1} \diamond [\gamma_1]_{-1}) \\ &= i_\omega([\gamma_0]_{-1}) + i_\omega([\gamma_1]_{-1}) \\ &= i_\omega(\gamma_0) + i_\omega(\gamma_1). \end{aligned}$$

Other cases of γ_0 and γ_1 being degenerate or not can be proved similarly. □

Theorem 7 *For $\omega \in \mathbf{U}$ and $\tau > 0$, the ω-index part of the index function defined on paths in* Sp(2*n*) *starting from the identity, i.e., the map $i_\omega : \cup_{n \in \mathbf{N}} \mathcal{P}_\tau(2n) \to \mathbf{Z}$, is uniquely characterized by the following five axioms:*

1° **(Homotopy invariant)** *For γ_0 and $\gamma_1 \in \mathcal{P}_\tau(2n)$, if $\gamma_0 \sim_\omega \gamma_1$ on $[0, \tau]$ in the sense of Definition* 5.0.3, *then*

$$i_\omega(\gamma_0) = i_\omega(\gamma_1). \tag{8}$$

2° **(Symplectic additivity)** *For any $\gamma_j \in \mathcal{P}_\tau(2n_j)$ with $j = 0$ and* 1,

$$i_\omega(\gamma_0 \diamond \gamma_1) = i_\omega(\gamma_0) + i_\omega(\gamma_1). \tag{9}$$

3° **(Clockwise continuity)** *For any $\gamma \in \mathcal{P}_\tau(2)$ satisfying $\gamma(\tau) = N_1(\omega, b)$ with $b = \pm 1$ or 0 if $\omega = \pm 1$, and $b = 0$ if $\omega \in (\mathbf{U} \backslash \mathbf{R})$, there exists a $\theta_0 > 0$ such that*

$$i_\omega([\gamma(\tau)\phi_{\tau,-\theta}] * \gamma) = i_\omega(\gamma), \quad \forall\, 0 < \theta \le \theta_0, \tag{10}$$

where $\phi_{\tau,-\theta}$ is defined by (5.1.1).

4° **(Counterclockwise jumping)** *For any* $\gamma \in \mathcal{P}_\tau(2n)$ *satisfying* $\gamma(\tau) = N_1(\omega, b)$ *with* $b = \pm 1$ *if* $\omega = \pm 1$, *and* $b = 0$ *if* $\omega \in (\mathbf{U}\backslash\mathbf{R})$, *there exists a* $\theta_0 > 0$ *such that*

$$i_\omega([\gamma(\tau)\phi_{\tau,\theta}] * \gamma) = i_\omega(\gamma) + 1, \quad \forall\, 0 < \theta \leq \theta_0. \tag{11}$$

5° **(Normality)** *For* $\hat{\alpha}_0(t) = D(1 + t/\tau)$,

$$i_\omega(\hat{\alpha}_0) = 0. \tag{12}$$

Proof: Without loss of generality, let $\tau = 1$. Clearly the index function $i_\omega : \cup_{n\in\mathbf{N}}\mathcal{P}_1(2n) \to \mathbf{Z}$ defined by Definition 5.4.3 satisfies these five conditions.

Suppose we have another index function $\mu : \cup_{n\in\mathbf{N}}\mathcal{P}_1(2n) \to \mathbf{Z}$ satisfying these five conditions with the given ω. Then by Theorems 1.9.3 and 2.4.1, for any $\gamma \in \mathcal{P}_1(2n)$, we can connect $\gamma(1)$ by a path $h : [0, 1] \to \Omega^0_\omega(\gamma(1))$ to

$$M = M_1 \diamond \cdots \diamond M_p \diamond M_0,$$

where p is a non-negative integer with $0 \leq p \leq n$, each $M_j = N_1(\omega_j, b_j)$ with $\omega_j = \omega$ or $\overline{\omega}$ and $b_j = \pm 1$ or 0 for $1 \leq j \leq p$, $M_0 = M^+_{n-p}$ or M^-_{n-p} when $p < n$. Then by 1°, 2°, and our argument in Case 2 of the proof of Theorem 5.4.1, $\mu(\gamma)$ is completely determined by its value on $\mathcal{P}_1(2)$. By 3° and 4°, μ's value is completely determined by its value on $\mathcal{P}^*_{1,\omega}(2)$. By 1°, the value of μ is completely determined by its value on all the standard paths $\hat{\alpha}_k$ in Sp(2) defined by (5.3.16) and (5.3.17). Then by 1°, 3°, 4°, and 5°, we obtain $\mu = i_\omega$ on $\hat{\alpha}_k$ for all $k \in \mathbf{Z}$. This completes the proof. □

Remark 8 It is easy to construct examples to show that the five axioms in Theorem 7 are independent from each other.

As a direct consequence of our study, for any end point free curve in the symplectic group, we can also define its index as follows.

Definition 9 Let $\omega \in \mathbf{U}$. For any curve $f \in C([a, b], \mathrm{Sp}(2n))$, choose $\xi \in \mathcal{P}_1(2n)$ so that $\xi(1) = f(a)$. Define $\eta(t) = \xi(2t)$ for $t \in [0, 1/2]$, and $\eta(t) = f(a + (2t - 1)(b - a))$ for $t \in [1/2, 1]$. Define

$$i(f) = i_1(\eta) - i_1(\xi). \tag{13}$$

By our above study of the index i_1 it is easy to see that the definition (13) only depends on f itself, and therefore is well defined. As a direct consequence of Theorem 7, we give a characterization of the index i below, whose proof is left to the readers.

Corollary 10 *The index i defined above for continuous curves in the symplectic groups is uniquely determined by the following five axioms:*

1° **(Homotopy invariant)** *Two continuous curves in* Sp(2n) *with the same initial and end points possess the same index if and only if they can be continuously deformed to each other with the end points fixed.*

2° **(Vanishing)** $i(f) = 0$ *if* $f \in C([a, b], \mathrm{Sp}(2n))$ *with* $\nu_1(f(t)) = constant$.

3° **(Symplectic additivity)** $i(f_0 \diamond f_1) = i(f_0) + i(f_1)$ *if* $f_i \in C([a,b], \mathrm{Sp}(2n_i))$ *with* $i = 0$ *and* 1.

4° **(Catenation)** $i(f) = i(f|_{[a,b]}) + i(f|_{[b,c]})$ *if* $f \in C([a,c], \mathrm{Sp}(2n))$ *with* $a < b < c$.

5° **(Normality)** *For* $f(t) = N_1(1,b)R(t/2)$ *with* $t \in [-1,1]$, $b = \pm 1$ *or* 0,

$$i(f|_{[-1,0]}) = 0 \quad \textit{and} \quad i(f|_{[0,1]}) = 2 - |b|.$$

Next we classify the exponential paths in Sp(2) defined before Lemma 1.7. We omit the subindex 1 when there is no confusion.

Lemma 11 (T. An and Y. Long [AnL2]). *Let* $\gamma \in \mathcal{P}_\tau(2)$ *be an exponential path. Then*

$$i_1(\gamma) \in \begin{cases} (2\mathbf{Z}+1) \cup \{0\}, & \textit{if} \ \ \nu_1(\gamma) = 0, \\ \{0, -1\}, & \textit{if} \ \ \nu_1(\gamma) = 1, \\ 2\mathbf{Z}+1, & \textit{if} \ \ \nu_1(\gamma) = 2. \end{cases} \tag{14}$$

Proof: For $\tau > 0$, let $\gamma(t) = \exp(tJB)$ for $0 \le t \le \tau$ be an exponential path with $B = \begin{pmatrix} a & b \\ b & c \end{pmatrix}$, for some a, b, and $c \in \mathbf{R}$. Then $1 \in \sigma(\gamma(\tau))$ if and only if

$$\sigma(JB) \cap \sqrt{-1}2\pi\mathbf{Z} \neq \emptyset.$$

By direct computation we obtain that the eigenvalues of JB are given by

$$\lambda^{\pm} = \pm\sqrt{b^2 - ac}.$$

Thus

$$\gamma \in \mathcal{P}^*_\tau(2) \quad \text{if and only if} \quad b^2 - ac \neq -4k^2\pi, \quad \forall k \in \mathbf{Z}.$$

Using notation from Chapters 4 to 6, we obtain

$$\frac{1}{2}(A-B)|_{E_m \oplus E_{-m}} = \mu\begin{pmatrix} 0 & -J \\ J & 0 \end{pmatrix} - \begin{pmatrix} B & 0 \\ 0 & B \end{pmatrix}, \quad \forall m \neq 0.$$

where $\mu = 2m\pi$ for $m \in \mathbf{Z}$. Then it possesses double eigenvalues

$$\lambda_m^{\pm} = \frac{1}{2}\left[-(a+c) \pm \sqrt{(a+c)^2 + 4(b^2-ac) + 4\mu^2}\right].$$

When $m = 0$ we have

$$(A-B)|_{E_0} = -B|_{\mathbf{R}^{2n}}.$$

Thus it possesses two simple eigenvalues

$$\lambda_0^{\pm} = \frac{1}{2}\left[-(a+c) \pm \sqrt{(a-c)^2 + 4(b^2 - ac)}\right].$$

To calculate the Morse index of the functional a in the saddle point reduction as studied in Section 2, it suffices to know the total multiplicity of negative eigenvalues of $(A - B)|_Z$. Recall that $2d = \dim Z$ and

$$Z = E_0 \oplus (\oplus_{m=1}^{n}(E_m \oplus E_{-m}),$$

where $n \in \mathbf{N}$ is large enough so that $\lambda_m^+ > 0$ and $\lambda_m^- < 0$ for $m > n$.

Suppose $b^2 - ac \neq -4k^2\pi$ for all $k \in \mathbf{Z}$. If $b^2 - ac > 0$, we obtain $\lambda_m^+ > 0$ and $\lambda_m^- < 0$ for all $m \in \mathbf{Z}$. So $m^+ = m^- = d$ and $m^0 = 0$. This implies $i_1(\gamma) = \nu_1(\gamma) = 0$. If $b^2 - ac < 0$, we obtain $\lambda_m^{\pm} \neq 0$ for all $m \in \mathbf{Z}\backslash\{0\}$. Since $0 \le b^2 < ac$, we must have $\lambda_0^{\pm} > 0$ or $\lambda_0^{\pm} < 0$. When $\lambda_0^{\pm} > 0$, it holds that $m^- = d + 2h - 1$, $m^+ = d - 2h + 1$, and $m^0 = 0$ for some $h \in \mathbf{Z}$. When $\lambda_0^{\pm} < 0$, it holds that $m^- = d + 2h + 1$, $m^+ = d - 2h - 1$, and $m^0 = 0$ for some $h \in \mathbf{Z}$. Thus we always have $i_1(\gamma) \in 2\mathbf{Z} + 1$ and $\nu_1(\gamma) = 0$.

When $b^2 - ac = -4k^2\pi$ for some $k \in \mathbf{Z}\backslash\{0\}$, similarly we obtain $i_1(\gamma) \in 2\mathbf{Z} + 1$ and $\nu_1(\gamma) = 2$.

When $b^2 - ac = 0$, we have $\lambda_m^+ > 0$ and $\lambda_m^- < 0$ for all $m \in \mathbf{Z}\backslash\{0\}$. If $a + c = 0$, we obtain $i_1(\gamma) = -1$ and $\nu_1(\gamma) = 2$. If $a + c > 0$ we obtain $i_1(\gamma) = 0$ and $\nu_1(\gamma) = 1$. If $a + c < 0$, we obtain $i_1(\gamma) = -1$ and $\nu_1(\gamma) = 1$.

Thus the proof is complete. □

Based on this result we have the following classification of exponential paths.

Theorem 12 (T. An and Y. Long [AnL2]). *An exponential path $\gamma(t) = \exp(tJB)$ for $t \ge 0$ with $B \in \mathcal{L}_s(\mathbf{R}^2)$ must be of one of the following three types:*

1° **fixed point type,** *i.e.* $\gamma(t) \equiv I_2$ *for all* $t \ge 0$.

2° **radiation type,** *i.e.* $I_2 \neq \gamma(t) \in \mathrm{Sp}(2)^0$ *for all* $t > 0$ *or* $\gamma(t) \in \mathrm{Sp}(2)^+$ *for all* $t > 0$.

3° **rotation type,** *i.e.* $\gamma(t)$ *rotates in the* S^1 *direction of* $\mathrm{Sp}(2)$ *within* $\mathrm{Sp}(2)^- \cup \{I_2\}$ *as* t *increases and intersects* $\mathrm{Sp}(2)^0$ *only at* I_2 *on discrete times.*

Proof: We use notation from the proof of Theorem 1.

If $b^2 - ac = 0$, we have $(tb)^2 - t^2ac = 0$ for all $t \ge 0$. By our discussion in the proof of Lemma 11, γ is either of the fixed point type or $\gamma(t) \in \mathrm{Sp}(2)^0$ is of the radiation type.

If $b^2 - ac > 0$, $\gamma(t) \in \mathrm{Sp}(2)^+$ for $t > 0$ is of radiation type.

If $b^2 - ac < 0$, γ lies in $\mathrm{Sp}(2)^- \cup \{I_2\}$ and there exists an increasing positive number sequence $\{t_k\}$ such that $t_k^2(b^2 - ac) = -4k^2\pi$. Then γ is of the rotation type. □

Corollary 13 *For any $\omega \in \mathbf{U}$, the homotopy class $\mathcal{P}^k_{\tau,\omega}(2)$ contains no exponential paths for any $k \in 2\mathbf{Z}\setminus\{0\}$.*

Remark 14 Corollary 13 explains why the standard paths in Section 5.3 should have the form of (5.3.10) when $k \in 2\mathbf{Z}\setminus\{0\}$ and the Case 2 in the proof of Lemma 1.7 is necessary.

Results in this section are due to [Lon13] and [Lon16] except Lemma 11 and Theorem 12 which were proved in [AnL2] by T. An and the author.

Chapter 7

Relations with other Morse indices

In this chapter, we study the relations between the index function theory on $\mathcal{P}_\tau(2n)$ and other Morse index theories for Hamiltonian or Lagrangian systems. In Section 7.1, we study the Galerkin approximation method for Hamiltonian systems. In Section 7.2, we give a simple proof for the coincidence of the Morse index of second order Hamiltonian systems with the 1-index defined in Chapter 5 of the corresponding linearized first order Hamiltonian systems. In Section 3, based on J. Duistermaat's work [Dui1], we study the general Lagrangian systems.

7.1 The Galerkin approximation

In this section we study the relation between the 1-index of periodic solutions of Hamiltonian systems and the Morse index of the corresponding functional defined via the Galerkin approximation method on the finite dimensional truncated space at its corresponding critical points.

Given a Hamiltonian function H satisfying the conditions (H1) and (H2) with $\tau > 0$ in Section 4.1, we consider the nonlinear Hamiltonian system,

$$\dot{x} = JH''(t, x), \qquad x \in \mathbf{R}^{2n}. \tag{1}$$

As in Section 4.1, we denote $W = W^{1/2,2}(S_\tau, \mathbf{R}^{2n})$ with $S_\tau = \mathbf{R}/(\tau\mathbf{Z})$ and the norm (4.1.4).

Fix a matrix function $B_\infty \in C(S_\tau, \mathcal{L}_s(\mathbf{R}^{2n}))$. For W_m given by (4.1.14), let $P_m : W \to W_m$ be the projection. Then $\{P_m : W \to W_m\}$ form a Galerkin approximation scheme with respect to $A - B_\infty$. Note that

$$\dim W_m = 2m + \nu_1(B_\infty). \tag{2}$$

Let f be the functional defined by (4.1.3) for the above Hamiltonian function H, and $f_m = f \circ P_m : W_m \to \mathbf{R}$ be its restriction on W_m. Our aim in this section is to study the Morse index of f_m at its critical point $x_m \in W_m$ in terms of the 1-index of B_∞ and $B = H''(\cdot, x(\cdot))$ defined via the linear system

$$\dot{y} = JB(t)y, \qquad y \in \mathbf{R}^{2n}. \tag{3}$$

For any bounded linear self-adjoint operator L defined on a subspace $\mathrm{dom}L$ of W, we denote by $E^*(L)$ with $* = -, 0$, and $+$, the eigenvector subspace of L belonging to the eigenvalues in $(-\infty, 0)$, $\{0\}$, and $(0, +\infty)$ respectively. For any $\delta > 0$ we denote by $E_\delta^*(L)$ with $* = -$, 0, and $+$, the eigenvector subspace of L belonging to the eigenvalues in $(-\infty, -\delta]$, $(-\delta, \delta)$, and $[\delta, +\infty)$ respectively. We also denote

$L^{\#} = (L|_{\mathrm{Im}L})^{-1}$ and for $\delta > 0$ let

$$\begin{cases} m^*(L) = \dim E^*(L), & \text{for } * = -, 0, +, \\ m^*_\delta(L) = \dim E^*_\delta(L), & \text{for } * = -, 0, +. \end{cases} \tag{4}$$

The following is the main result in this section.

Theorem 1 (G. Fei and Q. Qiu [FeQ2]). *Let $\{P_m : W \to W_m\}$ be the Galerkin approximation scheme defined above with respect to $A - B_\infty$ for $A = -Jd/dt$ and $B_\infty \in C(S_\tau, \mathcal{L}_s(\mathbf{R}^{2n}))$. Then there exists a sufficiently great $m_0 > 0$ depending on B_∞ and B such that for all $m \geq m_0$,*

$$m^+_\delta(P_m(A - B)P_m) = m + i_1(B_\infty) - i_1(B) + \nu_1(B_\infty) - \nu_1(B), \tag{5}$$

$$m^0_\delta(P_m(A - B)P_m) = \nu_1(B), \tag{6}$$

$$m^-_\delta(P_m(A - B)P_m) = m - i_1(B_\infty) + i_1(B), \tag{7}$$

where $\delta = \frac{1}{4}\|(A - B)^{\#}\|^{-1}$.

Remark 2 In general one can not hope that the left-hand sides of (5)–(7) may be replaced by $\dim E^*(P_m(A-B)P_m)$, because there exist B_∞ and $B \in C(S_\tau, \mathcal{L}_s(\mathbf{R}^{2n}))$ such that the equality in (4.1.15) does not hold. Since $P_m(A-B)P_m$ is an approximation of $A - B$, the m^*_δ index introduced by G. Fei and Q. Qiu in [FeQ2] is reasonable and reflects the possibility of perturbations of zero eigenvalues of $A - B$ to nearby positive or negative ones.

In order to prove this theorem, we need the following result from T. Kato's celebrated book [Kat1] whose proof is omitted here. Recall that $\mathcal{L}_b(W)$ and $\mathcal{L}_c(W)$ denote the sets of bounded and compact linear operators on W respectively. For any T and $T_k \in \mathcal{L}_b(W)$ for $k \in \mathbf{N}$, T is the strong limit of T_n and denoted by $s\text{-}\lim_{k\to+\infty} T_k$ if $T_k w \to Tw$ in W as $k \to +\infty$ for any $w \in W$.

Proposition 3 (Problem VI.2.36 and Lemma VI.5.6 of [Kat1]). *Let $P^+_B : W \to E^+(M + B)$ be the orthogonal projection for given $M \in \mathcal{L}_b(W)$ and $B \in \mathcal{L}_c(W)$. Suppose $M + B$ is invertible. Then*

$$P^+_B = \frac{1}{2}(U^2_B(0) + U_B(0)), \tag{8}$$

where

$$\begin{aligned} &U_{B,r,t}(0) \\ &\equiv \frac{2}{\pi}\int_r^t ((M + B)^2 + y^2)^{-1}(M + B)dy \\ &= U_{0,r,t}(0) + \frac{2}{\pi}\int_r^t ((M + B)^2 + y^2)^{-1}Bdy \\ &\quad -\frac{2}{\pi}\int_r^t ((M + B)^2 + y^2)^{-1}(B^2 + BM + MB)(M^2 + y^2)^{-1}Mdy, \end{aligned}$$

and

$$U_B(0) = s- \lim_{\substack{r\to 0 \\ t\to +\infty}} U_{B,r,t}(0).$$

Lemma 4 *Let $M \in \mathcal{L}_b(W)$ and $B \in \mathcal{L}_c(W)$ with $M + B$ being invertible. Using notation in Proposition* 3,

$$P_B^+ - P_0^+ \in \mathcal{L}_c(W). \tag{9}$$

Proof: Since $M + B$ is invertible and $B \in \mathcal{L}_c(W)$, there exist K_1 and $K_2 \in \mathcal{L}_c(W)$ such that

$$U_B(0) = U_0(0) + K_1 + K_2, \quad U_B^2(0) = U_0^2(0) + K_3,$$

where

$$K_3 = (K_1 + K_2)^2 + (K_1 + K_2)U_0(0) + U_0(0)(K_1 + K_2) \in \mathcal{L}_c(W).$$

Therefore

$$P_B^+ - P_0^+ = \frac{1}{2}(K_1 + K_2 + K_3) \in \mathcal{L}_c(W).$$

This proves the lemma. □

Definition 5 (G. Fei [Fei1], G. Fei and Q. Qiu [FeQ2]). For any $M \in \mathcal{L}_b(W)$ and $B_j \in \mathcal{L}_c(W)$ with $j = 1$ and 2, the relative Morse index of $(M - B_1, M - B_2)$ is defined by

$$\begin{aligned} I(M - B_1, M - B_2) &= \dim[E^+(M - B_1) \cap E^-(M - B_2)] \\ &- \dim[(E^-(M - B_1) \oplus E^0(M - B_1)) \cap (E^+(M - B_2) \oplus E^0(M - B_2))]. \end{aligned} \tag{10}$$

Theorem 6 (G. Fei and Q. Qiu [FeQ2]). *Let $M \in \mathcal{L}_b(W)$, B_j and $S_j \in \mathcal{L}_c(W)$ satisfy $E^0(M - B_j - S_j) = \{0\}$ and $\|S_j\| < \|(M - B_j)^{\#}\|^{-1}$ for $j = 1, 2$. Then*

$$\begin{aligned} &I(M - B_1 - S_1, M - B_2 - S_2) - \dim E^0(M - B_1) - \dim E^0(M - B_2) \\ &\le I(M - B_1, M - B_2) \le I(M - B_1 - S_1, M - B_2 - S_2). \end{aligned} \tag{11}$$

Proof: Let P_j^* and Q_j^* denote the orthogonal projections from W to $E^*(M - B_j)$ and $E^*(M - B_j - S_j)$ respectively for $j = 1, 2$ and $* = +, -, 0$. Let

$$\begin{aligned} T &= (P_2^+ + P_2^0)P_1^+ : P_1^+W \to P_2^+W \oplus P_2^0W, \\ T_s &= (P_2^+ + P_2^0)Q_1^+ : Q_1^+W \to P_2^+W \oplus P_2^0W. \end{aligned}$$

Then both T and T_s are Fredholm operators and their indices satisfy

$$\mathrm{ind}T = I(M-B_1, M-B_2), \quad \mathrm{ind}T_s = I(M-B_1-S_1, M-B_2).$$

Let

$$T_s' = (P_2^+ + P_2^0)Q_1^+P_1^+ : P_1^+W \to P_2^+W \oplus P_2^0W.$$

By Lemma 4, it holds that $Q_1^+ - P_1^+ \in \mathcal{L}_c(W)$. Noticing that

$$T_s(Q_1^+P_1^+) = T_s' = T + (P_2^+ + P_2^0)(Q_1^+ - P_1^+)P_1^+,$$

we obtain

$$\mathrm{ind}T = \mathrm{ind}T_s' = \mathrm{ind}T_s + \mathrm{ind}(Q_1^+P_1^+).$$

Since $E^0(M-B_1-S_1) = \{0\}$ and $\|S_1\| < \|(M-B_1)^\#\|^{-1}$, we get

$$-\dim E^0(M-B_1) \le \mathrm{ind}(Q_1^+P_1^+) \le 0.$$

Therefore we have

$$\begin{aligned} I(M-B_1-S_1, M-B_2) &- \dim E^0(M-B_1) \\ &\le I(M-B_1, M-B_2) \\ &\le I(M-B_1-S_1, M-B_2). \end{aligned} \tag{12}$$

Similarly we obtain

$$\begin{aligned} I(M-B_1-S_1, M-B_2-S_2) &- \dim E^0(M-B_2) \\ &\le I(M-B_1-S_1, M-B_2) \\ &\le I(M-B_1-S_1, M-B_2-S_2). \end{aligned} \tag{13}$$

Then (11) follows from (12) and (13). □

Theorem 7 (G. Fei [Fei1], G. Fei and Q. Qiu [FeQ2]). *It holds that*

$$I(A-B_1, A-B_2) = i_1(B_2) - i_1(B_1) - \nu_1(B_1), \tag{14}$$

for $B_j \in C(S_\tau, \mathcal{L}_s(\mathbf{R}^{2n}))$ with $j = 1, 2$.

Proof: Denote by γ_j the fundamental solution of the linear Hamiltonian system with coefficient $B_j(t)$ for $j = 1, 2$. We continue the proof in two cases.

Case 1 $\nu_1(B_1) = \nu_1(B_2) = 0$.

In this case, by our discussion in Section 5.3, there are two standard matrix functions $\hat{B}_{k_j}$ with $j = 1, 2$ defined by (5.3.8)–(5.3.15) such that the corresponding fundamental solutions satisfy $\hat{\beta}_{k_j} \sim_1 \gamma_j$. Denote the corresponding homotopy matrix functions by $B_{j,s}$ for $j = 1, 2$ and $s \in [0, 1]$. Then

$$\nu_1(B_{j,s}) = 0, \quad \forall s \in [0, 1], j = 1, 2.$$

This is

$$E^0(A - B_{j,s}) = \{0\}, \qquad \text{for } s \in [0, 1], j = 1, 2.$$

Thus by Theorem 6, the index $I(A - B_{1,s}, A - B_{2,s})$ is locally constant for $s \in [0, 1]$, and thus is constant on $[0, 1]$. Then we obtain

$$I(A - B_1, A - B_2) = I(A - \hat{B}_{k_1}, A - \hat{B}_{k_2}).$$

By direct verification, as in our discussion in Section 6.1, which is left to the readers, we obtain

$$I(A - \hat{B}_{k_1}, A - \hat{B}_{k_2}) = i_1(B_{k_2}) - i_1(B_{k_1}).$$

Then (14) follows in this case.

Case 2 $\nu_1(B_1) \neq 0$ or $\nu_1(B_2) \neq 0$.

In this case, let γ_j be the associated symplectic paths of B_j and $\gamma_{j,s}$ be the perturbation paths of γ_j defined by (5.4.6), and $B_{j,s}$ be coefficients of the corresponding Hamiltonian systems for $j = 1, 2$ and $s \in [-1, 1]$ respectively. Then

$$\nu_1(B_{j,s}) = 0, \quad \forall s \in [-1, 1]\backslash\{0\}, j = 1, 2,$$

and

$$\|B_{j,s} - B_j\| \to 0, \quad \text{as } s \to 0.$$

Therefore by Theorem 6 and the above Case 1, for $s > 0$ sufficiently close to 0 we obtain

$$\begin{aligned} I(A - B_1, A - B_2) &\leq I(A - B_{1,s}, A - B_{2,-s}) \\ &= i_1(\gamma_{2,-s}) - i_1(\gamma_{1,s}) \\ &= i_1(\gamma_2) - i_1(\gamma_1) - \nu_1(\gamma_1), \end{aligned}$$

$$\begin{aligned} &I(A - B_1, A - B_2) \\ &\geq I(A - B_{1,-s}, A - B_{2,s}) - \dim E^0(A - B_1) - \dim E^0(A - B_2) \\ &= i_1(\gamma_{2,s}) - i_1(\gamma_{1,-s}) - \nu_1(\gamma_1) - \nu_1(\gamma_2) \\ &= i_1(\gamma_2) - i_1(\gamma_1) - \nu_1(\gamma_1). \end{aligned}$$

Thus (14) holds. □

We need the following well-known algebraic lemma whose proof is left to the readers.

Lemma 8 *Suppose Z is a finite dimensional Hilbert space, X and Y are subspaces of Z. Then*

$$\dim(X \cap Y^{\perp}) - \dim(X^{\perp} \cap Y) = \dim X - \dim Y. \tag{15}$$

In order to approximate degenerate operators by non-degenerate ones, we need corresponding result on eigenprojections of bounded operators in a Banach space X. By Remark IV.2.16 of [Kat1], the topology on $\mathcal{L}_b(X)$ induced by the gap function $\hat{\delta}(\cdot,\cdot)$ is identical to the norm topology. Therefore Theorem IV.3.16 of [Kat1] implies the following theorem:

Theorem 9 (cf. Theorem IV.3.16 of [Kat1]). *Let X be a Banach space and $T \in \mathcal{L}_b(X)$ be a closed operator. Suppose $\sigma(T)$ contains a bounded part $\sigma_0(T)$ separated from the rest $\sigma_1(T)$ in such a way that $\sigma_0(T)$ consists of finitely many eigenvalues with finite multiplicities, and a rectifiable simple closed curve Γ in $\mathbf{C}$ can be drawn so as to enclose an open set containing $\sigma_0(T)$ in its interior. Let $X = M_0(T) \oplus M_1(T)$ be a decomposition of invariant subspaces of T such that $\sigma(T|_{M_i(T)}) = \sigma_i(T)$ for $i = 0$ and 1. Denote by $P_0(T): X \to M_0(T)$ the projection along $M_1(T)$. Then there exists a $\delta > 0$ such that any $S \in \mathcal{L}_b(X)$ that satisfies $\|S - T\| < \delta$ has a spectrum separated by Γ into two parts $\sigma_0(S)$ and $\Sigma_1(S)$, in the corresponding decomposition $X = M_0(S) \oplus M_1(S)$, $M_i(S)$ is isomorphic to $M_i(T)$, in particular* $\dim M_i(S) = \dim M_i(T)$*, and the projection $P_0(S): X \to M_0(S)$ along $M_1(S)$ tends to $P_0(T)$ in norm as $\|S - T\| \to 0$.*

Now we can give

Proof of Theorem 1: Without loss of generality, set $\tau = 1$. We divide the proof into three parts.

Case 1 $\nu_1(B_\infty) = \nu_1(B) = 0$.

Similar to Lemma 6.1.2, we obtain $\dim\ker(A - B) = \nu_1(B) = 0$ and $A - B$ is invertible. Then by Corollary 4.1.6 we obtain (6), i.e. there exists $m_1 > 0$ such that

$$m^0(P_m(A - B)P_m) = 0, \qquad \forall m \geq m_1. \tag{16}$$

Since the operators B and B_∞ on E are compact, there exists $m_2 \geq m_1$ such that for all $m \geq m_2$,

$$\begin{aligned} &\|(I - P_m)(B_\infty - B)\| + \|(B_\infty - B)(I - P_m)\| \\ &\leq \min\left\{2\delta, \frac{1}{4}\|A - B\|\right\}. \end{aligned} \tag{17}$$

Recall that $\delta = \frac{1}{4}\|(A - B)^{\#}\|^{-1}$. From

$$P_m(A - B)P_m = (A - B)P_m + (P_m - I)(B_\infty - B)P_m,$$

for any $m \geq m_2$,

$$\|P_m(A-B)P_m x\| \geq 2\delta\|x\|, \qquad \forall x \in W_m. \tag{18}$$

This implies

$$E_\delta^*(P_m(A-B)P_m) = E^*(P_m(A-B)P_m), \quad \text{for } * = -, 0, +, \tag{19}$$

for $m \geq m_2$. Therefore (6) holds by (16).

Let $Q_m = B_\infty + P_m(B - B_\infty)P_m$ and

$$F_m = (I - P_m)(B_\infty - B) + P_m(B_\infty - B)(I - P_m).$$

Then

$$A - B = A - Q_m + F_m.$$

By the definition of Q_m, there exists $m_3 \geq m_2$ such that for all $m \geq m_3$,

$$\|B - Q_m\| \leq \frac{1}{4}\|A - B\|.$$

This yields

$$\|A - Q_m\| \geq \|A - B\| - \|B - Q_m\| \geq \frac{1}{2}\|A - B\|.$$

Together with (17) this proves that $A - Q_m + sF_m$ for all $s \in [0, 1]$ defines a non-degenerate homotopy from $A - Q_m$ to $A - B$ provided $m \geq m_3$.

By the same homotopy argument as that in the proof of Case 1 of Theorem 7, we obtain

$$I(A - B, A - B_\infty) = I(A - Q_m, A - B_\infty). \tag{20}$$

By the definitions of $\{P_m\}$ and Q_m, when $m \to \infty$ we obtain

$$\begin{aligned}
\|P_m(A - B_\infty)P_m - (A - B_\infty)\|_{\mathcal{L}_b(W)} &\to 0,\\
\|P_m(A - Q_m)P_m - (A - Q_m)\|_{\mathcal{L}_b(W)} &\to 0.
\end{aligned}$$

Thus by the same homotopy argument in the proof of Theorem 7, there exists $m_4 \geq m_3$ such that for any $m \geq m_4$,

$$I(A - Q_m, A - B_\infty) = I(P_m(A - Q_m)P_m, P_m(A - B_\infty)P_m). \tag{21}$$

By the definition (4.1.14) of W_m, Definition 5, and Lemma 8, we then obtain

$$\begin{aligned}
&I(P_m(A - Q_m)P_m, P_m(A - B_\infty)P_m)\\
&\quad = m^+(P_m(A - B)P_m) - m^+(P_m(A - B_\infty)P_m)\\
&\quad = m^+(P_m(A - B)P_m) - m.
\end{aligned} \tag{22}$$

Together with (19), (20), (21), and Theorem 7, we obtain (5). Then (7) follows from (5) and (6).

Case 2 $\nu_1(B_\infty) > 0$ and $\nu_1(B) = 0$.

Note that by the same proof in Case 1, we obtain (16) and (19). Therefore (6) holds.

In order to prove (5) and (7), let γ be the fundamental solution of the linear Hamiltonian system corresponding to $B_\infty(t)$. For $s \in [-1, 1]$, let γ_s and $B_\infty^s(t)$ be the perturbation paths and matrix functions defined by the corresponding (5.4.6) and (6.1.38) respectively. By our study in Sections 5.4 and 6.1, specifically we have $\nu_1(B_\infty^s) = 0$ for $s \in [-1, 1]\backslash\{0\}$. Recall that we have chosen $\{P_m : W \to W_m\}$ to be a Galerkin approximation scheme with respect to $A - B_\infty$. Let $\{P_m^s : W \to W_m^s\}$ for $s \in [-1, 1]$ be a Galerkin approximation scheme with respect to $A - B_\infty^s$ given by Theorem 9 so that $\|P_m^s - P_m\|_W \to 0$ as $|s| \to 0$. Especially the space $W_m^s = P_m^s W$ satisfies

$$\dim W_m^s = \dim W_m.$$

By $\nu_1(B) = 0$ and (18), we can choose $m_5 > 0$ sufficiently great such that

(i) the non-zero eigenvalues of $A - B$ become non-zero eigenvalues of $P_m(A - B)P_m$ and they are away from 0 by at least $\epsilon > 0$ for any $m \geq m_5$ and some fixed $\epsilon > 0$.

Thus by our discussion at the beginning of Case 2, (6), and (18) again, there exists $t \in (0, 1]$ such that

(ii) $m^0(P_m^s(A - B)P_m^s) = 0$, positive eigenvalues of $P_m(A - B)P_m$ are perturbed to nearby positive eigenvalues of $P_m^s(A - B)P_m^s$, and all positive eigenvalues of $P_m^s(A - B)P_m^s$ are obtained by perturbation from positive eigenvalues of $P_m(A - B)P_m$ for any $m \geq m_5$ and $|s| \leq t$.

Now we choose $m_6 \geq m_5$ such that

(iii) Theorem 1 holds for $P_m^t(A - B)P_m^t$ when we replace P_m in (5)–(7) by P_m^t for any $m \geq m_6$.

By (ii) and a proof similar to that of Lemma 6.1.3 we obtain

$$m^+(P_m^s(A - B)P_m^s) = m^+(P_m^t(A - B)P_m^t), \tag{23}$$

for any $m \geq m_6$ and $|s| \leq t$.

When s leaves from $s = 0$, the zero eigenvalue of $A - B_\infty$ becomes positive or negative eigenvalues of $A - B_\infty^s$ whose total multiplicities are denoted by $\mu_+(s)$ and $\mu_-(s)$ respectively, and it holds that

$$\mu_+(s) + \mu_-(s) = \nu_1(B_\infty), \quad \forall 0 < |s| \leq 1. \tag{24}$$

Note that by Theorem 6.1.8, it holds that $\mu_+(s) = \mu_-(-s) = 0$ and $\mu_-(s) = \mu_+(-s) = \nu_1(B_\infty)$ for any $0 < s \leq 1$. But we do not need this result in our following proof.

By our above discussion, for any $m \geq m_6$ and $s = t$, we obtain

$$\begin{aligned} m^+(P_m(A-B)P_m) &= m^+(P_m^{-s}(A-B)P_m^{-s}) \\ &= m + \mu_+(-s) + i_1(B_\infty^{-s}) - i_1(B) \\ &\leq m + \nu_1(B_\infty) + i_1(B_\infty) - i_1(B), \end{aligned} \tag{25}$$

where we have used (23), Case 1 of Theorem 1, and Definition 5.4.2. Similarly we obtain

$$\begin{aligned} m^+(P_m(A-B)P_m) &= m^+(P_m^{s}(A-B)P_m^{s}) \\ &= m + \mu_+(s) + i_1(B_\infty^{s}) - i_1(B) \\ &= m + \mu_+(s) + \nu_1(B_\infty) + i_1(B_\infty) - i_1(B), \end{aligned} \tag{26}$$

where we have used (23), Case 1 of Theorem 1, Theorem 5.4.1, and Definition 5.4.2. Combining (25) and (26) together, we obtain $\mu_+(s) = 0$, $\mu_+(-s) = \nu_1(B_\infty)$, and by (19) we obtain (5). Then (7) follows from (5) and (6).

Case 3 $\nu_1(B) > 0$.

Note that in this case, (18) and then (19) may not hold anymore.

Let γ be the fundamental solution of the linear Hamiltonian system corresponding to $B(t)$. For $s \in [-1, 1]$, let γ_s and $B_s(t)$ be the perturbation paths and matrix functions defined by corresponding (5.4.6) and (6.1.38) respectively. Fix $s_0 \in (0, 1]$ such that $\|B - B_s\| < \delta/2$ for $0 < |s| \leq s_0$. By Cases 1 and 2, there exists $m_7 > 0$ such that for $m \geq m_7$, $0 < |s| \leq s_0$,

$$m^+(P_m(A-B_s)P_m) = m + i_1(B_\infty) - i_1(\gamma_s) + \nu_1(B_\infty), \tag{27}$$

$$m^0(P_m(A-B_s)P_m) = 0, \tag{28}$$

$$m^-(P_m(A-B_s)P_m) = m - i_1(B_\infty) + i_1(\gamma_s). \tag{29}$$

Let $m_8 \geq m_7$ be such that for $m \geq m_8$ it holds that $\dim P_m \ker(A-B) = \nu_1(B)$. Define a subspace Y_m of W_m by $W_m = P_m \ker(A-B) \oplus Y_m$. If there exists some $y \in E_\delta^0(P_m(A-B)P_m) \cap Y_m$ with $\|y\| = 1$, then

$$\|P_m(A-B)P_m y\| \leq \delta \|y\|. \tag{30}$$

But on the other hand, for $y \in Y_m$ we have

$$\begin{aligned} y &= (A-B)^\#(A-B)y \\ &= (A-B)^\#(P_m(A-B)P_m y + (P_m - I)(B - B_\infty)y). \end{aligned}$$

By (17) for $m \geq m_1$ this implies

$$\|y\| \leq \frac{1}{2\delta}\|P_m(A-B)P_m y\|.$$

This contradicts (30) and proves

$$E_{\delta}^{0}(P_m(A-B)P_m)\cap Y_m=\emptyset. \tag{31}$$

Thus by the definition of Y_m we obtain

$$m_{\delta}^{0}(P_m(A-B)P_m)\le \nu_1(B). \tag{32}$$

When $s>0$ is sufficiently small, $P_m(A-B_s)P_m$ and is a small perturbation of $P_m(A-B)P_m$ on the same space W. Note that positive eigenvalues of $P_m(A-B)P_m$ in $[\delta,+\infty)$ are perturbed to nearby positive eigenvalues of $P_m(A-B_s)P_m$ when $|s|>0$ is sufficiently small. Thus we can choose $m_9\ge m_8$ such that for all $m\ge m_9$ and $s>0$ small we obtain

$$\begin{aligned}m_{\delta}^{+}(P_m(A-B)P_m) &\le m^{+}(P_m(A-B_s)P_m)\\ &= m+i_1(B_\infty)+\nu_1(B_\infty)-i_1(B_s)\\ &\le m+i_1(B_\infty)+\nu_1(B_\infty)-i_1(B)-\nu_1(B),\end{aligned} \tag{33}$$

where we have used Cases 1 and 2, Theorem 5.4.1, and Definition 5.4.2.

Similarly for $m\ge m_9$ and $s>0$ sufficiently small, we obtain

$$\begin{aligned}&m_{\delta}^{+}(P_m(A-B)P_m)\\ &\ge m^{+}(P_m(A-B_{-s})P_m)-m_{\delta}^{0}(P_m(A-B)P_m)\\ &= m+i_1(B_\infty)+\nu_1(B_\infty)-i_1(B_{-s})-m_{\delta}^{0}(P_m(A-B)P_m)\\ &= m+i_1(B_\infty)+\nu_1(B_\infty)-i_1(B)-m_{\delta}^{0}(P_m(A-B)P_m,\end{aligned} \tag{34}$$

where we have used Cases 1 and 2, and Definition 5.4.2.

Combining (33), (34), and (32) we obtain (6) and (5). Then (7) follows from (5) and (6).

The proof is complete.

Corollary 10 (G. Fei [Fei1], G. Fei and Q. Qiu [FeQ2]). *Let $\{P_m\}$ be a Galerkin approximation scheme with respect to A. Using notation of Theorem 1, for any $B\in C(S_\tau,\mathcal{L}_s(\mathbf{R}^{2n}))$, there exists $\delta>0$ sufficiently small and $m_0>0$ such that for all $m\ge m_0$,*

$$m_{\delta}^{+}(P_m(A-B)P_m)=m+n-i_1(B)-\nu_1(B), \tag{35}$$

$$m_{\delta}^{0}(P_m(A-B)P_m)=\nu_1(B), \tag{36}$$

$$m_{\delta}^{-}(P_m(A-B)P_m)=m+n+i_1(B). \tag{37}$$

Proof: It suffices to note that $i_1(0)=-n$ and $\nu_1(0)=2n$. □

It is interesting to compare this corollary with Theorem 6.1.1.

Remark 11 Results in this section are due to G. Fei [Feil] of 1995, G. Fei and Q. Qiu [FeQ1] of 1996. Theorem 1 was proved by G. Fei and Q. Qiu in [FeQ2] of 1997. To the author's knowledge, the relative Morse index $I(A - B_1, A - B_2)$ of the above Definition 5 was introduced by G. Fei in [Feil] of 1995. The proof of Theorem 1 follows the idea in [FeQ2]. Here we give a somewhat different proof for Cases 1 and 2 in the proof of Theorem 1 which contains more details.

7.2 Second order Hamiltonian systems

For $\tau > 0$, we consider the periodic boundary problem for second order linear Hamiltonian systems

$$\begin{cases} \ddot{q}(t) + M(t)q(t) = 0, \\ q(\tau) = q(0), \\ \dot{q}(\tau) = \dot{q}(0), \end{cases} \tag{1}$$

where $M \in C(S_\tau, \mathcal{L}_s(\mathbf{R}^n))$ with $S_\tau = \mathbf{R}/(\tau\mathbf{Z})$, $\mathcal{L}(\mathbf{R}^n)$ denotes the set of $n \times n$ real matrices and $\mathcal{L}_s(\mathbf{R}^n)$ its subset of symmetric matrices. Define $H = W^{1,2}(S_\tau, \mathbf{R}^n)$ with inner product and norm:

$$\langle p, q \rangle = \int_0^\tau (\dot{p} \cdot \dot{q} + p \cdot q)dt,$$
$$\|q\| = \langle q, q \rangle^{1/2}.$$

Fix $M \in C(S_\tau, \mathcal{L}_s(\mathbf{R}^n))$. The solutions of (1) are in one to one correspondence with the critical points of the quadratic functional

$$\phi(q) = \int_0^\tau \frac{1}{2}\{|\dot{q}|^2 - M(t)q(t) \cdot q(t)\}dt, \quad \forall q \in H. \tag{2}$$

The space H has an orthogonal decomposition $H = H^+ \oplus H^0 \oplus H^-$ according to ϕ being positive, null, or negative definite respectively. As usual, the Morse index $m^-(M)$ and the nullity $m^0(M)$ of the second order linear Hamiltonian system (1) is defined by

$$m^-(M) = \dim H^-, \qquad m^0(M) = \dim H^0. \tag{3}$$

We have $0 \leq m^-(M) < \infty$, because $\phi'(q)$ is the difference of the identity and a compact operator.

Let $p = \dot{q}$ and $x = (p, q)$. For all $t \in \mathbf{R}$ define

$$B(t) = \begin{pmatrix} I & 0 \\ 0 & M(t) \end{pmatrix}. \tag{4}$$

Then the system (1) is transformed into the first order linear Hamiltonian system

$$\dot{x}(t) = JB(t)x(t). \tag{5}$$

Let γ be the fundamental solution of (5). Then our index theory in Chapter 5 assigns to system (5) through γ a pair of integers

$$(i_1(M), \nu_1(M)) = (i_1(\gamma), \nu_1(\gamma)) \in \mathbf{Z} \times \{0, 1, \ldots, 2n\}. \tag{6}$$

Our aim in this paper is to give a variational proof of the coincidence of the Morse indices of system (1) and the index of the corresponding system (5).

Theorem 1 (T. An and Y. Long [AnL1]). *Under the above conditions,*

$$m^-(M) = i_1(M), \qquad m^0(M) = \nu_1(M). \tag{7}$$

The remainder of this section is devoted to the proof of Theorem 1.

Fix $M \in C(S_\tau, \mathcal{L}_s(\mathbf{R}^n))$. We apply the saddle point reduction method in Section 4.3 to the system (5) with coefficient $B(t)$ of (4), and use notations there. Especially we obtain by Theorem 4.3.1 the functional $a_s \in C^2(Z_s, \mathbf{R})$ and the injection map $u_s \in C^2(Z_s, W)$.

Let $2d_s = \dim Z_s$. Denote the positive, null, and negative Morse indices of a_s at $z_s = 0$ by m_s^+, m_s^-, and m_s^0 respectively. By Theorem 6.1.1,

$$\begin{cases} m_s^+ = d_s - i_1(M) - \nu_1(M), \\ m_s^0 = \nu_1(M), \\ m_s^- = d_s + i_1(M). \end{cases} \tag{8}$$

For $x = (p, q) \in W$ with p and $q \in W^{1,2}(S_\tau, \mathbf{R}^n)$, the functional f correspondingly defined by (4.2.8) becomes

$$f(p, q) = \frac{1}{2}\int_0^\tau (\dot{q} \cdot p - \dot{p} \cdot q - |p|^2 - M(t)q \cdot q)dt.$$

The bilinear form Φ of f at 0 is defined by

$$\begin{aligned} \Phi(x, y) &= \frac{1}{2}(f(x + y) - f(x) - f(y)) \\ &= \int_0^\tau (\dot{x}_2 \cdot y_1 + \dot{y}_2 \cdot x_1 - x_1 \cdot y_1 - M(t)x_2 \cdot y_2)dt, \end{aligned} \tag{9}$$

for all $x = (x_1, x_2)$ and $y = (y_1, y_2) \in W$. Let X be the dense subspace of W defined by $X = H \times X_2$, where $X_2 = W^{2,2}(S_\tau, \mathbf{R}^n)$. Define two subspaces of X by

$$W_1 = \{(p, q) \in X \mid q = 0\}, \quad W_2 = \{(p, q) \in X \mid p = \dot{q}\}. \tag{10}$$

Lemma 2 *W_1 and W_2 are Φ-orthogonal and $X = W_1 \oplus W_2$.*

Proof: For any $x = (p, 0) \in W_1$ and $y = (\dot{q}, q) \in W_2$, by direct computation and (9) we obtain $\Phi(x, y) = 0$. On the other hand, for any $x = (p, q) \in X$, we have $x = (p - \dot{q}, 0) + (\dot{q}, q) \in W_1 \oplus W_2$. This proves the lemma. □

Define $Z_s^1 = Z_s \cap W_1$ and $Z_s^2 = Z_s \cap W_2$. Then $Z_s = Z_s^1 \oplus Z_s^2$.

Lemma 3 *The injection $u_s : Z_s \to W$ is linear, maps Z_s^1 and Z_s^2 into W_1 and W_2 respectively, and $u_s(Z_s) \subseteq X$.*

Proof: From the linearities of g_0', $P_s^{\pm}$, A_0, and the equation (4.5.13), we obtain the linearities of $x_s^{\pm}$. Then this yields the linearity of u_s.

We rewrite the equation (4.5.13) into the form

$$S_s^{\pm} x_s^{\pm}(R_s^{-1} z) = \pm (S_s^{\pm})^2 g_0'(u_s(z)). \tag{11}$$

Denote by $P_i^{\pm}$ the projection $P_s^{\pm}|_{V\cap W_i} : V \cap W_i \to Z_s^i$ for $i = 1$ and 2. For $z = (z_1, z_2) \in (W \times X_2) \cap Z_s$, denote $(p^+, q^+) = S_s^+ x_s^+(z)$ and $(p^-, q^-) = S_s^- x_s^-(z)$. Using (4.3.1), (4.3.9), (4.2.4), (4.3.13), and the projection P_0 defined at the beginning of Section 4.3, equation (11) yields

$$\dot{q}^+ + P_0 p^+ = P_1^+(I + P_0)(p^+ + p^- + z_1), \tag{12}$$

$$-\dot{p}^+ + P_0 q^+ = P_2^+(M + P_0)(q^+ + q^- + z_2), \tag{13}$$

$$\dot{q}^- + P_0 p^- = P_1^-(I + P_0)(p^+ + p^- + z_1), \tag{14}$$

$$-\dot{p}^- + P_0 q^- = P_2^-(M + P_0)(q^+ + q^- + z_2), \tag{15}$$

where M is the linear operator induced by $M(t)$. From (12) and (14), we obtain

$$\dot{q}^+ = p^+, \qquad \dot{q}^- = p^-. \tag{16}$$

For $u_s(z) = (u_s^1(z), u_s^2(z)) \in W \times X_2$,

$$u_s^1(z) = p^+(z) + p^-(z) + z_1, \qquad u_s^2(z) = q^+(z) + q^-(z) + z_2.$$

When $z \in Z_s^2$, we have $z_1 = \dot{z}_2$. Thus $u_s^1(z) = \dot{u}_s^2(z)$. Then $u_s(z) \in W_2$ whenever $z \in Z_s^2$.

From (12)–(16) we obtain

$$\begin{aligned} \dot{q}^+ &= p^+, \\ -\dot{p}^+ &= P_2^+ M(t)(q^+ + q^- + z_2), \\ \dot{q}^- &= p^-, \\ -\dot{p}^- &= P_2^- M(t)(q^+ + q^- + z_2). \end{aligned}$$

When $z_2 = 0$, by the uniqueness of this system, $p^+ = p^- = q^+ = q^- = 0$. Thus for $z = (z_1, 0)$, we have $S_s^{\pm} x_s^{\pm}(z) = 0$, and then $u_s(z) = z \in W_1$.

The proof is complete. □

Lemma 4 *For $s > c(M)$ and $s \notin \sigma(A)$, the Morse index of a_s on Z_s is the sum of the Morse indices of $a_s|_{Z_s^1}$ and $a_s|_{Z_s^2}$.*

Proof: Denote by Ψ_s the bilinear form of a_s. Then for $z_i \in Z_s^i$ with $i = 1, 2$, we obtain

$$\begin{aligned}\Psi_s(z_1, z_2) &= \frac{1}{2}(a_s(z_1 + z_2) - a_s(z_1) - a_s(z_2)) \\ &= \frac{1}{2}(f(u_s(z_1) + u_s(z_2)) - f(u_s(z_1)) - f(u_s(z_2))) \\ &= \Phi(u_s(z_1), u_s(z_2)) \\ &= 0.\end{aligned}$$

Here we have used Lemmas 2 and 3. This proves the lemma. □

Lemma 5 *For sufficiently large $s > c(M)$ and $s \notin \sigma(A)$, the Morse index $\hat{m}_s^-$ of $a_s|_{Z_s^2}$ satisfies $\hat{m}_s^- = i_1(M)$.*

Proof: Note that $\dim Z_s = 2d_s$. From direct computations on eigenvectors of A, we obtain $\dim Z_s^1 = d_s$. For any $z = (z_1, 0) \in Z_s \backslash \{0\}$, it holds that $\Psi_s(z, z) = a_s(z) = f(z) < 0$. So Ψ_s is negative definite on Z_s^1, and the Morse index of $a_s|_{Z_s^1}$ is equal to d_s. By (8) and Lemma 4, we obtain $\hat{m}_s^- = i_1(M)$. □

Lemma 6 *Let H be a Hilbert space with an orthonormal basis $\{e_i\}_{i\geq 1}$. Let P_n be the projection from H to $E_n = \operatorname{span}\{e_1, \dots, e_n\}$. Then for any finite dimensional subspace F of H, there exists $n_0 \geq 1$ such that $\dim P_n F = \dim F$ for all $n \geq n_0$.*

Proof: Let $\{v_1, \dots, v_k\}$ be an orthonormal basis of F. Fix $n_0 \geq 1$ such that $\| P_{n_0} v_i - v_i \|_H \leq 1/(k+1)$ for $1 \leq i \leq k$. Then for any $n \geq n_0$ and $(a_1, \dots, a_k) \in \mathbf{R}^k \backslash \{0\}$ we obtain

$$\begin{aligned}\left\| \sum_{i=1}^k a_i P_n v_i \right\|_H &\geq \left\| \sum_{i=1}^k a_i v_i \right\|_H - \sum_{i=1}^k \|a_i(P_n v_i - v_i)\|_H \\ &\geq \left(\sum_{i=1}^k |a_i|^2 \right)^{1/2} - \frac{k}{k+1} \max_{1\leq i\leq k} |a_i| \\ &> 0.\end{aligned}$$

This proves the lemma. □

Lemma 7 *Denote by m_f^- the Morse index of $f|_{W_2}$. Then $m_f^- = m^-(M)$.*

Proof: Note that $f(\dot{q}, q) = \phi(q)$ for all $(\dot{q}, q) \in W_2$. Since X_2 is dense in W and $\dim W_2 = \dim X_2$, we obtain $m_f^- = m^-(M)$. □

Now we can give the proof of our main result.

Proof of Theorem 1: Fix $M \in C(S_\tau, \mathcal{L}_s(\mathbf{R}^{2n}))$. By the Floquet theory, $\nu_1(M)$ is equal to the dimension of the solution space of (5) and then that of (1). Thus we obtain $m^0(M) = \nu_1(M)$.

For the first equality in (7), by Lemmas 5 and 7 it suffices to prove $m_f^- = \hat{m}_s^-$ for s large enough. Since $u_s : Z_s^2 \to W_2$ is a linear injection and $a_s(z) = f(u_s(z))$ for $z \in Z_s^2$, we obtain $\hat{m}_s^- \le m_f^-$ for large s. On the other hand, suppose f is negative definite on a subspace W_2^- of W_2. Since W_2^- is finite dimensional, there exists a constant $\delta > 0$ such that

$$f(x) \le -\delta < 0, \quad \forall x \in W_2^- \text{ with } \|x\| = 1.$$

By Proposition 4.5.3, there exists $s_0 > 0$ such that for $s \ge s_0$,

$$a_s(P_s x) = f(u_s(P_s x)) \le -\frac{1}{2}\delta < 0, \quad \forall x \in W_2^- \text{ with } \|x\| = 1. \tag{17}$$

Thus a_s is negative definite on $P_s(W_2^-)$ by its 2-homogeneity. By Lemma 6, $\dim P_s(W_2^-) = \dim W_2^-$ for large s. This proves $\hat{m}_s^- \ge m_f^-$ and Theorem 1. □

Remark 8 When $M(t) \equiv M$ is a constant symmetric matrix in (1), let $\alpha_1 \le \alpha_2 \le \cdots \le \alpha_n$ be the eigenvalues of M. Then Proposition 9.1 of [MaW1] yields the following formulae for the Morse index and nullity of the system (1):

$$m^-(M) = {}^\#\{k \mid \alpha_k > 0\} + 2\sum_{k=1}^{n} {}^\# \left\{ i \in \mathbf{N} \mid \frac{4\pi^2 i^2}{\tau^2} < \alpha_k \right\}, \tag{18}$$

$$m^0(M) = {}^\#\{k \mid \alpha_k = 0\} + 2\sum_{k=1}^{n} {}^\# \left\{ i \in \mathbf{N} \mid \frac{4\pi^2 i^2}{\tau^2} = \alpha_k \right\}. \tag{19}$$

Therefore $m^-(M)$ ranges over zero and all positive odd integers when $n = 1$ and $M \in \mathcal{L}_s(\mathbf{R}^n)$. Thus $m^-(M)$ ranges over all non-negative integers when $n \ge 2$ and $M \in \mathcal{L}_s(\mathbf{R}^n)$.

For any $\tau > 0$, by Theorem IV.5.3 of [Eke3] the superquadratic second order ordinary differential equation

$$\ddot{x}(t) + 4x^3(t) = 0, \quad x \in \mathbf{R}, \tag{20}$$

possesses a non-constant periodic solution x_1 with minimal period τ. For all $k \in \mathbf{N}$ define

$$x_k(t) = kx_1(kt), \qquad \forall t \in \mathbf{R}. \tag{21}$$

Then x_k is a non-constant periodic solution of (21) with minimal period τ/k. By Theorem 10.5 of [MaW1], for any $k \in \mathbf{N}$ it holds that $m^-(M_k) = 2k$, where $M_k(t) = 12x_k^2(t)$ for all $t \in \mathbf{R}$. Thus when $n = 1$, all the positive even integers are Morse indices of some non-autonomous second order linear Hamiltonian systems, and any second order linear Hamiltonian system possessing a positive even integer valued Morse index must be non-autonomous.

Remark 9 Theorem 1 was proved by T. An and the author in [AnL1] of 1998, which gives a rather simple proof for second order Hamiltonian systems as a special case of Theorem 3.1 below for general Lagrangian systems. It is interesting to point out that the method of Theorem 1 can be generalized to prove Theorem 3.1 below for general Lagrangian systems as having been done recently by A. Abbondandolo in [Abb1].

7.3 Lagrangian systems

In this section, we consider the periodic problem of the calculus of variations, i.e., for $\tau > 0$ finding the critical points of the functional

$$F(x) = \int_0^{\tau} L(t, x, \dot{x})dt, \quad \forall x \in W, \tag{1}$$

where $W = W^{1,2}(S_\tau, \mathbf{R}^n)$ with $S_\tau = \mathbf{R}/(\tau\mathbf{Z})$ is a Sobolev space with the usual inner product and norm

$$\langle x, y\rangle = \int_0^{\tau} (x \cdot y + \dot{x} \cdot \dot{y})dt,$$

$$\|x\| = \langle x, x\rangle^{1/2},$$

for all $x, y \in W$. In this section we suppose the Lagrangian function $L = L(t, x, p)$ satisfies

(L1) $L \in C^2(S_\tau \times \mathbf{R}^{2n}, \mathbf{R})$.

(L2) (The Legendre convexity condition)

$$L_{p,p}(x, p)z \cdot z > 0, \qquad \forall z \in \mathbf{R}^n \setminus \{0\}, \quad (x, p) \in \mathbf{R}^n \times \mathbf{R}^n.$$

Under these conditions, $F \in C^2(W, \mathbf{R})$. A critical point x of F, i.e., an extreme loop of F, corresponds to a solution of the periodic boundary value problem of the Lagrangian system

$$\begin{cases} \frac{d}{dt}L_p(t, x, \dot{x}) - L_x(t, x, \dot{x}) = 0. \\ x(\tau) = x(0), \quad \dot{x}(\tau) = \dot{x}(0). \end{cases} \tag{2}$$

Fix such an extremal loop x and define

$$\begin{cases} P(t) = L_{p,p}(t, x(t), \dot{x}(t)), \\ Q(t) = L_{x,p}(t, x(t), \dot{x}(t)), \\ R(t) = L_{x,x}(t, x(t), \dot{x}(t)), \end{cases} \tag{3}$$

where L_x and $L_{\dot{x}}$ denote the corresponding gradients. Note that in this case, P and $R \in C(S^1, \mathcal{L}_s(\mathbf{R}^n))$ and $Q \in C(S^1, \mathcal{L}(\mathbf{R}^n))$.

Suppose $x \in W$ is an extreme loop of the functional F. Then x is a τ-periodic solution of problem (2). The Hessian of F at x is given by

$$\langle F''(x)y, z\rangle_W = \int_0^1 \{(P\dot{y} + Qy) \cdot \dot{z} + Q^T\dot{y} \cdot z + Ry \cdot z\}dt, \tag{4}$$

for all y and $z \in W$, where P, Q, and R are defined in (3). Therefore the linearized system of (2) at x is given by the **Sturm system**:

$$-(P\dot{y} + Qy)^{\cdot} + Q^T\dot{y} + Ry = 0. \tag{5}$$

Denote the Morse index and nullity of F at x by $m^-(x)$ and $m^0(x)$, i.e., the total multiplicities of all the negative eigenvalues and zeros of $F''(x)$ respectively.

Note that the linear periodic Sturm system (5) corresponds to the linear Hamiltonian system

$$\dot{z} = JB(t)z, \qquad z \in \mathbf{R}^{2n}, \tag{6}$$

with

$$B(t) \equiv B_x(t) = \begin{pmatrix} P^{-1}(t) & -P^{-1}(t)Q(t) \\ -Q(t)^T P^{-1}(t) & Q(t)^T P^{-1}(t)Q(t) - R(t) \end{pmatrix}. \tag{7}$$

Denote the fundamental solution of this system (6)–(7) by γ_x, and its index by $(i_1(\gamma_x), \nu_1(\gamma_x))$. Then the following is the main result in this section.

Theorem 1 (C. Viterbo [Vit2], Y. Long and T. An [LoA1]). *Under the above conditions,*

$$m^-(x) = i_1(\gamma_x), \quad m^0(x) = \nu_1(\gamma_x). \tag{8}$$

The rest of this section is devoted to the proof of this theorem. We need the following results from [Dui1] of J. Duistermaat in 1976.

Let (E, ω) be a real symplectic vector space and Λ be the set of all Lagrangian subspaces of (E, ω). For $\alpha \in \Lambda$, let

$$\Lambda^k(\alpha) = \{\beta \in \Lambda \mid \dim \beta \cap \alpha k\}. \tag{9}$$

If $\alpha, \beta \in \Lambda$, $\alpha \cap \beta = \{0\}$, then any d-dimensional linear subspace γ of E with $\gamma \cap \beta = \{0\}$ can be written as

$$\gamma = \{u + Au \mid u \in \alpha\} \tag{10}$$

for a linear mapping $A : \alpha \to \beta$. Then the bilinear form

$$Q(\alpha, \beta; \gamma) : (u, v) \mapsto \omega(Au, v), \qquad \forall u, v \in \alpha, \tag{11}$$

is symmetric if and only if $\gamma \in \Lambda$, so

$$Q(\alpha, \beta) : \gamma \mapsto Q(\alpha, \beta; \gamma) \tag{12}$$

is a bijection from $\Lambda^0(\beta)$ to the space $S^2\alpha$ of symmetric bilinear forms on α.

Here note that in the actual computation the map A is obtained as a real matrix on E and is not unique in general. But the symmetric bilinear form $Q(\alpha, \beta, \gamma)|_\alpha$ is uniquely determined by α, β, and γ. We denote the negative and positive Morse indices of Q by $\mathrm{ind}^- Q$ and $\mathrm{ind}^+ Q$.

Note that by definition (9)–(10) for α, β, and $\gamma \in \Lambda$ satisfying $\alpha \cap \beta = \{0\}$ and $\alpha \cap \gamma = \{0\}$,

$$Q(\alpha, \beta; \gamma) = -Q(\alpha, \gamma; \beta). \tag{13}$$

Note that if $\beta \cap \gamma = \{0\}$ further holds, then

$$Q(\alpha, \beta; \gamma) = -Q(\beta, \gamma; \alpha). \tag{14}$$

Definition 2 (Definition 2.3 of [Dui1]). The index of a continuous curve $\Gamma \in \Lambda^0([a, b], \Lambda)$, not necessarily closed, is given by

$$\mathrm{ind}(\Gamma) = [\Gamma : \Psi] + \mathrm{ind}^- Q(\Gamma(b), \beta; \Gamma(a)), \tag{15}$$

for any $\Psi \in \Lambda^0(\Gamma(a)) \cap \Lambda^0(\Gamma(b))$, where $[\Gamma : \Psi]$ is the intersection number of Γ with Ψ. This definition is independent of the choice of Ψ in $\Lambda^0(\Gamma(a)) \cap \Lambda^0(\Gamma(b))$.

Let X be a C^2-manifold of dimension n and $L \in C^2(Z, \mathbf{R})$, where Z is an open subset of $\mathbf{R} \times TX$ and TX is the tangent bundle of X. Define

$$f(c) = \int_0^\tau L(t, c(t), \dot{c}(t))dt, \tag{16}$$

for all $c \in C^1([0, \tau], Z)$ with the usual uniform convergence topology of curves and their derivatives.

Proposition 3 (Proposition 4.6 of [Dui1]). *If c is a critical point of f with respect to the periodic boundary condition*

$$(c(a), c(\tau)) \in R = \{(x, y) \in X \times X \mid x = y\} \tag{17}$$

and Γ is a covering such that $D_y\Gamma(0, 0) = D_y\Gamma(\tau, 0)$, then the Morse index of f at c satisfies

$$i_R(c) = \mathrm{ind}(\varphi) - n, \tag{18}$$

where φ denotes the curve $t \mapsto$ $\mathrm{graph}\Phi(0, t)$*, t running from* 0 *to τ.*

To keep our discussion short, we refer readers to [Dui1] for details of the proofs of the above results.

Proof of Theorem 1: Without loss of generality, set $\tau = 1$. The second identity in (8) follows directly from the classical Floquet theory. The proof of the first identity in (8) is carried out below in six steps. In the following we denote by $V_{4n} \equiv (\mathbf{R}^{4n}, J \oplus (-J))$, Λ the set of Lagrangian subspaces in V_{4n}, and by $\Lambda^0(\Gamma)$ the set of all $\Psi \in \Lambda$ such that $\Psi \cap \Gamma = \{0\}$.

Step 1 By the above Proposition 3 and Definition 2,

$$\begin{aligned} m^-(x) &= \mathrm{ind}(\Gamma_x) - n \\ &\equiv [\Gamma_x : \Psi] + \mathrm{ind}^- Q(\Gamma_x(1), \Psi; \Gamma_x(0)) - n, \end{aligned} \tag{19}$$

where Γ_x is the graph of γ_x in the symplectic space V_{4n}, which forms a path in Λ, and $\Psi \in \Lambda^0(\Gamma_x(1)) \cap \Lambda^0(\Gamma_x(0))$. By (19), to get the first identity in (8), it suffices to show that for the graph Γ_x of γ_x,

$$i_1(\gamma_x) + n = [\Gamma_x : \Psi] + \mathrm{ind}^- Q(\Gamma_x(1), \Psi; \Gamma_x(0)), \tag{20}$$

for any $\Psi \in \Lambda^0(\Gamma_x(1)) \cap \Lambda^0(\Gamma_x(0))$.

Step 2 If γ_x is degenerate, for any sufficiently small $\epsilon > 0$, by our discussion in Chapters 5 and 6, we can construct a short path $\xi : [0,1] \to \mathrm{Sp}(2n)$ such that $\xi(0) = \gamma_x(1)$, $\xi(t) \in \mathrm{Sp}(2n)^*$ for all $0 < t \leq 1$, and $i_1(\xi * \gamma_x) = i_1(\gamma_x)$. Denote the graph of $\xi * \gamma_x$ by Ξ_x. On the other hand, since $\Psi \in \Lambda^0(\Gamma_x(1)) \cap \Lambda^0(\Gamma_x(0))$ is an open condition, for small enough $\epsilon > 0$ we still have $\Psi \in \Lambda^0(\Xi_x(1)) \cap \Lambda^0(\Xi_x(0))$, $[\Xi_x : \Psi] = [\Gamma_x : \Psi]$, and

$$\mathrm{ind}^- Q(\Xi_x(1), \Psi; \Xi_x(0)) = \mathrm{ind}^- Q(\Gamma_x(1), \Psi; \Gamma_x(0)).$$

Thus it suffices to prove (20) for the path $\xi * \gamma_x \in \mathcal{P}_1^*(2n)$ and its graph Ξ_x in Λ with a small enough $\epsilon > 0$.

Step 3 Since both Duistermaat's "ind(Γ)" and the index $i_1(\gamma)$ are homotopy invariant, we only need to compute the both sides of (20) with γ_x being replaced by a special path in each homotopy class $\mathcal{P}_{1,1}^*(2n)$ defined in Section 5.2. Then it suffices to prove (20) for each standard zigzag path $\phi_k \equiv \hat{\alpha}_k \in \mathcal{P}_{1,1}^k(2n)$ for $k \in \mathbf{Z}$ defined in Section 5.3 and its graph Φ_k with $k \in \mathbf{Z}$.

Step 4 Since the path ϕ_k is a $\diamond$-product of paths in $\mathcal{P}(2)^*$, both the Duistermaat "ind" and the index i_1 are symplectically additive, we only need to verify (20) for each $\diamond$-component of ϕ_k and its graph respectively. Thus without loss of generality, we suppose $\phi_k = \hat{\alpha}_k \in \mathcal{P}_1(2)$ is given by (5.3.16) or (5.3.17) with $n = 1$. We still denote the graph of $\phi_k \in \mathcal{P}_1(2)$ by Φ_k. From $i_1(\phi_k) = k$ of (5.3.18), to get (20) it suffices to prove the equalities

$$[\Phi_k : \Psi] + \mathrm{ind}^- Q(\Phi_k(1), \Psi; \Phi_k(0)) = k + 1, \tag{21}$$

for each $k \in \mathbf{Z}$ and some $\Psi \in \Lambda^0(\Phi_k(1)) \cap \Lambda^0(\Phi_k(0))$.

Step 5 The proof of (21) for odd integers k.

Fix $k \in 2\mathbf{Z} - 1$. Then Φ_k is a path in Λ, and

$$\Phi_k(1) = \Gamma_- \equiv \{(x_1, x_2, -2x_1, -x_2/2)^T \mid x_1, x_2 \in \mathbf{R}\}, \tag{22}$$

$$\Phi_k(0) = \Delta, \tag{23}$$

where Δ is the diagonal of V_4. Choose

$$\Psi = \Gamma_+ \equiv \{(x_1, x_2, 2x_1, x_2/2)^T \mid x_1, x_2 \in \mathbf{R}\} \in \Lambda. \tag{24}$$

Note that here we choose Ψ so that the three Lagrangian subspaces Ψ, Γ_-, and Γ_+ of V_4 are mutually transverse. By direct computation we obtain that

$$\Phi_k(t) \cap \Gamma_+ \neq \emptyset \quad \text{if and only if } \cos(tk\pi) = 4/5,$$

and the intersection number at such a time t is 1. When t runs from 0 to 1, there are precisely k such times. Therefore we obtain

$$[\Phi_k : \Psi] = k. \tag{25}$$

On the other hand,

$$Q(\Phi_k(1), \Phi; \Psi_k(0)) = Q(\Gamma_-, \Gamma_+; \Delta) = -Q(\Gamma_-, \Delta; \Gamma_+), \tag{26}$$

where the first equality follows from the definitions of Φ_k and Ψ, the second equality follows from (13). By a direct computation we obtain that the 4×4 matrix $A : \Gamma_- \to \Delta$ determined by Γ_+ via (10) can be chosen to have the form

$$A = \begin{pmatrix} 0 & 0 & 2 & 0 \\ 0 & 0 & 0 & -4 \\ -4 & 0 & 0 & 0 \\ 0 & 2 & 0 & 0 \end{pmatrix}.$$

By (11), the quadratic form $Q(\Gamma_-, \Delta; \Gamma_+)$ is defined to be the restriction of the quadratic form given by the 4×4 matrix $[J \oplus (-J)]A$ to the Lagrangian subspace Γ_- of V_4. By direct computation we obtain that for any $x = (x_1, x_2, -2x_1, -x_2/2)^T$ and $y = (y_1, y_2, -2y_1, -y_2/2)^T \in \Gamma_-$,

$$[J \oplus (-J)]Ax \cdot y = \begin{pmatrix} 0 & -6 \\ -6 & 0 \end{pmatrix} \begin{pmatrix} x_1 \\ x_2 \end{pmatrix} \cdot \begin{pmatrix} y_1 \\ y_2 \end{pmatrix}. \tag{27}$$

Thus we obtain

$$\operatorname{ind}^+ Q(\Gamma_-, \Delta; \Gamma_+) = \operatorname{ind}^- Q(\Gamma_-, \Delta; \Gamma_+) = 1. \tag{28}$$

Now (25), (26), and (28) yield (21) and then (8).

Step 6 The proof of (8) for even integers k.

Fix $k \in 2\mathbf{Z}$. Then (23) holds and $\Phi_k(1) = \Gamma_+$. Choose $\Psi = \Gamma_-$. By direct computation we obtain that

$$\Phi_k(t) \cap \Gamma_- \neq \emptyset \quad \text{if and only if } \cos(tk\pi) = -4/5,$$

and the intersection number at such a time t is 1. When t runs from 0 to 1, there are precisely k such times. Therefore (25) still holds.

On the other hand,

$$Q(\Phi_k(1), \Psi; \Phi_k(0)) = Q(\Gamma_+, \Gamma_-; \Delta) = -Q(\Gamma_-, \Delta; \Gamma_+), \tag{29}$$

where the first equality follows from the definitions of Φ_k and Ψ, the second equality follows from (14). Thus by (25), (28), and (29), we obtain (21) and then (8).

The proof of Theorem 1 is complete. □

Next we study the relationship between our index function theory and the index functions of R. Bott defined in [Bot1]. The proof of the following Theorem 4 needs some knowledge of the splitting numbers defined in Section 9.1 below. Note that the Bott index function is defined by the Morse index and nullity $(\Lambda(\omega), N(\omega))$ of the Lagrangian system in (2) with the boundary value conditions

$$x(\tau) = \omega x(0), \quad \dot{x}(\tau) = \omega \dot{x}(0). \tag{30}$$

As before, we denote by γ_x the fundamental solution of the linear Hamiltonian system (6) with $B(t)$ defined by (7).

Theorem 4 *Under the above assumptions,*

$$i_\omega(\gamma_x) = \Lambda(\omega), \quad \nu_\omega(\gamma_x) = N(\omega), \quad \forall \omega \in \mathbf{U}. \tag{31}$$

Proof: Note that the second equality holds by definition of the two sides. To prove the first equality, denote by $BS_M^{\pm}(\omega)$ the splitting numbers defined on page 180 of [Bot1]. Then by a direct verification we obtain that $(BS_M^+(\omega), BS_M^-(\omega))$ satisfies the four axioms in Theorem 9.1.10 below for all $M \in \cup_{n\geq 1}\mathrm{Sp}(2n)$. Therefore it coincides with our splitting numbers $(S_M^+(\omega), S_M^-(\omega))$ defined in Section 9.1 below. Thus we obtain

$$\Lambda(\omega) - i_\omega(\gamma_x) = \Lambda(1) - i_1(\gamma_x). \tag{32}$$

By Theorem 1 we obtain that the right-hand side of (32) is zero. Thus the first equality in (31) holds. □

Remark 5 Theorem 1 was first proved by C. Viterbo in [Vit2] of 1990. The proof here was given by T. An and the author in [LoA1] of 1998. Theorem 4 was proved in [Lon16] of 1999.

It is well known that there are infinitely many ways to define the index theory for symplectic paths in the non-degenerate case and degenerate case. The existing definitions in many other papers are different from ours. But in the sense of Theorems 1.1, 2.1, and 3.1 in this chapter, as well as Theorems 6.1.1 and 15.1.1, definitions of the index theory in Chapter 5 for both the non-degenerate case and the degenerate case are natural.

Part IV
Iteration Theory

As mentioned in the Introduction of this book, iteration theories of various Morse and Maslov type index theories were studied by many mathematicians for closed geodesics on Riemannian manifolds, periodic solutions of convex Hamiltonian systems and non-degenerate star-shaped Hamiltonian systems. In order to study general nonlinear Hamiltonian systems, we have established the iteration theory for the index of symplectic paths in our works since 1993. Part IV is devoted to this index iteration theory.

Part IV contains five chapters. In Chapter 8, we establish precise iteration formulae for any symplectic paths. In Chapter 9 we study splitting numbers and give its algebraic characterization, establish the Bott-type iteration formulae and abstract precise iteration formula. These two chapters form the basics for the iteration theory. In Chapter 10, we establish various iteration inequalities for any single symplectic path. In Chapter 11, we study the common index iteration properties for a finite family of symplectic paths. These two chapters give us powerful tools in applications. In Chapter 12, we show that this iteration theory also works for closed geodesics on Riemannian manifolds.

Chapter 8
Precise iteration formulae

For $\tau > 0$, let $x : S_\tau = \mathbf{R}/(\tau\mathbf{Z}) \to \mathbf{R}^{2n}$ be a τ-periodic solution of the nonlinear Hamiltonian system

$$\dot{x} = JH'(t, x). \tag{1}$$

Here we suppose H is C^2 along the orbit $\mathcal{O}(x) = x(\mathbf{R})$ and τ-periodic in time t. We define the iteration of the τ-periodic function x by

$$x^m(t) = x(t - j\tau), \quad j\tau \le t \le (j+1)\tau, \ \ 0 \le j \le m-1. \tag{2}$$

Then x^m becomes an $m\tau$-periodic solution of the system (1). But geometrically it is the same as x. To further our study we consider the linearized Hamiltonian system

$$\dot{y}(t) = JB(t)y(t), \tag{3}$$

where $B \equiv H''(\cdot, x(\cdot)) \in C(S_\tau, \mathcal{L}_s(\mathbf{R}^{2n}))$. As we have seen before, the fundamental solution $\gamma_x \in \mathcal{P}_\tau(2n)$ of (3) carries significant information about x. For this reason, in order to study the sequence $\{x^m\}_{m\in\mathbf{N}}$, for any $\tau > 0$ and $\gamma \in \mathcal{P}_\tau(2n)$, we extend the definition of γ to $[0, +\infty)$ by

$$\gamma(t) = \gamma(t - j\tau)\gamma(\tau)^j, \quad \forall j\tau \le t \le (j+1)\tau, \ j \in \mathbf{N}, \tag{4}$$

and define the **m-th iteration** γ^m of γ by

$$\gamma^m = \gamma|_{[0, m\tau]}. \tag{5}$$

If $\delta : [0,1] \times [0,\tau] \to \mathrm{Sp}(2n)$ is a homotopy in the sense of Definition 5.0.3, its iteration is defined to be the map $\delta : [0,1] \times [0, +\infty) \to \mathrm{Sp}(2n)$ such that for each $s \in [0,1]$ the iteration path $\delta(s, \cdot)$ is defined by (4) and (5). Note that if γ is the fundamental solution of the system (3) for some $B \in C(S_\tau, \mathcal{L}_s(\mathbf{R}^{2n}))$, then by the uniqueness of the initial value problem of (3), the path γ coincides with its iteration defined by (4) completely on $[0, \infty)$. Note that in general the iteration path $\gamma|_{[0,+\infty)}$ of (4) need not be C^1, even if the original path $\gamma|_{[0,\tau]}$ is.

In order to get more information on τ-periodic solutions of the system (1), we consider the index function sequence of $\{\gamma^m\}_{m\in\mathbf{N}}$.

Definition 1 For any $\tau > 0$, $\gamma \in \mathcal{P}_\tau(2n)$, $m \in \mathbf{N}$, and $\omega \in \mathbf{U}$, we define

$$i_\omega(\gamma, m) = i_\omega(\gamma^m), \qquad \nu_\omega(\gamma, m) = \nu_\omega(\gamma^m). \tag{6}$$

We also define

$$\hat{i}(\gamma, m) \equiv \lim_{k\to+\infty} \frac{i_1(\gamma, km)}{k}, \tag{7}$$

whenever this limit exists. By Theorems 8.3.1 and 9.3.5 it holds that $\hat{i}(\gamma, m)$ is always a finite real number. We call this number the **mean index** per $m\tau$ of $\gamma \in \mathcal{P}_\tau(2n)$. We also define

$$\hat{\nu}(\gamma, m) \equiv \lim_{k\to+\infty} \frac{\nu_1(\gamma, km)}{k} = 0. \tag{8}$$

Then for any τ-periodic solution x of (1), via its associated symplectic path γ_x we define

$$i_\omega(x, m) = i_\omega(\gamma_x, m), \quad \nu_\omega(x, m) = \nu_\omega(\gamma_x, m), \tag{9}$$

$$\hat{i}(x, m) = \hat{i}(\gamma_x, m), \quad \hat{\nu}(x, m) = \hat{\nu}(\gamma_x, m) = 0. \tag{10}$$

If the subindex $\omega = 1$, we simply write $(i(\gamma, m), \nu(\gamma, m))$, $(i(x, m), \nu(x, m))$ etc, and omit the subindex 1 when there is no confusion.

Note that (8) follows from the fact that $0 \le \nu(\gamma, m) \le 2n$ for any $m \in \mathbf{N}$.

In this chapter we establish the precise iteration formulae of the index theory for any path in $\mathcal{P}_\tau(2n)$. By homotopy methods and symplectic additivity of the index theory, this study is reduced to certain simple symplectic paths which end at basic normal forms introduced in Definition 1.8.9. These basic forms are in Sp(2) or Sp(4). In Section 1, we establish the iteration formulae for paths in Sp(2). In Section 2, we study the case for hyperbolic paths and elliptic paths in Sp(4). Then in Section 3, based on the basic normal form decomposition Theorem 1.8.10, we establish the iteration formula for any symplectic paths. Compared with the variational method used in the next Chapter 9, the method we used here is completely elementary.

8.1 Paths in Sp(2)

In this section, we study the indices for iterations of any symplectic path in Sp(2) ending in basic normal forms. Our study is based upon the results obtained in Chapters 1 and 2 on the topological structures of symplectic groups and their subsets. For any $a \in \mathbf{R}$, we define the functions

$$\begin{cases} [a] = \max\{k \in \mathbf{Z} \mid k \le a\}, & E(a) = \min\{k \in \mathbf{Z} \mid k \ge a\}, \\ \varphi(a) = E(a) - [a], & \{a\} = a - [a]. \end{cases} \tag{1}$$

Especially, $\varphi(a) = 0$ if $a \in \mathbf{Z}$, and $\varphi(a) = 1$ if $a \notin \mathbf{Z}$.

The following lemmas are frequently used in this chapter for our study on iterations.

Lemma 1 *Let* α *and* $\beta \in \mathcal{P}_\tau(2n)$. *Suppose*

$$\alpha(\tau) \in \Omega^0(\beta(\tau)), \quad i(\alpha, 1) = i(\beta, 1). \tag{2}$$

Then for any $m \in \mathbf{N}$,

$$i(\alpha, m) = i(\beta, m), \quad \nu(\alpha, m) = \nu(\beta, m). \tag{3}$$

Proof: Set $\tau = 1$. By (2), there is a path $f : [0, 1] \to \Omega^0(\beta(1))$ such that $f(0) = \alpha(1)$ and $f(1) = \beta(1)$. Thus by the inverse homotopy Theorem 6.2.4 and (2), it holds that $\alpha \sim \beta$ along f via a homotopy map $\delta : [0, 1]^2 \to \mathrm{Sp}(2n)$. Extending δ to $[0, 1 \times [0, m]$ yields a homotopy from α^m to β^m. Thus by the homotopy Theorem 6.2.3 we obtain (3). □

By direct computation or using the $\mathbf{R}^3$-cylinderical coordinate representation introduced in Section 5.1, we obtain the following result.

Lemma 2 *It holds that*

$$i(\phi_{\tau,\theta}, 1) = 2E\left(\frac{\theta}{2\pi}\right) - 1 = \begin{cases} 2[\theta/(2\pi)] + 1 & \text{if } \theta \neq 0 \bmod 2\pi, \\ 2[\theta/(2\pi)] - 1 & \text{if } \theta = 0 \bmod 2\pi, \end{cases} \tag{4}$$

where the rotation path $\phi_{\tau,\theta}$ is defined by (5.1.1).

Lemma 3 *Let $\gamma \in \mathcal{P}_\tau(2n)$ such that $\gamma(\tau) = M_1 \diamond M_2$ with $M_j \in \mathrm{Sp}(2n_j)$ and $n_1 + n_2 = n$. Suppose paths $\gamma_j \in \mathcal{P}_\tau(2n_j)$ with $j = 1$ and 2 satisfy $\gamma_j(\tau) = M_j$ and $i(\gamma, 1) = i(\gamma_1, 1) + i(\gamma_2, 1)$. Then for any $m \in \mathbf{N}$,*

$$i(\gamma, m) = i(\gamma_1, m) + i(\gamma_2, m), \tag{5}$$

$$\nu(\gamma, m) = \nu(\gamma_1, m) + \nu(\gamma_2, m). \tag{6}$$

Proof: Set $\tau = 1$. By the inverse homotopy Theorem 6.2.4, $\gamma \sim (\gamma_1 \diamond \gamma_2)$ with fixed points via a homotopy map $\delta : [0, 1]^2 \to \mathrm{Sp}(2n)$. Extending this δ to $[0, 1] \times [0, m]$ yields a homotopy from γ^m to $(\gamma_1 \diamond \gamma_2)^m = \gamma_1^m \diamond \gamma_2^m$. By Theorem 6.2.7, we then obtain (5) and (6). □

According to the end matrix $\gamma(\tau)$, we study the iteration of corresponding path $\gamma \in \mathcal{P}_\tau(2)$ in four cases. Here we use the notation in Section 2.1, and omit the subindex ω when $\omega = 1$.

Case 1 The degenerate case $\gamma \in \mathcal{P}_\tau^0(2)$.

In this case we have the following result.

Theorem 4 *For $\gamma \in \mathcal{P}_\tau^0(2)$, one of the following three cases must happen.*

1° *If $\gamma(\tau) \in \mathrm{Sp}(2)_-^0$, then*

$$i(\gamma, m) = m(i(\gamma, 1) + 1) - 1, \quad \nu(\gamma, m) = 1, \qquad \forall m \in \mathbf{N}, \tag{7}$$

$$(\gamma, 1) \in 2\mathbf{Z} + 1, \quad \hat{i}(\gamma, 1) = i(\gamma, 1) + 1. \tag{8}$$

2° *If $\gamma(\tau) = I$, then*

$$i(\gamma, m) = m(i(\gamma, 1) + 1) - 1, \quad \nu(\gamma, m) = 2, \qquad \forall m \in \mathbf{N}, \tag{9}$$

$$(\gamma, 1) \in 2\mathbf{Z} + 1, \quad \hat{i}(\gamma, 1) = i(\gamma, 1) + 1. \tag{10}$$

3° *If* $\gamma(\tau) \in \mathrm{Sp}(2)^0_+$, *then*

$$i(\gamma, m) = mi(\gamma, 1), \quad \nu(\gamma, m) = 1, \qquad \forall m \in \mathbf{N}, \tag{11}$$

$$i(\gamma, 1) \in 2\mathbf{Z}, \quad \hat{i}(\gamma, 1) = i(\gamma, 1). \tag{12}$$

Proof: We only prove 1°. The other two cases are similar and are left to the readers. Without loss of generality, set $\tau = 1$. Fix $\gamma \in \mathcal{P}^0_1(2)$ and extend it to $[0, +\infty]$. Define

$$\gamma_s(t) = \gamma(t)R(s\rho(t)\theta_0), \quad \forall t \in [0, 1], s \in [-1, 1] \tag{13}$$

where $\rho(t) = 3t^2 - 2t^3$ for $t \in [0, 1]$, and $\theta_0 > 0$ is so small such that $\gamma_s(1) \in \mathrm{Sp}(2)^*$ for all $s \in [-1, 1]\backslash\{0\}$. Then by our study in Chapter 6,

$$i(\gamma_s, 1) = i(\gamma_{-s}, 1) + 1 = i(\gamma, 1) + 1, \quad \forall s \in (0, 1], \tag{14}$$

$$\nu(\gamma_s, 1) = 0, \quad \forall s \in [-1, 1]\backslash\{0\}. \tag{15}$$

In the case of 1°, since $\gamma(1) \in \mathrm{Sp}(2)^0_-$, it holds that $\gamma_{-1}(1) \in \mathrm{Sp}(2)^-$. Thus $i_1(\gamma)$ must be odd, and by Theorem 1.7.3 there exists $P \in \mathrm{Sp}(2)$ such that $P^{-1}\gamma(1)P = N_1(1, 1)$. Therefore $\gamma(1)^m \in \mathrm{Sp}(2)^0_-$ and $\nu(\gamma, m) = 1$ for all $m \in \mathbf{N}$. Choose a smooth path $h : [0, 1] \to \mathrm{Sp}(2)^0$ such that $h(0) = I$, $h(1) = \gamma(1)$, and $h(s) \in \mathrm{Sp}(2)^0_-$ for all $s \in (0, 1]$. For $k = i(\gamma, 1) + 1$ and this path h, we define a new path $\delta_{k,h} : [0, +\infty) \to \mathrm{Sp}(2)$ by

$$\begin{cases} \delta_{k,h}(t) = h * \phi_{1,k\pi}(t), & \forall t \in [0, 1], \\ \delta_{k,h}(t) = \delta_{k,h}(t - i)\gamma(1)^i, & \forall i \le t \le i + 1, \ i \in \mathbf{N}. \end{cases} \tag{16}$$

Using Lemma 2 and Definition 5.1.3 for the case of $\gamma(1) \in \mathrm{Sp}(2)^0_-$, we obtain

$$\gamma(0) = I = \delta_{k,h}(0), \quad \gamma(1) = \delta_{k,h}(1),$$

$$i(\delta_{k,h}, 1) = 2\left[\frac{i(\gamma, 1) + 1}{2}\right] - 1 = i(\gamma, 1).$$

Thus by Lemma 1, this implies for any $m \in \mathbf{N}$:

$$i(\gamma, m) = i(\delta_{k,h}, m) = 2\left[m\frac{i(\gamma, 1) + 1}{2}\right] - 1 = m(i(\gamma, 1) + 1) - 1.$$

Thus (7) and (8) hold. □

In Figure 1, the idea of this proof is illustrated.

Case 2 The case $\gamma \in \mathcal{P}_\tau(2)$ with $\sigma(\gamma(\tau)) = \{-1\}$.

In this case we refer readers to Figure 2.1.2 for $\mathrm{Sp}(2)^0_{-1,\pm}$. We have the following result.

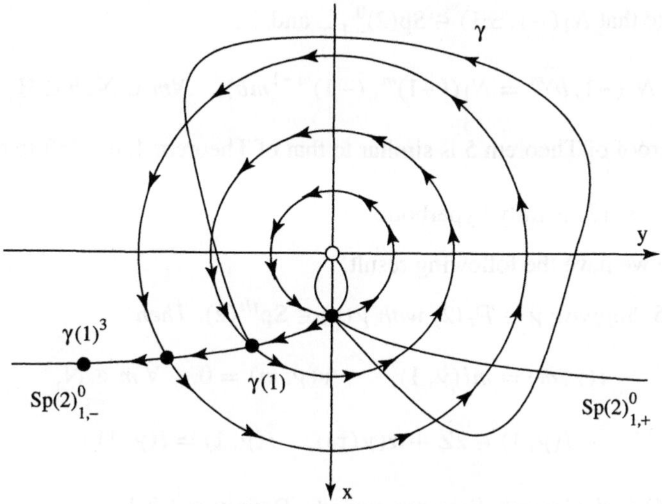

Figure 1: *Iteration of* $\delta_{k,h}$

Theorem 5 *For* $\gamma \in \mathcal{P}_\tau(2)$ *with* $\sigma(\gamma(\tau)) = \{-1\}$, *one of the following three cases must happen:*

1° *If* $\gamma(\tau) \in \mathrm{Sp}(2)^0_{-1,+}$, *then*

$$i(\gamma, m) = mi(\gamma, 1) - \frac{1+(-1)^m}{2}, \quad \nu(\gamma, m) = \frac{1+(-1)^m}{2}, \quad \forall m \in \mathbf{N}, \tag{17}$$

$$i(\gamma, 1) \in 2\mathbf{Z}+1, \quad \hat{i}(\gamma, 1) = i(\gamma, 1). \tag{18}$$

2° *If* $\gamma(\tau) = -I_2$, *then*

$$i(\gamma, m) = mi(\gamma, 1) - \frac{1+(-1)^m}{2}, \quad \nu(\gamma, 1) = 1 + (-1)^m, \quad \forall m \in \mathbf{N}, \tag{19}$$

$$i(\gamma, 1) \in 2\mathbf{Z}+1, \quad \hat{i}(\gamma, 1) = i(\gamma, 1). \tag{20}$$

3° *If* $\gamma(\tau) \in \mathrm{Sp}(2)^0_{-1,-}$, *then*

$$i(\gamma, m) = mi(\gamma, 1), \quad \nu(\gamma, m) = \frac{1+(-1)^m}{2}, \quad \forall m \in \mathbf{N}, \tag{21}$$

$$i(\gamma, 1) \in 2\mathbf{Z}+1, \quad \hat{i}(\gamma, 1) = i(\gamma, 1). \tag{22}$$

Proof: Note that $N_1(-1, \pm 1) \in \mathrm{Sp}(2)^0_{-1,\mp}$ and

$$N_1(-1, b)^m = N_1((-1)^m, (-1)^{m-1}mb), \quad \forall m \in \mathbf{N}, b \in \mathbf{R}.$$

Since the proof of Theorem 5 is similar to that of Theorem 4, it is left to the readers. □

Case 3 $\gamma \in \mathcal{P}_1(2)$ is truly hyperbolic.

In this case we have the following result.

Theorem 6 *Suppose $\gamma \in \mathcal{P}_\tau(2)$ with $\gamma(\tau) \in \mathrm{Sp}^{th}(2)$. Then*

$$i(\gamma, m) = mi(\gamma, 1), \qquad \nu(\gamma, m) = 0, \quad \forall\, m \in \mathbf{N}, \tag{23}$$

$$i(\gamma, 1) \in 2\mathbf{Z} + \alpha(\gamma(\tau)), \quad \hat{i}(\gamma, 1) = i(\gamma, 1), \tag{24}$$

where $\alpha(M)$ is the hyperbolic index given by Definition 1.8.1.

Proof: Set $\tau = 1$. There are two possibilities according to the sign of λ for $\sigma(\gamma(1)) = \{\lambda, \lambda^{-1}\}$. We only consider the case of $\lambda > 0$. The other case of $\lambda < 0$ is left to the readers.

In this case, $\gamma(1) \in \mathrm{Sp}(2)^+$. Thus $i(\gamma, 1)$ must be even and $\alpha(\gamma(1)) = 0$. By Theorem 1.7.3, there exists $P \in \mathrm{Sp}(2)$ such that $P^{-1}\gamma(1)P = D(\lambda)$. Therefore

$$\sigma(\gamma(m)) = \sigma(\gamma(1)^m) = \{\lambda^m, \lambda^{-m}\}, \quad \forall\, m \in \mathbf{N}. \tag{25}$$

Then we obtain $\gamma(1)^m \in \mathrm{Sp}(2)^+$ and $\nu_m(\gamma) = 0$ for all $m \in \mathbf{N}$. Choose a smooth path $h : [0, 1] \to \mathrm{Sp}(2)$ such that $h(0) = I$, $h(1) = \gamma(1)$, and $h(s) \in \mathrm{Sp}(2)^+$ for all $s \in (0, 1]$. For $k = i_1$ and this path h we define a new path $\delta_{k,h} : [0, +\infty) \to \mathrm{Sp}(2)$ by (16). We obtain

$$\gamma(0) = I = \delta_{k,h}(0), \quad \gamma(1) = \delta_{k,h}(1),$$

$$i(\delta_{k,h}, 1) = \left(2\left[\frac{i(\gamma, 1)}{2}\right] - 1\right) + 1 = i_1.$$

Then by Lemma 1 this implies

$$i(\gamma, m) = i(\delta_{k,h}, m) = \left(2\left[\frac{mi(\gamma, 1)}{2}\right] - 1\right) + 1 = mi(\gamma, 1), \quad \forall m \in \mathbf{N}.$$

Thus (23) and (24) hold. □

Case 4 $\gamma \in \mathcal{P}_\tau(2)$ is strongly elliptic.

In this case we have the following result.

Theorem 7 *Suppose $\gamma \in \mathcal{P}_\tau(2)$ satisfies $\sigma(\gamma(\tau)) = \{\omega, \omega^{-1}\}$ with $\omega = e^{\sqrt{-1}\theta}$ and $\theta \in (0, \pi) \cup (\pi, 2\pi)$. By Theorem* 1.7.3, *there exists $P \in \mathrm{Sp}(2)$ such that*

$$P^{-1}\gamma(\tau)P = R(\hat{\theta}) \quad \text{with } \hat{\theta} = \theta \text{ or } 2\pi - \theta. \tag{26}$$

Then for any $m \in \mathbf{N}$,

$$i(\gamma, m) = m(i(\gamma, 1) - 1) + 2E\left(\frac{m\hat{\theta}}{2\pi}\right) - 1, \tag{27}$$

$$\nu(\gamma, m) = 2 - 2\varphi\left(\frac{m\hat{\theta}}{2\pi}\right), \tag{28}$$

$$i(\gamma, 1) \in 2\mathbf{Z} + 1, \quad \hat{i}(\gamma, 1) = i(\gamma, 1) - 1 + \frac{\hat{\theta}}{\pi}. \tag{29}$$

Proof: Set $\tau = 1$. In this case, $\gamma(1) \in \mathrm{Sp}(2)^-$. Thus $i(\gamma, 1)$ must be odd. By (26), $\sigma(\gamma(m)) = \sigma(\gamma(1)^m) = \{\omega^m, \omega^{-m}\}$ for all $m \in \mathbf{N}$. Define a path $\beta : [0, +\infty) \to \mathrm{Sp}(2)$ by

$$\beta(t) = R(t(i(\gamma, 1) - 1)\pi + t\hat{\theta}), \qquad \forall\, t \geq 0.$$

We obtain $\gamma(0) = I_2 = \beta(0)$, and

$$i(\beta, 1) = 2[((i(\gamma, 1) - 1)\pi + \hat{\theta})/(2\pi)] + 1 = i(\gamma, 1).$$

By (26) there is a path $h : [0, 1] \to \Omega^0(\gamma(1))$ connecting $h(0) = \beta(1)$ to $h(1) = \gamma(1)$. Thus by Lemma 1, it suffices to prove (27)–(29) for the path β.

If $m\hat{\theta} \neq 0 \mod 2\pi$, (26) yields $\beta(m) = \beta(1)^m \notin \mathrm{Sp}(2)^0$. Thus we obtain $\nu(\beta, m) = 0$ and

$$\begin{aligned} i(\beta, m) &= 2\left[\frac{m(i(\beta, 1) - 1)\pi + \hat{\theta})}{2\pi}\right] + 1 \\ &= m(i(\beta, 1) - 1) + 2\left[\frac{m\hat{\theta}}{2\pi}\right] + 1, \quad \forall m \in \mathbf{N}. \end{aligned}$$

Noticing $E(a) - [a] = 1$ when $a \notin \mathbf{Z}$, this yields (27)–(29).

If $m\hat{\theta} = 0 \mod 2\pi$, (26) yields $\beta(m) = I_2$. We then obtain

$$\nu(\beta, m) = 2, \quad i(\beta, m) = 2\left[\frac{m((i(\beta, 1) - 1)\pi + \hat{\theta})}{2\pi}\right] - 1.$$

Thus (27)–(29) hold. □

Remark 8 Most of the results in Section 1 were proved in the author's ETH-Zürich 1994 preprint quoted by [Lon17] which was put into [Lon18] later.

8.2 Hyperbolic and elliptic paths

In this section, we establish the iteration formulae of the index theory for truly hyperbolic and hyperbolic paths. Then using these results we study the fifth basic case of elliptic paths in Sp(4).

Fix $\gamma \in \mathcal{P}_\tau^{th}(2n)$. By Lemma 2.3.5, there is a path $f \in C([0,1], \mathrm{Sp}^{th}(2n))$ such that $f(0) = \gamma(\tau)$ and

$$f(1) = M_n^+, \quad i(\gamma, 1) \in 2\mathbf{Z}, \quad \text{if} \quad \alpha(\gamma(\tau)) = 0, \tag{1}$$

$$f(1) = M_n^-, \quad i(\gamma, 1) \in 2\mathbf{Z} + 1, \quad \text{if} \quad \alpha(\gamma(\tau)) = 1 \tag{2}$$

where $\alpha(\gamma(\tau))$ is the hyperbolic index of $\gamma(\tau)$ given by Definition 1.8.1.

Theorem 1 *For any* $\gamma \in \mathcal{P}_\tau^{th}(2n)$,

$$i(\gamma, m) = mi(\gamma, 1), \quad \nu(\gamma, m) = 0, \quad \forall m \in \mathbf{N}, \tag{3}$$

and then $\hat{i}(\gamma, 1) = i(\gamma, 1)$.

Proof: Set $\tau = 1$. Since $\gamma(1)^m \in \mathrm{Sp}^{th}(2n)$, we obtain the second equality in (3). Define $\beta_\pm(t) = D(\pm(1+t))$ for $0 \le t \le 1$. Define

$$\psi = (\beta_+ * \phi_{1,k\pi}) \diamond \beta_+^{\diamond(n-1)}, \quad \text{if} \quad \alpha(\gamma(\tau)) = 0, \tag{4}$$

$$\psi = (\beta_- * \phi_{1,k\pi}) \diamond \beta_+^{\diamond(n-2)}, \quad \text{if} \quad \alpha(\gamma(\tau)) = 1, \tag{5}$$

where $k = i(\gamma, 1)$. Then by Lemma 1.1 we obtain

$$i(\gamma, m) = i(\psi, m), \qquad \forall m \in \mathbf{N}. \tag{6}$$

Together with Lemmas 1.1, 1.2, 1.3, and Theorems 1.6 and 6.2.7, we obtain the first equality in (3). □

Fix $\gamma \in \mathcal{P}_\tau^h(2n)$. By Lemma 2.3.6, when $n \ge 2$ there is a path $f \in C([0,1], \mathrm{Sp}^h(2n) \cap \Omega^0(\gamma(\tau)))$ such that $f(0) = \gamma(\tau)$ and

$$f(1) = N_1(1, b(\gamma(\tau)) \diamond M_{n-1}^+, \quad \text{if} \quad \alpha(\gamma(\tau)) = 0, \tag{7}$$

$$f(1) = N_1(1, b(\gamma(\tau)) \diamond M_{n-1}^-, \quad \text{if} \quad \alpha(\gamma(\tau)) = 1, \tag{8}$$

where $(\alpha(\gamma(\tau)), b(\gamma(\tau))$ is the hyperbolic character of $\gamma(\tau)$ given by Definition 2.3.7. Note that

$$i(\gamma, 1) \in 2\mathbf{Z}, \qquad \begin{array}{l} \text{if } b(\gamma(\tau)) < 0,\ \alpha(\gamma(\tau)) = 0, \\ \text{or } b(\gamma(\tau)) \ge 0,\ \alpha(\gamma(\tau)) = 1, \end{array} \tag{9}$$

$$i(\gamma, 1) \in 2\mathbf{Z} + 1, \qquad \begin{array}{l} \text{if } b(\gamma(\tau)) \ge 0,\ \alpha(\gamma(\tau)) = 0, \\ \text{or } b(\gamma(\tau)) < 0,\ \alpha(\gamma(\tau)) = 1. \end{array} \tag{10}$$

Theorem 2 *For any $\gamma \in \mathcal{P}_\tau^h(2n)$ and any $m \in \mathbf{N}$,*

$$b(\gamma(\tau)^m) = b(\gamma(\tau)), \quad \nu(\gamma, m) = 2 - |b(\gamma(\tau))|, \tag{11}$$

$$i(\gamma, m) = m(i(\gamma, 1) + 1) - 1, \quad \text{if } b(\gamma(\tau)) \in \{0, 1\}, \tag{12}$$

$$i(\gamma, m) = mi(\gamma, 1), \quad \text{if } b(\gamma(\tau)) = -1. \tag{13}$$

Proof: Set $\tau = 1$. When $n = 1$, the conclusions follow from Theorem 1.4. When $n \geq 2$, (11) follows from the fact that $b(\gamma(1)^m) =$ constant for all $m \in \mathbf{N}$. Define $g(t) = N_1(1, (1-t)b(\gamma(1))$ for $t \in [0, 1]$, and

$$\begin{cases} \psi = g\diamond(\beta^+ * \phi_{k\pi})\diamond\beta_+^{\diamond(n-2)}, & \text{if} \quad \alpha(\gamma(\tau)) = 0, \\ \psi = g\diamond(\beta_- * \phi_{k\pi})\diamond\beta_+^{\diamond(n-3)}, & \text{if} \quad \alpha(\gamma(\tau)) = 1, \end{cases} \tag{14}$$

where $k = i(\gamma, 1) + 1$ if $b(\gamma(1)) = 0$ or 1, $k = i(\gamma, 1)$ if $b(\gamma(1)) = -1$. Then by Lemma 1 we also get (11). Together with Theorems 1.6 and 6.2.7, we obtain (12) and (13). □

Based on Theorem 1 we further consider the following case for elliptic paths in addition to the four cases on $\mathcal{P}_\tau(2)$ studied in Section 1.

For any $a \in \mathbf{R}$, note that by definition (1.1), the function $\varphi(a) = E(a) - [a]$ satisfies $\varphi(a) = 0$ if $a \in \mathbf{Z}$, and $\varphi(a) = 1$ if $a \notin \mathbf{Z}$.

Case 5 $\gamma \in \mathcal{P}_\tau(4)$ with $\gamma(\tau) = N_2(\omega, b) \in \mathcal{M}_\omega(4)$ for $\omega \in \mathbf{U}\backslash\mathbf{R}$.

Using notation introduced in Lemma 1.9.2, let $\omega = \cos\theta + \sqrt{-1}\sin\theta$ with $\theta \in (0, \pi) \cup (\pi, 2\pi)$ and $b = \begin{pmatrix} b_1 & b_2 \\ b_3 & b_4 \end{pmatrix} \in \mathcal{L}(R^2)$. We have the following two theorems according to $N_2(\omega, b)$ being non-trivial or trivial.

Theorem 3 *Suppose $\gamma \in \mathcal{P}_\tau(4)$ satisfies $\gamma(\tau) = N_2(\omega, b) \in \mathcal{M}_\omega(4)$ with $\omega = e^{\sqrt{-1}\theta}$ and $\theta \in (0, \pi) \cup (\pi, 2\pi)$. Suppose $N_2(\omega, b)$ is non-trivial, i.e., $(b_2 - b_3)\sin\theta < 0$. Then using the function $\varphi(\cdot) \in \{0, 1\}$ defined by* (1.1), *for any $m \in \mathbf{N}$,*

$$i(\gamma, m) = mi(\gamma, 1) + 2\varphi\left(\frac{m\theta}{2\pi}\right) - 2, \quad \nu(\gamma, m) = 2 - 2\varphi\left(\frac{m\theta}{2\pi}\right). \tag{15}$$

Proof: Without loss of generality, let $\tau = 1$. By 8° of Lemma 1.9.2, there is a small perturbation path $f : [0, 1] \to \mathrm{Sp}(4)$ such that $f(0) = N_2(\omega, b)$ and $f(t)$ possesses two distinct pairs of eigenvalues on $\mathbf{U}$ sufficiently close to ω for all $t \in (0, 1]$. This then implies

$$i(\gamma, 1) = i(f * \gamma, 1) \in 2\mathbf{Z}, \qquad \nu(\gamma, 1) = 0. \tag{16}$$

To continue our study, we fix an $m \in \mathbf{N}\backslash\{1\}$.

Since $N_2(\omega, b)$ is non-trivial, it can be connected to $N_2(\omega, R(\theta))$ by a path $\eta : [0, 1] \to \Omega^0(N_2(\omega, b))$. Let $\beta = \eta * \gamma$. Then for $k = 1$ or m, we obtain

$$i(\gamma, k) = i(\beta, k), \quad \nu(\gamma, k) = \nu(\beta, k).$$

Thus it suffices to prove (15) for the path β. Note that

$$\beta(1)^m = N_2(\omega, R(\theta))^m = \begin{pmatrix} R(m\theta) & mR(m\theta) \\ 0 & R(m\theta) \end{pmatrix}, \qquad \forall m \in \mathbf{N}. \tag{17}$$

Thus we obtain

$$\begin{aligned} \beta(1)^m \in \mathrm{Sp}(2n)^0 \qquad &\text{if and only if} \qquad m\theta = 0 \bmod 2\pi, \\ &\text{if and only if} \qquad \nu(\beta, m) = 2. \end{aligned} \tag{18}$$

By the definition of φ, this implies the second equality of (15).

For small $\epsilon > 0$ we define

$$\xi(t) = N_2(\omega, R(\theta))R(t\epsilon)^{\diamond 2}, \qquad \forall t \in [-1, 1],$$

and let

$$\xi_+(t) = \xi(t), \quad \xi_-(t) = \xi(-t), \quad \forall t \in [0, 1]. \tag{19}$$

Then when ϵ is sufficiently small, by Theorem 6.2.7 we obtain

$$i(\beta, 1) = i(\xi_- * \beta, 1) = i(\xi_+ * \beta, 1), \tag{20}$$

$$i(\beta, m) = i(\xi_- * (\beta^m), 1) = i(\xi_- * \beta, m). \tag{21}$$

By the computations in Lemma 1.9.2, we obtain

$$\det(\xi(t) - \lambda I_4) = \lambda^4 - 4A_\xi(t)\lambda^3 + B_\xi(t)\lambda^2 - 4A_\xi(t)\lambda + 1. \tag{22}$$

By Lemma 1.9.1, the sign of the function

$$g(\sin(t\epsilon)) \equiv 4A_\xi^2(t) + 2 - B_\xi(t), \quad \forall t \in [-1, 1], \tag{23}$$

determines the situation of eigenvalues of $\xi(t)$. Let $s = \sin\theta$ and $s_1 = \sin(t\epsilon)$. Then for sufficiently small $\epsilon > 0$ and $t \in [-1, 1]$, by the proof of Lemma 1.9.2 we obtain

$$g(0) = 0, \qquad \frac{\partial g}{\partial s_1}(0) = s(b_2 - b_3).$$

Therefore by the condition $s(b_2 - b_3) < 0$ and Lemma 1.9.1, $\xi_+(t)$ possesses four eigenvalues outside $\mathbf{U}$ and $\xi_-(t)$ possesses four eigenvalues on $\mathbf{U}$ when $t \in (0, 1]$. Especially, $\xi_+ * \beta$ is a truly hyperbolic path in Sp(4).

Note that by the definition (5.4.6) of perturbation paths $(\beta^m)_{\pm 1}$ of β^m, we have

$$(\xi_+ * \beta)^m \sim_1 (\beta^m)_1, \quad (\xi_- * \beta)^m \sim_1 (\beta^m)_{-1}. \tag{24}$$

By Theorems 6.2.4 and 6.2.7, we then obtain

$$i((\xi_- * \beta)^m, 1) = i((\xi_+ * \beta)^m, 1) - \nu(\beta^m, 1). \tag{25}$$

Applying Theorem 1 to the truly hyperbolic path $\xi_+ * \beta$, by our above discussion we obtain

$$\begin{aligned} i(\beta, m) &= i((\xi_- * \beta)^m, 1) \\ &= i((\xi_+ * \beta)^m, 1) - \nu(\beta^m, 1) \\ &= mi(\xi_+ * \beta, 1) - \nu(\beta, m) \\ &= mi(\beta, 1) - \nu(\beta, m). \end{aligned} \tag{26}$$

By the second equality of (15), we then obtain the first one of (15) for the path β. □

Theorem 4 *Suppose $\gamma \in \mathcal{P}_\tau(4)$ satisfies $\gamma(\tau) = N_2(\omega, b) \in \mathcal{M}_\omega(4)$ with $\omega = e^{\sqrt{-1}\theta}$ and $\theta \in (0, \pi) \cup (\pi, 2\pi)$. Suppose $N_2(\omega, b)$ is trivial, i.e., $(b_2 - b_3)\sin\theta > 0$. Then*

$$i(\gamma, m) = mi(\gamma, 1), \quad \nu(\gamma, m) = 2 - 2\varphi\left(\frac{m\theta}{2\pi}\right). \tag{27}$$

Proof: Set $\tau = 1$. Since $N_2(\omega, b)$ is trivial, it can be connected within $\Omega^0(N_2(\omega, b))$ to $N_2(\omega, -R(\theta))$ by a path η. Let $\beta = \eta * \gamma$. Then instead of (17) we obtain

$$\beta(1)^m = N_2(\omega, -R(\theta))^m = \begin{pmatrix} R(m\theta) & -mR(m\theta) \\ 1 & R(m\theta) \end{pmatrix}, \quad \forall m \in \mathbf{N}. \tag{28}$$

Thus we obtain

$$\begin{aligned} \beta(m) = \beta(1)^m \in \mathrm{Sp}(2n)^0 \qquad &\text{if and only if} \qquad m\theta = 0 \bmod 2\pi, \\ &\text{if and only if} \qquad \nu(\beta, m) = 2. \end{aligned} \tag{29}$$

Thus the second equality in (27) holds. Using notations defined in the proof of Theorem 3, since $(b_2 - b_3)\sin\theta > 0$, the path $\xi_- * \beta$ is hyperbolic. Thus by (25), (26), and Theorem 1, we obtain

$$i_m(\beta^m) = i_m((\xi_- * \beta)^m) = mi_1(\xi_- * \beta) = mi_1(\beta).$$

This proves the first equality in (27). □

Remark 5 Result in this section were proved in [Lon18].

8.3 General symplectic paths

Based on our studies in Sections 1 and 2, we establish the precise iteration formula of the index theory for any paths in $\mathrm{Sp}(2n)$ based on the basic normal form decomposition Theorem 1.8.10.

Theorem 1 *Let $\gamma \in \mathcal{P}_\tau(2n)$. There is a path $f \in C([0,1], \Omega^0(\gamma(\tau))$ by Theorem* 1.8.10 *such that $f(0) = \gamma(1)$ and*

$$\begin{aligned} f(1) = {} & N_1(1,1)^{\diamond p_-} \diamond I_{2p_0} \diamond N_1(1,-1)^{\diamond p_+} \\ & \diamond N_1(-1,1)^{\diamond q_-} \diamond (-I_{2q_0}) \diamond N_1(-1,-1)^{\diamond q_+} \\ & \diamond R(\theta_1) \diamond \cdots \diamond R(\theta_r) \\ & \diamond N_2(\omega_1, u_1) \diamond \cdots \diamond N_2(\omega_{r_*}, u_{r_*}) \\ & \diamond N_2(\lambda_1, v_1) \diamond \cdots \diamond N_2(\lambda_{r_0}, v_{r_0}) \\ & \diamond M_0, \end{aligned} \tag{1}$$

where $N_2(\omega_j, u_j)$'s are non-trivial and $N_2(\lambda_j, v_j)$'s are trivial basic normal forms; $\sigma(M_0) \cap \mathbf{U} = \emptyset$; p_-, p_0, p_+, q_-, q_0, q_+, r, r_*, and r_0 are non-negative integers; $\omega_j = e^{\sqrt{-1}\alpha_j}$, $\lambda_j = e^{\sqrt{-1}\beta_j}$; $\theta_j, \alpha_j, \beta_j \in (0,\pi) \cup (\pi, 2\pi)$; these integers and real numbers are uniquely determined by $\gamma(\tau)$. Then using the functions defined in (1.1),

$$\begin{aligned} i(\gamma, m) = {} & m(i(\gamma,1) + p_- + p_0 - r) + 2\sum_{j=1}^{r} E\left(\frac{m\theta_j}{2\pi}\right) - r \\ & -p_- - p_0 - \frac{1+(-1)^m}{2}(q_0 + q_+) \\ & +2\left(\sum_{j=1}^{r_*} \varphi\left(\frac{m\alpha_j}{2\pi}\right) - r_*\right), \end{aligned} \tag{2}$$

$$\nu(\gamma, m) = \nu(\gamma, 1) + \frac{1+(-1)^m}{2}(q_- + 2q_0 + q_+) + 2\varsigma(m, \gamma(\tau)), \tag{3}$$

where we denote by

$$\begin{aligned} \varsigma(m, \gamma(\tau)) = {} & r - \sum_{j=1}^{r} \varphi\left(\frac{m\theta_j}{2\pi}\right) + r_* - \sum_{j=1}^{r_*} \varphi\left(\frac{m\alpha_j}{2\pi}\right) \\ & + r_0 - \sum_{j=1}^{r_0} \varphi\left(\frac{m\beta_j}{2\pi}\right). \end{aligned} \tag{4}$$

Recall that $\varphi(a) = E(a) - [a] \in \{0, 1\}$ is defined by (1.1) *for any $a \in \mathbf{R}$.*

Proof: By Theorems 1.8.10 and 6.2.7, Lemmas 1.1 and 1.3, summing the iteration formulae established for the five basic cases in Sections 1 and 2, we obtain (2) and (3). □

Corollary 2 1° *Using notation of Definition* 0.1 *and Theorem* 1,

$$\hat{i}(\gamma,1) = i(\gamma,1) + p_- + p_0 - r + \sum_{j=1}^{r} \frac{\theta_j}{\pi} \in \mathbf{R},$$
$$= 2 + (i(\gamma,1) - n) + (p_- + p_0 - 1) + (n - r - 1) + \sum_{j=1}^{r} \frac{\theta_j}{\pi}. \tag{5}$$

2° *Moreover, suppose that further,*

$$i(\gamma,1) \geq n \geq 2 \quad \text{and} \quad p_- + p_0 \geq 1. \tag{6}$$

Then

$$\hat{i}(\gamma,1) > 2. \tag{7}$$

Proof: 1° follows from Theorem 1. By the fact that $p_- + p_0 \geq 1$, we obtain $r \leq n-1$. Under condition (7), if $r = 0$, then $n - r - 1 \geq n - 1 > 0$. Thus (8) holds. If $r > 0$, then there exists at least one $\theta_j > 0$. Thus (8) still holds. □

Remark 3 Theorem 1 and Corollary 2 were proved in [Lon18]. Corollary 2 generalizes a corresponding result of I. Ekeland and H. Hofer obtained in [EkH2] of 1987 which proves that the mean index of any non-constant periodic solution of a convex Hamiltonian system is always greater than 2. Note that in this case condition (6) holds by Corollary 15.1.4 and Lemma 15.2.4 below.

Chapter 9

Bott-type iteration formulae

This chapter is devoted to the establishment and further developments of the Bott-type formula for the index theory of iterations of any paths in the symplectic group $\mathrm{Sp}(2n)$, starting from the identity matrix and studing its applications. We follow the ideas of [Bot1] and [Lon16], and base our study upon the index function theory introduced in Chapter 5. As a special case, such a formula also works for the fundamental solutions of general linear Hamiltonian systems with continuous symmetric and periodic coefficients, and thus for periodic solutions of any nonlinear Hamiltonian systems with Hamiltonian function periodic in time. As a consequence of this formula, new index invariants and iteration equalities and inequalities of the index theory for symplectic paths are established.

In Section 1, as the basis of this chapter, we study the splitting numbers and give an algebraic characterization for them. In Section 2, the Bott-type formula for the 1-index theory is proved for iterations of any symplectic paths. New invariants, L^2-index character and Fourier index coefficients, especially the mean index, of any symplectic path are introduced. In Section 3, based on our complete understanding of the splitting numbers and Bott-type formula, we establish an abstract precise iteration formula for our index theory and give some corollaries to it.

9.1 Splitting numbers

Fix $n \in \mathbf{N}$, $\tau > 0$ and a path $\gamma \in \mathcal{P}_\tau(2n)$. In this section, we study the properties of the index function $i_\omega(\gamma)$ of γ at ω as a function of $\omega \in \mathbf{U}$, and use the short hand notations

$$i(\omega) \equiv i_\omega(\gamma) \quad \text{and} \quad \nu(\omega) \equiv \nu_\omega(\gamma), \tag{1}$$

when there is no confusion. By Lemma 5.3.1, we obtain

$$i(\omega) = i(\overline{\omega}) \quad \text{and} \quad \nu(\omega) = \nu(\overline{\omega}), \quad \forall \omega \in \mathbf{U}. \tag{2}$$

Lemma 1 *$i(\cdot)$ is locally constant on $\mathbf{U}\setminus\sigma(\gamma(\tau))$, and thus is constant on each connected component of $\mathbf{U}\setminus\sigma(\gamma(\tau))$. It holds that*

$$\nu(\omega) = 0, \qquad \forall \omega \in \mathbf{U}\setminus\sigma(\gamma(\tau)). \tag{3}$$

Proof: Note that $\mathbf{U}\cap\sigma(\gamma(\tau))$ contains at most $2n$ points. For any $\omega_0 \in \mathbf{U}\setminus\sigma(\gamma(\tau))$, let $\mathcal{N}(\omega_0)$ be an open connected neighborhood of ω_0 in $\mathbf{U}\setminus\sigma(\gamma(\tau))$. By definition we obtain $\nu(\omega) = 0$ for all $\omega \in \mathcal{N}(\omega_0)$. Thus (3) holds and $\gamma \in \mathcal{P}^*_{\tau,\omega}(2n)$ for all $\omega \in \mathcal{N}(\omega_0)$. By Theorem 2.4.1, there is a path $\beta : [0, 1] \to \mathcal{P}^*_{\tau,\omega_0}(2n)$ connecting

$\gamma(\tau)$ to M_n^+ or M_n^-. By the compactness of the image of β and the openness of $\mathrm{Sp}(2n)^*_{\omega_0}$ in $\mathrm{Sp}(2n)$, we can further require $\mathcal{N}(\omega_0)$ to be so small that β is completely located within $\mathrm{Sp}(2n)^*_{\omega}$ for all $\omega \in \mathcal{N}(\omega_0)$. Then by definition, this implies $i(\omega) = i(\omega_0)$ for all $\omega \in \mathcal{N}(\omega_0)$. □

Corollary 2 *The discontinuity points of $i(\cdot)$ and $\nu_\tau(\cdot)$ are contained in $\mathbf{U} \cap \sigma(\gamma(\tau))$.*

By direct computations we obtain

Lemma 3 *For any $M \in \mathrm{Sp}(2n)$ and $\lambda \in \sigma(M)$, the algebraic and geometric multiplicities and the Krein type of $\lambda \in \sigma(N)$ for any $N \in [M]$ are the same as those of M, where $[M]$ is the conjugate set of M in* Sp(2*n*) *defined in Section* 1.8.

Note that by the above study, for a fixed path γ the index $i_\omega(\gamma)$ is a step function in $\omega \in \mathbf{U}$ with possible jumps only at eigenvalues of $M = \gamma(\tau)$ on $\mathbf{U}$. Next we introduce the splitting numbers $S_M^{\pm}(\omega)$ to measure the jumps between $i_\omega(\gamma)$ and $i_\lambda(\gamma)$ for $\lambda \in \mathbf{U}$ nearby from two sides of ω in $\mathbf{U}$.

Definition 4 For any $M \in \mathrm{Sp}(2n)$ and $\omega \in \mathbf{U}$, choosing $\tau > 0$ and $\gamma \in \mathcal{P}_\tau(2n)$ with $\gamma(\tau) = M$, we define

$$S_M^{\pm}(\omega) = \lim_{\epsilon \to 0^+} i_{\exp(\pm\epsilon\sqrt{-1})\omega}(\gamma) - i_\omega(\gamma). \tag{4}$$

They are called the **splitting numbers** of M at ω.

Lemma 5 *The splitting numbers $S_M^{\pm}(\omega)$ are well defined, i.e., are independent of the path $\gamma \in \mathcal{P}_1(2n)$ appearing in* (4). *For $\omega \in \mathbf{U}$ and $M \in \mathrm{Sp}(2n)$, $S_N^{\pm}(\omega)$ are constant for any $N \in \Omega^0(M)$ of Section* 1.8.

Proof: The theorem is proved in the following two claims.

Claim 1 *$S_M^{\pm}(\omega)$ is independent of the path γ in its definition.*

In fact, using (5.4.2) for γ in (4) and a sufficiently small $a > 0$, we define

$$\xi_a(t) = \gamma(1)Q(-ta, \ldots, -ta), \qquad \forall\, t \in [0, 1].$$

Then we obtain

$$i_\omega(\gamma) = i_\omega(\xi_a * \gamma),$$

and by Theorem 2.4.1 there is a path $\beta_0 : [0, 1] \to \mathrm{Sp}(2n)^*_\omega$ which connects $\xi_\alpha(1)$ to M_n^+ or M_n^-. Choose $\epsilon > 0$ to be small enough such that $\lambda(s) = \exp(s\epsilon\sqrt{-1})\omega \notin \sigma(M)$ for $0 < s \leq 1$. By Lemma 1, $i_{\lambda(s)}(\gamma) =$ constant for all $s \in (0, 1]$. Thus by Theorem 2.4.1, there exists a path $\beta_\epsilon : [0, 1] \to \mathrm{Sp}(2n)^*_{\lambda(1)}$ such that $\beta_\epsilon(0) = M$ and $\beta_\epsilon(1) = M_n^+$ or M_n^-. Thus we obtain

$$\begin{aligned} S_M^+(\omega) &= i_{\lambda(1)}(\gamma) - i_\omega(\gamma) \\ &= \frac{1}{\pi}\Delta_1(\beta_\epsilon * \gamma) - \frac{1}{\pi}\Delta_1(\beta_0 * \xi_a * \gamma) \\ &= \frac{1}{\pi}[\Delta_1(\beta_\epsilon) - \Delta_1(\beta_0) - \Delta_1(\xi_a)]. \end{aligned}$$

Then $S_M^+(\omega)$ is independent of the choice of γ. By Theorem 2.4.2, this number is also independent of the choices of the above $a > 0$ and paths β_ϵ and β_0, and then is completely determined by M and ω.

Claim 2 *$S_M^\pm(\omega) = S_N^\pm(\omega)$ for any $N \in \Omega^0(M)$.*

In fact, by the definition of $N \in \Omega^0(M)$, there exists a path $\xi : [0, 1] \to \Omega(M)$ such that $\xi(0) = M$ and $\xi(1) = N$. Choose any $\gamma \in \mathcal{P}_1(2n)$ such that $\gamma(1) = M$. Then by the definition of $\Omega(M)$, it holds that $\gamma \sim \xi * \gamma$ on $[0, 1]$ along ξ. Thus by Theorem 6.2.7, we obtain

$$\begin{aligned} S_M^\pm(\omega) &= \lim_{\epsilon\to 0^+} i_{\exp(\pm\epsilon\sqrt{-1})\omega}(\gamma) - i_\omega(\gamma) \\ &= \lim_{\epsilon\to 0^+} i_{\exp(\pm\epsilon\sqrt{-1})\omega}(\xi * \gamma) - i_\omega(\xi * \gamma) \\ &= S_N^\pm(\omega). \end{aligned}$$

The proof is complete. □

Lemma 6 *For $M \in \mathrm{Sp}(2n)$ and $\omega \in \mathbf{U}$,*

$$S_M^\pm(\omega) = 0, \qquad \text{if } \omega \notin \sigma(M), \tag{5}$$

$$S_M^+(\omega) = S_M^-(\overline{\omega}). \tag{6}$$

For any $M_i \in \mathrm{Sp}(2n_i)$ with $i = 0$ and 1,

$$S_{M_0\diamond M_1}^\pm(\omega) = S_{M_0}^\pm(\omega) + S_{M_1}^\pm(\omega), \qquad \forall\, \omega \in \mathbf{U}. \tag{7}$$

Proof: This is a direct consequence of Definition 4, Theorem 6.2.7, and Lemma 5. □

In order to give a complete explanation of the splitting number, we use the concept of ultimate type of any $\omega \in \mathbf{U}$ as an eigenvalue of $M \in \mathrm{Sp}(2n)$ introduced by Definition 1.8.12.

Theorem 7 *For any $\omega \in \mathbf{U}$ and $M \in \mathrm{Sp}(2n)$,*

$$S_M^+(\omega) = p \quad \text{and} \quad S_M^-(\omega) = q, \tag{8}$$

where (p, q) is the ultimate type of ω for M.

Proof: By Lemmas 5, 6, and Theorem 1.8.10, it suffices to prove (8) when M is one of basic normal forms $N_1(\pm 1, b)$ with $b = 1, 0, -1$, $N_1(\omega, b)$, and $N_2(\omega, b) \in \mathcal{M}_\omega^1(2n)$ for $\omega \in \mathbf{U}\backslash\mathbf{R}$ defined in Section 1.8. We continue the proof according to different basic normal forms.

Case 1 $M = N_1(1, b)$ with $b = \pm 1$ or 0.

Define $f_b(t) = N_1(1, tb)$ for $t \in [0, 1]$. The matrix $M = N_1(1, b)$ is nontrivial when $b = 1$ or 0 and possesses both the Krein type and the ultimate type $(1, 1)$ of the eigenvalue 1. The matrix $M = N_1(1, -1)$ is trivial, and possesses the Krein type $(1, 1)$ and the ultimate type $(0, 0)$ of the eigenvalue 1. Then using the $\mathbf{R}^3$ cylindrical coordinate representation introduced in Section 2.1, it is easy to see that for any $\epsilon > 0$ sufficiently small,

$$\begin{aligned} i_1(f_b) = -1, \quad & i_{\exp(\pm\epsilon\sqrt{-1})}(f_b) = 0, \quad \text{if } b = 1 \text{ or } 0, \\ \text{and} \quad i_1(f_{-1}) = 0, \quad & i_{\exp(\pm\epsilon\sqrt{-1})}(f_{-1}) = 0. \end{aligned}$$

Thus (8) holds when $\omega = 1$.

Case 2 $M = N_1(-1, b)$ with $b = \pm 1$ or 0.

The proof of the case of $\omega = -1$ is similar by using the $\mathbf{R}^3$-cylindrical coordinate representation and is left to the readers.

Case 3 $M = R(\theta)$ or $R(2\pi - \theta)$ with $\omega = e^{\theta\sqrt{-1}} \in \mathbf{U}\backslash\mathbf{R}$.

Fix $\omega = e^{\theta\sqrt{-1}} \in \mathbf{U}\backslash\mathbf{R}$. By Lemma 6 it suffices to prove (8) for $\theta \in (0, \pi) \cup (\pi, 2\pi)$ and $M = R(\theta)$. Fix a path $\gamma \in \mathcal{P}_1(2)$ in Definition 4 with $\gamma(1) = R(\theta)$.

When $\theta \in (0, \pi)$, note that $\sigma(R(\theta))$ consists of two simple eigenvalues ω and $\overline{\omega}$. By direct computation, $R(\theta)$ possesses $(p, q) = (0, 1)$ as the Krein type and ultimate type of ω.

On the other hand, let $\lambda(\pm\epsilon) = \exp(\pm\epsilon\sqrt{-1})\omega$ with $\epsilon > 0$ sufficiently small. When $\theta \in (0, \pi)$, the matrix $\gamma_{-1}(1)$ can be connected to $D(2)$ within $\mathrm{Sp}(2)^+_\omega$, where γ_{-1} is the perturbation path of γ defined by (5.4.6) with respect to ω. Note also that the matrix $R(\theta)$ can be connected to $D(2)$ within $\mathrm{Sp}(2)^+_{\lambda(\epsilon)}$, and the matrix $R(\theta)$ can be connected to $D(-2)$ within $\mathrm{Sp}(2)^-_{\lambda(-\epsilon)}$. Thus

$$i(\lambda(-\epsilon)) = i(\omega) + 1, \qquad i(\lambda(\epsilon)) = i(\omega). \tag{9}$$

Therefore we obtain (8).

When $\theta \in (\pi, 2\pi)$, similarly the matrix $\gamma_{-1}(1)$ can be connected to $D(-2)$ within $\mathrm{Sp}(2)^-_\omega$, the matrix $R(\theta)$ can be connected to $D(-2)$ within $\mathrm{Sp}(2)^-_{\lambda(\epsilon)}$, and the matrix $R(\theta)$ can be connected to $D(2)$ within $\mathrm{Sp}(2)^+_{\lambda(-\epsilon)}$. Thus (8) still hold.

Next we continue our study for $M = N_2(\omega, b) \in \mathcal{M}^1_\omega(4)$ with $\omega = e^{\theta\sqrt{-1}} \in \mathbf{U}\backslash\mathbf{R}$ in two cases. By Lemma 6, it suffices to consider the case of $N_2(\omega, b)$ with $\theta \in (0, \pi) \cup (\pi, 2\pi)$. In the following we always suppose $\theta \in (0, \pi)$. The case of $\theta \in (\pi, 2\pi)$ can be studied similarly and is left to the readers.

Case 4 $M = N_2(\omega, b) \in \mathcal{M}^1_\omega(4)$ and $(b_2 - b_3)\sin\theta < 0$.

Since $N_2(\omega, b)$ is non-trivial, by direct computation we obtain that ω as an eigenvalue of M possesses both the Krein type and the ultimate type $(1, 1)$.

Define

$$\xi(t) = MR((t-1)\alpha)^{\diamond 2}, \qquad \forall\, t \in [0,1], \tag{10}$$

for some small $\alpha \in (0, \theta/2]$. Note that by Lemma 1.9.2,

$$\det(\xi(t) - \lambda I_4) = \lambda^4 - 4A_\xi(t)\lambda^3 + B_\xi(t)\lambda^2 - 4A_\xi(t)\lambda + 1. \tag{11}$$

Then denoting $s = \sin\theta$, $c = \cos\theta$, $s_1 = \sin((t-1)\alpha)$ and $c_1 = \cos((t-1)\alpha)$ for $j = 1$ and 2, we obtain that for sufficiently small $\alpha > 0$ and $t \in [0,1)$,

$$D_\omega(\xi(t)) = 2s_1c_1s(b_2 - b_3) + o(|s_1|) > 0, \tag{12}$$

$$4A_\xi^2(t) + 2 - B_\xi(t) = 2s_1c_1s(b_2 - b_3) + o(|s_1|) > 0, \tag{13}$$

where $s_1 = \sin((t-1)\alpha) < 0$. Here the condition $s(b_2 - b_3) < 0$ is crucial. Thus by Lemma 1.9.2, for $0 \le t < 1$ the matrix $\xi(t)$ possesses two pairs of simple eigenvalues $\{\lambda_1(t), \overline{\lambda_1(t)}, \lambda_2(t), \overline{\lambda_2(t)}\}$ such that for $j = 1$ and 2, $\lambda_j \in C([0,1], \mathbf{U})$, $\lambda_j(1) = \omega$, $\lambda_j(t) \neq \omega$ and $|\lambda_j(t) - \omega|$ is sufficiently small provided $\alpha > 0$ is small enough for $t \in [0,1)$. Denote by $\lambda_j(t) = \exp(\beta_j(t)\sqrt{-1})$ with $\beta_j \in C([0,1], \mathbf{R})$ for $j = 1$ and 2, and $\beta_1(1) = \beta_2(1) = \theta$. Note that $\xi(t)$ is the fundamental solution of the system (8.0.3) with $B(t) = -J\dot{\xi}(t)\xi(t)^{-1}$ being positive definite. So by Theorem 1.3.11 when $0 \le t < 1$, $\lambda_1(t)$ and $\lambda_2(t)$ must have different Krein type. Without loss of generality, for $0 \le t < 1$ we assume that $\lambda_1(t)$ is Krein negative and $\lambda_2(t)$ is Krein positive. By Theorem 1.3.11 again, we obtain

$$\beta_1(t) > \theta > \beta_2(t), \qquad \forall\, 0 \le t < 1, \tag{14}$$

and $\beta_1(t)$ and $\beta_2(t)$ are strictly decreasing and increasing respectively when t increases to 1.

By (12), we have $\xi(0) \in \mathrm{Sp}(4)_\omega^-$. Thus there is a path $\eta : [0,1] \to \mathrm{Sp}(2n)_\omega^-$ connecting $\eta(0) = D(-2)\diamond D(2)$ to $\eta(1) = \xi(0)$. Denote by $\psi = \hat{\alpha}_1$ the standard zigzag path defined by (5.3.17). Then ψ connects I_4 to $D(-2)\diamond D(2)$ and $i_\lambda(\psi) = 1$ for all $\lambda \in \mathbf{U}$. Thus

$$i_\omega(\eta * \psi) = 1. \tag{15}$$

Define

$$\gamma = \xi * \eta * \psi. \tag{16}$$

Denote s and c as in (12). For any $b \in [-\alpha, 0]$, let $s_1 = \sin(-\alpha)$, $c_1 = \cos(-\alpha)$, $s_2 = \sin b$, $c_2 = \cos b$. By Lemma 1.9.2 we further require $\alpha > 0$ to be so small such that for all $b \in [-\alpha, 0]$,

$$\begin{aligned} D_\omega(M[R(-\alpha)\diamond R(b)]) \\ &= s_1[s(b_2-b_3)C_2+2b_1c(1-c_2)]+s_2[s(b_2-b_3)c_1 \\ &\quad +2b_4c(1-c_1)]+2s^2(2-c_1-c_2)+O(|s_1|^2) \\ &> 0, \end{aligned} \tag{17}$$

where $(b_2-b_3)sin\theta < 0$ is used. Thus $M[R(-\alpha)\diamond I_2]$ and $M[R(-\alpha)\diamond R(-\alpha)]$ can be connected within $\mathrm{Sp}(4)^*_\omega$. This proves $\gamma_{-1} \sim_\omega \eta*\psi$, where γ_{-1} is the perturbation path of γ defined by (5.4.6). Thus

$$i_\omega(\gamma)=i_\omega(\gamma_{-1})=i_\omega(\eta*\psi)=1. \tag{18}$$

Recall that we have supposed $\theta \in (0,\pi)$. So we can fix $\epsilon > 0$ sufficiently small so that for $\chi(\pm\epsilon)=e^{\pm\epsilon\sqrt{-1}}\omega$,

$$i_{\chi(\pm\epsilon)}(\gamma)=\lim_{s\to 0+} i_{\exp(\pm s\sqrt{-1})\omega}(\gamma), \tag{19}$$

$$S(\omega)\equiv\{\exp(a\sqrt{-1})\omega\in\mathbf{U}\mid |a|\le\epsilon/2\}\subset\mathbf{U}\backslash\mathbf{R}. \tag{20}$$

Choose $r \in (0,1)$ close enough to 1 so that

$$\sigma(\xi(t))\subset S(\omega)\cup S(\overline{\omega}), \qquad \forall t\in[r,1]. \tag{21}$$

Define $\hat{\xi}(t)=\xi(tr)$ for $t\in[0,1]$. Note that (21) implies

$$i_{\chi(\pm\epsilon)}(\gamma)=i_{\chi(\pm\epsilon)}(\hat{\xi}*\eta*\psi). \tag{22}$$

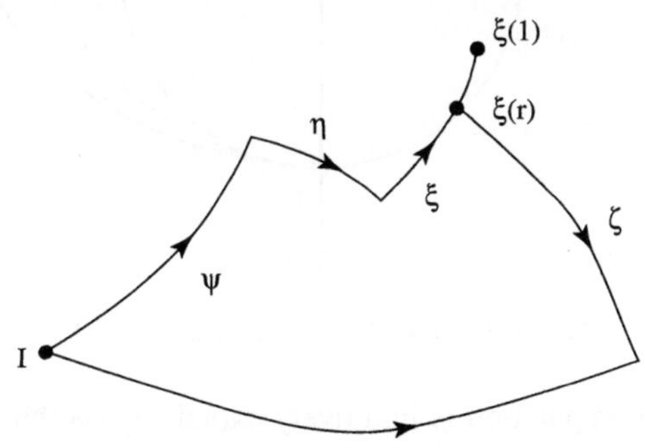

Figure 1: *Paths γ and f*

By (14) and the discussion there, $\lambda_1(r)$ and $\lambda_2(r)$ are Krein positive and negative respectively. By (20) and (21), both $\beta_1(r)$ and $\beta_2(r)$ are in $(0,\pi)$. Thus by Theorem 1.7.3 there exists $P\in \mathrm{Sp}(4)$ such that $P\xi(r)P^{-1} = R(2\pi-\beta_1(r))\diamond R(\beta_2(r))$. Let

$p \in \mathcal{P}_1(4)$ satisfy $p(1) = P$. Define $\zeta(t) = p(t)\xi(r)p(t)^{-1}$ for all $t \in [0, 1]$. Then the conjugate path ζ connects $\xi(r)$ to $R(2\pi - \beta_1(r))\diamond R(\beta_2(r))$. Define

$$f_1(t) = R(t(2\pi - \beta_1(r))), \quad f_2(t) = R(t\beta_2(r)),$$

$$f(t) = f_1\diamond f_2(t), \qquad \forall\, t \in [0, 1]. \tag{23}$$

By (14), it holds that $\beta_1(r) > \theta > \beta_2(r)$. Thus by Theorem 6.2.7 and a direct computation for paths in Sp(2), we obtain

$$i_\omega(f) = 1 + 0 = 1, \tag{24}$$

$$i_{\chi(\epsilon)}(f) = 2 + 0 = 2, \quad i_{\chi(-\epsilon)}(f) = 1 + 1 = 2. \tag{25}$$

By (15) and the definitions of $\hat{\xi}$ and ζ we obtain

$$i_\omega(\zeta * \hat{\xi} * \eta * \psi) = i_\omega(\eta * \psi) = 1. \tag{26}$$

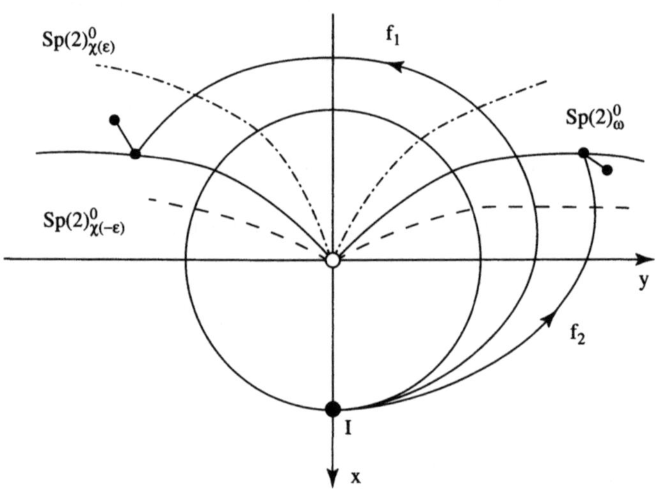

Figure 2: *An explanation of paths f_1 and f_2*

In Figure 2, this computation is intuitively explained via the $\mathbf{R}^3$-cylindrical representation of Sp(2) on the plane $\{z = 0\}$, where f_1 and f_2 are represented by two nearby paths with the same topological properties. Since $\zeta * \hat{\xi} * \eta * \psi(1) = f(1)$, by inverse homotopy Theorem 6.2.4, (24), and (26), we obtain $\zeta * \hat{\xi} * \eta * \psi \sim f$ on $[0, 1]$ with fixed end points. Since ζ is a conjugation path, this implies $\hat{\xi} * \eta * \psi \sim_\mu f$ on $[0, 1]$ along ζ for any $\mu \in \mathbf{U}$. Then by Theorem 6.2.3 and (25) we obtain

$$i_{\chi(\pm\epsilon)}(\hat{\xi} * \eta * \psi) = i_{\chi(\pm\epsilon)}(f) = 2. \tag{27}$$

Together with (22) we obtain

$$i_{\chi(\pm\epsilon)}(\gamma) = 2. \tag{28}$$

By (18), (19), and (28), we obtain

$$S_M^+(\omega) = S_M^-(\omega) = 1, \tag{29}$$

i.e., (8) holds in this case.

Case 5 $M = N_2(\omega, b) \in \mathcal{M}_\omega^1(4)$ and $(b_2 - b_3)\sin\theta > 0$.

Since in this case M is trivial, by direct computation we obtain that ω as an eigenvalue of M possesses the Krein type (1, 1) and the ultimate type (0, 0).

Define the path $\xi : [0, 1] \to \mathrm{Sp}(4)$ by (10) for some small $\alpha \in (0, \theta/2]$. By the notation of Case 4, we obtain that for sufficiently small $\alpha > 0$ and $t \in [0, 1)$ as in (12) and (13),

$$D_\omega(\xi(t)) < 0, \quad 4A_\xi^2(t) + 2 - B_\xi(t) < 0, \tag{30}$$

where $s_1 = \sin((t-1)\alpha) < 0$. Here the condition $s(b_2 - b_3) > 0$ is crucial. By Lemma 1.9.1, this implies

$$\sigma(\xi(t)) \cap (\mathbf{U} \cup \mathbf{R}) = \emptyset, \qquad \forall\, t \in [0, 1). \tag{31}$$

Similar to our study in Case 1, by (30) we obtain that M_2^+ can be connected to $\xi(0)$ by a path $\eta : [0, 1] \to \mathrm{Sp}(2n)$ satisfying

$$\sigma(\eta(t)) \cap \mathbf{U} = \emptyset, \qquad \forall\, t \in [0, 1]. \tag{32}$$

Define $\psi = \hat{\alpha}_0$ of Section 5.3. Then ψ connects I_4 to M_2^+ and satisfies $i_\lambda(\psi) = 0$ for all $\lambda \in \mathbf{U}$. Together with (32),

$$i_\omega(\eta * \psi) = 0. \tag{33}$$

Similar to the proof of (18), using Lemma 1.9.2, we get that for $\gamma = \xi * \eta * \psi$,

$$i_\omega(\gamma) = i_\omega(\eta * \psi) = 0. \tag{34}$$

Fix $\epsilon > 0$ so small that (19) holds for $\chi(\pm\epsilon) = e^{\pm\epsilon\sqrt{-1}}\omega \in \mathbf{U}\backslash\mathbf{R}$. By (33) and (34), $\chi(\pm\epsilon) \notin \sigma(\gamma(t))$ for $0 \le t \le 1$. Since γ passes through M_2^+, we obtain

$$i_{\chi(\pm\epsilon)}(\gamma) = 0. \tag{35}$$

Then together with (19) and (34) we obtain

$$S_M^+(\omega) = S_M^-(\omega) = 0, \tag{36}$$

i.e., (8) holds in this case.

The proof is complete. □

Now we obtain the following results immediately.

Corollary 8 *If $\omega \in \mathbf{U} \cap \sigma(\gamma(\tau))$ is of Krein type (p, q), then*

$$\lim_{\epsilon \to 0^+} (i(e^{\epsilon\sqrt{-1}}\omega) - i(e^{-\epsilon\sqrt{-1}}\omega)) = p - q. \tag{37}$$

Corollary 9 *For any $\omega \in \mathbf{U}$ and $M \in \mathrm{Sp}(2n)$,*

$$0 \le S_M^{\pm}(\omega) \le \nu_\omega(M). \tag{38}$$

Theorem 10 *The integer valued splitting number pair $(S_M^+(\omega), S_M^-(\omega))$ defined for all $(\omega, M) \in \mathbf{U} \times \cup_{n\ge 1}\mathrm{Sp}(2n)$ are uniquely determined by the following axioms:*

1° (Homotopy invariant) $S_M^{\pm}(\omega) = S_N^{\pm}(\omega)$ *for all $N \in \Omega^0(M)$.*
2° (Symplectic additivity) $S_{M_1 \diamond M_2}^{\pm}(\omega) = S_{M_1}^{\pm}(\omega) + S_{M_2}^{\pm}(\omega)$ *for all $M_i \in \mathrm{Sp}(2n_i)$ with $i = 1$ and 2.*
3° (Vanishing) $S_M^{\pm}(\omega) = 0$ *if $\omega \notin \sigma(M)$.*
4° (Normality) $(S_M^+(\omega), S_M^-(\omega))$ *coincides with the ultimate type of ω for M when M is any basic normal form given by Definition* 1.8.9.

Our Theorems 7 and 10 give a complete understanding of the splitting numbers for all $M \in \mathrm{Sp}(2n)$. Consequently, the ultimate type is also characterized by Theorem 10.

Proposition 11 *For any $\omega \in \mathbf{U}$ and $M \in \mathrm{Sp}(2n)$, denote the Krein type of $\omega \in \sigma(M)$ by $(p_\omega(M), q_\omega(M))$. Then*

$$0 \le \nu_\omega(M) - S_M^-(\omega) \le p_\omega(M), \tag{39}$$

$$0 \le \nu_\omega(M) - S_M^+(\omega) \le q_\omega(M). \tag{40}$$

Proof: The left inequality follows from Corollary 9. To prove the right inequality in (39), note that by Theorems 1.8.10 and 10, it suffices to prove (39) for basic normal forms. Then (39) follows from the direct verifications for basic normal forms as in our study in the cases of the proof for Theorem 7, which are left to the readers. □

List 12 For the reader's convenience, we list the splitting numbers below for all basic normal forms given by Theorem 7. Then via Theorem 1.8.10, Lemma 6, Theorem 10, and this list, we can get splitting numbers of any $M \in \mathrm{Sp}(2n)$ at any $\omega \in \mathbf{U}$.

⟨1⟩ $(S_M^+(1), S_M^-(1)) = (1, 1)$ for $M = N_1(1, b)$ with $b = 1$ or 0.
⟨2⟩ $(S_M^+(1), S_M^-(1)) = (0, 0)$ for $M = N_1(1, -1)$.
⟨3⟩ $(S_M^+(-1), S_M^-(-1)) = (1, 1)$ for $M = N_1(-1, b)$ with $b = -1$ or 0.
⟨4⟩ $(S_M^+(-1), S_M^-(-1)) = (0, 0)$ for $M = N_1(-1, 1)$.
⟨5⟩ $(S_M^+(e^{\sqrt{-1}\theta}), S_M^-(e^{\sqrt{-1}\theta})) = (0, 1)$ for $M = R(\theta)$ with $\theta \in (0, \pi) \cup (\pi, 2\pi)$.

⟨6⟩ $(S_M^+(\omega), S_M^-(\omega)) = (1, 1)$ for $M = N_2(\omega, b)$ being non-trivial with $\omega = e^{\sqrt{-1}\theta} \in \mathbf{U}\backslash\mathbf{R}$.

⟨7⟩ $(S_M^+(\omega), S_M^-(\omega)) = (0, 0)$ for $M = N_2(\omega, b)$ being trivial with $\omega = e^{\sqrt{-1}\theta} \in \mathbf{U}\backslash\mathbf{R}$.

⟨8⟩ $(S_M^+(\omega), S_M^-(\omega)) = (0, 0)$ for any $\omega \in \mathbf{U}$ and $M \in \mathrm{Sp}(2n)$ satisfying $\sigma(M) \cap \mathbf{U} = \emptyset$.

Remark 13 Splitting numbers for symplectic matrices and results in this section were introduced and obtained in [Lon16] of 1999. Proposition 11 is new. These results generalize those obtained by R. Bott in [Bot1] and I. Ekeland in [Eke3].

9.2 Bott-type formulae

Based on the index function theory of Chapter 5, we are able to establish the following Bott-type formula of the index theory for iterations of any path $\gamma \in \mathcal{P}_\tau(2n)$.

Theorem 1 *For any* $\tau > 0$, $\gamma \in \mathcal{P}_\tau(2n)$, $z \in \mathbf{U}$, *and* $m \in \mathbf{N}$,

$$i_z(\gamma, m) = \sum_{\omega^m = z} i_\omega(\gamma), \tag{1}$$

$$\nu_z(\gamma, m) = \sum_{\omega^m = z} \nu_\omega(\gamma). \tag{2}$$

Proof: We only prove the case $z = 1$. Other cases can be proved similarly. Based on the results of Chapters 5 and 6, we prove Theorem 1 in two steps. Without loss of generality, we set $\tau = 1$. We consider first the case for the fundamental solution $\gamma \in \hat{\mathcal{P}}_1(2n)$ of (6.1.4) for any linear Hamiltonian system (6.1.1) with symmetric continuous and periodic coefficients. Then the case of any path $\gamma \in \mathcal{P}_1(2n)$ is reduced to the case of $\hat{\mathcal{P}}_1(2n)$. We use the notation introduced in Chapters 5 and 6.

Step 1 Let m and $k \in \mathbf{N}$ with $k \le m$, $\tau > 0$ and $\omega \in \mathbf{U}$. We define L_m to be the Hilbert space $L^2([0, m\tau], \mathbf{C}^{2n})$ with the usual L^2 inner product $\langle \cdot, \cdot \rangle_L$, and

$$W_m(k, \omega) = \{y \in W^{1,2}([0, m\tau], \mathbf{C}^{2n}) \mid y(k\tau) = \omega y(0)\} \tag{3}$$

with the usual $W^{1,2}$ norm which forms a subspace of L_m. We write $W_m = W_m(m, 1)$.

Fix $B \in C(S_1, \mathcal{L}_s(\mathbf{R}^{2n}))$. Using the operators A and B introduced in Section 6.1, we define

$$f_{m,k}^\omega(y) = \frac{1}{2}\langle (A - B)y, y \rangle_L, \quad \forall y \in W_m(k, \omega), \tag{4}$$

and we write $f_m = f_{m,m}^1$ for simplicity. Then $f_{m,k}^\omega$ is smooth in the induced topology of L_m.

Let $\gamma : [0, +\infty) \to \mathrm{Sp}(2n)$ be the fundamental solution of (8.0.3) with B given above. The bilinear form ϕ_m on W_m corresponding to the functional f_m is given by

$$\phi_m(x, y) = \frac{1}{2}\langle (A - B)x, y\rangle_L, \qquad \forall x, y \in \mathrm{dom} A = W_m \subset L_m.$$

Two subspaces F_1 and F_2 of W_m are ϕ_m**-orthogonal** and is denoted by $F_1 \perp F_2$, if it holds that $\phi_m(x, y) = 0$ for all $x \in F_1$ and $y \in F_2$. We denote by $F_1 \oplus F_2$ the ϕ_m-orthogonal product of F_1 and F_2.

We define

$$\omega_p = \exp\left(\sqrt{-1}\frac{2p\pi}{m}\right), \qquad \forall 0 \le p \le m. \tag{5}$$

Then $\omega_p^m = 1$.

Lemma 2 *For $m \in \mathbf{N}$ and $z \in \mathbf{U}$,*

$$W_m(1, \omega_p) \perp W_m(1, \omega_q), \qquad \text{if } 0 \le p \ne q \le m, \tag{6}$$

$$W_m(m, 1) = \oplus_{\omega^m=1} W_m(1, \omega). \tag{7}$$

Proof: For $x \in W_m(1, \omega_p)$ and $y \in W_m(1, \omega_q)$ with $0 \le p \ne q \le m$, we have

$$\phi_m(x, y) = \phi_1(x, y)\sum_{i=0}^{m-1}(\omega_p\overline{\omega_q})^i = 0.$$

Thus (6) holds, and then (7) follows. □

Continuation of the proof of Theorem 1: For any $y_i \in W_m(1, \omega_i)$ with $\omega_i^m = 1$ and $0 \le i \le m - 1$,

$$f_m\left(\sum_{i=0}^{m-1} y_i\right) = \sum_{i=0}^{m-1} f_{m,1}^{\omega_i}(y_i) = m\sum_{i=0}^{m-1} f_{1,1}^{\omega_i}(y_i).$$

By a slight abuse of notation, this yields

$$f_m|_{W_m} = \sum_{i=0}^{m-1} f_{m,1}^{\omega}|_{W_m(1,\omega_i)} = m\sum_{i=0}^{m-1} f_{1,1}^{\omega_i}|_{W_1(1,\omega_i)}. \tag{8}$$

Now we carry out the saddle point reduction for f_m on W_m, and obtain the functional $a_m = f_m \circ u_m$ defined on Z_m. Simultaneously this induces saddle point reductions for $f_{m,1}^{\omega_i}$ on $W_m(1, \omega_i)$ for $0 \le i \le m - 1$, and yields the functional $a_{m,1}^{\omega_i} = f_{m,1}^{\omega} \circ u_{m,1}^{\omega_i}$ defined on $Z_m(1, \omega_i)$ which is the truncated subspace of $W_m(1, \omega_i)$. Denote the Morse indices of a_m at the origin by $m^*(m)$ and the Morse indices of $a_{m,1}^{\omega}$ at the origin by $m^*(m,\omega)$ for $* = +, 0$, and $-$ respectively. Then $m^*(m,\omega)$ coincides with $m^*(1,\omega)$. Denote the dimensions of Z_m and $Z_m(1,\omega)$ by $2d_m$ and $2d_{m,\omega}$ respectively.

Then from (7) and (8) we obtain

$$\begin{aligned} m^*(m) &= \sum_{\omega^m=1} m^*(m,\omega) \\ &= \sum_{\omega^m=1} m^*(1,\omega), \qquad \text{for } * = +, 0, -. \end{aligned} \tag{9}$$

Note that the dimensions of the spaces Z_m and $Z_m(1,\omega)$ satisfy

$$d_m = \sum_{\omega^m=1} d_{m,\omega}. \tag{10}$$

Combining (9), (10), and Theorem 6.1.1, we have proved the special case of Theorem 1 whenever $\gamma \in \hat{\mathcal{P}}_1(2n)$.

Step 2 The case of $\gamma \in \mathcal{P}_1(2n)$.

Fix $\gamma \in \mathcal{P}_1(2n)$. By the method of Step 2 in the proof of Theorem 6.1.8, we can choose a path $\psi \in \hat{\mathcal{P}}_1(2n)$ such that $\psi \sim \gamma$ with fixed end points. This homotopy can be extended to $[0,1] \times [0,m]$ for any $m \in \mathbf{N}$. Thus we obtain

$$\psi^m \sim \gamma^m, \qquad \forall m \in \mathbf{N}. \tag{11}$$

Specially, by Theorem 6.2.7, for all $m \in \mathbf{N}$ and $\omega \in \mathbf{U}$ we obtain

$$i_\omega(\gamma, m) = i_\omega(\psi, m), \qquad \nu_\omega(\gamma, m) = \nu_\omega(\psi, m). \tag{12}$$

Together with (1) and (2) for ψ, we obtain them for γ and complete the proof of Theorem 1. □

Remark 3 Theorem 1 can also be proved by firstly perturbing γ to a nearby path α such that $\alpha(\tau)$ possesses only simple eigenvalues. This reduces the proof of Theorem 1 to symplectic paths in Sp(2) and Sp(4). Then using our method in Chapter 8, the formulae (1) and (2) can be established directly.

As a direct consequence of Theorem 1, we obtain the Bott-type formulae for splitting numbers.

Corollary 4 (C. Liu-Y. Long-C. Zhu [LLZ1]). *For any* $M \in \mathrm{Sp}(2n)$, $m \in \mathbf{N}$, *and* $z \in \mathbf{U}$,

$$S^{\pm}_{M^m}(z) = \sum_{\omega^m=z} S^{\pm}_M(\omega). \tag{13}$$

Proof: Let $\gamma \in \mathcal{P}_1(2n)$ with $\gamma(1) = M$. Define $\theta \in [0, 2\pi)$ such that $z = e^{\sqrt{-1}\theta}$. Extend γ to $[0, +\infty]$ by (8.0.4). By Definition 1.4 of the splitting numbers and Theorem 1, we obtain

$$\begin{aligned}
S^+_{M^m}(z) &= \lim_{\epsilon\to 0+} i_{e^{\sqrt{-1}(\theta+\epsilon)}}(\gamma, m) - i_{e^{\sqrt{-1}\theta}}(\gamma, m) \\
&= \lim_{\epsilon\to 0+} \sum_{k=1}^{m} i_{e^{\sqrt{-1}(\theta+\epsilon+2k\pi)/m}}(\gamma) - \sum_{k=1}^{m} i_{e^{\sqrt{-1}(\theta+2k\pi)/m}}(\gamma) \\
&= \sum_{k=1}^{m} \left(\lim_{\epsilon\to 0+} i_{e^{\sqrt{-1}(\theta+\epsilon+2k\pi)/m}}(\gamma) - i_{e^{\sqrt{-1}(\theta+2k\pi)/m}}(\gamma) \right) \\
&= \sum_{\omega^m = z} S^+_M(\omega).
\end{aligned}$$

This proves (13) for $S^+_{M^m}(z)$. The proof for $S^-_{M^m}(z)$ is similar, and thus omitted. □

As a direct consequence of Theorem 1, the mean index per τ for any $\gamma \in \mathcal{P}_\tau(2n)$ can be further characterized as follows.

Theorem 5 *For any $\tau > 0$ and $\gamma \in \mathcal{P}_\tau(2n)$,*

$$\hat{i}(\gamma, 1) \equiv \lim_{m\to+\infty} \frac{i(\gamma, m)}{m} = \frac{1}{2\pi} \int_0^{2\pi} i_{e^{\sqrt{-1}\theta}}(\gamma) d\theta, \tag{14}$$

$$\hat{\nu}(\gamma, 1) \equiv \lim_{m\to+\infty} \frac{\nu(\gamma, m)}{m} = \frac{1}{2\pi} \int_0^{2\pi} \nu_{e^{\sqrt{-1}\theta}}(\gamma) d\theta = 0. \tag{15}$$

Especially, $\hat{i}(\gamma, 1)$ is always a finite real number as we have seen in Corollary 8.3.2.

Proof: By Theorem 1,

$$\frac{i(\gamma, m)}{m} = \frac{1}{2\pi} \sum_{\omega^m=1} i_\omega(\gamma) \frac{2\pi}{m}, \tag{16}$$

$$\frac{\nu(\gamma, m)}{m} = \frac{1}{2\pi} \sum_{\omega^m=1} \nu_\omega(\gamma) \frac{2\pi}{m}. \tag{17}$$

By Lemma 1.1, the function $i_\omega(\gamma)$ is locally constant and $\nu_\omega(\gamma)$ is zero in $\omega \in \mathbf{U}$ except at finitely many points. Therefore the right-hand side of (16) and (17) are Riemannian sums, and converge to the corresponding integrals as $m \to \infty$. □

By our study on $i_\omega(\gamma)$ as a function on $\omega \in \mathbf{U}$ for fixed $\gamma \in \mathcal{P}_\tau(2n)$, it is a step function with only finitely many jumps. Suggested by the mean index $\hat{i}(\gamma, 1)$, we introduce the following sequence of new invariants for elements in $\mathcal{P}_\tau(2n)$.

Definition 6 For $\tau > 0$ and $\gamma \in \mathcal{P}_\tau(2n)$, we define

$$I_{\gamma,m}(\omega) = i_\omega(\gamma, m), \qquad \forall \omega \in \mathbf{U},\ m \in \mathbf{N}. \tag{18}$$

Then as an $L^2(\mathbf{U}, \mathbf{R})$-function, we call $I_{\gamma,1}$ the **L^2-index character** of γ per τ. For any k and $m \in \mathbf{N}$, we further define

$$C_k(\gamma, m) = \frac{1}{2\pi}\int_0^{2\pi} I_{\gamma,m}(e^{\sqrt{-1}\theta})e^{-\sqrt{-1}k\theta}\,d\theta, \quad \forall k \in \mathbf{Z}, \tag{19}$$

$$A_0(\gamma, m) = 2C_0(\gamma, m), \tag{20}$$

$$A_k(\gamma, m) = C_k(\gamma, m) + C_{-k}(\gamma, m), \tag{21}$$

$$B_k(\gamma, m) = \sqrt{-1}\left(C_k(\gamma, m) - C_{-k}(\gamma, m)\right). \tag{22}$$

We call these $C_k(\gamma, 1)$, $A_k(\gamma, 1)$, and $B_k(\gamma, 1)$ the **k-th Fourier index coefficients** of γ per τ.

Corollary 7 1° *Index functions of Definition* 5.4.3 *and the concepts introduced in Definition* 6 *are all* $\mathbf{U}$*-homotopy invariant in the sense of Definition* 5.0.3, *i.e. invariant if* $\gamma_0 \sim_\omega \gamma_1$ *for all* $\omega \in \mathbf{U}$.

2° *For* $\tau > 0$ *and* $\gamma \in \mathcal{P}_\tau(2n)$, *the* L^2*-index character* $I_{\gamma,1}$ *of* γ *per* τ *is completely determined by the Fourier index coefficients of* γ *per* τ.

3° *It holds that*

$$A_0(\gamma, 1) = \hat{i}(\gamma, 1), \tag{23}$$

$$C_k(\gamma, m) = mC_k(\gamma, 1), \qquad \forall k \in \mathbf{Z},\ m \in \mathbf{N}. \tag{24}$$

The equality (24) *also hold for* $A_k(\gamma, m)$ *and* $B_k(\gamma, m)$. *Especially,*

$$\hat{i}(\gamma, m) = m\hat{i}(\gamma, 1), \qquad \forall m \in \mathbf{N}. \tag{25}$$

Proof: 1° follows from Theorem 6.2.7. 2° follows from the completeness of the space L^2. 3° follows from Theorem 5 and Definition 6. □

Note that the new concepts introduced in Definition 6 can be defined naturally for any τ-periodic solution x of the nonlinear Hamiltonian system (8.0.1) with the Hamiltonian function $H \in C^2$ along $x(\mathbf{R})$ and τ-periodic in t through the fundamental solution γ_x of the linearized Hamiltonian system (8.0.3) with $B(t) = H''(t, x(t))$.

Remark 8 Bott-type index iteration formulae and the mean index of any symplectic matrix path, as well as results contained in this section, were introduced or obtained in [Lon16] except Corollary 4 which is due to [LLZ1]. The L^2-index character and Fourier index coefficients given by Definition 6 are new and further generalize the mean index. So far, only the mean index has been used in our applications to nonlinear problems. We believe that the other concepts in Definition 6 should also be useful, since they in fact characterize greatly Morse-type information of the given periodic solution.

9.3 Abstract precise iteration formulae

For any $M \in \mathrm{Sp}(2n)$, we define

$$C(M) = \sum_{0<\theta<2\pi} S_M^-(e^{\sqrt{-1}\theta}). \tag{1}$$

Since $S_M^-(\omega) = 0$ if $\omega \notin \sigma(M)$, then (1) is a finite sum.

The main result in this section is the following abstract precise iteration formula.

Theorem 1 (Y. Long and C. Zhu [LZu3]). *For any path $\gamma \in \mathcal{P}_\tau(2n)$, let $M = \gamma(\tau)$. We extend γ to $[0, +\infty)$ by (8.0.4). Then for any $m \in \mathbf{N}$ we have*

$$\begin{aligned} i(\gamma, m) &= m(i(\gamma, 1) + S_M^+(1) - C(M)) \\ &\quad + 2\sum_{\theta\in(0,2\pi)} E\left(\frac{m\theta}{2\pi}\right) S_M^-(e^{\sqrt{-1}\theta}) - (S_M^+(1) + C(M)). \end{aligned} \tag{2}$$

Proof: Set $\tau = 1$ and fix a path $\gamma \in \mathcal{P}_1(2n)$. Let $M = \gamma(1)$. As mentioned in Section 1, the splitting numbers $S_M^\pm(\omega)$ measure the jumps between $i_\omega(\gamma)$ and $i_\lambda(\gamma)$ with $\lambda \in \mathbf{U}$ near ω from two sides of ω in $\mathbf{U}$. Therefore for any $\omega_0 = e^{\sqrt{-1}\theta_0} \in \mathbf{U}$ with $0 \le \theta_0 < 2\pi$, we denote by ω_j with $1 \le j \le p_0$ the eigenvalues of M on $\mathbf{U}$ which are distributed counterclockwise from 1 to ω_0 and located strictly between 1 and ω_0. Then we have

$$\begin{aligned} i_{\omega_0}(\gamma) &= i(\gamma, 1) + S_M^+(1) + \sum_{j=1}^{p_0}(-S_M^-(\omega_j) + S_M^+(\omega_j)) - S_M^-(\omega_0) \\ &= i(\gamma, 1) + \sum_{0\le\theta<\theta_0} S_M^+(e^{\sqrt{-1}\theta}) - \sum_{0<\theta\le\theta_0} S_M^-(e^{\sqrt{-1}\theta}). \end{aligned} \tag{3}$$

Thus by the Bott-type formula Theorem 2.1, for any $m \in \mathbf{N}$ we have

$$\begin{aligned} i(\gamma, m) &= \sum_{\omega^m=1} i_\omega(\gamma) \\ &= i(\gamma, 1) + \sum_{k=1}^{m-1}\left(i(\gamma, 1) + \sum_{0\le\theta<\frac{2k\pi}{m}} S_M^+(e^{\sqrt{-1}\theta}) - \sum_{0<\theta\le\frac{2k\pi}{m}} S_M^-(e^{\sqrt{-1}\theta})\right) \end{aligned}$$

$$
\begin{aligned}
&= i(\gamma,1) + (m-1)(i(\gamma,1) + S_M^+(1)) \\
&\quad + \sum_{\theta\in(0,2\pi)} \left(\sum_{\frac{m\theta}{2\pi}<k\le m-1} S_M^+(e^{\sqrt{-1}\theta}) - \sum_{\frac{m\theta}{2\pi}\le k\le m-1} S_M^-(e^{\sqrt{-1}\theta}) \right) \qquad (4)\\
&= i(\gamma,1) + (m-1)(i(\gamma,1) + S_M^+(1)) \\
&\quad + \sum_{\theta\in(0,2\pi)} \left(\left(m - 1 - \left[\frac{m\theta}{2\pi}\right]\right) S_M^+(e^{\sqrt{-1}\theta}) \right. \\
&\qquad \left. - \left[\frac{m(2\pi-\theta)}{2\pi}\right] S_M^-(e^{\sqrt{-1}\theta}) \right).
\end{aligned}
$$

Here to get the last equality, we have used the fact that $[m-a] = m - [a]$ for $a \in \mathbf{Z}$ and $[m-a] = m-[a]-1$ for $a \notin \mathbf{Z}$ to count the number of ks in the given intervals.

By Lemma 1.6 on splitting numbers, regrouping terms in the last expression of (4) according to $S_M^-(\omega)$ with $\omega \in \mathbf{U}$, we get

$$
\begin{aligned}
i(\gamma,m) &= i(\gamma,1) + (m-1)(i(\gamma,1) + S_M^+(1)) \\
&\quad + \left(m - 1 - 2\left[\frac{m}{2}\right]\right) S_M^-(-1) \\
&\quad + \sum_{\theta\in(0,\pi)} \left(\left(m-1-\left[\frac{m(2\pi-\theta)}{2\pi}\right]\right) S_M^-(e^{\sqrt{-1}\theta}) \right. \\
&\qquad \left. + \left(m-1-\left[\frac{m\theta}{2\pi}\right]\right) S_M^-(e^{\sqrt{-1}(2\pi-\theta)}) \right) \qquad (5)\\
&\quad - \sum_{\theta\in(0,\pi)} \left(\left[\frac{m(2\pi-\theta)}{2\pi}\right] S_M^-(e^{\sqrt{-1}\theta}) + \left[\frac{m\theta}{2\pi}\right] S_M^-(e^{\sqrt{-1}(2\pi-\theta)}) \right) \\
&= i(\gamma,1) + (m-1)(i(\gamma,1) + S_M^+(1)) + \left(m - 2\left[\frac{m}{2}\right] - 1\right) S_M^-(-1) \\
&\quad + \sum_{\theta\in(0,\pi)} \left(\left(m-1-2\left[\frac{m\theta}{2\pi}\right]\right) S_M^-(e^{\sqrt{-1}(2\pi-\theta)}) \right. \\
&\qquad \left. + \left(m-1-2\left[\frac{m(2\pi-\theta)}{2\pi}\right]\right) S_M^-(e^{\sqrt{-1}\theta}) \right).
\end{aligned}
$$

Since $E(a) + [b] = a + b$ if $a, b \in \mathbf{R}$ and $a + b \in \mathbf{Z}$, from (5) we obtain

$$
\begin{aligned}
i(\gamma,m) &= i(\gamma,1) + (m-1)(i(\gamma,1) + S_M^+(1)) \\
&\quad + \left(2E\left(\frac{m}{2}\right) - (m+1)\right) S_M^-(-1)
\end{aligned}
$$

$$
\begin{aligned}
&+\sum_{\theta\in(0,\pi)}\left(\left(2E\left(\frac{m(2\pi-\theta)}{2\pi}\right)-(m+1)\right)S_M^-(e^{\sqrt{-1}(2\pi-\theta)})\right. \\
&\qquad\left.+\left(2E\left(\frac{m\theta}{2\pi}\right)-(m+1)\right)S_M^-(e^{\sqrt{-1}\theta})\right) \\
&=i(\gamma,1)+(m-1)(i(\gamma,1)+S_M^+(1))-(m+1)C(M) \\
&\quad+\sum_{\theta\in(0,2\pi)}2E\left(\frac{m\theta}{2\pi}\right)S_M^-(e^{\sqrt{-1}\theta}).
\end{aligned}\tag{6}
$$

Here to get the last equality in (6), we have used definition (1) of $C(M)$. This yields (2). □

Corollary 2 *For any $\gamma\in\mathcal{P}_\tau(2n)$, setting $M=\gamma(\tau)$, we have*

$$
\hat{i}(\gamma,1)=i(\gamma,1)+S_M^+(1)-C(M)+\sum_{\theta\in(0,2\pi)}\frac{\theta}{\pi}S_M^-(e^{\sqrt{-1}\theta}).\tag{7}
$$

Remark 3 It is easy to verify that the iteration formulae (8.3.2) and the above (2) coincide, and that the formulae (8.3.5) and the above (7) of the mean index coincide. Note that 2° of Corollary 8.3.2 can also be proved similarly by the above Corollary 2.

The following formula gives the parity of $i(\gamma,m)$ in terms of $i(\gamma,1)=i_1(\gamma)$, $i_{-1}(\gamma)$, and the iteration time m.

Theorem 4 (X. Hu and Y. Long [HuL1]). *For any path $\gamma\in\mathcal{P}_\tau(2n)$, let $M=\gamma(\tau)$. We extend γ to $[0,+\infty)$ by (8.0.4). Then for any $m\in\mathbf{N}$ we have*

$$
\begin{aligned}
i(\gamma,m)&=i(\gamma,1)+(m-1)i_{-1}(\gamma) \\
&\quad+2\sum_{0<\theta<\pi}(S_M^-(e^{\sqrt{-1}\theta})-S_M^+(e^{\sqrt{-1}\theta}))\left[\frac{m\theta}{2\pi}\right] \\
&\quad-2\sum_{0<\theta<\pi}\left[1-\left\{\frac{m\theta}{2\pi}\right\}\right]S_M^-(e^{\sqrt{-1}\theta}) \\
&\quad+2\left(m-\left[\frac{m}{2}\right]-1\right)S_M^-(-1).
\end{aligned}\tag{8}
$$

Especially, if $\nu(\gamma,m)=0$ for all $m\in\mathbf{N}$, then

$$
\begin{aligned}
i(\gamma,m)&=i(\gamma,1)+(m-1)i_{-1}(\gamma) \\
&\quad+2\sum_{0<\theta<\pi}(S_M^-(e^{\sqrt{-1}\theta})-S_M^+(e^{\sqrt{-1}\theta}))\left[\frac{m\theta}{2\pi}\right].
\end{aligned}\tag{9}
$$

Proof: Note that

$$\begin{aligned} C(M) &= \sum_{0<\theta<2\pi} S_M^-(e^{\sqrt{-1}\theta}) \\ &= \sum_{0<\theta<\pi} (S_M^-(e^{\sqrt{-1}\theta}) + S_M^+(e^{\sqrt{-1}\theta})) + S_M^-(-1), \end{aligned} \tag{10}$$

and

$$\begin{aligned} i_{-1}(\gamma) &= i(\gamma, 1) + S_M^+(1) \\ &\quad + \sum_{0<\theta<\pi} (S_M^+(e^{\sqrt{-1}\theta}) - S_M^-(e^{\sqrt{-1}\theta})) - S_M^-(-1) \\ &= i(\gamma, 1) + S_M^+(1) - C(M) + 2 \sum_{0<\theta<\pi} S_M^+(e^{\sqrt{-1}\theta}). \end{aligned} \tag{11}$$

By the fact that $E(a) + [b] = a + b$ for $a, b \in \mathbf{R}$ and $a + b \in \mathbf{Z}$, we obtain

$$\begin{aligned} &\sum_{0<\theta<2\pi} E\left(\frac{m\theta}{2\pi}\right) S_M^-(e^{\sqrt{-1}\theta}) \\ &= \sum_{0<\theta<\pi} \left(m - \left[\frac{m\theta}{2\pi}\right]\right) (S_M^-(e^{\sqrt{-1}\theta}) + S_M^+(e^{\sqrt{-1}\theta})) \\ &\quad + \left(m - \left[\frac{m}{2}\right]\right) S_M^-(-1) \\ &= \sum_{0<\theta<\pi} \left[\frac{m\theta}{2\pi}\right] (S_M^-(e^{\sqrt{-1}\theta}) - S_M^+(e^{\sqrt{-1}\theta})) \\ &\quad - \sum_{0<\theta<\pi} \left[-\left\{\frac{m\theta}{2\pi}\right\}\right] S_M^-(e^{\sqrt{-1}\theta}) \\ &\quad + m \sum_{0<\theta<\pi} S_M^+(e^{\sqrt{-1}\theta}) + \left(m - \left[\frac{m}{2}\right]\right) S_M^-(-1). \end{aligned} \tag{12}$$

Therefore by (2), (10), (11), and (12), we obtain

$$\begin{aligned} i(\gamma, m) &= i(\gamma, 1) + (m-1)i_{-1}(\gamma) - 2(m-1) \sum_{0<\theta<\pi} S_M^+(e^{\sqrt{-1}\theta}) \\ &\quad + 2 \sum_{0<\theta<2\pi} E\left(\frac{m\theta}{2\pi}\right) S_M^-(e^{\sqrt{-1}\theta}) - 2C(M) \\ &= i(\gamma, 1) + (m-1)i_{-1}(\gamma) + 2 \sum_{0<\theta<\pi} S_M^+(e^{\sqrt{-1}\theta}) \\ &\quad - 2 \sum_{0<\theta<\pi} \left[\frac{m\theta}{2\pi}\right] (S_M^+(e^{\sqrt{-1}\theta}) - S_M^-(e^{\sqrt{-1}\theta})) \end{aligned}$$

$$
\begin{aligned}
&-2\sum_{0<\theta<\pi}\left[-\left\{\frac{m\theta}{2\pi}\right\}\right] S_M^-(e^{\sqrt{-1}\theta}) \\
&+2\left(m-\left[\frac{m}{2}\right]\right) S_M^-(-1) - 2C(M) \\
= {}& i(\gamma,1)+(m-1)i_{-1}(\gamma) \\
&-2\sum_{0<\theta<\pi}\left[\frac{m\theta}{2\pi}\right](S_M^+(e^{\sqrt{-1}\theta}) - S_M^-(e^{\sqrt{-1}\theta})) \\
&-2\sum_{0<\theta<\pi}\left(1+\left[-\left\{\frac{m\theta}{2\pi}\right\}\right]\right) S_M^-(e^{\sqrt{-1}\theta}) \\
&+2\left(m-\left[\frac{m}{2}\right]-1\right) S_M^-(-1).
\end{aligned}
$$

This yields (8).

When $\nu(\gamma, m) = 0$ for all $m \in \mathbf{N}$, the last two terms in (8) vanish. Therefore (9) holds. The proof is complete. □

Remark 5 The abstract iteration formula Theorem 1 was established by C. Zhu and the author in [LZu3]. Theorem 4 is proved in [HuL1] of Xijun Hu and the author, which generalizes Proposition 6.14 of Viterbo in [Vit3] of 1989, where the iteration formula (9) was proved for indices of iterations of any periodic solution of star-shaped Hamiltonian systems under the condition that the periodic solution and all its iterations are non-degenerate.

Chapter 10

Iteration inequalities

Usually in applications to differential equations of the index theory, only certain iteration inequalities are needed. In this chapter, based upon our results in Chapters 8 and 9, we establish various iteration inequalities.

In fact, our precise iteration formula (8.3.2) yields a simple way to detect any proposed iteration inequality of the index theory, i.e., plugging the ten formulae obtained in Theorems 8.1.4–8.1.7 and 8.2.3–8.2.4 for the ten basic normal forms to see if the inequality holds. These ten formulae suggest corrections one should make if the proposed inequality does not hold, and further improvements if the proposed inequality does hold. The abstract precise iteration formula (9.3.1) together with the complete understanding of Theorem 9.1.7 on splitting numbers also can play the same role. In this sense, the iteration theory of the index defined in Chapter 5 is completely established.

In Section 1, we derive inequalities on the increasing of $i(\gamma, m)$ in terms of mean index, or initial index, and iteration times. In Section 2, we derive estimates on successive iterations. In Section 3, we derive estimates on the iteration times via known information on indices. These inequalities and estimates will be applied to various problems on periodic solutions of nonlinear Hamiltonian systems in Part V of this book.

10.1 Estimates via mean index and initial index

Based on our studies of the index function theory in Chapters 5 and 6, and the Bott-type formula in Chapter 9, in this section we establish various iteration inequalities of the index theory for symplectic paths. In the following we denote by $\mathbf{U}^+$ and $\mathbf{U}^-$ the closed upper and lower semi-circles in $\mathbf{U}$ respectively.

Theorem 1 (C. Liu and Y. Long [LLo1], [LLo3]). 1° *For any* $\gamma \in \mathcal{P}_\tau(2n)$ *and* $\omega \in \mathbf{U}\setminus\{1\}$, *it always holds that*

$$i_1(\gamma) + \nu_1(\gamma) - n \le i_\omega(\gamma) \le i_1(\gamma) + n - \nu_\omega(\gamma). \qquad (1)$$

2° *The left equality in* (1) *holds for some* $\omega \in \mathbf{U}^+\setminus\{1\}$ (*or* $\mathbf{U}^-\setminus\{1\}$) *if and only if* $I_{2p}\diamond N_1(1,-1)^{\diamond q}\diamond K \in \Omega^0(\gamma(\tau))$ *for some non-negative integers p and q satisfying* $0 \le p+q \le n$ *and* $K \in \mathrm{Sp}(2(n-p-q))$ *with* $\sigma(K) \subset \mathbf{U}\setminus\{1\}$ *satisfying the condition that all eigenvalues of K located within the arc between* 1 *and* ω *including* ω *in* $\mathbf{U}^+\setminus\{1\}$(*or* $\mathbf{U}^-\setminus\{1\}$) *possess total multiplicity* $n-p-q$. *If* $\omega \ne -1$, *all eigenvalues of K are in* $\mathbf{U}\setminus\mathbf{R}$ *and those in* $\mathbf{U}^+\setminus\mathbf{R}$ (*or* $\mathbf{U}^+\setminus\mathbf{R}$) *are all Krein-negative* *(or Krein-positive) definite. If $\omega = -1$, it holds that $-I_{2s}\diamond N_1(-1,1)^{\diamond t}\diamond H \in \Omega^0(K)$ for some non-negative integers s and t satisfying $0 \le s+t \le n-p-q$, and some $H \in \mathrm{Sp}(2(n-p-q-s-t))$ statisfying $\sigma(H) \subset \mathbf{U}\backslash\mathbf{R}$ and that all elements in $\sigma(H)\cap\mathbf{U}^+$ (or $\sigma(H)\cap\mathbf{U}^-$) are all Krein-negative (or Krein-positive) definite.*

3° The left equality in (1) holds for all $\omega \in \mathbf{U}\backslash\{1\}$ if and only if $I_{2p}\diamond N_1(1,-1)^{\diamond(n-p)} \in \Omega^0(\gamma(\tau))$ for some integer $p \in [0,n]$. Especially in this case, all the eigenvalues of $\gamma(\tau)$ are equal to 1 and $\nu_1(\gamma) = n+p \ge n$.

4° The right equality in (1) holds for some $\omega \in \mathbf{U}^+\backslash\{1\}$ (or $\mathbf{U}^-\backslash\{1\}$) if and only if $I_{2p}\diamond N_1(1,1)^{\diamond r}\diamond K \in \Omega^0(\gamma(\tau))$ for some non-negative integers p and r satisfying $0 \le p+r \le n$ and $K \in \mathrm{Sp}(2(n-p-r))$ with $\sigma(K) \subset \mathbf{U}\backslash\{1\}$ satisfying the condition that all eigenvalues of K located within the closed arc between 1 and ω in $\mathbf{U}^+\backslash\{1\}$ (or $\mathbf{U}^-\backslash\{1\}$) possess total multiplicity $n-p-r$. If $\omega \ne -1$, all eigenvalues in $\sigma(K)\cap\mathbf{U}^+$ (or $\sigma(K)\cap\mathbf{U}^-$) are all Krein positive (or negative) definite. If $\omega = -1$, it holds that $-I_{2s}\diamond N_1(-1,1)^{\diamond t}\diamond H \in \Omega^0(K)$ for some non-negative integers s and t satisfying $0 \le s+t \le n-p-r$, and some $H \in \mathrm{Sp}(2(n-p-r-s-t))$ satisfying $\sigma(H) \subset \mathbf{U}\backslash\mathbf{R}$ and that all elements in $\sigma(H)\cap\mathbf{U}^+$ (or $\sigma(H)\cap\mathbf{U}^-$) are all Krein positive (or negative) definite.

5° The right equality in (1) holds for all $\omega \in \mathbf{U}\backslash\{1\}$ if and only if $I_{2p}\diamond N_1(1,1)^{\diamond(n-p)} \in \Omega^0(\gamma(\tau))$ for some integer $p \in [0,n]$. Especially in this case, all the eigenvalues of $\gamma(\tau)$ must be 1, and $\nu_1(\gamma) = n+p \ge n$.

6° Both equalities in (1) hold for all $\omega \in \mathbf{U}\backslash\{1\}$ if and only if $\gamma(\tau) = I_{2n}$.

Proof: We only consider the case for $\omega \in \mathbf{U}^+\backslash\{1\}$. The proof for $\omega \in \mathbf{U}^-\backslash\{1\}$ is similar and omitted. Without loss of generality, we set $\tau = 1$. Fix $\gamma \in \mathcal{P}_1(2n)$ and $\omega \in \mathbf{U}^+$. It is proved in Section 1.8 that $\gamma(1)$ can be connected within its homotopy component $\Omega^0(\gamma(1))$ to a matrix M which is a $\diamond$-product of normal forms,

$$M = I_{2p}\diamond N_1(1,-1)^{\diamond q}\diamond N_1(1,1)^{\diamond r}\diamond K, \tag{2}$$

where p, q and r are non-negative integers satisfying $p+q+r \le n$, $N_1(1,b)$ is a normal form belonging to the eigenvalue 1 with $b = \pm 1$ defined in Section 1.4, and $K \in \mathrm{Sp}(2(n-p-q-r))$ satisfies $1 \notin \sigma(K)$. We continue our proof in six steps.

Step 1 The proof of the left inequality in (1).

Note that $S^+_{\gamma(1)}(1) = S^-_{\gamma(1)}(1)$. Denote by $\omega_1, \ldots, \omega_k \in \mathbf{U}$ all the eigenvalues of $\gamma(1)$ that lie on the open arc between 1 and ω in $\mathbf{U}^+$. By the definition of the splitting numbers we obtain

$$i_\omega(\gamma) = i_1(\gamma) + S^+_{\gamma(1)}(1) - \sum_{j=1}^{k}[S^-_{\gamma(1)}(\omega_j) - S^+_{\gamma(1)}(\omega_j)] - S^-_{\gamma(1)}(\omega). \tag{3}$$

We continue the proof in two cases.

Case 1 $q = 0$ in (2).

In this case, by our discussion on splitting numbers in Section 9.1, we have

$$2S^+_{\gamma(1)}(1) = 2p + 2r \geq 2p + r = \nu_1(\gamma). \tag{4}$$

By (3), (4), and our study in Section 9.1, we obtain

$$\begin{aligned} i_\omega(\gamma) &\geq i_1(\gamma) + S^+_{\gamma(1)}(1) - \sum_{j=1}^k S^-_{\gamma(1)}(\omega_j) - S^-_{\gamma(1)}(\omega) \\ &\geq i_1(\gamma) + \nu_1(\gamma) - S^+_{\gamma(1)}(1) - \sum_{j=1}^k S^-_{\gamma(1)}(\omega_j) - S^-_{\gamma(1)}(\omega) \\ &\geq i_1(\gamma) + \nu_1(\gamma) - n. \end{aligned} \tag{5}$$

Thus the left inequality of (1) holds in this case.

Case 2 $q > 0$ in (2).

Suppose $q = 1$. By our study in Chapters 5, 6, and (2), there are paths $\gamma_1 \in \mathcal{P}_1(2)$ and $\gamma_2 \in \mathcal{P}_1(2n-2)$ such that $\gamma \sim \gamma_1 \diamond \gamma_2$ with $\gamma_1(1) = N_1(1, -1)$ and $\gamma_2(1) = I_{2p} \diamond N_1(1, 1)^{\diamond r} \diamond K$. Thus $\nu_1(\gamma_1) = 1$, $i_1(\gamma_1) \in 2\mathbf{Z}$, and

$$i_\omega(\gamma_1) = i_1(\gamma_1) = i_1(\gamma_1) + \nu_1(\gamma_1) - 1.$$

By symplectic additivity of i_ω and i_1, the proof of the left inequality in (1) is reduced to

$$i_\omega(\gamma_2) \geq i_1(\gamma_2) + \nu_1(\gamma_2) - (n-1).$$

Now we can apply Case 1 to the path γ_2 to get (1).

When $q \geq 2$, by induction and our study above we obtain the left inequality of (1).

Step 2 The proof of the right inequality in (1).

Similar to Step 1, by (3) and our study in Section 9.1

$$\begin{aligned} i_\omega(\gamma) + \nu_\omega(\gamma) \leq i_1(\gamma) &+ S^+_{\gamma(1)}(1) + \sum_{j=1}^k S^+_{\gamma(1)}(\omega_j) \\ &+ [\nu_\omega(\gamma) - S^-_{\gamma(1)}(\omega)]. \end{aligned} \tag{6}$$

If $\omega \neq -1$, from (6), $\nu_\omega(\gamma) = \nu_{\bar{\omega}}(\gamma)$, and our study in Section 9.1, we obtain

$$\begin{aligned} i_\omega(\gamma) + \nu_\omega(\gamma) &\leq i_1(\gamma) + S^+_{\gamma(1)}(1) + \sum_{j=1}^k S^+_{\gamma(1)}(\omega_j) + \nu_\omega(\gamma) \\ &\leq i_1(\gamma) + n. \end{aligned} \tag{7}$$

If $\omega = -1$, similar to (2) it holds that

$$(-I_{2s}) \diamond N_1(-1, 1)^{\diamond t} \diamond N_1(-1, -1)^{\diamond u} \diamond H \in \Omega^0(K), \tag{8}$$

for some non-negative integers s, t, and u satisfying $s+t+u \le n$, and some $H \in \mathrm{Sp}(2(n-p-q-r-s-t-u))$ satisfying $-1 \notin \sigma(H)$. From this representation and our discussion in Section 9.1, we obtain

$$0 \le \nu_{-1}(\gamma) - S^-_{\gamma(1)}(-1) \le s+t+u. \tag{9}$$

Together with (6), we obtain the right inequality of (1). Thus the right inequality of (1) always holds.

Note that the proofs of sufficiency parts of 2° to 5° follow by direct verifications, and therefore are omitted.

Step 3 The proof of 2°.

Suppose the left equality of (1) holds for the given $\omega \in \mathbf{U}^+\backslash\{1\}$. When $q=0$ in (2), all equalities must hold in (5). This then implies

$$\nu_1(\gamma) = 2S^+_{\gamma(1)}(1) = 2p, \tag{10}$$

$$\sum_{j=1}^{k} S^+_{\gamma(1)}(\omega_j) = 0, \tag{11}$$

$$S^+_{\gamma(1)}(1) + \sum_{j=1}^{k} S^-_{\gamma(1)}(\omega_j) + S^-_{\gamma(1)}(\omega) = n. \tag{12}$$

Thus 2° holds. When $q \ge 1$, by an induction argument as in the proof of Case 2 in Step 1, we get the necessity part of 2°.

Step 4 The proof of 3°.

When $q=0$ in (2), since ω can range over all of $\mathbf{U}^+\backslash\{1\}$, by 2° all the eigenvalues of $\gamma(1)$ must be equal to 1 and (10) to (12) must hold. This implies $\nu_1(\gamma) = 2n$. Thus $\gamma(1) = I_{2n}$. When $q \ge 1$, by induction we get the necessity part of 3°.

Step 5 The proof of 4°.

Suppose the right equality of (1) holds for the given $\omega \in \mathbf{U}^+\backslash\{1\}$.
If $\omega \ne -1$, all equalities must hold in (6) and (7). This yields

$$\sum_{j=1}^{k} S^-_{\gamma(1)}(\omega_j) = 0, \tag{13}$$

$$S^-_{\gamma(1)}(\omega) = 0, \tag{14}$$

$$S^+_{\gamma(1)}(1) + \sum_{j=1}^{k} S^+_{\gamma(1)}(\omega_j) + \nu_\omega(\gamma) = n. \tag{15}$$

By our discussion in Section 9.1, (15) implies $q = 0$ in (2). Together with our study in Section 9.1, we obtain the necessity part of 4° in this case.

If $\omega = -1$, all equalities must hold in (6). This yields (13) and

$$S^+_{\gamma(1)}(1) + \sum_{j=1}^{k} S^+_{\gamma(1)}(\omega_j) + \nu_\omega(\gamma) - S^-_{\gamma(1)}(\omega) = n. \tag{16}$$

By our discussion in Section 9.1, (16) implies $q = 0$ in (2) and $u = 0$ in (8). Together with our study in Section 9.1, we obtain the necessity part of 4° in this case.

Step 6 The proof of 5°.

Since ω can range over all of $\mathbf{U}^+\backslash\{1\}$, by 4° all the eigenvalues of $\gamma(1)$ must be 1 and equalities in (6) must hold. This implies $S^+_{\gamma(1)}(1) = n$. Thus in (2) it must hold that $p + r = n$ and $q = 0$, i.e., the necessity part of 5° holds.

6° is a direct consequence of 3° and 5°. The proof is complete. □

Theorem 2 (C. Liu and Y. Long [LLo1], [LLo3]). 1° *For any $\gamma \in \mathcal{P}_\tau(2n)$ and $m \in \mathbf{N}$,*

$$m\hat{i}(\gamma, 1) - n \le i(\gamma, m) \le m\hat{i}(\gamma, 1) + n - \nu(\gamma, m). \tag{17}$$

2° *The right equality in* (17) *holds for all $m \in \mathbf{N}$ if and only if $I_{2p}\diamond N_1(1, -1)^{\diamond(n-p)} \in \Omega^0(\gamma(\tau))$ for some integer $p \in [0, n]$. Especially in this case, all the eigenvalues of $\gamma(\tau)$ are equal to* 1 *and $\nu(\gamma, 1) = n + p \ge n$.*

3° *The left equality in* (17) *holds for all $m \in \mathbf{N}$ if and only if $I_{2q}\diamond N_1(1, 1)^{\diamond(n-q)} \in \Omega^0(\gamma(\tau))$ for some integer $q \in [0, n]$. Especially in this case, all the eigenvalues of $\gamma(\tau)$ are equal to* 1 *and $\nu(\gamma, 1) = n + q \ge n$.*

4° *Both equalities in* (17) *hold for all $m \in \mathbf{N}$ if and only if $\gamma(\tau) = I_{2n}$.*

Proof: Set $\tau = 1$. By Theorem 9.2.5, integrating (1) on $\mathbf{U}$ we obtain

$$i(\gamma, 1) + \nu(\gamma, 1) - n \le \hat{i}(\gamma, 1) \le i(\gamma, 1) + n. \tag{18}$$

Replacing $\tau = 1$ by m in (18), (9.2.25) yields (17). The right (left) equality in (17) holds if and only if the left (right) equality in (1) holds for all $\omega \in \mathbf{U}\backslash(\sigma(\gamma(\tau)) \cup \{1\})$. Thus 2°–4° follow from Theorem 1. □

Note that by Theorems 8.1.4–8.1.7 and 8.2.3–8.2.4, the necessary and sufficient conditions in terms of the basic normal form decomposition of $\gamma(\tau)$ can be given for the right or left equality in (17) to hold for some $m \in \mathbf{N}$. The details are left to the readers.

Theorem 3 (C. Liu and Y. Long [LLo1], [LLo3]). 1° *For any $\gamma \in \mathcal{P}_\tau(2n)$ and $m \in \mathbf{N}$,*

$$\begin{aligned} & m(i(\gamma, 1) + \nu(\gamma, 1) - n) + n - \nu(\gamma, 1) \\ & \le i(\gamma, m) \\ & \le m(i(\gamma, 1) + n) - n - (\nu(\gamma, m) - \nu(\gamma, 1)). \end{aligned} \tag{19}$$

2° *The left equality of* (19) *holds for some* $m \geq 2$ *if and only if* $I_{2p} \diamond N_1(1,-1)^{\diamond q} \diamond K \in \Omega^0(\gamma(\tau))$ *for some non-negative integres* p *and* q *satisfying* $p+q \leq n$ *and some* $K \in \mathrm{Sp}(2(n-p-q))$ *satisfying the condition that* $\sigma(K) \subset \mathbf{U}\backslash\{1\}$. *If* $m=2$ *and* $r=n-p-q>0$, *then* $N_1(-1,-1)^{\diamond r} \in \Omega^0(K)$. *If* $m \geq 3$ *and* $r=n-p-q>0$, *then* $R(\theta_1)\diamond\cdot\diamond R(\theta_r) \in \Omega^0(K)$ *for some* $\theta_j \in (0,\pi)$ *satisfying the condition that* $0 < m\theta_j\backslash(2\pi) \leq 1$ *with* $1 \leq j \leq r$. *In this case, all eigenvalues of* K *on* $\mathbf{U}^+\backslash\{1\}$ *(or* $\mathbf{U}^-\backslash\{1\}$*) are located on the arc between* 1 *and* $\exp(2\pi\sqrt{-1}/m)$*(or* $\exp(-2\pi\sqrt{-1}/m)$*) in* $\mathbf{U}^+$ *(in* $\mathbf{U}^-$*) and are all Krein-negative (or Krein-positive) definite.*

3° *The right equality of* (19) *holds for some* $m \geq 2$ *if and only if* $I_{2p} \diamond N_1(1,-1)^{\diamond r} \diamond K \in \Omega^0(\gamma(\tau))$ *for some non-negative integres* p *and* r *satisfying* $p+r \leq n$ *and some* $K \in \mathrm{Sp}(2(n-p-r))$ *with* $\sigma(K)=\{-1\}$ *satisfying the following conditions:*

If $m=2$ *and* $n-p-r>0$, *it holds that* $N_1(-1,-1)^{\diamond s} \diamond N_1(-1,1)^{\diamond t} \in \Omega^0(K)$ *for some non-negative integers* s *and* t *satisfying* $s+t=n-p-r$.

If $m>2$, *we must have* $n=p+r$.

4° *The two equalities of* (19) *hold for some* m_1 *and* $m_2 \geq 2$ *respectively if and only if* $\gamma(\tau)=I_{2p}\diamond N_1(-1,-1)^{\diamond(n-p)}$ *for some non-negative integres* $p \leq n$. *Here* $p<n$ *happens only when* $m_1=m_2=2$.

Proof: Set $\tau=1$. By Theorem 9.2.1 with $z=1$, summing (1) over all m-th roots of unity, we obtain

$$\begin{aligned} m(i(\gamma,1)+\nu(\gamma,1)-n)+i(\gamma,1) &\leq i(\gamma,m) \\ &\leq (m-1)(i(\gamma,1)+n)+i(\gamma,1)-(\nu(\gamma,m)-\nu(\gamma,1)). \end{aligned}$$

This yields (19). The equality conditions follow from Theorems 6.2.7, 8.1.4–8.1.7, and 8.2.3–8.2.4. □

Remark 4 Note that $\nu(\gamma,m) \geq \nu(\gamma,1)$ for any $\gamma \in \mathcal{P}_\tau(2n)$ and $m \in \mathbf{N}$.

Remark 5 I. Ekeland proved the following iteration inequality for his index theory in Theorem I.5.1 of [Eke3] for any $\gamma \in \mathcal{E}_\tau(2n)$ defined in Remark 15.1.5 below:

$$i_m^E(\gamma) \geq m(i_1^E(\gamma)+\nu_1^E(\gamma))-\nu_1^E(\gamma). \tag{20}$$

Note that by Theorem 15.1.6 proved in [Lon16], this inequality follows from the right-hand side inequality in 1° of our above Theorem 3 immediately. Note that the left-hand side in 1° of Theorem 3 can also be used to prove (20).

Note that by Theorem 7.3.1, the above Theorem 2 improves an estimate of H.B. Rademacher in [Rad1] of 1989.

Remark 6 Both of the first parts of Theorems 1 and 2 were established by C. Liu and the author in [LLo1] of 1997. The necessary and sufficient conditions for the equalities in Theorems 1–3 in this section were proved by C. Liu and the author in [LLo3] of 2000. Here we have made some necessary corrections for certain errors in [LLo3].

10.2 Successive estimates

The following estimate is basic for our iteration inequalities below.

Lemma 1 *Let $A \subset \mathbf{U}$ and $A' = \mathbf{U} \backslash A$. Suppose A is symmetric with respect to* **R**, *i.e., $\omega \in \mathcal{A}$ implies $\overline{\omega} \in A$. Then for any $M \in \mathrm{Sp}(2n)$,*

$$\sum_{\omega \in A} (\nu_\omega(M) - S_M^-(\omega)) + \sum_{\omega \in A'} S_M^-(\omega) \le \frac{e(M)}{2}, \tag{1}$$

where $e(M)$ is the elliptic height of M defined by the total algebraic multiplicity of all eigenvalues of M on **U** *given by Definition* 1.8.1.

Proof: Fix $M \in \mathrm{Sp}(2n)$. Let $e(M, A)$ denote the total algebraic multiplicity of all eigenvalues in $A \cap \sigma(M)$. For any $\omega \in \sigma(M) \cap \mathbf{U}$, denote its Krein type by $(p_\omega(M), q_\omega(M))$. Since A is symmetric with respect to **R**, so is A'. Therefore by Lemma 1.3.8, we have

$$\sum_{\omega \in A} p_\omega(M) = \frac{e(M, A)}{2}, \quad \sum_{\omega \in A'} q_\omega(M) = \frac{e(M, A')}{2}. \tag{2}$$

Together with Proposition 9.1.11, we obtain

$$\sum_{\omega \in A} (\nu_\omega(M) - S_M^-(\omega)) \le \frac{e(M, A)}{2}.$$

Together with Lemma 1.8.14 and Theorem 9.1.7, we obtain

$$\sum_{\omega \in A'} S_M^-(\omega) \le \frac{e(M, A')}{2}.$$

Thus (1) holds. □

Next we establish some iteration inequalities of the index theory for symplectic paths.

Theorem 2 (Y. Long and C. Zhu [LZu3]). *For any $\gamma \in \mathcal{P}_\tau(2n)$, set $M = \gamma(\tau)$ and extend γ to $[0, +\infty)$ by (8.0.4). Then for any m_1 and $m_2 \in \mathbf{N}$,*

$$\begin{aligned} &\nu(\gamma, m_1) + \nu(\gamma, m_2) - \nu(\gamma, (m_1, m_2)) - \frac{e(M)}{2} \\ &\le i(\gamma, m_1 + m_2) - i(\gamma, m_1) - i(\gamma, m_2) \\ &\le \nu(\gamma, (m_1, m_2)) - \nu(\gamma, m_1 + m_2) + \frac{e(M)}{2}, \end{aligned} \tag{3}$$

where (m_1, m_2) is the greatest common divisor of m_1 and m_2.

Proof: Fix m_1 and $m_2 \in \mathbf{N}$. We define

$$\begin{aligned}
\Psi_{m_1,m_2}(\theta) &= E\left(\frac{(m_1+m_2)\theta}{2\pi}\right) - E\left(\frac{m_1\theta}{2\pi}\right) - E\left(\frac{m_2\theta}{2\pi}\right), \quad \forall \theta \in [0,2\pi), \\
A &= \left\{\theta \in [0,2\pi) \mid \frac{m_1\theta}{2\pi} \in \mathbf{Z} \text{ or } \frac{m_2\theta}{2\pi} \in \mathbf{Z}\right\}, \\
B &= \left\{\theta \in [0,2\pi) \mid \frac{(m_1+m_2)\theta}{2\pi} \in \mathbf{Z} \text{ and } \frac{m_1\theta}{2\pi} \notin \mathbf{Z}\right\}, \\
D &= \{\theta \in [0,2\pi) \mid \Psi_{m_1,m_2}(\theta) = 0\}, \\
D' &= \{\theta \in (0,2\pi) \mid \Psi_{m_1,m_2}(\theta) \neq 0\}.
\end{aligned}$$

Note that we always have $\Psi_m(\theta) = 0$ or -1, $0 \in A \subset D$ and $B \subset D'$.

By Theorem 9.3.1, we have

$$\begin{aligned}
&i(\gamma, m_1+m_2) - i(\gamma, m_1) - i(\gamma, m_2) \\
&= (S_M^+(1) + C(M)) + 2\sum_{\theta\in(0,2\pi)} \Psi_{m_1,m_2}(\theta) S_M^-(e^{\sqrt{-1}\theta}) \\
&= S_M^-(1) + \left(\sum_{\theta\in D\setminus\{0\}} S_M^-(e^{\sqrt{-1}\theta}) + \sum_{\theta\in D'} S_M^-(e^{\sqrt{-1}\theta})\right) \qquad (4) \\
&\quad -2\sum_{\theta\in D'} S_M^-(e^{\sqrt{-1}\theta}) \\
&= \sum_{\theta\in D} S_M^-(e^{\sqrt{-1}\theta}) - \sum_{\theta\in D'} S_M^-(e^{\sqrt{-1}\theta}).
\end{aligned}$$

By Lemma 1, noting that both A and B are symmetric with respect to $\mathbf{R}$, we obtain

$$\begin{aligned}
&\sum_{\theta\in A} \nu_{e^{\sqrt{-1}\theta}}(M) - \sum_{\theta\in D} S_M^-(e^{\sqrt{-1}\theta}) + \sum_{\theta\in D'} S_M^-(e^{\sqrt{-1}\theta}) \\
&\quad \le \sum_{\theta\in A} (\nu_{e^{\sqrt{-1}\theta}}(M) - S_M^-(e^{\sqrt{-1}\theta}) + \sum_{\theta\in A'} S_M^-(e^{\sqrt{-1}\theta}) \qquad (5) \\
&\quad \le \frac{e(M)}{2},
\end{aligned}$$

$$\begin{aligned}
&\sum_{\theta\in D} S_M^-(e^{\sqrt{-1}\theta}) + \sum_{\theta\in B} \nu_{e^{\sqrt{-1}\theta}}(M) - \sum_{\theta\in D'} S_M^-(e^{\sqrt{-1}\theta}) \\
&\quad \le \sum_{\theta\in B'} S_M^-(e^{\sqrt{-1}\theta}) + \sum_{\theta\in B} (\nu_{e^{\sqrt{-1}\theta}}(M) - S_M^-(e^{\sqrt{-1}\theta})) \qquad (6) \\
&\quad \le \frac{e(M)}{2}.
\end{aligned}$$

By the definitions of A and B, we have

$$\nu(\gamma, m_1) + \nu(\gamma, m_2) - \nu(\gamma, (m_1, m_2)) = \sum_{\theta \in A} \nu_{e^{\sqrt{-1}\theta}}(\gamma), \tag{7}$$

$$\nu(\gamma, (m_1, m_2)) - \nu(\gamma, m_1 + m_2) = \sum_{\theta \in B} \nu_{e^{\sqrt{-1}\theta}}(\gamma). \tag{8}$$

By (4) we have

$$\begin{aligned} & i(\gamma, m_1 + m_2) - i(\gamma, m_1) - i(\gamma, m_2) \\ & = \sum_{\theta \in D} S_M^-(e^{\sqrt{-1}\theta}) - \sum_{\theta \in D'} S_M^-(e^{\sqrt{-1}\theta}) \\ & = \sum_{\theta \in A} \nu_{e^{\sqrt{-1}\theta}}(M) \\ & \quad - \left(\sum_{\theta \in A} \nu_{e^{\sqrt{-1}\theta}}(M) - \sum_{\theta \in D} S_M^-(e^{\sqrt{-1}\theta}) + \sum_{\theta \in D'} S_M^-(e^{\sqrt{-1}\theta}) \right). \end{aligned} \tag{9}$$

By (5) and (7), this implies the first inequality in (3).

Similarly, we have

$$\begin{aligned} & i(\gamma, m_1 + m_2) - i(\gamma, m_1) - i(\gamma, m_2) \\ & = \sum_{\theta \in D} S_M^-(e^{\sqrt{-1}\theta}) - \sum_{\theta \in D'} S_M^-(e^{\sqrt{-1}\theta}) \\ & = \sum_{\theta \in B} \nu_{e^{\sqrt{-1}\theta}}(M) \\ & \quad + \left(\sum_{\theta \in D} S_M^-(e^{\sqrt{-1}\theta}) + \sum_{\theta \in B} \nu_{e^{\sqrt{-1}\theta}}(M) - \sum_{\theta \in D'} S_M^-(e^{\sqrt{-1}\theta}) \right). \end{aligned} \tag{10}$$

Thus by (6) and (8), we obtain the second inequality in (3).

The proof is complete. □

The following result is a direct corollary of Theorem 2. We give an independent proof to show how to use precise iteration formulae obtained in Chapter 8 to detect and prove possible iteration inequalities.

Theorem 3 (Y. Long and C. Zhu [LZu3]). *For $\gamma \in \mathcal{P}_\tau(2n)$, set $M = \gamma(\tau)$ and extend γ to $[0, +\infty)$ by (8.0.4). Then for any $m \in \mathbf{N}$,*

$$\begin{aligned} \nu(\gamma, m) - \frac{e(M)}{2} & \le i(\gamma, m+1) - i(\gamma, m) - i(\gamma, 1) \\ & \le \nu(\gamma, 1) - \nu(\gamma, m+1) + \frac{e(M)}{2}. \end{aligned} \tag{11}$$

Proof: Without loss of generality, we set $\tau = 1$.

By Theorem 1.8.10, there exist $f \in C([0,1], \Omega^0(M))$ and basic normal forms $M_1, \ldots, M_p \in \mathrm{Sp}(2)$ and $N_1, \ldots, N_q \in \mathrm{Sp}(4)$ such that $f(0) = M$ and

$$f(1) = M_1 \diamond \cdots \diamond M_p \diamond N_1 \diamond \cdots \diamond N_q. \tag{12}$$

Since $\mathrm{Sp}(2n)$ is path connected, we can pick up paths $\xi_i \in \mathcal{P}_1(2)$ and $\eta_j \in \mathcal{P}_1(4)$ such that $\xi_i(1) = M_i$ and $\eta_j(1) = N_j$ for $1 \le i \le p$ and $1 \le j \le q$.

For $k \in \mathbf{Z}$, we define a new path $\beta \in \mathcal{P}_1(2n)$ by

$$\beta = \begin{cases} (\xi_1 * \phi_{1,2k\pi}) \diamond \xi_2 \diamond \cdots \xi_p \diamond \eta_1 \diamond \cdots \diamond \eta_q, & \text{if } p \ge 1, \\ (\eta_1 * (\phi_{1,2k\pi} \diamond I_2)) \diamond \eta_2 \diamond \cdots \diamond \eta_q, & \text{if } p = 0, \end{cases} \tag{13}$$

where $\phi_{1,\theta}$ is defined by (5.1.1). By this construction and the definition of $f(1)$, we have

$$\nu(\beta, 1) = \nu(\gamma, 1), \qquad e(\beta(1)) = e(f(1)) \le e(M). \tag{14}$$

Since the curve f is in $\Omega^0(M)$, we can fix a $k \in \mathbf{Z}$ in (13) such that

$$i(\beta, 1) = i(\gamma, 1). \tag{15}$$

Thus by the inverse homotopy Theorem 6.2.4,

$$\beta \sim_1 \gamma \quad \text{on } [0,1] \text{ along } f. \tag{16}$$

Extending this homotopy to $[0,1] \times [0,m]$ for any $m \in \mathbf{N}$, we have

$$\beta \sim_1 \gamma \quad \text{on } [0,m] \text{ along } f^m, \tag{17}$$

where $f^m(s) = f(s)^m$ for any $s \in [0,1]$. Then by Theorem 6.2.3, we obtain

$$i(\beta, m) = i(\gamma, m), \quad \nu(\beta, m) = \nu(\gamma, m), \quad \forall m \in \mathbf{N}. \tag{18}$$

So by (14) and (18), it suffices to prove (11) for the path β.

Note that all terms in the two inequalities in (11), except the elliptic height $e(\gamma(1))$, are symplectic additive and homotopy invariant. When we consider the path β, the elliptic height is also additive in terms of the decomposition of $\beta(1) = f(1)$ in (12). Thus by (12), (14), and Theorem 6.2.7, the proofs of both inequalities in (11) for the path β are reduced to those for each component path of β in (13), i.e., any path in $\mathcal{P}_1(2)$ or $\mathcal{P}_1(4)$ ending at a basic normal form of symplectic matrices defined in Section 1.8.

Next we continue our proof for paths ending at these normal forms in eight cases according to all the different patterns of iteration formulae.

Case 1 $\gamma \in \mathcal{P}_1(2)$ and $M \equiv \gamma(1) = N_1(1, b)$ with $b = 1$ or 0.

In this case by Theorem 8.3.1 (or Theorem 8.1.4), we have

$$e(M) = 2, \quad \nu(\gamma, 1) = 2 - b, \tag{19}$$

$$i(\gamma, m) = m(i(\gamma, 1) + 1) - 1, \quad \nu(\gamma, m) = \nu(\gamma, 1), \quad \forall m \in \mathbf{N}. \tag{20}$$

Thus we have

$$\begin{aligned} \nu(\gamma, m) - \frac{e(M)}{2} &= 1 - b, \\ i(\gamma, m+1) - i(\gamma, m) - i(\gamma, 1) &= 1, \\ \nu(\gamma, 1) - \nu(\gamma, m+1) + \frac{e(M)}{2} &= 1. \end{aligned}$$

Then both inequalities in (11).

Here we notice that when $M = N_1(1, 1)$, the first inequality in (11) can be improved by 1. We shall see this point in Theorem 4 below.

Case 2 $\gamma \in \mathcal{P}_1(2)$ and $M \equiv \gamma(1) = N_1(1, -1)$.

In this case by Theorem 8.3.1 (or Theorem 8.1.4), we have

$$\begin{aligned} e(M) &= 2, \quad \nu(\gamma, 1) = 1, \\ i(\gamma, m) &= mi(\gamma, 1), \quad \nu(\gamma, m) = \nu(\gamma, 1), \quad \forall m \in \mathbf{N}. \end{aligned}$$

Thus we have

$$\begin{aligned} \nu(\gamma, m) - \frac{e(M)}{2} &= 0, \\ i(\gamma, m+1) - i(\gamma, m) - i(\gamma, 1) &= 0, \\ \nu(\gamma, 1) - \nu(\gamma, m+1) + \frac{e(M)}{2} &= 1. \end{aligned}$$

Then both inequalities in (11) hold.

Case 3 $\gamma \in \mathcal{P}_1(2)$ and $M \equiv \gamma(1) = N_1(-1, -b)$ with $b = 1$ or 0.

In this case by Theorem 8.3.1 (or Theorem 8.1.5), we have

$$\begin{aligned} e(M) &= 2, \quad \nu(\gamma, 1) = 0, \\ i(\gamma, m) &= mi(\gamma, 1) - \frac{1 + (-1)^m}{2}, \quad \nu(\gamma, m) = \frac{1 + (-1)^m}{1 + b}, \quad \forall m \in \mathbf{N}. \end{aligned}$$

Thus we have

$$\begin{aligned} \nu(\gamma, m) - \frac{e(M)}{2} &= \frac{(-1)^m - b}{1 + b}, \\ i(\gamma, m+1) - i(\gamma, m) - i(\gamma, 1) &= (-1)^m, \end{aligned}$$

$$\nu(\gamma,1)-\nu(\gamma,m+1)+\frac{e(M)}{2} \;=\; \frac{(-1)^m+b}{1+b}.$$

Then both inequalities in (11) hold.

Case 4 $\gamma \in \mathcal{P}_1(2)$ and $M \equiv \gamma(1) = N_1(-1,1)$.

In this case by Theorem 8.3.1 (or Theorem 8.1.5), we have

$$\begin{aligned} e(M) &= 2, \quad \nu(\gamma,1)=0, \\ i(\gamma,m) &= mi(\gamma,1), \quad \nu(\gamma,m)=\frac{1+(-1)^m}{2}, \quad \forall m\in\mathbf{N}. \end{aligned}$$

Thus we have

$$\begin{aligned} \nu(\gamma,m)-\frac{e(M)}{2} &= \frac{(-1)^m-1}{2}\le 0, \\ i(\gamma,m+1)-i(\gamma,m)-i(\gamma,1) &= 0, \\ \nu(\gamma,1)-\nu(\gamma,m+1)+\frac{e(M)}{2} &= \frac{1-(-1)^m}{2}\ge 0. \end{aligned}$$

Then both inequalities in (11) hold.

Case 5 $\gamma \in \mathcal{P}_1(2)$ and $M \equiv \gamma(1) = D(2)$ or $D(-2)$.

In this case by Theorem 8.3.1 (or Theorem 8.1.6), we have

$$\begin{aligned} e(M) &= 0, \quad \nu(\gamma,1)=0, \\ i(\gamma,m) &= mi(\gamma,1), \quad \nu(\gamma,m)=0, \quad \forall m\in\mathbf{N}. \end{aligned}$$

Thus we have

$$\begin{aligned} \nu(\gamma,m)-\frac{e(M)}{2} &= 0, \\ i(\gamma,m+1)-i(\gamma,m)-i(\gamma,1) &= 0, \\ \nu(\gamma,1)-\nu(\gamma,m+1)+\frac{e(M)}{2} &= 0. \end{aligned}$$

Then both inequalities in (11) hold.

Case 6 $\gamma \in \mathcal{P}_1(2)$ and $M \equiv \gamma(1) = R(\theta)$ for some $\theta \in (0,\pi)\cup(\pi,2\pi)$.

In this case by Theorem 8.3.1 (or Theorem 8.1.7), we have

$$\begin{aligned} e(M) &= 2, \quad \nu(\gamma,1)=0, \\ i(\gamma,m) &= m(i(\gamma,1)-1)+2E\left(\frac{m\theta}{2\pi}\right)-1, \\ \nu(\gamma,m) &= 2-2\varphi\left(\frac{m\theta}{2\pi}\right), \quad \forall m\in\mathbf{N}. \end{aligned}$$

Thus we have

$$\nu(\gamma, m) - \frac{e(M)}{2} = 1 - 2\varphi\left(\frac{m\theta}{2\pi}\right),$$

$$i(\gamma, m+1) - i(\gamma, m) - i(\gamma, 1) = 2\left(E\left(\frac{(m+1)\theta}{2\pi}\right) - E\left(\frac{m\theta}{2\pi}\right)\right) - 1,$$

$$\nu(\gamma, 1) - \nu(\gamma, m+1) + \frac{e(M)}{2} = 2\varphi\left(\frac{(m+1)\theta}{2\pi}\right) - 1.$$

Note that

$$\begin{aligned}
&2\left(E\left(\frac{(m+1)\theta}{2\pi}\right) - E\left(\frac{m\theta}{2\pi}\right)\right) - 1 \\
&= 1 - 2\varphi\left(\frac{m\theta}{2\pi}\right) + 2\left(E\left(\frac{(m+1)\theta}{2\pi}\right) - \left[\frac{m\theta}{2\pi}\right] - 1\right) \\
&= 2\varphi\left(\frac{(m+1)\theta}{2\pi}\right) - 1 - 2\left(E\left(\frac{m\theta}{2\pi}\right) - \left[\frac{(m+1)\theta}{2\pi}\right]\right).
\end{aligned}$$

Note that also

$$E\left(\frac{(m+1)\theta}{2\pi}\right) - \left[\frac{m\theta}{2\pi}\right] - 1 \geq 0,$$

$$E\left(\frac{m\theta}{2\pi}\right) - \left[\frac{(m+1)\theta}{2\pi}\right] \geq 0.$$

Thus by the above estimates, we obtain both inequalities in (11).

Case 7 $\gamma \in \mathcal{P}_1(4)$ and $M \equiv \gamma(1) = N_2(\omega, b)$ for some $\omega \in \mathbf{U}\backslash\mathbf{R}$ is non-trivial.

In this case by Theorem 8.3.1 (or Theorem 8.2.3), we have

$$\begin{aligned}
e(M) &= 4, \quad \nu(\gamma, 1) = 0, \\
i(\gamma, m) &= mi(\gamma, 1) + 2\varphi\left(\frac{m\theta}{2\pi}\right) - 2, \quad \nu(\gamma, m) = 2 - 2\varphi\left(\frac{m\theta}{2\pi}\right), \quad \forall m \in \mathbf{N}.
\end{aligned}$$

Thus we have

$$\nu(\gamma, m) - \frac{e(M)}{2} = -2\varphi\left(\frac{m\theta}{2\pi}\right),$$

$$i(\gamma, m+1) - i(\gamma, m) - i(\gamma, 1) = 2\varphi\left(\frac{(m+1)\theta}{2\pi}\right) - 2\varphi\left(\frac{m\theta}{2\pi}\right),$$

$$\nu(\gamma, 1) - \nu(\gamma, m+1) + \frac{e(M)}{2} = 2\varphi\left(\frac{(m+1)\theta}{2\pi}\right).$$

Then both inequalities in (11) hold.

Case 8 $\gamma \in \mathcal{P}_1(4)$ and $M \equiv \gamma(1) = N_2(\omega, b)$ for some $\omega \in \mathbf{U}\backslash\mathbf{R}$ is trivial.

In this case by Theorem 8.3.1 (or Theorem 8.2.4), we have

$$\begin{aligned} e(M) &= 4, \quad \nu(\gamma, 1) = 0, \\ i(\gamma, m) &= mi(\gamma, 1), \quad \nu(\gamma, m) = 2 - 2\varphi\left(\frac{m\theta}{2\pi}\right), \quad \forall m \in \mathbf{N}. \end{aligned}$$

Thus we have

$$\begin{aligned} \nu(\gamma, m) - \frac{e(M)}{2} &= -2\varphi\left(\frac{m\theta}{2\pi}\right), \\ i(\gamma, m+1) - i(\gamma, m) - i(\gamma, 1) &= 0, \\ \nu(\gamma, 1) - \nu(\gamma, m+1) + \frac{e(M)}{2} &= 2\varphi\left(\frac{(m+1)\theta}{2\pi}\right). \end{aligned}$$

Then both inequalities in (11) hold.

The proof is complete. □

Suggested by the study in Case 1 with $b = 1$ of the proof of Theorem 3, we have the following result, which is useful for the study of closed characteristics on convex as well as star shaped hypersurfaces in $\mathbf{R}^{2n}$.

Theorem 4 (Y. Long and C. Zhu [LZu3]). *For any $\gamma \in \mathcal{P}_\tau(2n)$, set $M = \gamma(\tau)$ and extend γ to $[0, +\infty)$ by (8.0.4). Suppose that there exist $P \in \mathrm{Sp}(2n)$ and $Q \in \mathrm{Sp}(2n-2)$ such that*

$$M = P^{-1}(N_1(1, 1)\diamond Q)P. \tag{21}$$

Then for any $m \in \mathbf{N}$, we have

$$\begin{aligned} \nu(\gamma, m) - \frac{e(M)}{2} + 1 &\le i(\gamma, m+1) - i(\gamma, m) - i(\gamma, 1) \\ &\le \nu(\gamma, 1) - \nu(\gamma, m+1) + \frac{e(M)}{2}. \end{aligned} \tag{22}$$

Proof: Without loss of generality, we set $\tau = 1$.

Using the path $\phi_{1,\theta}$ given in (5.1.1), for $k \in \mathbf{Z}$ we define

$$\zeta(t) = N_1(1, t), \quad \xi(t) = \zeta * \phi_{1,2k\pi}(t), \quad \forall t \in [0, 1]. \tag{23}$$

Then by Theorems 6.2.7 and 8.1.4, we have

$$i(\xi, 1) = 2k - 1, \tag{24}$$

$$i(\xi, m) = 2mk - 1, \quad \nu(\xi, m) = 1, \quad \forall m \in \mathbf{N}. \tag{25}$$

Since $\mathrm{Sp}(2n-2)$ is path connected, there is a path $\eta \in \mathcal{P}_1(2n-2)$ such that $\eta(1) = Q$. By (23), we can fix a $k \in \mathbf{Z}$ so that

$$2k - 1 + i(\eta, 1) = i(\xi, 1) + i(\eta, 1) = i(\gamma, 1). \tag{26}$$

Pick up a path $g \in \mathcal{P}_1(2n)$ such that $g(1) = P$. Define

$$h(s) = g(s)^{-1} M g(s), \qquad \forall t \in [0, 1].$$

Then h connects $M = \gamma(1)$ to $N_1(1, 1) \diamond Q = (\xi \diamond \eta)(1)$ within $\Omega^0(M)$. Thus by the inverse homotopy Theorem 6.2.4, we have

$$\gamma \sim_1 (\xi \diamond \eta) \quad \text{on} \quad [0, 1] \quad \text{along} \quad h.$$

By extending this homotopy map to $[0, 1] \times [0, m]$ via iteration, we obtain

$$\gamma \sim_1 (\xi \diamond \eta) \quad \text{on} \quad [0, m] \quad \text{along} \quad h^m,$$

where $h^m(s) = h(s)^m$. Therefore by Theorem 6.2.7 and (26), for any $m \in \mathbf{N}$ we obtain

$$e(M) = e(N_1(1, 1)) + e(Q) = 2 + e(Q), \tag{27}$$

$$i(\gamma, m) = i(\xi, m) + i(\eta, m) = 2mk - 1 + i(\eta, m), \tag{28}$$

$$\nu(\gamma, m) = \nu(\xi, m) + \nu(\eta, m) = 1 + \nu(\eta, m). \tag{29}$$

Thus plugging (27)–(29) into (22), they become

$$\begin{aligned} \nu(\eta, m) - \frac{e(Q)}{2} &\le i(\eta, m+1) - i(\eta, m) - i(\eta, 1) \\ &\le \nu(\eta, 1) - \nu(\eta, m+1) + \frac{e(Q)}{2}, \end{aligned}$$

which follows from Theorem 3 for the path η. □

Remark 5 All the results in this section were established by C. Zhu and the author in the recent preprint [LZu3] of 1999.

10.3 Controlling iteration numbers via indices

For given $\tau > 0$, we study the number of iterations for symplectic paths via their indices. The following Lemma 1 is obtained from Theorems 1.2 and 1.3 immediately.

Lemma 1 *For any $\gamma \in \mathcal{P}_\tau(2n)$ and $m \in \mathbf{N}$,*

$$m \le 1 + \frac{i(\gamma, m) - i(\gamma, 1)}{i(\gamma, 1) + \nu(\gamma, 1) - n}, \quad \text{if } i(\gamma, 1) + \nu(\gamma, 1) - n > 0, \tag{1}$$

$$m \le 1 + \frac{i(\gamma, m) + \nu(\gamma, m) - \nu(\gamma, 1) - i(\gamma, 1)}{i(\gamma, 1) + n}, \quad \text{if } i(\gamma, 1) + n < 0, \tag{2}$$

$$m \le \frac{i(\gamma, m) + n}{\hat{i}(\gamma, 1)}, \quad \text{if } \hat{i}(\gamma, 1) > 0, \tag{3}$$

$$m \le \frac{i(\gamma, m) + \nu(\gamma, m) - n}{\hat{i}(\gamma, 1)}, \quad \text{if } \hat{i}(\gamma, 1) < 0. \tag{4}$$

A direct consequence of inequality (1) is the following

Corollary 2 *Suppose for $\gamma \in \mathcal{P}_\tau(2n)$ and integers $m, p \in \mathbf{N}$, $q \in \mathbf{Z}$,*

$$i(\gamma, m) \le n + q, \quad i(\gamma, 1) + \nu(\gamma, 1) \ge n + p. \tag{5}$$

Then

$$m \le \frac{\nu(\gamma, 1) + q}{p} \le \frac{2n + q}{p}. \tag{6}$$

Specially, if $p = q = 1$, we obtain $m \le 2n + 1$.

Theorem 3 (C. Liu and Y. Long [LLo3]). 1° *Suppose for $\gamma \in \mathcal{P}_\tau(2n)$ and integers $m, h \in \mathbf{N}$ and $l, s \in \mathbf{Z}$,*

$$i(\gamma, m) \le l, \quad i(\gamma, 1) \ge s, \quad i(\gamma, 1) + \nu(\gamma, 1) \ge h \ge n + 1, \tag{7}$$

$$l - s < 2(h - n). \tag{8}$$

Then

$$1 \le m \le 2 \quad \text{and} \quad l - s \ge 0. \tag{9}$$

2° *Moreover, suppose further that*

$$l - s \le h - n. \tag{10}$$

Then $m = 2$ only if

$$I_{2p} \diamond N_1(1, -1)^{\diamond q} \diamond N_1(-1, -1)^{\diamond (n-p-q)} \in \Omega^0(\gamma(\tau)), \tag{11}$$

for some non-negative integers p and q satisfying $0 \le p+q \le n$. In this case there exists an integer $k \ge 1$ such that

$$i(\gamma, 1) = 2k + n - 2p - q = s, \tag{12}$$

$$\nu(\gamma, 1) = 2p + q = h - s, \tag{13}$$

$$i(\gamma, 2) = 4k + n - 2p - q = l, \tag{14}$$

$$h - n = l - s = 2k \ge 2. \tag{15}$$

3° *There exists a path $\gamma \in \mathcal{P}_\tau(2n)$ satisfying conditions in 2° such that (7) and (10) hold with $m = 2$.*

Proof: 1° By (7) and the first inequality of (1.19), we obtain

$$\begin{aligned} l &\ge i(\gamma, m) \\ &\ge m(i(\gamma, 1) + \nu(\gamma, 1) - n) + n - \nu(\gamma, 1) \\ &= (m-1)(i(\gamma, 1) + \nu(\gamma, 1) - n) + i(\gamma, 1) \\ &\ge (m-1)(h-n) + s. \end{aligned} \tag{16}$$

Thus by (8) we obtain

$$m \le 1 + \left[\frac{l-s}{h-n}\right] \le 2, \qquad 0 \le l - s. \tag{17}$$

2° If $m = 2$ holds, (10) and (17) imply $l - s = h - n$. Then all equalities in (16) must hold, and we obtain

$$i(\gamma, 2) = l, \quad i(\gamma, 1) = s, \quad i(\gamma, 1) + \nu(\gamma, 1) = h. \tag{18}$$

Especially this implies that the first equality of (1.19) holds. We then obtain (11) by 2° of Theorem 1.3. By our earlier discussions, there are paths $\alpha \in \mathcal{P}_\tau(2p)$, $\beta \in \mathcal{P}_\tau(2q)$, $\psi \in \mathcal{P}_\tau(2(n-p-q))$ such that $\gamma \sim \alpha \diamond \beta \diamond \psi$, $\alpha(\tau) = I_{2p}$, $\beta(\tau) = N_1(1, -1)^{\diamond q}$, and $\psi(\tau) = N_1(-1, -1)^{\diamond (n-p-q)}$. Therefore by 2° of Theorem 1.3, there are integers $k_1, k_2, k_3 \in \mathbf{Z}$ such that

$$i(\alpha, 1) = 2k_1 - p, \quad \nu(\alpha, 1) = 2p, \quad i(\alpha, 2) = 4k_1 - p, \tag{19}$$

$$i(\beta, 1) = 2k_2, \quad \nu(\beta, 1) = q, \quad i(\beta, 2) = 4k_2, \tag{20}$$

$$\begin{aligned} i(\psi, 1) &= 2k_3 + (n-p-q), \quad \nu(\psi, 1) = 0, \\ i(\psi, 2) &= 4k_3 + (n-p-q). \end{aligned} \tag{21}$$

Thus by Theorem 6.2.7 and (18), we obtain (12)–(14) with $k = k_1 + k_2 + k_3$. Since $h \ge n+1$, from (12) and (13) we obtain $k \ge 1$. Then (15) follows from (12)–(14).

3° The path γ can be constructed using our study in Chapters 2, 5, and 6. The details are left to the readers.

The proof is complete. □

A direct consequence of Theorem 3 is the following

Corollary 4 *Given* $\gamma \in \mathcal{P}_\tau(2n)$, *suppose for some* $m, h \in \mathbf{N}$ *and* $l, s \in \mathbf{Z}$, (7) *and* (10) *hold, and at least one of* (11) *to* (15) *does not hold. Then* $m = 1$.

Corollary 5 (G. Fei, S.-K. Kim, and T. Wang [FKW1]). *Suppose for* $\gamma \in \mathcal{P}_\tau(2n)$ *and some* $m \in \mathbf{N}$ *and* $l \in \mathbf{Z}$,

$$l \geq i(\gamma, m), \quad i(\gamma, 1) \geq l - 1, \quad i(\gamma, 1) + \nu(\gamma, 1) \geq n + 1. \tag{22}$$

Then $m = 1$.

Proof: Let $s = l - 1$ and $h = n + 1$. Then (7) and (10) follow from (22), and (15) does not hold. Thus $m = 1$ by Corollary 4. □

Corollary 6 (D. Dong and Y. Long [DoL1]). *Suppose for* $\gamma \in \mathcal{P}_\tau(2n)$ *and some* $m \in \mathbf{N}$,

$$n + 1 \geq i(\gamma, m), \quad i(\gamma, 1) \geq n, \quad \nu(\gamma) \geq 1. \tag{23}$$

Then $m = 1$.

Proof: Letting $l = n + 1$ we get (22) from (23). Thus $m = 1$ by Corollary 5. □

Next we consider the case of paths in Sp(2). Note that in this case the iteration number can be determined by $i(\gamma, m)$ and $\nu(\gamma, m)$ only.

Theorem 7 (D. Dong and Y. Long [DoL1]). *Suppose for* $\tau > 0$ *and* $\gamma \in \mathcal{P}_\tau^0(2)$ *there exists an* $m \in \mathbf{N}$ *such that*

$$2 - \nu(\gamma, m) \leq i(\gamma, m) \leq 2. \tag{24}$$

Then $m = 1$.

Proof: Without loss of generality, we suppose $\tau = 1$. According to $\gamma(1)$ we continue our study in two cases.

Case 1 $\gamma(1) \in \mathrm{Sp}(2)_+^0$.

In this case, $i(\gamma, 1)$ must be even and $\nu(\gamma, m) = 1$. From (24) and Theorem 8.1.4 we obtain

$$1 \leq i(\gamma, m) = m i_1(\gamma) \leq 2.$$

Thus $i(\gamma, 1) \geq 2$ and then $m = 1$.

Case 2 $\gamma(1) \in \mathrm{Sp}(2)^0_- \cup \{I_2\}$.

In this case, $i(\gamma, 1)$ must be odd and $1 \le \nu(\gamma, m) \le 2$. From (24) and Theorem 8.1.4 we obtain

$$0 \le i(\gamma, m) = m(i(\gamma, 1) + 1) - 1 \le 2.$$

Therefore

$$1 \le m(i(\gamma, 1) + 1) \le 3.$$

Thus $i(\gamma, 1) \ge 1$ and then $m = 1$. □

It is natural to ask whether Theorem 7 can be generalized to higher dimensional cases. Unfortunately, the answer is negative in general, because of the following example.

Example 8 (D. Dong and Y. Long [DoL1]). Fix any integers $n \ge 2$ and $m \ge 2$. We choose $\theta \in (0, 2\pi)$ such that $m\theta = 2\pi + \chi\pi$ for some $\chi \in (0, 2)$. Let $k = [n/2]$. We define a path $\gamma \in \mathcal{P}_1(2n)$ by

$$\gamma(t) = F_1(t) \diamond \cdots \diamond F_k(t) \diamond Q_1(t) \diamond \cdots \diamond Q_k(t) \diamond D(2^{-t}) \diamond \cdots \diamond D(2^{-t}), \qquad (25)$$

where $F_i(t) = R(2\pi t)$ and $Q_i(t) = R((\theta - 2\pi)t)$ for $1 \le i \le k$. By results in Section 9.1, we obtain

$$i(\gamma, 1) = k(2-1) + k(1-2) = 0, \quad \nu(\gamma, 1) = 2k \ge 2, \qquad (26)$$

$$i(\gamma, m) = k(2m-1) + k(2+1-2m) = 2k, \quad \nu(\gamma, m) = 2k. \qquad (27)$$

Thus

$$n + p_n - \nu(\gamma, m) \le i(\gamma, m) \le n + 1, \qquad (28)$$

$$\nu(\gamma, 1) \ge 2\left[\frac{n}{2}\right], \qquad (29)$$

where we define $p_n = n$ if n is even, and $p_n = n - 2$ if n is odd. Therefore when $n \ge 2$, the condition (28)–(29) does not yield any bound on m in terms of n in general.

Remark 9 The idea of controlling the iteration number by corresponding indices started from the study of D. Dong and the author in [DoL1] of 1997 whose first preprint version appeared in 1993, where Corollary 6 was established via iteration theory developed in [DoL1]. The current proof is due to [LLo3]. In the papers

[LLo1] and [LLo3] by C. Liu and the author, this idea was further developed to get more information on the iterations. All the other results in this section were proved in [LLo3]. Special cases of Corollary 5 were proved by G. Fei, S.-K. Kim, and T. Wang in [FKW1] by using iteration inequalities of [DoL1]. Theorem 7 was proved in [DoL1] of 1997. Here we give a simpler proof. Example 8 was given in [DoL1].

Chapter 11

The common index jump theorem

Given a finite family of symplectic paths, the relationship among their iterated index sequences is very important in the study of periodic solutions of nonlinear Hamiltonian systems; such a relationship is specially crucial in distinguishing Hamitonian solution orbits geometrically. For every symplectic path, we define its index jumps by certain iterated index intervals. The common index jump theorem claims that for any finite family of symplectic paths in $\mathrm{Sp}(2n)$, their index jumps always contain infinitely many common non-empty intervals provided that their mean indices are positive and their initial indices are not less than n. This theorem turns out to be a powerful tool for applications to problems on nonlinear Hamiltonian systems.

Results of this chapter were established by C. Zhu and the author in [LZu3] of 1999. To the author's knowledge, no analogous results exist on common properties of indices of a finite family of symplectic paths before this work in any related areas, such as closed geodesics. Because of the precise iteration formulae derived in [Lon18] (cf. Chapter 8), it is possible to classify the problem into finitely many cases according to all possible different iteration patterns. The discovery of the common index jump theorem is based upon the classification given in [Lon18] for all possible eight iteration patterns of indices of closed characteristics on convex compact hypersurfaces in $\mathbf{R}^4$, and a further study on such a classification of all possible iteration patterns for paths in $\mathrm{Sp}(6)$. An illustration for this interesting common property is given in Example 15.3.7 below via weakly non-resonant ellipsoids. We refer the readers to Chapter 15 for further discussions.

In Section 1, we derive a common selection theorem using certain properties of dynamics on tori. In Section 2, we prove the common index jump theorem and derive its consequences. This theorem plays a crucial role in our study in Chapter 15 below.

11.1 A common selection theorem

We need the following common selection theorem. Recall the functions $E(\cdot)$, $[\cdot]$, $\varphi(\cdot)$, and $\{\cdot\}$ defined by (8.1.1).

Theorem 1 (Y. Long and C. Zhu [LZu3]). *Fix an integer $q > 0$. Let $\mu_i \geq 0$ and β_i be integers for all $i = 1, \ldots, k$. Let $\alpha_{i,j}$ be positive numbers for $j = 1, \ldots, \mu_i$ and $i = 1, \ldots, q$. Let $\delta \in (0, \frac{1}{2})$ satisfying*

$$\delta \max_{1\leq i\leq q} \mu_i < \frac{1}{2}. \tag{1}$$

Set

$$D_i = \beta_i + \sum_{j=1}^{\mu_i} \alpha_{i,j}, \quad \textit{for } i = 1, \dots, q. \tag{2}$$

Suppose

$$D_i > 0, \qquad \forall i = 1, \dots, q. \tag{3}$$

Then there exist infinitely many $(N, m_1, \dots, m_q) \in \mathbf{N}^{q+1}$ *such that*

$$m_i\beta_i + \sum_{j=1}^{\mu_i} E(m_i\alpha_{i,j}) = N + \Delta_i, \qquad \forall i = 1, \dots, q, \tag{4}$$

$$\min\{\{m_i\alpha_{i,j}\},\ 1 - \{m_i\alpha_{i,j}\}\} < \delta, \qquad \forall j = 1, \dots, \mu_i, i = 1, \dots, q, \tag{5}$$

$$m_i\alpha_{i,j} \in \mathbf{N}, \qquad \textit{if } \alpha_{i,j} \in \mathbf{Q}, \tag{6}$$

where

$$\Delta_i = \sum_{0<\{m_i\alpha_{i,j}\}<\delta} 1 \tag{7}$$

for all $i = 1, \dots, q$.

Proof: Firstly, we reduce the claims (4)–(6) to a dynamical problem on a torus. Then we prove the existence of numbers $(N, m_1, \dots, m_q)$ by the properties of the closed additive subgroups of some high dimensional torus. The proof is carried out in two steps.

Step 1 The reduction to a problem on a torus

We consider the left-hand side of (4) first. Using the function $\phi(\cdot)$ defined by (8.3.2), we obtain

$$\begin{aligned} m_i\beta_i &+ \sum_{j=1}^{\mu_i} E(m_i\alpha_{i,j}) \\ &= m_i D_i + \sum_{j=1}^{\mu_i} (\varphi(m_i\alpha_{i,j}) - \{m_i\alpha_{i,j}\}). \end{aligned} \tag{8}$$

To handle all the rational rotators, for $1 \le i \le q$ we require that each m_i has a factor $M \in \mathbf{N}$ such that $M\alpha_{i,j} \in \mathbf{N}$ whenever $\alpha_{i,j} \in \mathbf{Q}$ for $j = 1, \dots, \mu_i$ and $i = 1, \dots, q$. Let $M = 1$ if no such $\alpha_{i,j}$ exists.

To get the common integer $N \in \mathbf{N}$, we replace $m_i D_i$ in (8) by $\frac{N}{MD_i} MD_i$. To make m_i be an integer, we subtract the decimal part of $\frac{N}{MD_i}$ from itself. Because of the requirement of our later torus problem and to keep more flexibility on our choice of N, we add a term $\chi_i \in \{0, 1\}$ to it. This yields

$$\begin{aligned} m_i D_i &= \frac{N}{MD_i} MD_i - \left\{ \frac{N}{MD_i} \right\} MD_i + \chi_i MD_i \\ &= N + \left(\chi_i - \left\{ \frac{N}{MD_i} \right\} \right) MD_i. \end{aligned} \tag{9}$$

Here we define

$$m_i = \left(\left[\frac{N}{MD_i} \right] + \chi_i \right) M, \quad \forall i = 1, \ldots, q, \tag{10}$$

where $\chi_i = 0$ or 1 for $1 \le i \le q$ will be determined later. By this choice, (8) becomes

$$\begin{aligned} & m_i \beta_i + \sum_{j=1}^{\mu_i} E(m_i \alpha_{i,j}) \\ &= N + \left(\chi_i - \left\{ \frac{N}{MD_i} \right\} \right) MD_i + \sum_{j=1}^{\mu_i} (\varphi(m_i \alpha_{i,j}) - \{m_i \alpha_{i,j}\}). \end{aligned} \tag{11}$$

Now the claims (5) and (6) require that only the following three possibilities for each term $\{m_i \alpha_{i,j}\}$ can happen:

(A) $\{m_i \alpha_{i,j}\} = 0$ if $\alpha_{i,j} \in \mathbf{Q}$,
(B) $0 < \{m_i \alpha_{i,j}\} < \delta$ if $\alpha_{i,j} \in \mathbf{R} \setminus \mathbf{Q}$, or
(C) $1 - \delta < \{m_i \alpha_{i,j}\} < 1$ if $\alpha_{i,j} \in \mathbf{R} \setminus \mathbf{Q}$.

Note that (A) already holds by our choice of m_is of (10). Suppose now that all these three requirements are fulfilled by our choice of m_is. Then we have

$$\begin{aligned} & \sum_{j=1}^{\mu_i} (\varphi(m_i \alpha_{i,j}) - \{m_i \alpha_{i,j}\}) \\ &= \sum_A + \sum_B + \sum_C (\varphi(m_i \alpha_{i,j}) - \{m_i \alpha_{i,j}\}) \\ &= \sum_{0 < \{m_i \alpha_{i,j}\} < \delta} 1 - \sum_{0 < \{m_i \alpha_{i,j}\} < \delta} \{m_i \alpha_{i,j}\} \\ &\quad + \sum_{0 < 1 - \{m_i \alpha_{i,j}\} < \delta} (1 - \{m_i \alpha_{i,j}\}) \\ &= \Delta_i - \sum_{0 < \{m_i \alpha_{i,j}\} < \delta} \{m_i \alpha_{i,j}\} + \sum_{0 < 1 - \{m_i \alpha_{i,j}\} < \delta} (1 - \{m_i \alpha_{i,j}\}). \end{aligned} \tag{12}$$

Thus (11) becomes

$$\begin{aligned} m_i\beta_i + \sum_{j=1}^{\mu_i} E(m_i\alpha_{i,j}) &= N + \left(\chi_i - \left\{\frac{N}{MD_i}\right\}\right) MD_i \\ &\quad + \Delta_i - \sum_{0<\{m_i\alpha_{i,j}\}<\delta} \{m_i\alpha_{i,j}\} + \sum_{0<1-\{m_i\alpha_{i,j}\}<\delta} (1 - \{m_i\alpha_{i,j}\}). \end{aligned} \tag{13}$$

Therefore

$$\left| m_i\beta_i + \sum_{j=1}^{\mu_i} E(m_i\alpha_{i,j}) - N - \Delta_i \right| \le \left|\chi_i - \left\{\frac{N}{MD_i}\right\}\right| MD_i + \mu_i\delta. \tag{14}$$

Here to get (4), by (1) we need

$$\left|\chi_i - \left\{\frac{N}{MD_i}\right\}\right| MD_i < \frac{1}{2}. \tag{15}$$

Next we estimate $\{m_i\alpha_{i,j}\}$. By our choice of m_i in (10), we have

$$\begin{aligned} \{m_i\alpha_{i,j}\} &= \left\{ M\left(\left[\frac{N}{MD_i}\right] + \chi_i\right)\alpha_{i,j} \right\} \\ &= \left\{ \frac{N\alpha_{i,j}}{D_i} + \left(\chi_i - \left\{\frac{N}{MD_i}\right\}\right) M\alpha_{i,j} \right\} \\ &= \{A_{i,j}(N) + B_{i,j}(N)\}, \end{aligned} \tag{16}$$

where in (16), we set

$$A_{i,j}(N) = \left\{\frac{N\alpha_{i,j}}{D_i}\right\} - \chi_{i,j}, \quad B_{i,j}(N) = \left(\chi_i - \left\{\frac{N}{MD_i}\right\}\right) M\alpha_{i,j}, \tag{17}$$

and $\chi_{i,j} = 0$ or 1 will be determined later.

Assume now that we can choose $N \in \mathbf{N}$ such that

$$\left| \left\{\frac{N\alpha_{i,j}}{D_i}\right\} - \chi_{i,j} \right| = |A_{i,j}(N)| < \frac{\delta_1}{3}, \tag{18}$$

$$\left| \left(\chi_i - \left\{\frac{N}{MD_i}\right\}\right) M\alpha_{i,j} \right| = |B_{i,j}(N)| < \frac{\delta_1}{3}, \tag{19}$$

for a given δ_1 satisfying $0 < \delta_1 < \delta < 1/2$. Then if the sum $A_{i,j}(N) + B_{i,j}(N) \geq 0$, we obtain

$$\{m_i \alpha_{i,j}\} = A_{i,j}(N) + B_{i,j}(N) < \delta_1 < \delta.$$

If $A_{i,j}(N) + B_{i,j}(N) < 0$, we obtain

$$\{m_i \alpha_{i,j}\} = A_{i,j}(N) + B_{i,j}(N) - (-1).$$

That is

$$0 < 1 - \{m_i \alpha_{i,j}\} = -(A_{i,j}(N) + B_{i,j}(N)) < \delta_1 < \delta.$$

Therefore to prove (4)–(6), by (15), (18), and (19), it suffices to prove that we can choose integers χ_i and $\chi_{i,j}$ to be 0 or 1 and choose infinitely many integers $N \in \mathbf{N}$ such that all the quantities

$$\left| \left\{ \frac{N\alpha_{i,j}}{D_i} \right\} - \chi_{i,j} \right| \quad \text{and} \quad \left| \left\{ \frac{N}{MD_i} \right\} - \chi_i \right| \tag{20}$$

can be made simultaneously as small as we want.

Let $n = q + \sum_{i=1}^{q} \mu_i$, and

$$v = \left(\frac{1}{MD_1}, \ldots, \frac{1}{MD_q}, \frac{\alpha_{1,1}}{D_1}, \frac{\alpha_{1,2}}{D_1}, \ldots, \frac{\alpha_{1,\mu_1}}{D_1}, \frac{\alpha_{2,1}}{D_2}, \ldots, \frac{\alpha_{q,\mu_q}}{D_q} \right) \in \mathbf{R}^n. \tag{21}$$

Then the problem becomes for any given small $\epsilon \in (0, \min\{\delta, 1/3\})$ to find a vertex χ of the cube $[0, 1]^n$ and infinitely many integers $N \in \mathbf{N}$ such that

$$|\{Nv\} - \chi| < \epsilon. \tag{22}$$

This is a problem of the dynamics on the standard torus $T^n = (\mathbf{R}/\mathbf{Z})^n$.

Step 2 Dynamics on a torus

To solve problem (22), it suffices to note that the closure $\overline{G}$ of the set $G = \{\{mv\} \mid m \in \mathbf{Z}\}$ in T^n forms a closed additive subgroup of the torus T^n. Thus $\overline{G}$ is a product of a possibly lower dimensional torus and a cyclic group, and G contains the identity element of T^n.

Define a dynamical system $f : T^n \to T^n$ by $f(x) = x + \pi(v)$ for all $x \in T^n$. We consider the α and ω-limit sets of f which are defined as usual by

$$\alpha(x) = \bigcap_{N \in \mathbf{N}} \overline{\{f^k(x) \mid k \leq -N\}}, \quad \omega(x) = \bigcap_{N \in \mathbf{N}} \overline{\{f^k(x) \mid k \geq N\}},$$

for any $x \in T^n$.

If $v \in \mathbf{Q}^n$, G must be the finite cyclic group generated by $\pi(v)$. So we have $\alpha(0) = \omega(0) = G = \overline{G}$.

If $v \in \mathbf{R}^n \setminus \mathbf{Q}^n$, then we have $^{\#}G = +\infty$ and hence the limit sets $\alpha(0)$ and $\omega(0)$ are non-empty. By the definition of the limit sets and the fact $v \in \mathbf{R}^n \setminus \mathbf{Q}^n$, we have

$$\alpha(0) + G \subset \alpha(0), \qquad \omega(0) + G \subset \omega(0).$$

Since the limit sets are closed, we have

$$\alpha(0) + \overline{G} \subset \alpha(0), \quad \omega(0) + \overline{G} \subset \omega(0).$$

Thus by the facts that $\alpha(0) \cup \omega(0) \subset \overline{G}$ and that $\overline{G}$ is a closed additive subgroup of T^n, we also obtain

$$\alpha(0) = \omega(0) = \overline{G}. \tag{23}$$

Therefore there always exist infinitely many $N \in \mathbf{N}$ such that the point $\{Nv\}$ is located in the open ball in $\mathcal{T}^n$ centered at its identity element with radius $\epsilon < 1/3$. More precisely, we have proved that in the product space $X = [0, 1]^n$, there exist one vertex x of X and infinitely many $N \in \mathbf{N}$ such that

$$\{Nv\} \in B_\epsilon(x) \cap X, \tag{24}$$

where $B_\epsilon(x) = \{y \in \mathbf{R}^n \mid |y - x| < \epsilon\}$. Now we define $\chi = x$, and then (22) holds.

The proof is complete. □

In our later applications, we need to know more about the possible choices of the vector χ in Theorem 1. The following result gives this information.

Theorem 2 (Y. Long and C. Zhu [LZu3]). *Fix $v = (v_1, \dots, v_n) \in \mathbf{R}^n$. Let H be the closure of the subset $\{\{mv\} \mid m \in \mathbf{N}\}$ in $\mathcal{T}^n$ and $V = T_0\pi^{-1}H$ be the tangent space of $\pi^{-1}H$ at the origin in $\mathbf{R}^n$, where $\pi : \mathbf{R}^n \to \mathcal{T}^n$ is the projection map. Define*

$$A(v) = V \setminus \cup_{v_k \in \mathbf{R} \setminus \mathbf{Q}} \{x = (x_1, \dots, x_n) \in V \mid x_k = 0\}. \tag{25}$$

Define $\psi(x) = 0$ when $x \geq 0$ and $\psi(x) = 1$ when $x < 0$. Then for any $a = (a_1, \dots, a_n) \in A(v)$, the vector

$$\chi = (\psi(a_1), \dots, \psi(a_n)) \tag{26}$$

makes (22) hold for infinitely many $N \in \mathbf{N}$.

Moreover, this set $A(v)$ possesses the following properties.

(a) $A(v) \neq \emptyset$.

(b) *When $v \in \mathbf{Q}^n$, it holds that $V = A(v) = \{0\}$.*

(c) *When* $v \in \mathbf{R}^n \backslash \mathbf{Q}^n$, *it holds that* $\dim V \geq 1$, $0 \notin A(v) \subset V$, $A(v) = -A(v)$, *and that* $A(v)$ *is open in* V.

(d) *When* $\dim V = 1$, *then* $A(v) = V \backslash \{0\}$.

(e) *When* $\dim V \geq 2$, $A(v)$ *is obtained from* V *by deleting all the coordinate hyperplanes with dimension strictly smaller than* $\dim V$ *from* V, *and especially* $\dim A(v) = \dim V$.

Proof: If $v \in \mathbf{Q}^n$, H is a cyclic subgroup of $\mathcal{T}^n$. We then have $V = \{0\}$. Thus $A(v) = V = \{0\}$.

Suppose $v = (v_1, \ldots, v_n) \in \mathbf{R}^n \backslash \mathbf{Q}^n$. Then H is a closed additive subgroup of $\mathcal{T}^n$ which is a product of a torus with a cyclic subgroup of $\mathcal{T}^n$. By the definition of V, we have $\dim V \geq 1$. For any $a = (a_1, \ldots, a_n) \in V$,

$$a_i \neq 0 \quad \text{only if } v_i \in \mathbf{R} \backslash \mathbf{Q}.$$

When $v_i \in \mathbf{R} \backslash \mathbf{Q}$, there exists $a = (a_1, \ldots, a_n) \in V$ with $a_i \neq 0$. Thus for any $a \in A(v)$, the point χ defined by (26) gives a vertex χ of $X = [0, 1]^n$ which is in the closure of $\{\{mv\} \mid m \in \mathbf{N}\}$ in X. Therefore (22) holds for infinitely many $N \in \mathbf{N}$ and this χ.

The claims (a) to (e) follow from this argument, and their proofs are therefore omitted. □

11.2 The common index jump theorem

The following is the main result of this section.

Theorem 1 (Y. Long and C. Zhu [LZu3]). *Let* $\gamma_k \in \mathcal{P}_{\tau_k}(2n)$ *for* $k = 1, \ldots, q$ *be a finite collection of symplectic paths. Let* $M_k = \gamma(\tau_k)$. *We extend* γ_k *to* $[0, +\infty)$ *by iteration via* (8.0.4) *for* $k = 1, \ldots, q$. *Suppose*

$$\hat{i}(\gamma_k, 1) > 0, \quad \forall k = 1, \ldots, q. \tag{1}$$

Then there exist infinitely many $(N, m_1, \ldots, m_q) \in \mathbf{N}^{q+1}$ *such that*

$$\nu(\gamma_k, 2m_k - 1) = \nu(\gamma_k, 1), \tag{2}$$

$$\nu(\gamma_k, 2m_k + 1) = \nu(\gamma_k, 1), \tag{3}$$

$$\begin{aligned} i(\gamma_k, &2m_k - 1) + \nu(\gamma_k, 2m_k - 1) \\ &= 2N - (i(\gamma_k, 1) + 2S^+_{M_k}(1) - \nu(\gamma_k, 1)), \end{aligned} \tag{4}$$

$$i(\gamma_k, 2m_k + 1) = 2N + i(\gamma_k, 1), \tag{5}$$

$$i(\gamma_k, 2m_k) \geq 2N - \frac{e(M_k)}{2} \geq 2N - n, \tag{6}$$

$$i(\gamma_k, 2m_k) + \nu(\gamma_k, 2m_k) \leq 2N + \frac{e(M_k)}{2} \leq 2N + n, \tag{7}$$

for every $k = 1, \ldots, q$.

Proof: We complete the proof in four steps.

Step 1 Application of Theorem 1.1

Set

$$\delta_0 = \min_{1\leq k\leq q} \left\{\frac{1}{2}, \frac{\theta}{2\pi}, 1 - \frac{\theta}{2\pi} \mid \theta \in (0, 2\pi) \text{ and } e^{\sqrt{-1}\theta} \in \sigma(M_k)\right\}, \tag{8}$$

$$C_k = \sum_{\theta\in(0,2\pi)} S_{M_k}^-(e^{\sqrt{-1}\theta}), \tag{9}$$

$$\rho_k = i(\gamma_k, 1) + S_{M_k}^+(1) - C_k, \tag{10}$$

$$I(k, m) = m\rho_k + \sum_{\theta\in(0,2\pi)} E\left(\frac{m\theta}{\pi}\right) S_{M_k}^-(e^{\sqrt{-1}\theta}), \tag{11}$$

for $k = 1, \ldots, q$ and $m \in \mathbf{N}$.

By the definition of $I(k, m)$, we rewrite it as

$$I(k, m) = m\rho_k + \sum_{\theta\in(0,2\pi)} \sum_{j=1}^{S_{M_k}^-(e^{\sqrt{-1}\theta})} E\left(\frac{m\theta}{\pi}\right). \tag{12}$$

Note that by (1) and Corollary 9.3.2, for every $k = 1, \ldots, q$, we have

$$0 < \hat{i}(\gamma_k, 1) = \rho_k + \sum_{\theta\in(0,2\pi)} \frac{\theta}{\pi} S_{M_k}^-(e^{\sqrt{-1}\theta}). \tag{13}$$

Now in the statement of Theorem 1.1, other than the integer q, we set

$$\begin{cases} \delta \in (0, \delta_0), \quad \beta_i = \rho_i, \quad D_i = \hat{i}(\gamma_i, 1), \\ \mu_i = \sum_{\theta\in(0,2\pi)} S_{M_i}^-(e^{\sqrt{-1}\theta}), \quad \forall 1 \leq i \leq q, \end{cases} \tag{14}$$

$$\alpha_{i,j} = \frac{\theta_j}{\pi}, \quad \text{where } e^{\sqrt{-1}\theta_j} \in \sigma(M_i), \quad \forall 1 \leq j \leq \mu_i,\ 1 \leq i \leq q. \tag{15}$$

Note that μ_i is a non-negative integer by Corollary 9.1.9.

By (13), the condition (1.3) of Theorem 1.1 holds. Applying this theorem to $I(k,m)$s for $k=1,\ldots,q$ and any $\delta\in(0,\delta_0)$, there exist infinitely many $(N,m_1,\ldots,m_q)\in\mathbf{N}^{q+1}$ such that

$$I(k,m_k)=N+\Delta_k, \tag{16}$$

$$\min\left\{\left\{\frac{m_k\theta}{\pi}\right\},1-\left\{\frac{m_k\theta}{\pi}\right\}\right\}<\delta,\quad \text{if } S^-_{M_k}(e^{\sqrt{-1}\theta})>0, \tag{17}$$

$$\frac{m_k\theta}{\pi}\in\mathbf{Z},\quad \text{if } \frac{\theta}{\pi}\in\mathbf{Q}\cap(0,2)\ \text{ and } S^-_{M_k}(e^{\sqrt{-1}\theta})>0, \tag{18}$$

$$\Delta_k=\sum_{0<\{m_k\theta/\pi\}<\delta}S^-_{M_k}(e^{\sqrt{-1}\theta}), \tag{19}$$

for $k=1,\ldots,q$.

Step 2 Verifications of (2) and (3).

Whenever $e^{\sqrt{-1}\theta}\in\sigma(M_k)$ and $\frac{\theta}{\pi}\in\mathbf{Q}\cap(0,2)$, by (18), we always have $2m_k\theta\in 2\pi\mathbf{Z}$. Thus for any such a θ,

$$2m_k\theta\pm\theta\notin 2\pi\mathbf{Z}.$$

Since the change of the nullity happens only when iterations of some eigenvalues in $\mathbf{U}\backslash\{1\}$ hit 1, we obtain

$$\nu(\gamma_k,2m_k-1)=\nu(\gamma_k,2m_k+1)=\nu(\gamma_k,1).$$

Thus (2) and (3) hold.

Step 3 Verifications of (4) and (5).

By Theorem 9.3.1, (16), (2), and the definition of $I(k,m)$, we have

$$\begin{aligned}
&i(\gamma_k,2m_k-1)+\nu(\gamma_k,2m_k-1)\\
&\quad=2I(k,m_k)-(i(\gamma_k,1)+S^+_{M_k}(1)-C_k)-(S^+_{M_k}(1)+C_k)\\
&\qquad-2\sum_{\theta\in(0,2\pi)}\xi_-(m_k,\theta)S^-_{M_k}(e^{\sqrt{-1}\theta})+\nu(\gamma_k,1) \qquad\qquad (20)\\
&\quad=2(N+\Delta_k)-(i(\gamma_k,1)+2S^+_{M_k}(1))\\
&\qquad-2\sum_{\theta\in(0,2\pi)}\xi_-(m_k,\theta)S^-_{M_k}(e^{\sqrt{-1}\theta})+\nu(\gamma_k,1),
\end{aligned}$$

where we define

$$\begin{aligned}
\xi_-(m_k, \theta) &= E\left(\frac{2m_k\theta}{2\pi}\right) - E\left(\frac{(2m_k-1)\theta}{2\pi}\right) \\
&= E\left(\left[\frac{m_k\theta}{\pi}\right] + \left\{\frac{m_k\theta}{\pi}\right\}\right) - E\left(\left[\frac{m_k\theta}{\pi}\right] + \left\{\frac{m_k\theta}{\pi}\right\} - \frac{\theta}{2\pi}\right) \quad (21) \\
&= E\left(\left\{\frac{m_k\theta}{\pi}\right\}\right) - E\left(\left\{\frac{m_k\theta}{\pi}\right\} - \frac{\theta}{2\pi}\right).
\end{aligned}$$

Note that $\xi_-(m_k, \theta)$ takes only the value 0 or 1. To evaluate $\xi_-(m_k, \theta)$ for any $\theta \in (0, 2\pi)$ and $e^{\sqrt{-1}\theta} \in \sigma(M_k)$ with some $k \in \{1, \dots, q\}$, we consider the following three cases:

Case 1 $\{\frac{m_k\theta}{\pi}\} = 0$.

By (21) and the definition of the function $E(\cdot)$, we have $\xi_-(m_k, \theta) = 0$.

Case 2 $\{\frac{m_k\theta}{\pi}\} \in (0, \delta)$.

Since $\delta \in (0, \delta_0)$, by the definition (8) of δ_0, we obtain

$$\left\{\frac{m_k\theta}{\pi}\right\} - \frac{\theta}{2\pi} < \delta_0 - \frac{\theta}{2\pi} \leq 0.$$

Thus we have $\xi_-(m_k, \theta) = 1$.

Case 3 $\{\frac{m_k\theta}{\pi}\} \geq \delta$ and $S_{M_k}^-(e^{\sqrt{-1}\theta}) > 0$.

By (17) it must hold that

$$1 - \delta < \left\{\frac{m_k\theta}{\pi}\right\} < 1.$$

Thus by the fact $\delta \in (0, \delta_0)$ and the definition (8) of δ_0, we obtain

$$\left\{\frac{m_k\theta}{\pi}\right\} - \frac{\theta}{2\pi} > 1 - \delta - \frac{\theta}{2\pi} > 0.$$

So in this case we have $\xi_-(m_k, \theta) = 0$.

Therefore only in the above Case 2, the term $\xi_-(m_k, \theta)$ makes a contribution. Together with (20) and (21), we obtain

$$\begin{aligned}
&i(\gamma_k, 2m_k - 1) + \nu(\gamma_k, 2m_k - 1) \\
&\quad = 2(N + \Delta_k) - (i(\gamma_k, 1) + 2S_{M_k}^+(1)) \\
&\qquad -2\Delta_k + \nu(\gamma_k, 1) \\
&\quad = 2N - (i(\gamma_k, 1) + 2S_{M_k}^+(1) - \nu(\gamma_k, 1)).
\end{aligned}$$

This proves (4).

Similarly we have

$$\begin{aligned} & i(\gamma_k, 2m_k+1) \\ & \quad = 2I(k, m_k) + (i(\gamma_k, 1) + S^+_{M_k}(1) - C_k) - (S^+_{M_k}(1) + C_k) \\ & \quad +2 \sum_{\theta\in(0,2\pi)} \xi_+(m_k, \theta) S^-_{M_k}(e^{\sqrt{-1}\theta}), \end{aligned} \tag{22}$$

where

$$\xi_+(m_k, \theta) = E\left(\frac{(2m_k+1)\theta}{2\pi}\right) - E\left(\frac{2m_k\theta}{2\pi}\right).$$

Similarly to our above discussion, we have $\xi_+(m_k, \theta) = 1$ in the above Cases 1 and 3, and $\xi_+(m_k, \theta) = 0$ in the above Case 2. Thus from (22) we obtain

$$\begin{aligned} i(\gamma_k, 2m_k+1) &= 2(N + \Delta_k) + (i(\gamma_k, 1) - 2C_k) + 2(C_k - \Delta_k) \\ &= 2N + i(\gamma_k, 1). \end{aligned} \tag{23}$$

This proves (5).

Step 4 Verifications of (6) and (7).

By the second inequality in (10.2.11) of Theorem 10.2.3 and (3) as well as (5), we have

$$\begin{aligned} i(\gamma_k, 2m_k) &\geq i(\gamma_k, 2m_k+1) - i(\gamma_k, 1) - \frac{e(M_k)}{2} \\ &= 2N - \frac{e(M_k)}{2} \\ &\geq 2N - n. \end{aligned}$$

This proves (6).

By the first inequality in (10.2.11) of Theorem 10.2.3 as well as (5), we have

$$\begin{aligned} i(\gamma_k, 2m_k) + \nu(\gamma_k, 2m_k) &\leq i(\gamma_k, 2m_k+1) - i(\gamma_k, 1) + \frac{e(M_k)}{2} \\ &= 2N + \frac{e(M_k)}{2} \\ &\leq 2N + n. \end{aligned} \tag{24}$$

This proves (7).

The proof is complete. □

The following consideration is useful for the study of closed characteristics in our applications.

Theorem 2 (Y. Long and C. Zhu [LZu3]). *In Theorem* 1, *further suppose*

$$M_k = P_k^{-1}(N_1(1,1)\diamond G_k)P_k \qquad (25)$$

holds for some $P_k \in \mathrm{Sp}(2n)$ *and* $G_k \in \mathrm{Sp}(2n-2)$ *with* $1 \le k \le q$. *Then* (7) *can be improved to*

$$i(\gamma_k, 2m_k) + \nu(\gamma_k, 2m_k) \le 2N + \frac{e(M_k)}{2} - 1 \le 2N + n - 1. \qquad (26)$$

Proof: In Step 4 of the proof of Theorem 1, we use the first inequality in (10.2.22) of Theorem 10.2.4 as well as (5), and obtain

$$\begin{aligned} i(\gamma_k, 2m_k) + \nu(\gamma_k, 2m_k) &\le i(\gamma_k, 2m_k + 1) - i(\gamma_k, 1) + \frac{e(M_k)}{2} - 1 \\ &= 2N + \frac{e(M_k)}{2} - 1 \\ &\le 2N + n - 1. \end{aligned}$$

This proves (26). □

Next we introduce the index jump for any symplectic paths.

Definition 3 (Y. Long and C. Zhu [LZu3]). For any $\gamma \in \mathcal{P}_\tau(2n)$ and $m \in \mathbf{N}$, we define the m**-th index jump** $\mathcal{G}_m(\gamma)$ of γ by the open interval

$$\mathcal{G}_m(\gamma) = (i(\gamma, m) + \nu(\gamma, m) - 1, i(\gamma, m+2)). \qquad (27)$$

A direct consequence of Theorem 2 is the following

Corollary 4 (Y. Long and C. Zhu [LZu3]). *Let* $\gamma_k \in \mathcal{P}_{\tau_k}(2n)$ *for* $k = 1, \ldots, q$ *be a finite collection of symplectic paths. We extend* γ_k *to* $[0, +\infty)$ *by iteration via* (8.0.4) *for* $k = 1, \ldots, q$. *Define*

$$\kappa_1 = \min_{1\le k\le q} (i(\gamma_k, 1) + 2S^+_{\gamma_k(\tau_k)}(1) - \nu(\gamma_k, 1)), \qquad (28)$$

$$\kappa_2 = \min_{1\le k\le q} (i(\gamma_k, 1) - 1). \qquad (29)$$

For all $k = 1, \ldots, q$, *suppose condition* (1) *holds, i.e., the mean index of each* γ_k *is positive. Suppose*

$$\kappa_1 + \kappa_2 \ge 0. \qquad (30)$$

Then there exist infinitely many $(N, m_1, \ldots, m_q) \in \mathbf{N}^{q+1}$ *such that*

$$\emptyset \neq [2N - \kappa_1, 2N + \kappa_2] \subset \bigcap_{k=1}^{q} \mathcal{G}_{2m_j-1}(\gamma_k), \tag{31}$$

We call $[2N - \kappa_1, 2N + \kappa_2]$ the **common index jump** of these γ_ks. It will play a crucial role in Chapter 15.

Chapter 12

Index iteration theory for closed geodesics

Morse index theory for closed geodesics on Riemannian manifolds has a very long history, for details of which we refer the readers to the original book [Mor1] of M. Morse, and the excellent survey works [Bot2] of R. Bott and [Ban2] of V. Bangert.

Let M be an n-dimensional Riemannian manifold with metric g. Denote the Levi-Civita connection of (M, g) by ∇. A closed geodesic c on (M, g) is a map $c : \mathbf{R}/\mathbf{Z} \to M$, which satisfies the following equation on M:

$$\nabla_{\dot{c}(t)}\dot{c}(t) = 0, \qquad \forall t \in \mathbf{R}. \tag{1}$$

Locally geodesics are shortest curves between any two given points in M. Let ΛM denote the Hilbert manifold of closed curves in (M, g) under the $W^{1,2}(\mathbf{R}/\mathbf{Z}, M)$ norm. It is well known (cf. [Kli1] and [Kli2]) that a closed geodesic is a critical point of the energy functional $\mathcal{E} : \Lambda M \to \mathbf{R}$:

$$\mathcal{E}(c) = \frac{1}{2}\int_0^1 g(\dot{c}(t), \dot{c}(t))\, dt, \qquad \forall c \in \Lambda M. \tag{2}$$

In the 1920s, M. Morse assigned to each closed geodesic c a pair of non-negative integers $(\mathrm{ind}(c), \mathrm{nul}(c))$, which are now called **Morse index and nullity** of c, and discovered a significant relation between Morse index theory and the base manifold M.

For a closed geodesic c, its q-th iterate c^q is defined by $c^q(t) = c(qt)$ for $q \in \mathbf{N}$. Note that each c^q is also a critical point of $\mathcal{E}$ but possessing the same orbit set in M with c. In order to distinguish critical points of $\mathcal{E}$ corresponding to different geodesic orbit sets in M, one way is to study the index sequence $\{(\mathrm{ind}(c^q), \mathrm{nul}(c^q))\}$ of iterations $\{c^q\}$ of c. The goal of this chapter is to prove that the iteration theory established in previous chapters in Part IV for index theory of symplectic paths holds also for Morse index theory of closed geodesics.

In the following Section 1, we briefly review the Morse index theory for closed geodesics. In Section 2, we study the splitting numbers of closed geodesics and prove that they coincide with the splitting numbers of symplectic matrices via Definition 9.1.4. This then implies that our iteration theory also works for Morse index theory of closed geodesics on Riemannian manifolds.

Most of the results in Section 1 are taking from [Bot1] of R. Bott and [BTZ1] of W. Ballmann, G. Thobergsson, and W. Ziller. Especially Theorem 1.1 was proved in [Bot1] and [BTZ1]. Theorems 1.2 and 1.3 were proved in [BTZ1]. Theorems 2.1, 2.2, 2.3, and Corollary 2.4 were proved in the recent paper [LLo4].

12.1 Morse index theory

We first recall some classical notations and definitions on geodesics. Let (M, g) be an n-dimensional complete connected C^∞ Riemannian manifold. Let $c : \mathbf{R} \to M$ be a closed geodesic satisfying $c(t+1) = c(t)$ for all $t \in \mathbf{R}$ and let $p = c(0) = c(1)$. A Jacobi field U along c is a vector field U transversal to c and satisfies

$$\nabla_{\dot{c}(t)}\nabla_{\dot{c}(t)}U(t) + R_{\dot{c}(t)U(t)}\dot{c}(t) = 0.$$

Let E be the orthogonal complement of $v = \dot{c}(0) = \dot{c}(1)$ in T_pM. Then the linearized Poincaré map of c is defined by $P : E \oplus E \to E \oplus E$ with $P : (X, Y) \mapsto (U(1), \nabla U(1))$, where U is the unique Jacobi field along c determined by $U(0) = X$ and $\nabla U(0) = Y$. The map P preserves the symplectic structure φ on $E \oplus E$ defined by

$$\varphi((X_1, Y_1), (X_2, Y_2)) = \langle X_1, Y_2\rangle - \langle Y_1, X_2\rangle,$$

where $\langle\cdot, \cdot\rangle$ denotes the usual inner product in $E \cong \mathbf{R}^{n-1}$.

Note that we can choose an orthonormal basis such that $E \cong \mathbf{R}^{n-1}$ and $E \oplus E \cong \mathbf{R}^{2(n-1)}$. With this equivalence, the linearized Poincaré map P corresponds to a symplectic matrix which is still denoted by P, and the above symplectic structure can be given on $\mathbf{R}^{2(n-1)}$ by

$$\varphi(X, Y) = -\langle XJ, Y\rangle$$

for any $X, Y \in \mathbf{R}^{2n-2}$. Here J is the standard symplectic matrix introduced in Section 1.1.

Let $\Lambda = \Lambda(M)$ be the space of piecewise C^∞ curves $c : I = [0, 1] \to M$ such that $c(0) = c(1)$, which is equipped with the compact-open topology induced by the distance $\rho(c, c') = \max_{t\in I} d(c(t), c'(t))$. Here the distance $d(p, q)$ for any two points p and q on M is defined by the infimum of the length of any curve on M connecting p and q induced by the Riemannian metric. The critical points of the energy functional $\mathcal{E} : \Lambda \to \mathbf{R}$ defined by (0.2) are the closed geodesics and the point curves.

For any closed geodesic $c \in \Lambda$, let $V_\Lambda(c)$ be the space of piecewise C^∞ vector fields X along c such that $\langle X, \dot{c}(t)\rangle = 0$ for all t and such that $X(0) = X(1)$. The index form $H : V_\Lambda(c) \times V_\Lambda(c) \to \mathbf{R}$ of c is given by

$$H(X, Y) = \int_0^1 (\langle\nabla X, \nabla Y\rangle - \langle R(X, \dot{c})\dot{c}, Y\rangle)\, dt, \tag{1}$$

where $\langle\cdot, \cdot\rangle$ is the inner product induced by the Riemannian metric of the manifold M. Note that we have abused the symbol $\langle\cdot, \cdot\rangle$ in two different senses. We hope that it will not cause any confusion. The **Morse index** $\mathrm{ind}(c)$ of c is defined to be the index of H, i.e., the total multiplicity of negative eigenvalues of H. The **nullity** $\mathrm{nul}(c)$ of c is defined by

$$\mathrm{nul}(c) = \dim_{\mathbf{C}} \ker_{\mathbf{C}}(H|V_{\Lambda(c)}). \tag{2}$$

Similarly the index $\mathrm{ind}_\Omega(c)$ of a geodesic segment $c : [0, 1] \to M$ is defined to be the index of the form H on the space $V_\Omega(c)$ of piecewise C^∞ vector fields X along c such that $\langle X, \dot{c}(t)\rangle = 0$ for all t and that $X(0) = X(1) = 0$. The nullity of c is defined by

$$\mathrm{nul}_\Omega(c) = \dim_{\mathbf{C}} \ker_{\mathbf{C}}(H | V_\Omega(c)). \tag{3}$$

It is well known that $\mathrm{nul}(c) = \dim_{\mathbf{C}} \ker_{\mathbf{C}}(P - id)$ and $\mathrm{ind}(c) \geq \mathrm{ind}_\Omega(c)$. For $t > 0$, let $\mu(t)$ be the number of linearly independent Jacobi fields Y along c with $Y(0) = Y(t) = 0$. Then the Morse index theorem (cf. [Mor1]) for geodesic segments states that

$$\mathrm{ind}_\Omega(c) = \sum_{0<t<1} \mu(t), \quad \mathrm{nul}_\Omega(c) = \mu(1).$$

The concavity $C(c)$ of the closed geodesic c is defined by

$$C(c) = \mathrm{ind}(c) - \mathrm{ind}_\Omega(c). \tag{4}$$

Then $0 \leq C(c) \leq n - 1$.

Let $c : \mathbf{R}/\mathbf{Z} \to M$ be a closed geodesic. Two points $c(s)$ and $c(t)$ are **conjugate** if there is a non-zero Jacobi field U along c such that $U(s) = U(t) = 0$. Choose points $0 = t_0 < t_1 < \cdots < t_k = 1$ such that there are no pairs of conjugate points on $c|_{[t_i, t_{i+1}]}$. Denote by $\mathcal{J}$ the vector space of continuous vector fields X along c such that $X|_{[t_i+k, t_{i+1}+k]}$ is a Jacobi field orthogonal to $\dot{c}$ for all i and all $k \in \mathbf{Z}$. Denote the complexification of $\mathcal{J}$ by $\mathcal{J}_{\mathbf{C}}$. For $q \in \mathbf{N}$ and $\omega \in \mathbf{U}$, set

$$\mathcal{J}_\omega = \{X \in \mathcal{J}_{\mathbf{C}} \mid X(t+1) = \omega X(t) \ \ \forall t \in \mathbf{R}\}.$$

$\mathcal{J}_\omega$ is a finite dimensional subspace of $\mathcal{J}_{\mathbf{C}}$. A symmetric form H_ω is defined on $\mathcal{J}_\omega$ by

$$H_\omega(X, Y) = \int_0^1 (\langle \nabla X, \nabla Y\rangle - \langle R(X, \dot{c})\dot{c}, Y\rangle)\, dt,$$

where we have extended $\langle \cdot, \cdot \rangle$ to a Hermitian form and R to a complex linear tensor of $TM \otimes \mathbf{C}$. We define

$$I(\omega) = \mathrm{ind} H_\omega, \qquad N(\omega) = \dim_{\mathbf{C}} \ker_{\mathbf{C}} H_\omega,$$

and $I_0(\omega) = I(\omega) + N(\omega)$. Note that $I(1) = \mathrm{ind}(c)$, $N(1) = \mathrm{nul}(c)$ and $I_0(1) = \mathrm{ind}_0(c) = \mathrm{ind}(c) + \mathrm{nul}(c)$. Denote the complexification of E by $E_C \cong \mathbf{C}^{n-1}$ and extend the linear Poincaré map P to a complex linear automorphism of $E_C \oplus E_C \cong \mathbf{C}^{n-1} \oplus \mathbf{C}^{n-1}$. Here the matrix P acts from the right on row vectors.

We define the q fold iteration of the closed geodesic c by $c^q(t) = c(qt)$ for all $t \in \mathbf{R}$. The **splitting numbers** of the closed geodesic c at $\omega \in \mathbf{U}$ are defined by

$$\mathcal{S}_P^\pm(\omega) = \lim_{\theta \to \pm 0} I(\omega e^{\sqrt{-1}\theta}) - I(\omega). \tag{5}$$

Theorem 1 (R. Bott [Bot1], W. Ballmann, G. Thobergsson, and W. Ziller [BTZ1]). *For* $\omega \in \mathbf{U}$,

(i) $\operatorname{ind}(c^q) = \sum_{\omega^q=1} I(\omega)$, $\operatorname{nul}(c^q) = \sum_{\omega^q=1} N(\omega)$.

(ii) $N(\omega) = \dim_{\mathbf{C}} \ker_{\mathbf{C}}(P - \omega I)$, *where whenever there is no confusion, I denotes the identity matrix.*

(iii) *I is locally constant near the points at which $N(\omega) = 0$, and the splitting numbers of c at z satisfy* $0 \leq \mathcal{S}_P^{\pm}(\omega) \leq N(\omega)$.

(iv) $I(\omega) = I(\bar{\omega})$, $N(\omega) = N(\bar{\omega})$.

Proof: Note that when ω is perturbed to $\omega e^{\sqrt{-1}\theta}$ with $|\theta| > 0$ sufficiently small, there are at most $N(\omega)$ of 0 eigenvalues that can be changed to negative eigenvalues. Therefore $\mathcal{S}_P^{\pm}(\omega) \leq N(\omega)$ follows. Proofs for other parts are similar to our study in Section 9.2, and therefore are omitted. □

We define a Hermitian form φ_h on $E_C \oplus E_C \cong \mathbf{C}^{n-1} \oplus \mathbf{C}^{n-1}$ by

$$\begin{aligned}\varphi_h((X_1, Y_1), (X_2, Y_2)) &= \sqrt{-1}(\langle X_1, Y_2\rangle - \langle Y_1, X_2\rangle) \\ &= -\sqrt{-1}\langle (X_1, Y_2)J, (X_2, Y_2)\rangle.\end{aligned}$$

On the subspace $W \subset \mathbf{C}^{n-1} \oplus \mathbf{C}^{n-1}$ defined by

$$\begin{aligned}W &= (P - \omega I)^{-1}(0, E_C) \\ &= \{X \in \mathbf{C}^{n-1} \oplus \mathbf{C}^{n-1} \mid X(P - \omega I) \in (0, \mathbf{C}^{n-1})\},\end{aligned} \tag{6}$$

we define

$$\tilde{H}_{P,\omega}(X, Y) = \sqrt{-1}\varphi_h((P\bar{\omega} - I)X, Y) = \langle X(P\bar{\omega} - I)J, Y\rangle.$$

Set

$$\begin{aligned}C_P(\omega) &= \operatorname{ind}\tilde{H}_{P,\omega} + \dim_{\mathbf{C}} \ker_{\mathbf{C}} \tilde{H}_{P,\omega} - \dim_{\mathbf{C}} \ker_{\mathbf{C}}(P - \omega I) \\ &= (\operatorname{ind} + \dim_{\mathbf{C}} \ker_{\mathbf{C}})\tilde{H}_{P,\omega} - \dim_{\mathbf{C}} \ker_{\mathbf{C}}(P - \omega I).\end{aligned} \tag{7}$$

Theorem 2 (W. Ballmann, G. Thobergsson, and W. Ziller [BTZ1]).

(i) $I(\omega) = \operatorname{ind}_{\Omega}(c_0) + C_P(\omega)$.

(ii) *$C_P(\omega)$ depends only on P and ω. The splitting numbers $\mathcal{S}_P^{\pm}(\omega)$ defined by (5) satisfy*

$$\mathcal{S}_P^{\pm}(\omega) = \lim_{\theta \to \pm 0} C_P(\omega e^{\sqrt{-1}\theta}) - C_P(\omega), \tag{8}$$

and depend only on P and ω.

(iii) $0 \leq C_P(\omega) \leq n - 1$.

Proof: (i) Let A be a symmetric form on a finite dimensional real vector space V (or a Hermitian form on a complex vector space). For any subspace $W \subset V$,

$$\mathrm{ind}A = \mathrm{ind}A|_W + \mathrm{ind}A|_{W^\perp} + \dim(W \cap W^\perp) - \dim(W \cap \ker A), \quad (9)$$

$$\dim\ker A = \dim\ker A|_{W^\perp} - \dim(W \cap W^\perp) + \dim(W \cap \ker A). \quad (10)$$

From (9) and (10) we obtain (i). (ii) follows from the definitions of C_P, splitting numbers, and (i). (iii) follows from the second equation of (7). □

The space $V = E_C \oplus E_C \cong \mathbf{C}^{n-1} \oplus \mathbf{C}^{n-1}$ is a symplectic space with the standard symplectic form φ. $L = (0, E_C) \cong (0, \mathbf{C}^{n-1})$ is a Lagrangian subspace of V. Considering another Lagrangian subspace L' of V with

$$L \cap L' = 0, \qquad PL \cap L' = 0, \quad (11)$$

for $z \in \mathbf{U}$, define the following form on V:

$$\begin{aligned} B_\omega(X, Y) &\equiv B_\omega(P, L, L')(X, Y) \\ &= \sqrt{-1}\varphi_h((P\bar{\omega} - I)X, Y) \\ &\quad + \sqrt{-1}\varphi_h((P\bar{\omega} - I)X, \pi(P\bar{\omega} - I)Y), \end{aligned} \quad (12)$$

where π is the projection of V onto L such that $\ker \pi = L'$.

Theorem 3 (W. Ballmann, G. Thobergsson, and W. Ziller [BTZ1]).

(i) $\ker B_\omega = \ker(P - \omega I)$.
(ii) $\mathcal{S}_P^\pm(\omega) = \lim_{\theta\to\pm 0} \mathrm{ind}B_{\omega e^{\sqrt{-1}\theta}} - \mathrm{ind}B_\omega$.
(iii) $\mathrm{ind}B_\omega(P, L, L') - \mathrm{ind}B_\omega(P, L, L'_1)$ *is independent of* ω *if* (L, L') *and* (L, L'_1) *satisfy* (11).

Proof: (i) By definition, we have $\ker(P - \omega I) \subset \ker B_\omega$. To prove

$$\ker B_\omega \subset \ker(P - \omega I), \quad (13)$$

note that

$$\begin{aligned} B_\omega(X, Y) &= \sqrt{-1}\varphi_h(\pi(P\bar{\omega} - I)X, Y) \\ &\quad + \sqrt{-1}\varphi_h(\pi'(P\bar{\omega} - I)X, P\bar{\omega}Y), \end{aligned} \quad (14)$$

where π is the projection of V onto L such that $\ker \pi = L'$, and π' is the projection of V onto L' such that $\ker \pi' = L$. Let $X \in \ker B_\omega$. For every $Y \in L$,

$$B_\omega(X, Y) = \sqrt{-1}\varphi_h(\pi'(P\bar{\omega} - I)X, P\bar{\omega}Y) = 0. \quad (15)$$

Thus $\pi'(P\bar{\omega} - I)X \in PL$. By (11) this implies $\pi'(P\bar{\omega} - I)X = 0$. Therefore we obtain

$$B_\omega(X, Y) = \sqrt{-1}\varphi_h(\pi(P\bar{\omega} - I)X, Y) = 0,$$

for all $Y \in V$. This then implies $\pi(P\bar{\omega} - I)X = 0$. Thus $(P\bar{\omega} - I)X = 0$ and (13) holds.

(ii) By (14), for $X, Y \in L$ we get

$$B_\omega(X, Y) = \sqrt{-1}\varphi_h(\pi'(P\bar{\omega} - I)X, P\bar{\omega}Y) = \sqrt{-1}\varphi_h(\pi' PX, PY).$$

Thus (ii) holds.

(iii) By (9) we get

$$\mathrm{ind} B_\omega = \mathrm{ind} B_\omega|_L + \mathrm{ind} B_\omega|_{L^\perp} + \dim(L \cap L^\perp) - \dim(L \cap \ker B_\omega),$$

where $L^\perp$ is the orthogonal complement of L with respect to B_ω. By (i) we get $\ker B_\omega = \ker(P - \omega I) \cap V$. Thus it is only left to prove $L^\perp = (P - \omega I)^{-1} L \cap V$. By (14), $X \in L^\perp$ if and only if (15) holds for all $Y \in L$. By $PL \cap L' = 0$, (15) is equivalent to $\pi'(P\bar{\omega} - I)X = 0$. And the latter is equivalent to $(P\bar{\omega} - I)X \in L$. Therefore (iii) holds. □

12.2 Splitting numbers

Let $c : \mathbf{R}/\mathbf{Z} \to M$ be a closed geodesic on an n-dimensional Riemannian manifold (M, g). Denote its Poincaré map by P as in Section 1. By Theorem 1.2, the splitting numbers $\mathcal{S}_P(\omega)$ of the closed geodesic c at ω defined by (1.5) depend only on $P \in \mathrm{Sp}(2n-2)$ and $\omega \in \mathbf{U}$. In this section, we prove that the splitting numbers of c coincide with the splitting numbers of its Poincaré map P viewed as a symplectic matrix $P \in \mathrm{Sp}(2n-2)$ given by Definition 9.1.4. We first establish the homotopy invariance (on the homotopy component of P via Definition 1.8.5) and symplectic additivity for splitting numbers of closed geodesics. Then we compute them explicitly when P corresponds to basic normal forms of symplectic matrices. These computation results coincide with the splitting numbers $S_P^\pm(\omega)$ given by Theorem 9.1.7. Therefore the index iteration theory we have established in the previous chapters works also for closed geodesics.

Theorem 1 (C. Liu and Y. Long [LLo4]). *For $P \in \mathrm{Sp}(2n-2)$, $M \in \Omega^0(P)$, and $\omega \in \mathbf{U}$,*

$$\mathcal{S}_P^\pm(\omega) = \mathcal{S}_P^\mp(\bar{\omega}), \tag{1}$$

and

$$\mathcal{S}_P^\pm(\omega) = \mathcal{S}_M^\pm(\omega). \tag{2}$$

Proof: (1) follows from the fact $I(\omega) = I(\bar{\omega})$ and the properties of eigenvalues of symplectic matrices.

To prove (2) we fix an $M \in \Omega^0(P)$. Pick up a smooth path $f : [0,1] \to \Omega^0(P)$ such that $f(0) = P$ and $f(1) = M$. For any $\omega \in \mathbf{U}$ and $t \in [0,1]$,

$$\dim_C \ker_C(P - \omega I) = \dim_C \ker_C(f(t) - \omega I) = \dim_C \ker_C(M - \omega I). \quad (3)$$

Set $L = (0, E_C)$. It is a Lagrangian subspace of $V = E_C \oplus E_C$. For any $t_0 \in [0,1]$, by the continuity of f, there exist $\epsilon(t_0) > 0$ and a subspace L_{t_0} of V such that

$$L \cap L_{t_0} = 0, \quad f(t)L \cap L_{t_0} = 0, \quad \forall t \in (t_0 - \epsilon(t_0), t_0 + \epsilon(t_0)).$$

By the compactness of $[0,1]$, there exist $k-1$ numbers $0 = t_0 < t_1 < \cdots < t_k = 1$ and k Lagrangian subspaces L_i' for $i = 1, 2, \ldots, k$ such that

$$L \cap L_i' = 0, \quad f(t)L \cap L_i' = 0, \quad \forall t \in [t_{i-1}, t_i], \ i = 1, \cdots, k. \quad (4)$$

Thus similarly to (1.11), by (1.12) and (4) we can define the form $B_\omega(f(t), L, L_i')$ for $t \in [t_{i-1}, t_i]$ and $i = 1, \ldots, k$. By (i) of Theorem 1.3,

$$\dim_{\mathbf{C}} \ker_{\mathbf{C}} B_\omega(f(t), L, L_i') = \dim_{\mathbf{C}} \ker_{\mathbf{C}}(f(t) - \omega I) = \dim_{\mathbf{C}} \ker_{\mathbf{C}}(P - \omega I),$$

which does not depend on $t \in [t_{i-1}, t_i]$. Thus $\operatorname{ind} B_\omega(f(t), L, L_i')$ is constant for all $t \in [t_{i-1}, t_i]$, and

$$\operatorname{ind} B_\omega(f(t_{i-1}), L, L_i') = \operatorname{ind} B_\omega(f(t_i), L, L_i').$$

By (iii) of Theorem 1.3, we have that

$$\operatorname{ind} B_\omega(f(t_i), L, L_i') - \operatorname{ind} B_\omega(f(t_i), L, L_{i+1}')$$

is constant in the variable $\omega \in \mathbf{U}$. Thus for each $i = 0, 1, \ldots, k-1$, the function

$$\operatorname{ind} B_\omega(f(t_i), L, L_i') - \operatorname{ind} B_\omega(f(t_{i+1}), L, L_{i+1}')$$

is constant. This implies that

$$\operatorname{ind} B_\omega(P, L, L_0') - \operatorname{ind} B_\omega(M, L, L_k')$$

does not depend on $\omega \in \mathbf{U}$. By (ii) of Theorem 1.3,

$$\begin{aligned} S_P^{\pm}(\omega) &= \lim_{\theta \to \pm 0} \operatorname{ind} B_{\omega e^{\sqrt{-1}\theta}}(P, L, L_0') - \operatorname{ind} B_\omega(P, L, L_0') \\ &= \lim_{\theta \to \pm 0} \operatorname{ind} B_{\omega e^{\sqrt{-1}\theta}}(M, L, L_k') - \operatorname{ind} B_\omega(M, L, L_k') \\ &= S_M^{\pm}(\omega). \end{aligned}$$

This completes the proof. □

For two vectors $X = (x_1, x_2) \in \mathbf{C}^i \oplus \mathbf{C}^i$ and $Y = (y_1, y_2) \in \mathbf{C}^j \oplus \mathbf{C}^j$, we define $X\diamond Y = (x_1, y_1, x_2, y_2) \in \mathbf{C}^{i+j} \oplus \mathbf{C}^{i+j}$.

Theorem 2 (C. Liu and Y. Long [LLo4]). *For $P \in \mathrm{Sp}(2n-2)$ and $P_i \in \mathrm{Sp}(2n_i)$ with $i = 1, 2$, and $n_1 + n_2 = n - 1$ such that $P = P_1\diamond P_2$,*

$$\mathcal{S}_P^{\pm}(\omega) = \mathcal{S}_{P_1}^{\pm}(\omega) + \mathcal{S}_{P_2}^{\pm}(\omega), \qquad \forall \omega \in \mathbf{U}. \tag{5}$$

Proof: Let $E = E_C \oplus E_C \cong \mathbf{C}^{n-1} \oplus \mathbf{C}^{n-1}$ and $E_i = \mathbf{C}^{n_i-1}$ with $i = 1, 2$. For $X \in E_1 \oplus E_1$ and $Y \in E_2 \oplus E_2$, it holds that $(X\diamond Y)(P_1\diamond P_2) = (XP_1)\diamond(YP_2)$ and

$$P_1\diamond P_2 - \omega I = (P_1 - \omega I)\diamond(P_2 - \omega I), \quad \forall \omega \in \mathbf{U}.$$

By the definition (1.6), we have

$$(P - \omega I)^{-1}(0, E_c) = (P_1 - \omega I)^{-1}(0, E_1)\diamond(P_2 - \omega I)^{-1}(0, E_2).$$

Denote by J_i the standard symplectic matrix on $\mathbf{R}^{2n_i}$ for $i = 1$ and 2. Then $J = J_1\diamond J_2$. For any X and $Y \in (P - \omega I)^{-1}(0, E_C)$ with $X = X_1\diamond X_2$ and $Y = Y_1\diamond Y_2$, X_i and $Y_i \in (P_i - \omega I)^{-1}(0, E_i)$ for $i = 1$ and 2. By the definition below (1.6),

$$\begin{aligned}
\tilde{H}_{P,\omega}(X, Y) &= \langle X(P\bar{\omega} - I)J, Y\rangle \\
&= \langle (X_1\diamond X_2)((P_1\bar{\omega} - I)J_1\diamond(P_2\bar{\omega} - I)J_2), Y_1\diamond Y_2\rangle \\
&= \langle (X_1(P_1\bar{\omega} - I)J_1)\diamond(X_2(P_2\bar{\omega} - I)J_2), Y_1\diamond Y_2\rangle \\
&= \langle X_1(P_1\bar{\omega} - I)J_1, Y_1\rangle + \langle X_2(P_2\bar{\omega} - I)J_2, Y_2\rangle \\
&= \tilde{H}_{P_1,\omega}(X_1, Y_1) + \tilde{H}_{P_2,\omega}(X_2, Y_2).
\end{aligned}$$

It is clear that

$$\dim_{\mathbf{C}} \ker_{\mathbf{C}}(P_1\diamond P_2 - \omega I) = \dim_{\mathbf{C}} \ker_{\mathbf{C}}(P_1 - \omega I) + \dim_{\mathbf{C}} \ker_{\mathbf{C}}(P_2 - \omega I).$$

By the definition of $C_P(\omega)$,

$$\begin{aligned}
C_P(\omega) &= (\mathrm{ind} + \dim_{\mathbf{C}} \ker_{\mathbf{C}})\tilde{H}_{P,\omega} - \dim_{\mathbf{C}} \ker_{\mathbf{C}}(P - \omega I) \\
&= (\mathrm{ind} + \dim_{\mathbf{C}} \ker_{\mathbf{C}})\tilde{H}_{P_1,\omega} - \dim_{\mathbf{C}} \ker_{\mathbf{C}}(P_1 - \omega I) \\
&\quad +(\mathrm{ind} + \dim_{\mathbf{C}} \ker_{\mathbf{C}})\tilde{H}_{P_2,\omega} - \dim_{\mathbf{C}} \ker_{\mathbf{C}}(P_2 - \omega I) \\
&= C_{P_1}(\omega) + C_{P_2}(\omega).
\end{aligned}$$

Thus by Theorem 1.2,

$$\begin{aligned}
\mathcal{S}_P^{\pm}(z) &= \lim_{\theta\to\pm 0} C_P(\omega e^{\sqrt{-1}\theta}) - C_P(\omega) \\
&= \lim_{\theta\to\pm 0} C_{P_1}(\omega e^{\sqrt{-1}\theta}) - C_{P_1}(\omega) + \lim_{\theta\to\pm 0} C_{P_2}(\omega e^{\sqrt{-1}\theta}) - C_{P_2}(\omega) \\
&= \mathcal{S}_{P_1}^{\pm}(\omega) + \mathcal{S}_{P_2}^{\pm}(\omega).
\end{aligned}$$

This completes the proof. □

For splitting numbers of geodesics with basic normal forms defined in Section 1.8, we have the following result.

Theorem 3 (C. Liu and Y. Long [LLo4]). (i) *If* $P = N_1(1, b)$ *with* $b = 0$ *or* 1, *then* $\mathcal{S}_P^{\pm}(1) = 1$.

(ii) *If* $P = N_1(1, -1)$, *then* $\mathcal{S}_P^{\pm}(1) = 0$.

(iii) *If* $P = N_1(-1, b)$ *with* $b = 0$ *or* -1, *then* $\mathcal{S}_P^{\pm}(-1) = 1$.

(iv) *If* $P = N_1(-1, 1)$, *then* $\mathcal{S}_P^{\pm}(-1) = 0$.

(v) *If* $P = R(\theta)$ *with* $\theta \in (0, \pi) \cup (\pi, 2\pi)$, *then* $\mathcal{S}_P^{+}(e^{\sqrt{-1}\theta}) = 0$ *and* $\mathcal{S}_P^{-}(e^{\sqrt{-1}\theta}) = 1$.

(vi) *If* $P = N_2(\omega, b)$ *with* $\omega = e^{\sqrt{-1}\theta}$ *and* $\theta \in (0, \pi) \cup (\pi, 2\pi)$ *is non-trivial, i.e.,* $(b_2 - b_3) \sin\theta < 0$, *then* $\mathcal{S}_P^{\pm}(e^{\sqrt{-1}\theta}) = 1$.

(vii) *If* $P = N_2(\omega, b)$ *with* $\omega = e^{\sqrt{-1}\theta}$ *and* $\theta \in (0, \pi) \cup (\pi, 2\pi)$ *is trivial, i.e.,* $(b_2 - b_3) \sin\theta > 0$, *then* $\mathcal{S}_P^{\pm}(e^{\sqrt{-1}\theta}) = 0$.

Proof: (vi) Note that in this case, we must have $n = \dim M = 3$ and $E_C = \mathbf{C}^2$. For $z = e^{\sqrt{-1}\varphi}$ with $\varphi \neq \theta$ and φ being very close to θ, the matrix $(R(\theta) - zI)$ is invertible. If $(x_1, y_1, x_2, y_2)(P - zI) = (0, 0, *, *) \in (0, \mathbf{C}^2)$, then $x_1 = y_1 = 0$. This implies that $(P - zI)^{-1}(0, E_C) = (0, E_C)$. For any $X = (0, 0, y_1, y_2)$ and $Y = (0, 0, u_1, u_2) \in (P - zI)^{-1}(0, E_C)$,

$$\begin{aligned}\tilde{H}_{P,z}(X, Y) &= \langle X(P\bar{z} - I)J, Y\rangle \\ &= \bar{z}\langle (0, 0, *, *)J, (0, 0, u_1, u_2)\rangle \\ &= \bar{z}\langle (*, *, 0, 0), (0, 0, u_1, u_2)\rangle \\ &= 0.\end{aligned}$$

Here and below $*$ denotes all complex numbers that we do not care about. Thus $(\mathrm{ind} + \dim_{\mathbf{C}} \ker_{\mathbf{C}})\tilde{H}_{P,z} = 2$ and $\dim_{\mathbf{C}} \ker_{\mathbf{C}}(P - zI) = 0$. It implies $C_P(z) = 2$.

Note that

$$(P - \omega I)^{-1}(0, E_C) = \{(x_1, \sqrt{-1}x_1, y_1, y_2) \mid x_1,\ y_1,\ y_2 \in \mathbf{C}\}.$$

In fact,

$$(x_1, x_2, y_1, y_2)(P - \omega I) = (0, 0, *, *) \in (0, \mathbf{C}^2),$$

if and only if $(x_1, x_2)(R(\theta) - \omega I) = 0$. The latter implies $x_2 = \sqrt{-1}x_1$.

For $X = (x_1, \sqrt{-1}x_1, y_1, y_2)$ and $Y = (x_2, \sqrt{-1}x_2, u_1, u_2)$, by direct computation,

$$\begin{aligned}\tilde{H}_{P,\omega}(X, Y) &= \langle X(P\bar{\omega} - I)J, Y\rangle \\ &= \bar{\omega}(x_1\bar{x}_2 b_1 + \sqrt{-1}x_1\bar{x}_2 b_3 - \sqrt{-1}x_1\bar{x}_2 b_2 + x_1\bar{x}_2 b_4) \\ &= x_1\bar{x}_2\bar{\omega}(b_1 + b_4 - \sqrt{-1}(b_2 - b_3)).\end{aligned}$$

Denote $P \in \mathrm{Sp}(2n-2)$ by (1.9.6). By Lemma 1.9.2 and notations there, we have

$$(b_2 - b_3)\cos\theta + (b_1 + b_4)\sin\theta = 0.$$

Thus

$$\begin{aligned}\tilde{H}_{P,\omega}(X,Y) &= x_1\bar{x}_2\bar{\omega}\left(-\frac{(b_2-b_3)\cos\theta}{\sin\theta} - \sqrt{-1}(b_2-b_3)\right)\\ &= -x_1\bar{x}_2(b_2-b_3)\left(\frac{\cos\theta}{\sin\theta}(\cos\theta - \sqrt{-1}\sin\theta)\right.\\ &\quad \left.+\sqrt{-1}\cos\theta + \sin\theta\right)\\ &= -\frac{x_1\bar{x}_2}{\sin^2\theta}(b_2-b_3)\sin\theta,\end{aligned}$$

and

$$\tilde{H}_{P,\omega}(X,X) = -\frac{|x_1|^2}{\sin^2\theta}(b_2-b_3)\sin\theta. \tag{6}$$

Since P is non-trivial, i.e., $(b_2-b_3)\sin\theta < 0$, it holds that $\dim_{\mathbf{C}}\ker_{\mathbf{C}}\tilde{H}_{P,\omega} = 2$, $\dim_{\mathbf{C}}\ker_{\mathbf{C}}(P-\omega I) = 1$, and $\mathrm{ind}\tilde{H}_{P,\omega} = 0$. This implies $C_P(\omega) = 1$. By definition, $\mathcal{S}_P^{\pm}(\omega) = 2 - 1 = 1$. This proves (vi).

(vii) By (6) similarly to the last paragraph in the above proof of (vi), since P is trivial, i.e., $(b_2-b_3)\sin\theta > 0$, it holds that $\mathrm{ind}\tilde{H}_{P,\omega} = 1$. Thus $C_P(\omega) = 2$. By Theorem 1.2, $\mathcal{S}_P^{\pm}(\omega) = 2 - 2 = 0$.

Since the proofs for (i)–(v) are similar to those for (vi) and (vii), we are rather sketchy in the rest of the proofs.

(i) If $P = N_1(1,0)$, in this case $n = 2$. It is easy to see that

$$\begin{aligned}(P - zI)^{-1}(0, E_C) &= E_C \oplus E_C, && \text{if } z = 1,\\ (P - zI)^{-1}(0, E_C) &= \{0\} \oplus E_C, && \text{if } z \in \mathbf{U}\setminus\{1\}.\end{aligned}$$

By a direct computation,

$$\begin{aligned}&(\mathrm{ind} + \dim_{\mathbf{C}}\ker_{\mathbf{C}})\tilde{H}_{P,z} = 2, && \text{if } z = 1,\\ &(\mathrm{ind} + \dim_{\mathbf{C}}\ker_{\mathbf{C}})\tilde{H}_{P,z} = 1, && \text{if } z \in \mathbf{U}\setminus\{1\}.\end{aligned}$$

Thus by definition, $C_P(z) = 1$ for $z \neq 1$, and $C_P(1) = 0$. Thus $\mathcal{S}_P^{\pm}(1) = 1$ follows from Theorem 1.2.

When $P = N_1(1,1)$, the proof is similar to the above case, and is thus omitted. The proofs of (ii)–(iv) are similar to that of (i), and are also omitted.

(v) Suppose $P = R(\theta)$ with $0 < \theta < \pi$, and $z = e^{\sqrt{-1}\varphi} = \cos\varphi + \sqrt{-1}\sin\varphi$. Then by a direct computation we have

$$W = (P - zI)^{-1}(0, E_C) = \left\{\left(x, -\frac{\cos\theta - z}{\sin\theta}x\right) \in \mathbf{C} \oplus \mathbf{C} \,|\, x \in \mathbf{C}\right\}.$$

For $X = (x, -\frac{\cos\theta - z}{\sin\theta}x)$ and $Y = (y, -\frac{\cos\theta - z}{\sin\theta}y)$,

$$\tilde{H}_{P,z}(X, Y) = x\bar{y}\frac{2(\cos\theta - \cos\varphi)}{\sin\theta}.$$

From this we get

$$\dim_{\mathbf{C}} \ker_{\mathbf{C}}(P - zI) = \begin{cases} 0, & \text{if} \quad 0 < \varphi < \theta < \pi, \\ 0, & \text{if} \quad 0 < \theta < \varphi < \pi, \\ 1, & \text{if} \quad 0 < \theta = \varphi < \pi, \end{cases}$$

and

$$(\text{ind} + \dim_{\mathbf{C}} \ker_{\mathbf{C}})\tilde{H}_{P,z} = \begin{cases} 1, & \text{if} \quad 0 < \varphi < \theta < \pi, \\ 0, & \text{if} \quad 0 < \theta < \varphi < \pi, \\ 1, & \text{if} \quad 0 < \theta = \varphi < \pi. \end{cases}$$

Thus by definition,

$$C_P(e^{\sqrt{-1}\varphi}) = \begin{cases} 1, & \text{if} \quad 0 < \varphi < \theta < \pi, \\ 0, & \text{if} \quad 0 < \theta < \varphi < \pi, \\ 0, & \text{if} \quad 0 < \theta = \varphi < \pi. \end{cases}$$

Therefore by Theorem 1.2, we obtain $\mathcal{S}_P^+(e^{\sqrt{-1}\theta}) = 0$ and $\mathcal{S}_P^-(e^{\sqrt{-1}\theta}) = 1$ for $P = R(\theta)$ with $0 < \theta < \pi$. The proof for $P = R(\theta)$ with $\pi < \theta < 2\pi$ is similar, and thus is omitted.

The proof is complete. □

Now by the above Theorems 1 to 3 and discussions in Section 9.1, we obtain

Corollary 4 *For any* $P \in \mathrm{Sp}(2n-2)$ *and* $\omega \in \mathbf{U}$,

$$\mathcal{S}_P^{\pm}(\omega) = S_P^{\pm}(\omega), \tag{7}$$

where the left-hand side of (7) *is defined by* (1.5) *and the right-hand side of* (7) *is defined by* (9.1.4).

Remark 5 As we mentioned at the beginning of this section, Corollary 4 implies that our index iteration theory established in the previous chapters also holds for Morse index theory of closed geodesics on any n-dimensional Riemannian manifold (M, g). Here we should point out a major difference of index iteration formulae between the Morse index theory of closed geodesics and the index theory for closed characteristics on convex compact hypersurfaces in $\mathbf{R}^{2n}$ studied in our Chapter 15 below. Note that the linearized Poincaré map P of a closed geodesic c on (M, g) is a $(2n-2)\times(2n-2)$

symplectic matrix. By (1.1), the Morse index theory $(\mathrm{ind}(c), \mathrm{nul}(c))$ is the index of the form H defined on vector fields which are orthogonal to the tangent vector field $\dot{c}(t)$ along c. Therefore comparing our study in Chapter 15, the term $N_1(1,1)$ appearing in Lemma 15.2.4 for the end point $\gamma_x(\tau)$ of the associated symplectic path for closed characteristic (τ, x) has no corresponding part in the Poincaré map P for closed geodesic c. Correspondingly, all the results in our Chapters 8 to 11, without using the condition (10.2.21)

$$\gamma(\tau) = P^{-1}(N_1(1,1) \diamond Q)P,$$

hold also for Morse index theory of closed geodesics, provided we replace n by $n-1$.

We also point out that our index iteration theory also works for closed geodesics on Finsler manifolds.

Part V
Applications

Chapter 13

The Rabinowitz conjecture

For the nonlinear autonomous Hamiltonian system

$$\dot{x} = JH'(x), \qquad x \in \mathbf{R}^{2n}, \tag{1}$$

in his pioneering work [Rab2] of 1978, P. Rabinowitz proved the following famous result via the variational method.

Theorem 1 (P. Rabinowitz [Rab2]). *Suppose H satisfies the conditions:*

(R1) $H \in C^1(\mathbf{R}^{2n}, \mathbf{R})$.

(R2) *There exist constants $\mu > 2$ and $r_0 > 0$ such that*

$$0 < \mu H(x) \leq H'(x) \cdot x, \qquad \forall |x| \geq r_0.$$

(R3) $H(x) = o(|x|^2)$ *at* $x = 0$.

(R4) $H(x) \geq 0$ *for all* $x \in \mathbf{R}^{2n}$.

Then for any $\tau > 0$, the system (1) *possesses a non-constant τ-periodic solution.*

Because a τ/k-periodic function is also a τ-periodic function, in [Rab2] Rabinowitz proposed a conjecture that "We suspect however that under the hypotheses of Theorem 2.1 (i.e., the above Theorem 1) there is a non-constant solution having any prescribed minimal period". Since 1978, there have appeared many contributions to this conjecture. F. Clarke and I. Ekeland in their paper [ClE1] of 1980 made the first breakthrough, where they solved the case for subquadratic Hamiltonian systems. Then a great number of contributions to this minimal period problem followed from many mathematicians (cf. [AmM1], [AmC1], [ClE1,2], [Den1], [Eke3], [GiM1,2,3,4], [Hof1], etc.). Among them, significant progress was made by I. Ekeland and H. Hofer in their celebrated paper [EkH1] of 1985, where they proved Rabinowitz's conjecture for strictly convex Hamiltonian systems. In [DoL1] of 1997 by D. Dong and the author, Corollary 10.3.6 was proved, which discovered the intrinsic relationship between the minimal period and the indices of a solution. This result in fact unified all the results obtained before under various convexity conditions. Note also that in [Lon8], [Lon9], and [Lon12], the first result on this conjecture for second order Hamiltonian systems under precisely Rabinowitz's original structure conditions was obtained.

Among all these results, four kinds of methods are used to determine the minimality of the period of a solution.

1° *A priori estimate method.* This method estimates the difference of critical values between minimal period solutions and non-minimal ones. Because the critical values increase according to iterations in many cases, when the critical value of a particular solution is small enough, it must have minimal period. Usually this method requires the Hamiltonian function to be close to monomials (cf. [AmM1], [ClE1,2], [Den1], [Eke3], [GiM1,2], [Lon10]).

2° *Convexity method.* This method uses the dual action principle of convex Hamiltonian systems, a Bott-type iteration formula and the iteration inequality of Ekeland index theory, and Hofer's topological characterization of mountain-pass points. This method was introduced by Ekeland and Hofer in their celebrated paper [EkH1] of 1985, where a special case of Corollary 4.3 was proved. This method was applied by many other authors to various convex Hamiltonian systems before 1993 (cf. [AmC1], [Eke3], [GiM3–4]).

3° *Iteration inequalities of symmetric Morse indices.* This method depends on the iteration inequalities of symmetric Morse indices for second order Hamiltonian systems. It has no need of any further requirement on the potential function. This method was introduced in [Lon8] in 1993 and [Lon9] in 1994, and has been used to give a natural lower bound for the minimal period of solutions found by variational methods (cf. [Lon8], [Lon9], and [Lon12]).

4° *Iteration theory method of the index function theory.* This method was first introduced by D. Dong and the author in [DoL1] of 1997, which had already appeared in 1993 as a Nankai Institute preprint. It has been further developed by C. Liu and the author in [LLo1] and [LLo3]. In fact, the main reason for our interest or developing the iteration theory of the index for symplectic paths is to find such a new way to study Rabinowitz's conjecture. Based upon the work [DoL1], G. Fei, Q. Qiu, T. Wang and others have applied this method to various problems of Rabinowitz's conjecture (cf. [FeQ1], [FKW1]).

In this chapter, the last two methods above are introduced. We use our iteration theory of the index for symplectic paths established in Chapters 8 to 10 to reveal the intrinsic connections between the minimal period and the index of periodic solutions. Especially, Theorem A in the Introduction of this book is proved in Section 2 below. In Section 3 and 4, we study this conjecture for asymptotically linear and superquadratic Hamiltonian systems. We also include some results on the same conjecture for second order Hamiltonian systems in Section 5 under precisely Rabinowitz's original structure conditions, as well as some results on subharmonics of Hamiltonian systems in Section 6. In Section 1, we briefly review some basic results on the minimax principle which are used in this chapter.

13.1 Minimax principle preparations

The minimax principle is a very powerful tool in finding solutions of nonlinear problems. In this section, we introduce some concepts and minimax theorems which form a part of the basis for our study in Chapters 12 and 13 of this book. Because these results are well known and can be found in many books, we shall only list the theorems, indicate necessary modifications, for our applications, and refer readers for the details of proofs to the celebrated books [Rab7] of P. Rabinowitz for minimax principles, and [Cha6] of K.C. Chang and [Gho2] of N. Ghossoub for information on the Morse indices of critical points found by minimax arguments. In Section 14.3, we shall give another minimax theorem for our study in Chapter 14.

In the rest of this section, all the homologies are singular homology, we fix the coefficient field of all homological groups to be $\mathbf{R}$, and the working space X to be a real Banach space.

As in Section 7 of [Rab7], Section II.4 of [Cha6], and Section 7.1 of [Gho2], we let G be a compact Lie group acting on X via one of its representations which we denote by $g * x \in X$ for $x \in X$ and $g \in G$. A subset A of X is **G-invariant** if $G * A = A$. A function $f : X \to \mathbf{R}$ is **G-invariant** on X if

$$f(g * x) = f(x), \quad \forall x \in X, \; g \in G.$$

Let Y be another separable Banach space with a G-action represented by $g * y \in Y$ for any $y \in Y$ and $g \in G$. A map $\varphi : X \to Y$ is **G-equivariant** on X if

$$\varphi(g * x) = g * \varphi(x), \quad \forall x \in X, \; g \in G.$$

We denote by $C_G(X, Y)$ the set of all continuous and G-equivariant maps from X to Y. In the rest of this book, we shall be interested in the cases of $G = \{\mathrm{id}\}$, $\mathbf{Z}_2$, and S^1. For any $f \in C^1(X, \mathbf{R})$, we denote $f_c = \{x \in X \mid f(x) \le c\}$, $\mathcal{K}_c(f) = \{x \in X \mid f'(x) = 0, \; f(x) = c\}$ for any $c \in \mathbf{R}$, and $\mathcal{K}(f) = \{x \in X \mid f'(x) = 0\}$.

Definition 1 (cf. pp. 83–85 of [Cha6]). Let D be a k-dimensional ball in X, and S be a subset of X satisfying $\partial D \cap S = \emptyset$. We say that ∂D and S **homotopically link,** if $\varphi(D) \cap S \neq \emptyset$ for any $\varphi \in C(D, X)$ satisfying $\varphi|_{\partial D} = \mathrm{id}|_{\partial D}$. We say that ∂D and S **homologically link,** if $|\tau| \cap S \neq \emptyset$ for any singular k-chain τ with $\partial \tau = \partial D$ where $|\tau|$ is the support of τ.

Definition 2 (cf. pp. 166–167 of [Gho2]). A family $\mathcal{F}$ of G-invariant subsets of X is said to be **G-homotopic of dimension n with boundary B** if there exists a compact G-invariant subset D of $\mathbf{R}^n$ containing a closed subset D_0 and a G-equivariant surjective $\sigma \in C(D_0, B)$ such that

$$\mathcal{F} = \{A \subset X \mid A = f(D) \text{ for some } f \in C_G(D, X) \text{ with } f = \sigma \text{ on } D_0\}. \quad (1)$$

A family $\mathcal{F}(\alpha)$ of subsets of X is said to be a **homological family of dimension n with boundary B** if for some non-trivial class α in $H_n(X, B)$, the family $\mathcal{F}(\alpha)$ is defined by

$$\mathcal{F}(\alpha) = \{A \subset X \mid A \text{ is compact}, \; B \subset A, \; \alpha \in \mathrm{Im}(i_*^A)\}, \quad (2)$$

where i_*^A is the homomorphism from $H_n(A, B) \to H_n(X, B)$ induced by the immersion $i : A \to X$.

Definition 3 (cf. [Rab7]). As usual, suppose $f \in C^1(X, \mathbf{R})$. f satisfies the **Palais–Smale condition for** $c \in \mathbf{R}$ ((PS)$_c$ for short), if every (PS)$_c$ sequence $\{x_k\} \subset X$, i.e., it satisfies the condition that

$$f'(x_k) = 0, \quad f(x_k) \to c, \quad \text{as} \quad k \to \infty, \quad (3)$$

possesses a convergent subsequence in X. f satisfies the **(PS) condition**, if it satisfies the $(\mathrm{PS})_c$ condition for any $c \in \mathbf{R}$.

Definition 4 (cf. p. 32 of [Cha6]). For $x \in \mathcal{K}_c(f)$ the q-th critical group $C_q(f, x)$ of f at x is defined by

$$C_q(f, x) = H_q(f_c \cap U_x, (f_c \backslash \{x\}) \cap U_x), \tag{4}$$

where U_x is any small open neighborhood of x in X such that $\mathcal{K}_c(f) \cap U_x = \{x\}$.

Next we give some powerful minimax theorems which will be used later.

Theorem 5 (cf. Theorem 10.4, Corollaries 10.5 and 10.16 of [Gho2]). *As in Definition 2, let $\mathcal{F}(\alpha)$ be a homological family of dimension q with boundary D. Suppose that $f \in C^2(X, \mathbf{R})$ satisfies the (PS) condition. Define*

$$c \equiv c(f, \mathcal{F}(\alpha)) = \inf_{A \in \mathcal{F}(\alpha)} \sup_{x \in A} f(x). \tag{5}$$

Suppose that

$$\sup_{x \in D} f(x) < c \tag{6}$$

and $f''(x)$ is Fredholm for $x \in \mathcal{K}_c(f)$. Then there exists an $x \in \mathcal{K}_c(f)$ such that the Morse index $m^-(x)$ and the nullity $m^0(x)$ of the functional f at x, which are defined to be the dimensions of the largest subspaces of X on which $f''(x_0)$ are negative definite or null respectively, satisfy

$$m^-(x) \leq q \leq m^-(x) + m^0(x). \tag{7}$$

Moreover, if f is S^1-invariant under an S^1-action on X, and the set $S^1 * x$ is not a single point, (7) can be improved to

$$m^-(x) \leq q \leq m^-(x) + m^0(x) - 1. \tag{8}$$

Remark 6 This theorem is a slight modification of Theorem 10.4, Corollaries 10.5 and 10.16 of [Gho2]. The -1 in the right-hand side of (8) comes from the S^1-invariance of f. In this case, this critical orbit makes at least one non-trivial element of the null space not to turn into a negative definite element in the proofs of [Gho2].

Next we describe the saddle point theorem.

Theorem 7 (cf. [Rab7], [Ni1], [Cha2], and [Gho2]). *Let X be a real Hilbert space with orthogonal decomposition $X = Z \oplus Y$, where* $\dim Z < \infty$. *Suppose $f \in C^2(X, \mathbf{R})$ satisfies the (PS) condition, and the following conditions:*

(F1) *There exist ρ and $\alpha > 0$ such that*

$$f(x) \geq \alpha \quad \forall x \in \partial B_\rho(0) \cap Y.$$

(F2) *There exist* $e \in \partial B_1(0) \cap Y$ *and* $R > \rho$ *such that*

$$f(x) < \alpha, \qquad \forall x \in \partial Q,$$

where $Q = (\overline{B_R(0)} \cap Z) \oplus \{re \mid 0 \le r \le R\}$. *Then*

1° f *possesses a critical value c satisfying* $\alpha \le c \le \max\{f(x) \mid x \in Q\}$.

2° *If* $f''(x)$ *is Fredholm for* $x \in \mathcal{K}_c(f)$, *then there exists an element* $x_0 \in \mathcal{K}_c(f)$, *such that the negative Morse index* $m^-(x_0)$ *and nullity* $m^0(x_0)$ *of* f *at* x_0 *satisfy*

$$m^-(x_0) \le \dim Z + 1 \le m^-(x_0) + m^0(x_0). \tag{9}$$

3° *Suppose that there is an* S^1 *action on* X, f *is* S^1*-invariant, and for* x_0 *found in* 2° *the set* $S^1 * x_0$ *is not a single point. Then* (9) *can be further improved to*

$$m^-(x_0) \le \dim Z + 1 \le m^-(x_0) + m^0(x_0) - 1. \tag{10}$$

Sketch of the Proof: 1° is the well-known Theorem 5.3 of [Rab7]. Next we sketch the proof of 2° and 3° based on Theorem 5. Since Q is a strong deformation retract of the whole space X, it holds that $H_q(Q, \partial Q) \cong H_q(Z, \partial Q)$, where ∂Q is the boundary of Q in $Z \oplus (\mathbf{R}e)$. We have

$$H_q(Q, \partial Q) \cong H_{q-1}(S^{q-1}) \cong \mathbf{R}.$$

Denote by $i : Q \to Z$ the inclusion map. Let $\alpha = [Q] \in H_q(Q, \partial Q)$ be a generator. Then $i_*\alpha$ is non-trivial in $H_q(Z, \partial Q)$, and $\mathcal{F}(i_*\alpha)$ defined by (2) is a homological family of dimension $\dim Z + 1$ with boundary ∂Q and $Q \in \mathcal{F}(i_*\alpha_k)$. Define

$$c = \inf_{A \in \mathcal{F}(i_*\alpha)} \sup_{x \in A} f(x). \tag{11}$$

Note that by a topological degree argument as in Proposition 5.9 of [Rab7], under the conditions (F1) and (F2) for $R > \rho$, the sets ∂Q and $\partial B_\rho(0) \cap Y$ homotopically link in X. Thus by (F1) and (F2), we obtain

$$\sup_{x \in \partial Q} f(x) < \alpha \le c \le \max_{x \in Q} f(x). \tag{12}$$

Since f satisfies the (PS) condition, and f' is Fredholm, by Theorem 5 we obtain 2° and 3°.

Remark 8 On the minimax principle, we also refer to [Ni1] of W.-M. Ni. As far as the author knows, the study on the topological information as well as the Morse index of critical points found by minimax arguments started from the works [Hof1] of H. Hofer and [Tia1] of G. Tian independently, as well as [LaS1] of A. Lazer and S. Solimini and [Sol1] of S. Solimini in the 1980s. We also refer readers to [Cha6], [Gho1], [BaL1], [Tan1], and [Wzq2] for further related results.

13.2 Controlling the minimal period via indices

For given $\tau > 0$, let $S_\tau = \mathbf{R}\backslash(\tau\mathbf{Z})$. In order to study the minimal period, say τ/m for some $m \in \mathbf{N}$, of a non-constant τ-periodic solution x of a nonlinear autonomous Hamiltonian system (0.1), we depend on the iteration theory of the index established in previous chapters. Using the index given by Definitions 5.4.3 and 8.0.1, in the following, we denote

$$(i(x, 1/m), \nu(x, 1/m)) = (i_1(x|_{[0,\tau/m]}), \nu_1(x|_{[0,\tau/m]})). \tag{1}$$

Theorem 1 *For $\tau > 0$, let $x_0 \in C^2(S_\tau, \mathbf{R}^{2n})$ be a τ-periodic solution of the system* (0.1) *with minimal period τ/m for some $m \in \mathbf{N}$. Suppose H satisfies the condition*

(HX0) *H is C^2 along the orbit $\mathcal{O}(x) = \{x(t) \mid t \in \mathbf{R}\}$.*

If there exists some $k \in \mathbf{Z}$ such that the indices of x_0 satisfy the following conditions:

(X1) $i(x_0, 1) \leq k + 1$.

(X2) $i(x_0, 1/m) \geq k$.

(X3) $i(x_0, 1/m) + \nu(x_0, 1/m) \geq n + 1$.

Then $m = 1$, i.e., the solution x_0 possesses minimal period τ.

Proof: This is a direct consequence of Corollary 10.3.5 with $l = k + 1$. □

Corollary 2 (D. Dong and Y. Long [DoL1]). *Let $x_0 \in C^2(S_\tau, \mathbf{R}^{2n})$ be a τ-periodic solution of the system* (0.1) *with minimal period τ/m for some $m \in \mathbf{N}$ satisfying* (HX0). *Let the indices of x_0 satisfy the following conditions:*

(X4) $i(x_0, 1) \leq n + 1$.

(X5) $i(x_0, 1/m) \geq n$.

Then $m = 1$, i.e., the solution x_0 possesses minimal period τ.

Proof: Since x_0 is a non-constant τ-periodic solution of the autonomous system (0.1), the function $\dot{x}_0$ is a non-trivial τ-periodic solution of the corresponding linearized system

$$\dot{y} = JH''(x_0(t))y. \tag{2}$$

Thus $\nu(x_0, 1/m) \geq 1$. Now we can apply Theorem 1 with $k = n$ to conclude $m = 1$. □

Proposition 3 (D. Dong and Y. Long [DoL1]). *For $\tau > 0$, let $x_0 \in C^2(S_\tau, \mathbf{R}^{2n})$ be a non-constant τ-periodic solution of the system* (0.1) *which satisfies* (HX0) *and the following conditions:*

(HX1) *$H''(x_0(t)) \geq 0$, i.e., $H''(x_0(t))$ possesses only non-negative eigenvalues for every $t \in \mathbf{R}$.*

(HX2) *$\int_0^\tau H''(x_0(t))dt$ is positive definite.*

Then

$$i(x_0, 1) \geq n. \tag{3}$$

If x_0 *further satisfies* (X4), *it then possesses minimal period* τ.

Proof: For the saddle point reduction described in Section 4.3 with respect to the Hamiltonian function $K = \frac{1}{2}H''(x_0(t))x \cdot x$, the space $L = L^2(S_\tau, \mathbf{R}^{2n})$ possesses an orthogonal decomposition

$$L = L^+ \oplus Z \oplus L^-, \tag{4}$$

where $L^\pm = P^\pm L$ and $Z = PL$ are defined by (4.3.4) via the projections $P^\pm$ and P defined by (4.3.2). Let g be the functional defined by (4.2.4) with H replaced by K, and $a : Z \to \mathbf{R}$ be the functional defined by Theorem 4.3.1. Then by Theorem 4.5.1, we obtain

$$a(z) \leq \frac{1}{2}\langle A(P^- w(z) + z), P^- w(z) + z\rangle - g(P^- w(z) + z), \quad \forall z \in Z, \tag{5}$$

where P^- and $w(z)$ are given by (4.3.2) and (4.3.22) respectively. Note that Z possesses a decomposition according to the positive, null, and negative definiteness of $A = -Jd/dt$:

$$Z = Z^+ \oplus Z^0 \oplus Z^-, \quad Z^0 = \mathbf{R}^{2n}. \tag{6}$$

By (5), we obtain that for any $z = z_- + z_0 \in (Z^- \oplus Z^0)\backslash\{0\}$,

$$\begin{aligned} a(z) &\leq -\frac{1}{2}\|P^- w(z) + z_-\|^2 - g(P^- w(z) + z_- + z_0) \\ &\leq -\frac{1}{2}(\|P^- w(z)\|^2 + \|z_-\|^2) - g(P^- w(z) + z_- + z_0), \end{aligned} \tag{7}$$

where the orthogonality between $P^- w(z)$ and z_- has been used.

Thus if $\|P^- w(z)\|^2 + \|z_-\|^2 > 0$, by (HX1) we have $a(z) < 0$. On the other hand if $\|P^- w(z)\|^2 + \|z_-\|^2 = 0$, (7) becomes

$$a(z) \leq -\frac{1}{2}\left(\int_0^\tau H''(x_0(t))dt\right) z_0 \cdot z_0 < 0.$$

Here in the last step we have used (HX2). Thus the negative Morse index m^- of the functional a at $z = 0$ satisfies

$$m^- \geq \dim Z^- + \dim Z^0 = (d - n) + 2n = d + n, \tag{8}$$

where $2d = \dim Z$. Combining this with Theorem 6.1.1, we obtain (3).

Denote the minimal period of x_0 by τ/m for some $m \in \mathbf{N}$. Then (3) with τ replaced by the period τ/m implies (X5). Now by (X4) we can apply Corollary 2 to conclude $m = 1$, and complete the proof. □

Remark 4 Corollary 2 and Proposition 3 were proved by D. Dong and the author in [DoL1]. Note that Corollary 2 actually contains all the known results on the Rabinowitz conjecture under various convexity conditions. Theorem 1 is a slight generalization of Corollary 2.

13.3 Asymptotically linear Hamiltonian systems

In this section we apply results obtained earlier to any autonomous asymptotically linear Hamiltonian system (0.1) defined on $\mathbf{R}^{2n}$. The main result in this section is the following theorem.

Theorem 1 (D. Dong and Y. Long [DoL1]). *Let $\tau > 0$. Suppose the Hamiltonian function H satisfies the conditions* (R3), (R4), *and*

(HC1) $H \in C^2(\mathbf{R}^{2n}, \mathbf{R})$.

(HB0) *There exists a positive definite matrix $B \in \mathcal{L}_s(\mathbf{R}^{2n})$ such that*

$$H'(x) = Bx + o(|x|) \quad as \quad |x| \to \infty. \tag{1}$$

(HB1) $\nu(B|_{[0,\tau]}, 1) = 0$.

(HB2) $i(B|_{[0,\tau]}, 1) > n$.

Then there exists a τ-periodic non-constant solution x_0 of (0.1) *such that*

$$n + 2 - \nu(x_0, 1) \le i(x_0, 1) \le n + 1. \tag{2}$$

Moreover, suppose the conditions (HX1) *and* (HX2) *hold for this x_0. Then the minimal period of x_0 is τ.*

Proof: We carry out the proof in several steps.

Step 1 Truncated Hamiltonian functions.

In order to use the saddle point reduction method, we need to truncate the function H suitably to get the boundedness of $\|H''\|_C$.

We note that the condition (HB0) implies the fact

$$H(x) = \frac{1}{2}Bx \cdot x + o(|x|^2) \quad \text{as} \quad |x| \to \infty, \tag{3}$$

and the existence of constants $\Lambda_0 \ge \lambda_0 > 0$ such that

$$\Lambda_0 I \ge B \ge \lambda_0 I. \tag{4}$$

Thus there exists a constant $k_0 \ge 5$ such that

$$2\Lambda_0|x|^2 \ge H(x) \ge \frac{\lambda_0}{4}|x|^2, \quad \forall |x| \ge k_0. \tag{5}$$

Claim 1 *For any integer $k \geq k_0$, there exist a constant $b(k) \geq 1$ and a function $\chi_k \in C^2([0, +\infty), [0, 1])$ such that*

$$\begin{cases} \chi_k(r) = 1, & \textit{for } 0 \leq r \leq k, \\ \chi_k(r) = 0, & \textit{for } k + b(k) \leq r, \\ 0 \leq -\chi_k'(r) \leq \frac{2}{r}, & \textit{for } k < r < k + b(k). \end{cases} \tag{6}$$

In fact, we first define $\chi_k = 1$ for $r \in [0, k]$. Then by induction on integers $m \geq 0$, we can assume that χ_k has been extended smoothly to $[0, k + m]$ such that

$$\begin{cases} 0 \leq \chi_k(k+m) \leq \max\{0, 1 - \sum_{i=1}^{m} \frac{1}{k+i}\} \equiv M(k, m), \\ 0 \leq -\chi_k'(k+m) < \frac{2}{k+m+1}, \\ 0 \leq -\chi_k'(r) < \frac{2}{k+m} \leq \frac{2}{r}, \quad \text{for } k+m-1 < r \leq k+m. \end{cases} \tag{7}$$

Since $\sum_{i \geq 1} \frac{1}{k+i} = +\infty$, there exists an integer $b(k) \geq 3$ so that

$$M(k, b(k) - 1) = 0 \quad \text{and} \quad M(k, b(k) - 2) > 0.$$

Then we slightly modify χ_k near $k + b(k) - 1$ and extend it to $(k + b(k) - 1, k + b(k)]$ so that (7) holds for $m = k + b(k)$ and

$$\chi_k(k + b(k)) = \chi_k'(k + b(k)) = \chi_k''(k + b(k)) = 0.$$

Finally we extend χ_k to $(k + b(k), +\infty)$ by the second equation of (6), and obtain the claim.

For $k \geq k_0$ we define

$$H_k(x) = \chi_k(|x|)H(x) + \frac{1}{2}(1 - \chi_k(|x|))Bx \cdot x, \quad \forall x \in \mathbf{R}^{2n}. \tag{8}$$

Then by (4), (5), and (6) for any $k \geq k_0$,

$$2\Lambda_0|x|^2 \geq H_k(x) \geq \frac{\lambda_0}{4}|x|^2, \quad \forall |x| \geq k_0, \tag{9}$$

$$2\Lambda_0|x|^2 + M_0 \geq H_k(x) \geq \frac{\lambda_0}{4}|x|^2 - M_0, \quad \forall x \in \mathbf{R}^{2n}, \tag{10}$$

and

$$\begin{aligned} &|H_k'(x) - Bx| \\ &\leq |\chi_k(|x|)(H'(x) - Bx)| + \left|\chi_k'(|x|)\left(H(x) - \frac{1}{2}Bx \cdot x\right)\frac{x}{|x|}\right| \\ &\leq |H'(x) - Bx| + \frac{2}{|x|}\left|H(x) - \frac{1}{2}Bx \cdot x\right|, \quad \forall |x| > 0, \end{aligned} \tag{11}$$

where $M_0 = \max_{|x|\le k_0} H(x)$. Thus by (HB0) and (1) for H and (11) the function H_k satisfies (HB0) for the same $B \in \mathcal{L}_s(\mathbf{R}^{2n})$, and $H_k'(x)$ converges to Bx as $|x| \to \infty$ uniformly for all $k \ge k_0$. Note that for each $k \ge k_0$, the function H_k satisfies (HC1), (HB0), (R3), (R4), (HB1), (HB2), and

$$\|H_k''(x)\|_{C(\mathbf{R}^{2n})} < +\infty. \tag{12}$$

Step 2 An a priori estimate.

Fix a $k \ge k_0$. By (HC1) and (12), we apply the saddle point reduction method described in Chapter 4 to the system

$$\dot{x} = JH_k'(x), \tag{13}$$

to get the spaces L, $W = \mathrm{dom} A$ with $A = -Jd/dt$, Z_k, the map $u_k : Z_k \to W$, and the functionals f_k, $a_k = f_k \circ u_k$.

Next we derive an a priori estimate on the upper bound of $\|x\|_C = \max_{0\le t\le\tau} |x(t)|$ for any critical point x of f_k.

Claim 2 *There exists a constant $k_1 \ge k_0$ such that for any $k \ge k_0$ if x_k is a τ-periodic solution of the system* (13), *then*

$$\|x_k\|_{C^1(S_\tau)} \le k_1. \tag{14}$$

In fact, by (HB1) there exists a constant $\alpha > 0$ such that

$$\|(A - B)y\|_{L^2} \ge 2\alpha\|y\|_{L^2}, \qquad \forall y \in W. \tag{15}$$

Fix $k \ge k_0$. Let

$$g_k(y) = \int_0^\tau H_k(y(t))dt, \qquad \forall y \in L.$$

Then g_k is C^1 on L. By (HB0) and (11), we obtain

$$\|g_k'(y) - By\|_L / \|y\|_L \to 0, \tag{16}$$

for $\|y\|_L \to \infty$ and $y \in L$ uniformly for all $k \ge k_0$. Thus there exists a constant $M_1 \ge k_0$ such that when $k \ge k_0$,

$$\|g_k'(y) - By\|_L \le \alpha\|y\|_L, \qquad \forall \|y\|_L \ge M_1, y \in L. \tag{17}$$

Combining (15) and (17) we obtain

$$\begin{aligned} \|f_k'(y)\|_L &\ge \|(A - B)y\|_L - \|g_k'(y) - By\|_L \\ &\ge \alpha\|y\|_L, \qquad \forall \|y\|_L \ge M_1, \quad y \in W. \end{aligned} \tag{18}$$

Thus for any $k \geq k_0$ and any critical point $x_k \in W$ of f_k it must hold that

$$\|x_k\|_L \leq M_2, \tag{19}$$

for some $M_2 > 0$ independent of k. Then by (10), (13), and (19), we obtain that for every $t \in \mathbf{R}$,

$$\begin{aligned} \tau H_k(x_k(t)) &= \int_0^\tau H_k(x_k(s))ds \\ &\leq \tau M_0 + 2\Lambda_0 \int_0^\tau |x_k(s)|^2 ds \\ &\leq \tau M_0 + 2\Lambda_0 M_2^2. \end{aligned} \tag{20}$$

Combining (20) with (10) again yields

$$\frac{\tau\lambda_0}{4}|x_k(t)|^2 \leq 2\tau M_0 + 2\Lambda_0 M_2^2, \quad \forall t \in \mathbf{R}. \tag{21}$$

This yields a uniform estimate of C-norm for all τ-periodic solutions of (13) with $k \geq k_0$:

$$\|x_k(\cdot)\|_{C(S_\tau)} \leq 1 + \sqrt{\frac{8}{\tau\lambda_0}(\tau M_0 + \Lambda_0 M_2^2)}. \tag{22}$$

Now together with (11), we obtain an integer k_1 such that (14) holds for any $k \geq k_0$ and any critical point x of f_k. The claim is proved.

Claim 3 *a_k satisfies the (PS) condition on Z_k.*

In fact, by Claim 2, the critical points of f_k are uniformly bounded in the balls in L and W centered at the origin with radius $2k_1^2$. By the definition of the injection $u_k : Z_k \to W$ given in Section 4.3 by (4.3.17)–(4.3.18), the map u_k and its inverse map u_k^{-1} are bounded maps, where u_k^{-1} is defined on the image of u_k. Thus the critical points of a_k are bounded. Therefore the claim follows from the finiteness of the dimension of Z_k.

Step 3 The existence of a solution x_k of (13).

In this step, for each $k \geq k_1$ we prove the existence of a special τ-periodic solution of the Hamiltonian system (13). For notational simplicity, in this step, we shall omit all the subindex k.

By (R3) and Proposition 4.5.2, we obtain

$$\begin{aligned} a(z) &\geq \frac{1}{2}\langle A(P^+w(z) + z), P^+w(z) + z\rangle_L - o(\|z\|^2), \\ &\quad \text{as} \quad z \to 0 \text{ in } Z. \end{aligned} \tag{23}$$

Thus there exists $\rho > 0$ small enough so that

$$\begin{aligned} a(z) &\geq \frac{1}{2}\|P^+w(z)+z\|^2 - o(\|z\|^2) \\ &= \frac{1}{2}(\|P^+w(z)\|^2 + \|z\|^2) - o(\|z\|^2) \\ &\geq \frac{1}{4}\|z\|^2 = \frac{1}{4}\rho^2 > 0, \quad \forall z \in \partial B_\rho(0, Z^+), \end{aligned} \tag{24}$$

where Z^+ is defined by (2.6), and $\partial B_\rho(0, Z^+)$ is the boundary of the open ball $B_\rho(0, Z^+)$ in Z^+ centered at the origin with radius ρ. Thus the condition (F1) of Theorem 1.7 holds.

On the other hand, by (HB2) and Theorem 6.1.1, there exists at least an element $y \in Z^+$ with $\|y\| = 1$ and $\lambda_1 > 0$ such that

$$(A - B)y = -\lambda_1 y \qquad \text{in } L, \tag{25}$$

where $\langle Bx, x\rangle_{L^2} = \int_0^\tau Bx(t)\cdot x(t)dt$ for $x \in W$. For large $R > 0$ to be determined later, we define

$$\begin{aligned} Q = \{z = ry + z_0 + z_- \in Z | z_0 + z_- \in Z^0 \oplus Z^-, \\ \|z_0 + z_-\| \leq R,\ 0 \leq r \leq R\}. \end{aligned} \tag{26}$$

By (2.7) and (R4), we obtain

$$a(z) \leq 0, \qquad \forall z = z_0 + z_- \in Z^0 \oplus Z^-. \tag{27}$$

For $z = ry + z_0 + z_- \in \partial Q$ by (3) and (2.5) we have

$$\begin{aligned} a(z) &\leq \frac{1}{2}\langle A(P^-w(z) + ry + z_-), P^-w(z) + ry + z_-\rangle_L \\ &\quad -\frac{1}{2}\int_0^\tau B(P^-w(z)+z)\cdot(P^-w(z)+z)dt \\ &\quad +o(\|P^-w(z)+z\|_{L^2}^2) \\ &\leq -\frac{1}{2}\|P^-w(z)+z_-\|^2 + \frac{r^2}{2}\langle (A-B)y\cdot y\rangle_L \\ &\quad -r\int_0^\tau By\cdot(P^-w(z)+z_0+z_-)dt \\ &\quad -\frac{1}{2}\int_0^\tau B(P^-w(z)+z_0+z_-)\cdot(P^-w(z)+z_0+z_-)dt \\ &\quad +o(\|P^-w(z)+z\|_L^2). \end{aligned} \tag{28}$$

Note that $x = P^- w(z) + z_0 + z_-$ is orthogonal to y and Ay in L^2. We obtain

$$\begin{aligned}\int_0^\tau By \cdot x dt &= -\int_0^\tau (A-B)y \cdot x dt + \int_0^\tau Ay \cdot x dt \\ &= \lambda_1 \int_0^\tau y \cdot x dt = 0.\end{aligned} \tag{29}$$

By (28) and (29) we have

$$\begin{aligned} a(z) &\le -\frac{1}{2}\|P^- w(z) + z_-\|^2 - \lambda_1 \frac{r^2}{2}\|y\|_L^2 \\ &\quad -\frac{\lambda_0}{2}\|P^- w(z) + z_0 + z_-\|^2 + o(\|P^- w(z) + z\|_{L^2}^2) \\ &\le -\frac{1}{2}\min\{1, \lambda_0, \lambda_1\}\|P^- w(z) + z\|_L^2 + o(\|P^- w(z) + z\|_L^2). \end{aligned} \tag{30}$$

Thus by taking $R > 0$ to be large enough we obtain

$$a(z) \le 0, \tag{31}$$

for all $z = ry + z_0 + z_- \in \partial Q$ with $\|z_0 + z_-\| = R$ or $r = R$. Therefore the condition (F2) of Theorem 1.7 holds. Note that here by Theorem 6.1.1 with $\omega = 1$, for $2d = \dim Z$ we have

$$\dim Q = \dim(Z^- \oplus Z^0) + 1 = (d-n) + 2n + 1 = d + n + 1. \tag{32}$$

Note that there is a natural S^1 action on W and Z defined by

$$\theta * x(t) = x(t+\theta), \qquad \forall x \in W, \theta \in [0,\tau]/\{0,\tau\} = S^1. \tag{33}$$

Since a satisfies the (PS) condition, is S^1-invariant on Z, and $a''(z)$ is Fredholm if $a'(z) = 0$, by Theorem 1.7 we obtain a critical point $z \in Z$ of a whose Morse index $m^-(z)$ and nullity $m^0(z)$ satisfy

$$m^-(z) \le d + n + 1 \le m^-(z) + m^0(z) - 1. \tag{34}$$

By Theorem 6.1.1 with $\omega = 1$, this implies that the index and nullity defined in Chapter 5 of $x = u(z)$ as a solution of (13) satisfies (2). Note that by (24), we have $f(x) > 0$. By (8), (R4), and (HB0) of H_k, this x is not a constant function.

Step 4 Solutions of the system (0.1).

Now for each $k \ge k_0$, the above Step 3 yields a non-constant τ-periodic solution x_k of the system (13) according to H_k, which satisfies (2).

By our choice of k_1 in (14) and the definition of χ_k in (6), any τ-periodic solution x_k with $k \ge k_1$ is actually a τ-periodic solution of (0.1) with the given Hamiltonian function H.

Step 5 The above proof yields a non-constant τ-periodic solution x_0 of (0.1), which satisfies (2). Thus (X4) holds for x_0. Then by (HX1) and (HX2) for this x_0 and Proposition 2.3, the solution x_0 possesses minimal period τ.

The proof is complete. □

The following corollary gives more accessible sufficient conditions for the existence of solutions with prescribed minimal period.

Corollary 2 (D. Dong and Y. Long [DoL1]). *For $\tau > 0$, suppose the Hamiltonian function H satisfies* (HC1), (HB0), (HB1), (HB2), (R3), *and the conditions:*

(HC2) $H''(x) \geq 0$, *i.e., $H''(x)$ possesses no negative eigenvalues for all $x \in \mathbf{R}^{2n}$.*

(HC3) *The set $D = \{x \in \mathbf{R}^{2n} | H'(x) \neq 0, \ 0 \in \sigma(H''(x))\}$ is hereditarily disconnected, i.e., every connected component of D contains only one point.*

Then the system (0.1) possesses a τ-periodic solution x with τ as its minimal period.

Proof: Note that (R4) follows from (HC1), (R3), and (HC2). Thus (0.1) possesses at least one non-constant τ-periodic solution x_0 satisfying (2) and thus (X4) by the first part of Theorem 1.

Now (HX1) for this x_0 follows from (HC2). Then $H'(x_0(t)) \neq 0$ for all $t \in \mathbf{R}$. Thus $\{x_0(t) \,|\, t \in \mathbf{R}, \ 0 \in \sigma(H''(x_0(t)))\}$ is a subset of D, hence is hereditarily disconnected. This implies (HX2) for x_0. Now we can apply Proposition 2.3 to complete the proof.

□

Remark 3 To the author's knowledge, the study on periodic solutions of asymptotically linear Hamiltonian systems (ALHS for short) in the global sense started from the celebrated paper [AmZ1] of H. Amann and E. Zehnder of 1980, who proved the existence of non-trivial 1-periodic solutions for such systems with constant symmetric coefficients non-degenerate at zero and infinity under a twist condition. Since then many contributions on this problem have appeared, for example, the papers [Cha1] of K.C. Chang in 1981 and [LiL1] of S. Li and J.-Q. Liu in 1989. In [CoZ2] of 1984, C. Conley and E. Zehnder extended these results to ALHS which are non-degenerate at zero and infinity with continuous periodic symmetric coefficients in $\mathbf{R}^{2n}$ for $n \geq 2$. In [LZe1] of 1990, E. Zehnder and the author extended this result to the case of $n = 1$. This problem was further studied in [Lon5] of 1990 and the non-degeneracy condition at zero was removed. In 1997, in [CLL1] of K.C. Chang, J.-Q. Liu, and M.-J. Liu and in [FeQ2] of G. Fei and Q. Qiu independently, the non-degeneracy condition at infinity was removed. Note that these mentioned results are closely related to the famous Poincaré–Birkhoff theorem on the existence of fixed points of area preserving homeomorphisms on an annulus under twist conditions on the boundary. The following is one of the theorems on ALHS proved in [CLL1] and [FeQ2]:

Suppose the function H satisfies the following conditions:

(AL1) $H \in C^2(\mathbf{R} \times \mathbf{R}^{2n}, \mathbf{R})$ is 1-periodic in time t and satisfies

$$|H''(t,x)| \le a(1+|x|^{\mu}), \qquad \forall (t,x) \in \mathbf{R} \times \mathbf{R}^{2n}, \tag{35}$$

where $a > 0$ and $\mu > 1$ are constants.

(AL2) There exists 1-periodic matrix function $B_0 \in C(\mathbf{R}, \mathcal{L}_s(\mathbf{R}^{2n}))$ such that

$$H'(t,x) = B_0(t)x + o(|x|) \quad \text{as} \quad |x| \to 0 \text{ uniformly in } t. \tag{36}$$

(AL3) There exists 1-periodic matrix function $B_\infty \in C(\mathbf{R}, \mathcal{L}_s(\mathbf{R}^{2n}))$ such that

$$H'(t,x) = B_\infty(t)x + o(|x|) \quad \text{as} \quad |x| \to \infty \text{ uniformly in } t. \tag{37}$$

(AL4) For $h(t,x) = H(t,x) - B_\infty(t)x$ with $(t,x) \in \mathbf{R} \times \mathbf{R}^{2n}$, either

$$h(t,x) \to 0, \quad |h'(t,x)| \to 0, \quad \text{as} \quad |x| \to \infty \text{ uniformly in } t, \tag{38}$$

or

$$h(t,x) \to \pm\infty, \quad |h'(t,x)| = O(1), \quad \text{as} \quad |x| \to \infty \text{ uniformly in } t, \tag{39}$$

Theorem 4 (K.C. Chang, J.-Q. Liu, and M.-J. Liu [CLL1], G. Fei and Q. Qiu [FeQ2]). *Suppose H satisfies the conditions* (AL1)–(AL4). *Then the problem*

$$\dot{x} = JH'(t,x), \tag{40}$$

$$x(1) = x(0), \tag{41}$$

possesses a non-trivial 1*-periodic solution, provided that indices of the corresponding linear Hamiltonian systems satisfy*

$$[i_1(B_0), i_1(B_0) + \nu_1(B_0)] \cap [i_1(B_\infty), i_1(B_\infty) + \nu_1(B_\infty)] = \emptyset. \tag{42}$$

Here the index theory is used to measure the twist of the Poincaré map of the nonlinear problem (40)–(41) between the origin and infinity. We refer readers to [CLL1] and [FeQ2] for the proof of Theorem 4 and further results on ALHS.

13.4 Superquadratic Hamiltonian systems

In this section we study autonomous Hamiltonian systems

$$-J\dot{x} = \hat{H}'(x), \qquad x \in \mathbf{R}^{2n}, \tag{1}$$

where $\hat{H}(x) = \frac{1}{2}Bx \cdot x + H(x)$. Here we suppose B satisfies the following condition:

(B) $B \in \mathcal{L}_s(\mathbf{R}^{2n})$ *is semi-positive definite, i.e., B possesses no negative eigenvalues. The operator norm of B is denoted by $\omega = \|B\|$.*

The following is our main result in this section.

Theorem 1 (G. Fei and Q. Qiu [FeQ1], C. Liu and Y. Long [LLo3]). *Suppose the matrix B satisfies the condition* (B), *and the function H satisfies conditions* (HC1), (R2), (R3), and (R4). *Then for every $0 < \tau < \frac{2\pi}{\omega}$, the system* (1) *possesses a non-constant τ-periodic solution x_0 satisfying*

$$n + 2 - \nu(x_0, 1) \le i(x_0, 1) \le n + 1. \tag{2}$$

Moreover, suppose this solution x_0 satisfies the conditions (HX1) *and* (HX2). *Then τ is the minimal period of x_0.*

Proof: The proof is carried out in several steps.

Step 1 The truncation of the Hamiltonian function.

For $k \ge 1$ to be determined later, following Rabinowitz's pioneering work [Rab2] we choose $\chi_k \in C^\infty(\mathbf{R}, \mathbf{R})$ such that $\chi_k(t) = 1$ if $t \le k$, $\chi_k(t) = 0$ if $t \ge k + 1$, and $\chi_k'(t) < 0$ if $t \in (k, k+1)$. Set

$$\hat{H}_k(x) = \frac{1}{2} Bx \cdot x + H_k(x), \quad \forall x \in \mathbf{R}^{2n}, \tag{3}$$

and

$$H_k(x) = \chi_k(|x|)H(x) + (1 - \chi_k(|x|))R_k|x|^4,$$

where the constant R_K is chosen to satisfy

$$R_k \ge \max\left\{ \frac{H(x)}{|x|^4} \mid k \le |x| \le k+1 \right\}.$$

In the following we denote by a_i or K_i the positive constants which are independent or dependent on k respectively without further explanations.

Then this H_k still satisfies conditions (HC1), (R2) with μ replaced by the number $\eta = \min\{\mu, 4\}$, (R3), and (R4). Consequently there exist positive constants a_1 and a_2 such that for any integer $k \ge r_0$,

$$\begin{cases} 0 < \eta a_1 |x|^\eta \le \eta H_k(x) \le H_k'(x) \cdot x, & \forall |x| \ge r_0, \\ H_k(x) \ge a_1 |x|^\eta - a_2, & \forall x \in \mathbf{R}^{2n}. \end{cases} \tag{4}$$

Note that for $k \ge r_0$ there exist positive constants K_1, K_2, and K_3 such that

$$R_k \eta |x|^4 - K_1 \le \eta H_k(x) \le H_k'(x) \cdot x + K_2 \le 4R_k|x|^4 + K_3, \tag{5}$$

for all $x \in \mathbf{R}^{2n}$.

As in Section 4.1, let $W = W^{1/2,2}(S_\tau, \mathbf{R}^{2n})$ with the norm (4.1.5). For any $k \geq 1$, we define a functional f_k on W by

$$f_k(x) = \frac{1}{2}\int_0^\tau (\dot{x} \cdot Jx - \hat{H}_k(x))dt, \quad \forall x \in W. \tag{6}$$

According to the operator A induced by $-J\frac{d}{dt}$ via (4.1.6) being positive, null, and negative definite respectively, W possesses a decomposition

$$W = W^+ \oplus W^0 \oplus W^- \tag{7}$$

with $W^0 = \mathbf{R}^{2n}$.

For $m \in \mathbf{N}$, let W_m be the space defined by (4.1.14) for A defined by $-Jd/dt$ via (4.1.6) and $B_\infty = 0$, and $\{P_m : W \to W_m\}$ be the corresponding Galerkin approximation scheme with respect to A. Let $W_m^* = W_m \cap W^*$ for $* = +, 0$, and $-$. Let $f_{k,m} = f_k|_{W_m}$.

Step 2 The Palais–Smale conditions.

For $k \geq r_0$ we need the following claims.

Claim 1 *Each f_k always satisfies the (PS) condition on W_m for any $m \in \mathbf{N}$.*

Claim 2 *f_k satisfies the $(PS)_c^*$ condition on W for $c \in \mathbf{R}$, i.e., any sequence $\{x_m\}$ satisfying $x_m \in W_m$, $f'_{k,m}(x_m) \to 0$ and $f_k(x_m) \to c$ as $m \to +\infty$ always possesses a convergent subsequence in W.*

In fact, we fix such a sequence $\{x_m\}$ and prove Claim 2 in the following. The similar proof of Claim 1 is left to the readers. In the following we denote by b_i or K_i the positive constants which are independent or dependent on k respectively without further explanations. The proof is given in two substeps.

(A) *Boundedness of $\{x_m\}$ in W.*

By the definition of $\{x_m\}$ and (4),

$$\begin{aligned} a_3 + \|x_m\| &\geq f_{k,m}(x_m) - \frac{1}{2}\langle f'_{k,m}(x_m), x_m\rangle \\ &= \int_0^\tau \left(\frac{1}{2}H'_k(x_m) \cdot x_m - H_k(x_m)\right) dt \\ &\geq \left(\frac{1}{2} - \frac{1}{\eta}\right) \int_0^\tau H'_k(x_m) \cdot x_m \, dt - a_4 \\ &\geq \left(\frac{\eta}{2} - 1\right) \int_0^\tau H_k(x_m)\, dt - a_4 \\ &\geq a_5\|x_m\|_{L^\eta}^\eta - a_6. \end{aligned} \tag{8}$$

Writing $x_m = x_m^+ + x_m^0 + x_m^-$ according to the decomposition (7), then (8) implies

$$|x_m^0| \leq a_7(1 + \|x_m\|^{1/\eta}). \tag{9}$$

For $m > 0$ great enough, letting $\zeta = x_m^+$ in

$$\left|\langle Ax_m, \zeta\rangle - \int_0^\tau (Bx_m + H_k'(x_m)) \cdot \zeta dt\right|$$
$$= |\langle f_{k,m}'(x_m), \zeta\rangle| \le \|\zeta\|, \quad \forall \zeta \in W_m,$$

and using the Hölder inequality and the inequality (4.1.5), by (5) we obtain

$$\begin{aligned}
2\|x_m^+\|^2 &= \langle Ax_m, x_m^+\rangle \\
&\le \int_0^\tau |Bx_m + H_k'(x_m)||x_m^+|dt + \|x_m^+\| \\
&\le K_4 \int_0^\tau (1 + |x_m|^3)|x_m^+|dt + \|x_m^+\| \\
&\le K_5(1 + \|x_m\|_{L^4}^3)\|x_m^+\|.
\end{aligned}$$

Thus together with (5), (8), and Sobolev embedding theorem (4.1.5),

$$\begin{aligned}
\|x_m^+\| &\le K_6 \left(1 + \left(\int_0^\tau |x_k|^4 dt\right)^{3/4}\right) \\
&\le K_7(1 + \|x_m\|^{3/4}).
\end{aligned} \tag{10}$$

Similarly, we obtain an estimate on $\|x_m^-\|$ as in (10). Together with (9), we then have

$$\|x_m\| \le K_8(1 + \|x_m\|^{3/4} + \|x_m\|^{1/\eta}). \tag{11}$$

Therefore $\{x_m\}$ is bounded in W.

(B) *The convergence of some subsequence.*

The condition $f_{k,m}'(x_m) \to 0$ as $m \to +\infty$ implies

$$-J\dot{x}_m - P^m \hat{H}_k'(x_m) = \epsilon_m, \tag{12}$$

with

$$\|\epsilon_m\|_{W_m'} \to 0, \quad \text{as } m \to +\infty.$$

By Step (A),

$$\|x_m\| = \|x_m\|_{W_m} \le K_9, \quad \forall m \in \mathbf{N}.$$

Thus by passing to a subsequence, we may assume that

$$\begin{aligned}
x_m &\to x \quad &&\text{in} \quad W \text{ weakly,} \\
x_m &\to x \quad &&\text{in} \quad L^p \text{ strongly for any } 1 \le p < +\infty, \\
x_m^0 &\to x^0 \quad &&\text{in} \quad \mathbf{R}^{2n}.
\end{aligned}$$

By (5), this implies $\hat{H}_k'(x_m) \to \hat{H}_k'(x)$ strongly in L^2. Thus $P_m \hat{H}_k'(x_m) \to \hat{H}_k'(x)$ strongly in L^2, and thus in W'. Therefore (12) implies that

$$\dot{x}_m = \xi_m + \epsilon_m, \tag{13}$$

holds in W', where $\xi_m \to \xi = J\hat{H}_k'(x)$ in L^2. This then implies

$$\dot{x} = \xi, \tag{14}$$

in W'. Since $\xi \in L^2$, then $x \in W^{1,2}$ and thus $x \in C^2$, i.e., (14) holds in the classical sense. Because W_m is a subspace of the Hilbert space W with the reduced norm,

$$\|x_m - P_m x\|_{W_m}^2 = \|\dot{x}_m - P_m \dot{x}\|_{W_m'}^2 + |x_m^0 - x^0|^2.$$

Then

$$\|x_m - P_m x\|_{W_m}^2 \le (\|\xi_m - P_m \xi\|_{W_m'} + \|\epsilon_m\|_{W_m'})^2 + |x_m^0 - x^0|^2.$$

From

$$\|\xi_m - P_m \xi\|_{W_m'} \le M \|\xi_m - P_m \xi\|_{L^2} \to 0,$$

for some $M > 0$ independent of m, we then obtain

$$\|x_m - P_m x\|_W = \|x_m - P_m x\|_{W_m} \to 0.$$

This proves $x_m \to x$ in W strongly. The proof of Claim 2 is completed.

Step 3 The existence of a critical point x_k of f_k on W with $k \in \mathbf{N}$.

We prove first that $f_{k,m}$ possesses a critical point $x_{k,m}$ in W_m satisfying the index estimate for $m > n$.

In fact, note that

$$\|x\|_{L^2}^2 \le \frac{\tau}{2\pi} \|x\|^2, \quad \forall x \in W^+. \tag{15}$$

Firstly by the conditions (B), (R3) for H, and the definition of H_k, for any $\epsilon \in (0, 1]$ there exists $\delta > 0$ such that

$$H_k(x) \le \epsilon |x|^2, \qquad \forall |x| \le \delta.$$

Thus there exists $K_{10}(\epsilon) > 0$ depending on k and ϵ such that

$$H_k(x) \le \epsilon |x|^2 + K_{10}(\epsilon)|x|^4, \quad \forall x \in \mathbf{R}^{2n}. \tag{16}$$

Therefore for any $x \in W^+$, by (16) we obtain

$$f_k(x) \ge \frac{1}{2}\|x\|^2 - \frac{\omega}{2}\|x\|_{L^2}^2 - \epsilon \|x\|_{L^2}^2 - K_{10}(\epsilon)\|x\|_{L^4}^4.$$

When $\omega > 0$, by (15) we then obtain

$$f_k(x) \geq \left(\frac{\omega}{4\pi}\left(\frac{2\pi}{\omega} - \tau\right) - \epsilon C_2^2 - K_{10}(\epsilon) C_2^4 \|x\|^2\right) \|x\|^2,$$

where C_2 is the Sobolev embedding constant given by (4.1.5). Thus by the definition of τ, we can fix $\epsilon > 0$ sufficiently small independent of k first and then choose $\rho(k) \in (0, 1]$ and $c_k > 0$ small and depending on k such that

$$f_{k,m}(x) \geq c_k > 0, \quad \forall x \in \partial B_{\rho(k)}(0) \cap W_m^+. \tag{17}$$

When $\omega = 0$, $B = 0$. Thus from (R3) for H and (16), we also get (17). Therefore the condition (F1) of Theorem 1.7 holds.

Secondly, for $m > n$, set

$$Q_m = \{re \,|\, 0 \leq r \leq r_1\} \oplus (B_{r_2}(0) \cap (W_m^- \oplus W_m^0)\},$$

where $e \in \partial B_1(0) \cap W_m^+$, r_1 and r_2 are positive constants to be determined later. By (4) for $r \in [0, r_1]$ and $x = re + x^0 + x^- \in \partial Q_m$ with $x^- \in W^-$ and $x^0 \in W^0$, we obtain

$$\begin{aligned} f_{k,m}(x) &= \frac{r^2}{2}\langle Ae, e\rangle + \frac{1}{2}\langle Ax^-, x^-\rangle - \int_0^\tau \hat{H}_k(x)\,dt \\ &\leq \frac{r^2}{2} - \frac{\|x^-\|^2}{2} - \frac{1}{2}\langle Bx, x\rangle - a_1 \int_0^\tau |x|^\eta dt + a_2. \end{aligned} \tag{18}$$

By Jensen's inequality (Theorem 204 of [HLP1]) and the mutual L^2 orthogonality among x^0, x^-, and e, we have

$$\begin{aligned} \int_0^\tau |x|^\eta dt &\geq \left(\int_0^\tau |x|^2 dt\right)^{\eta/2} \\ &= (\|x^-\|_{L^2}^2 + \tau |x^0|^2 + r^2 \|e\|_{L^2}^2)^{\eta/2}, \\ &\geq a_7(\|x^-\|_{L^2}^2 + |x^0|^2 + r^2)^{\eta/2}, \end{aligned} \tag{19}$$

for some $a_7 > 0$ independent of m and k. Combining (18) and (19) together yields

$$f_{k,m}(x) \leq \frac{r^2}{2} - \frac{1}{2}\|x^-\|^2 - a_8 \|x\|_{L^2}^2 - a_8(\|x^-\|^2 + |x^0|^2 + r^2)^{\eta/2}, \tag{20}$$

for some $a_8 > 0$ independent of m and k. Thus we can choose large enough r_1 and $r_2 > 1$ independent of m and k such that

$$f_{k,m}(x) \leq 0, \quad \forall x \in \partial Q_m. \tag{21}$$

Therefore the condition (F2) of Theorem 1.7 holds.

Note that by our above discussion, there exists a constant $a_9 > 0$ independent of m and k such that

$$f_{k,m}(x) \le a_9, \quad \forall x \in Q_m. \tag{22}$$

Now since $f''_{k,m}(x)$ is Fredholm at any critical point x, and $f_{k,m}$ is S^1-invariant under the S^1-action (3.33) on W_m, by Claim 1, (17), (21), and (22), we can apply Theorem 1.7 and obtain a critical point $x_{k,m} \in W_m$ of $f_{k,m}$ for each m, which satisfies

$$a_9 \ge f_{k,m}(x_{k,m}) \ge c_k > 0, \quad \forall k, m \in \mathbf{N}, \tag{23}$$

and

$$\begin{aligned} m^-(x_{k,m}) &\le \dim(W_m^- \oplus W_m^0) + 1 \\ &= m + 2n + 1 \\ &\le m^-(x_{k,m}) + m^0(x_{k,m}) - 1. \end{aligned} \tag{24}$$

This completes our study of the critical point $x_{k,m}$ of $f_{k,m}$.

Now by Claim 2 and (23), $\{x_{k,m}\}$ possesses a subsequence which converges to some x_k in W, which is a critical point of f_k. Thus it is a non-constant solution of (1) with H replaced by H_k, and then we obtain $x_k \in C^1(S_\tau, \mathbf{R}^{2n})$. By (23) this x_k further satisfies

$$0 < c_k \le f_k(x_k) \le a_9, \quad \forall k \in \mathbf{N}. \tag{25}$$

Step 4 The energy estimate and existence.

By (4), similarly to (8) we obtain

$$\begin{aligned} a_9 &\ge f_k(x_k) - \frac{1}{2}\langle f_k'(x_k), x_k\rangle \\ &\ge \left(\frac{\eta}{2} - 1\right)\int_0^\tau H_k(x_k(t))dt - a_{10}, \end{aligned} \tag{26}$$

for some $a_{10} > 0$ independent of k.

On the other hand, by (4) we can choose $r_3 \ge r_0$ sufficiently large such that

$$\frac{1}{2}Bx \cdot x \le a_1|x|^\eta \le H_k(x), \quad \forall |x| \ge r_3.$$

Let $D_k = \{t \in [0,\tau] \mid |x_k(t)| \ge r_3\}$ and $D_k' = [0,\tau]\backslash D_k$. Then from (3) for $k \ge r_3$ we obtain

$$\begin{aligned} \tau \hat{H}_k(x_k(t)) &= \left(\int_{D_k'} + \int_{D_k}\right)\hat{H}_k(x_k(t))dt \\ &\le a_{11} + 2\int_{D_k} H_k(x_k(t))dt \\ &\le a_{11} + 2\int_0^\tau H_k(x_k(t))dt, \end{aligned} \tag{27}$$

for some $a_{11} > 0$ independent of k. Together with (26), we then obtain a constant $a_{12} > 0$ independent of k such that

$$\hat{H}_k(x_k(t)) = \frac{1}{2}Bx_k(t) \cdot x_k(t) + H_k(x_k(t)) \le a_{12},$$

for any $t \in [0, \tau]$. By the semi-positive definiteness of B given by (B) and (4), we then obtain a constant $a_{13} > 0$ independent of k such that

$$|x_k(t)| \le a_{13}, \quad \forall t \in [0, \tau], \ k \ge r_3. \tag{28}$$

Then choose $k \ge a_{13} + 1 + r_3 \equiv b_1$, and let $x_0 = x_k$. By (28) and the definition (3) of $\hat{H}_k$, this x_0 is a non-constant classical solution of (1).

Step 5 The index and nullity estimates.

For any $k \ge b_1$, let $B_k(t) = \hat{H}_k''(x_k(t))$ and B_k be the linear operator defined by (4.1.10) corresponding to $B_k(t)$. Then we have

$$\dim\ker(A - B_k) = \nu(x_k, 1). \tag{29}$$

Let $\delta = \delta(k) = \frac{1}{4}\|(A - B_k)^{\#}\|^{-1}$ with $(A - B_k)^{\#} = [(A - B_k)|_{\mathrm{Im}(A-B_k)}]^{-1}$. As we have pointed out in Section 4.1, the value 0 is isolated in the set $\{0\} \cup \sigma(A - B_k)$. Therefore for any $k \ge b_1$, there exists $m_1 = m_1(k)$ depending only on k such that for any $m \ge m_1$ and any $u \in E^-_{\delta(k)}(P_m(A - B_k)P_m)$,

$$\|P_m(A - B_k)P_m u\| \ \ge\ 3\delta\|P_m u\|, \tag{30}$$

$$\langle P_m(A - B_k)P_m u,\, P_m u\rangle \ \le\ -3\delta\|P_m u\|^2. \tag{31}$$

Since $f_k''(x_k) = A - B_k$, we have

$$\|f_k''(z) - (A - B_k)\| \to 0 \ \text{ as } \|z - x_k\| \to 0. \tag{32}$$

Thus there exists a $r_4 > 0$ sufficiently small such that

$$\|f_k''(z) - (A - B_k)\| < \delta, \ \ \forall z \in \overline{B_{r_4}(x_k)}.$$

Hence there exists $m_2 = m_2(k) \ge m_1$ such that for all $m \ge m_2$,

$$\|f_{k,m}''(z) - P_m(A - B_k)P_m\| < 2\delta, \quad \forall z \in \overline{B_{r_4}(x_k)} \cap W_m. \tag{33}$$

Then using notations in (7.1.4), and $u \in E^-_\delta(P_m(A - B_k)P_m)\backslash\{0\}$ for $m \ge m_2$, by (31) and (33) we obtain

$$\begin{aligned}
&\langle f_{k,m}''(x_{k,m})u, u\rangle \\
&\quad \le \langle P_m(A - B_k)P_m u,\, P_m u\rangle + \|f_{k,m}''(x_{k,m}) \\
&\qquad -P_m(A - B_k)P_m\|\,\|P_m u\|^2 \\
&\quad \le -\delta\|P_m u\|^2 \\
&\quad < 0, \qquad \forall u \in E^-_\delta(P_m(A - B_k)P_m).
\end{aligned} \tag{34}$$

Thus

$$m^-(x_{k,m}) \geq m_\delta^-(P_m(A - B_k)P_m), \quad \forall m \geq m_2. \tag{35}$$

By Corollary 7.1.10, (24) and (35), for all $m \geq m_2$ we obtain

$$\begin{aligned} 2n + m + 1 &= \dim Q_m \\ &\geq m^-(x_{k,m}) \\ &\geq m_\delta^-(P_m(A - B_k)P_m) \\ &= m + n + i(x_k, 1). \end{aligned} \tag{36}$$

This implies $i(x_k, 1) \leq n + 1$. Thus the right-hand side inequality of (2) holds for x_0 chosen in Step 4.

Similarly to the proof of (35), we obtain $m_3 = m_3(k) \geq m_2$ such that

$$m^+(x_{k,m}) \geq m_\delta^+(P_m(A - B_k)P_m), \quad \forall m \geq m_3. \tag{37}$$

Thus by Corollary 7.1.10, (24) and (37), for all $m \geq m_3$ we obtain

$$\begin{aligned} 2n + m + 1 &= \dim Q_m \\ &\leq m^-(x_{k,m}) + m^0(x_{k,m}) - 1 \\ &= \dim W_m - m^+(x_{k,m}) - 1 \\ &\leq 2(m + n) - m_\delta^+(P_m(A - B_k)P_m) - 1 \\ &= 2(m + n) - (m + n - i(x_k, 1) - \nu(x_k, 1)) - 1. \end{aligned} \tag{38}$$

This implies $n + 1 \leq i(x_k, 1) + \nu(x_k, 1) - 1$. Thus the left-hand side inequality of (2) holds for x_0 chosen in Step 4.

Step 6 The minimal period of x_0.

Denote the minimal period of this solution x by τ/m for some $m \in \mathbf{N}$. By the conditions (HX1), (HX2) and (B), we can use Proposition 2.3 to conclude $m = 1$ and complete the proof. □

Remark 2 If $B = 0$, we can choose $\omega = 0$, and then $2\pi/\omega = +\infty$. Thus in this case for every $\tau > 0$, Theorem 1 holds.

The following corollary gives more accessible sufficient conditions for the existence of solutions with prescribed minimal period.

Corollary 3 *Suppose the conditions* (B), (HC1), (R2), (R3), (R4), (HC2), and (HC3) *hold. Then for every* $0 < \tau < \frac{2\pi}{\omega}$, *the system* (1) *possesses a non-constant* τ*-periodic solution* x_0 *with* τ *as its minimal period.*

Proof: The conclusion follows from Theorem 1 and the second paragraph in the proof of Corollary 3.2. □

At the end of this section we apply the iteration theory to Hamiltonian systems defined on $\mathbf{R}^2$ with no convexity type conditions.

Theorem 4 *Suppose the Hamiltonian function H satisfies the condition* (HC1) *on* $\mathbf{R}^2$. *For $\tau > 0$, let $x \in C^2(S_\tau, \mathbf{R}^2)$ be a non-constant τ-periodic solution of the system* (0.1). *If the index of x satisfies the condition*

(X6) $2 - \nu(x, 1) \leq i(x, 1) \leq 2$,

then the solution x possesses minimal period τ.

Proof: Let $\alpha = \tau/m$ be the minimal period of x for some $m \in \mathbf{N}$. Then (X6) implies (10.3.24). Since x is a non-constant α-periodic solution of the autonomous system (0.1), we obtain $\nu_\alpha(x) \geq 1$. Thus Theorem 10.3.7 yields $m = 1$. □

Corollary 5 *When $n = 1$, the τ-periodic non-constant solution x_0 found by the first part of the above Theorem* 1 *or* 3.1 *possesses minimal period τ.*

Remark 6 Based on the iteration results and the idea of relating the indices to the minimal periods of solutions of [DoL1], G. Fei and Q. Qiu using the Galerkin approximation method obtained Theorem 1 with $B \equiv 0$ in [FeQ1]. The full Theorem 1 was proved by C. Liu and the author in [LLo3] which generalized the result in [FeQ1]. Note that the proof of the $(\mathrm{PS})^*_c$ condition in Step 2 of the proof of Theorem 1 uses an idea of A. Bahri and H. Berestycki in [BaB2].

13.5 Second order systems

In this section, we consider the existence of non-constant periodic solutions with prescribed minimal period for the autonomous second order Hamiltonian systems,

$$\ddot{x} + V'(x) = 0, \qquad \forall x \in \mathbf{R}^n, \tag{1}$$

where $V : \mathbf{R}^n \to \mathbf{R}$ is a function, and V' denotes its gradient. Rabinowitz's conjecture also includes the system (1) when V satisfies conditions similar to (R1)–(R4). Our main results in this section are the following theorems which give partial answers to Rabinowitz's conjecture under precisely his structure conditions (V2)–(V4) as well as (V5).

Theorem 1 *Suppose V satisfies the following conditions.*

(V1) $V \in C^2(\mathbf{R}^n, \mathbf{R})$.

(V2) *There exist constants* $\mu > 2$ *and* $r_0 > 0$ *such that*

$$0 < \mu V(x) \leq V'(x) \cdot x, \qquad \forall |x| \geq r_0.$$

(V3) $V(x) \geq 0, \quad \forall x \in \mathbf{R}^n$.

(V4) $V(x) = o(|x|^2)$, *at* $x = 0$.

Then for every $\tau > 0$, *the system* (1) *possesses a non-constant* τ*-periodic even solution with minimal period* τ/m *for some integer* m *satisfying* $1 \leq m \leq n+1$.

Theorem 2 *Suppose* V *satisfies conditions* (V1)–(V3) *and the condition*

(V5) *There exist constants* $\omega > 0$ *and* $r_1 > 0$ *such that*

$$V(x) \leq \frac{\omega}{2}|x|^2, \qquad \forall |x| \leq r_1.$$

Then for every positive $0 < \tau < \pi/\sqrt{\omega}$, *the conclusion of Theorem* 1 *holds.*

Our proof depends on an inequality, Theorem 4 below, of iterated Morse indices for the functional corresponding to (1) defined on even function spaces.

Let $E_\tau = W^{1,2}(S_\tau, \mathbf{R}^n)$, with the norm

$$\|x\|_\tau = \left(\int_0^\tau |\dot{x}|^2 dt + \tau|x(0)|^2\right)^{1/2}, \quad \forall x \in E_\tau.$$

The functional corresponding to the system (1) is defined by

$$f_\tau(x) = \int_0^\tau \left(\frac{1}{2}|\dot{x}|^2 - V(x)\right) dt, \quad \forall x \in E_\tau. \tag{2}$$

Define

$$SE_\tau = \{x \in E_\tau \mid x(-t) = x(t), \ \forall t \in \mathbf{R}\}.$$

Because

$$f_\tau(\overline{x}) = f_\tau(x), \quad \forall x \in E_\tau, \tag{3}$$

where $\overline{x}(t) = x(-t)$, it is easy to prove (cf. [Lon9]) that the critical points of f_τ restricted to SE_τ are in one to one correspondence to τ-periodic even solutions of the system (1).

For any critical point x of $f_\tau|_{SE_\tau}$ the following bilinear form is defined by $f_\tau''(x)$ on SE_τ:

$$\phi_\tau(y, z) = \int_0^\tau \{\dot{y} \cdot \dot{z} - A(t)y \cdot z\} dt, \quad \forall y, z \in SE_\tau, \tag{4}$$

where $A(t) = V''(x(t))$ satisfies the condition

(AS) $A \in C(S_\tau, \mathcal{L}_s(\mathbf{R}^n))$ *and* $A(t)$ *is even about* $t = 0$.

Note that ϕ_τ corresponds to the linear second order Hamiltonian system

$$\ddot{y} + A(t)y = 0, \quad \forall y \in \mathbf{R}^n. \tag{5}$$

We write $\phi_\tau(x) = \phi_\tau(x, x)$. Note that under (AS), SE_τ possesses a ϕ_τ-orthogonal decomposition

$$SE_\tau = SE_\tau^+ \oplus SE_\tau^0 \oplus SE_\tau^-, \tag{6}$$

according to ϕ_τ being positive, null, and negative definite respectively.

If x is a non-constant critical point of f_τ in SE_τ, then $\dot{x}$ is a non-trivial solution of (5) with $A(t) = V''(x(t))$, and $\dot{x} \in E_\tau$ is odd about $t = 0$. Thus we define the space of such odd solutions of (5) by

$$OE_\tau^0 = \{y \in E_\tau \mid y \text{ is an odd solution of (5)}\}. \tag{7}$$

Definition 3 We define

$$si_\tau = \dim SE_\tau^- \quad \text{and} \quad ov_\tau = \dim OE_\tau^0,$$

and call them the symmetric index and odd nullity of x.

It is an exercise for the readers that the symmetric index is always finite. The following estimate describes the iteration property of Morse indices.

Theorem 4 *Suppose the condition* (AS) *holds. Then*

$$si_{m\tau} \geq m \min\{ov_\tau, 1\}, \quad \forall m \in \mathbf{N}. \tag{8}$$

Proof: We carry out the proof in two steps.

Step 1 Suppose $ov_\tau \geq 1$ and fix an integer $m \in \mathbf{N}$. Fix $u \in OE_T^0 \setminus \{0\}$. We define new functions u_+ and $u_- \in C(S_{m\tau}, \mathbf{R}^n)$ by

$$u_+(t) = \begin{cases} u(t), & \text{if } 0 \leq t \leq \tau/2, \\ 0, & \text{if } \tau/2 \leq t \leq m\tau, \end{cases}$$

and

$$u_-(t) = \begin{cases} u(t + \tau/2), & \text{if } 0 \leq t \leq \tau/2, \\ 0, & \text{if } \tau/2 \leq t \leq m\tau. \end{cases}$$

Define a $\tau/2$-translation operator $\eta : C(S_{m\tau}, \mathbf{R}^n) \to C(S_{m\tau}, \mathbf{R}^n)$ by

$$\eta v(t) = v(t - \tau/2), \quad \forall v \in C(S_{m\tau}, \mathbf{R}^n). \tag{9}$$

Now we define a sequence of functions $\{u_i\}$ for $1 \leq i \leq m$ based on u by:

$$u_i = \eta^{i-1} u_+ - \eta^{2m-i} u_-, \quad \text{if } i \in 2\mathbf{N} - 1, \tag{10}$$

$$u_i = -\eta^{i-1} u_- + \eta^{2m-i} u_+, \quad \text{if } i \in 2\mathbf{N}, \tag{11}$$

and define $N = \text{span}\{u_i \mid 1 \leq i \leq m\}$.

Since all the u_is have mutually disjoint supports, they are linearly independent. Note that each $u_i \in E_{m\tau}$ is even, and

$$\dim N = m, \quad N \subset SE_{m\tau}, \tag{12}$$

$$\phi_{m\tau}(u_i, u_j) = 0, \quad \text{if} \quad i \neq j. \tag{13}$$

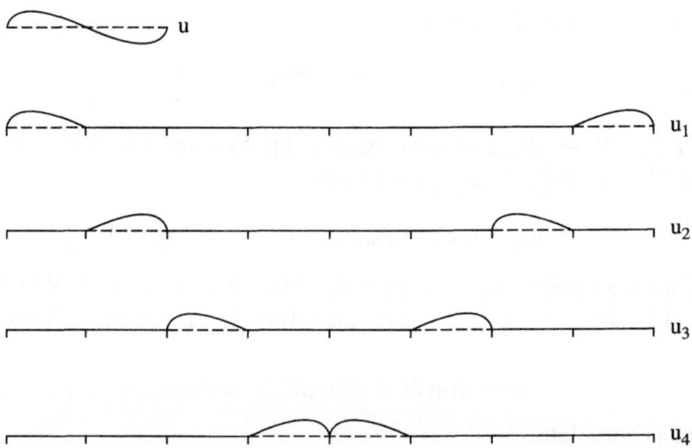

Figure 1: *Functions in N defined by* (10) *and* (11) *for the case of* $m = 4$

Since $u \in OE_\tau^0 \backslash \{0\}$, we obtain $\dot{u}(0) = \dot{u}(\tau) \neq 0$ and $\dot{u}(\tau/2) \neq 0$. Therefore each u_i is not C^1, and therefore does not belong to $\ker \phi_\tau$ in E_τ. Since u_m is not C^1 at $t = m\tau/2$, any function in $N \backslash \{0\}$ is not C^1. Therefore we obtain

$$N \cap \ker \phi_{m\tau} = \{0\}. \tag{14}$$

If i is odd, from the definition of u_i, we obtain

$$\begin{aligned}
\phi_{m\tau}(u_i, u_i) &= \int_{(i-1)\tau/2}^{i\tau/2} (|\dot{u}_i|^2 - A(t)u_i \cdot u_i)dt \\
&\quad + \int_{(2m-i)\tau/2}^{(2m-i+1)\tau/2} (|\dot{u}_i|^2 - A(t)u_i \cdot u_i)dt \\
&= \int_0^{\tau/2} (|\dot{u}_+|^2 - A(t)u_+ \cdot u_+)dt \\
&\quad + \int_0^{\tau/2} (|\dot{u}_-|^2 - A(t + \tau/2)u_- \cdot u_-)dt \\
&= \int_0^{\tau} (|\dot{u}|^2 - A(t)u \cdot u)dt \\
&= 0.
\end{aligned} \tag{15}$$

If i is even, from the definition of u_i, we obtain similarly

$$\begin{aligned}\phi_{m\tau}(u_i, u_i) &= \int_0^{\tau/2} (|\dot{u}_-|^2 - A(t+\tau/2)u_- \cdot u_-)dt \\ &\quad + \int_0^{\tau/2} (|\dot{u}_+|^2 - A(t)u_+ \cdot u_+)dt \\ &= 0.\end{aligned} \tag{16}$$

Note that (13), (15), and (16) imply

$$\phi_{m\tau}(v_1, v_2) = 0, \quad \forall v_1, v_2 \in N. \tag{17}$$

Step 2 Let $P_{m\tau}^- : N \to SE_{m\tau}^-$ be the orthogonal projection. For $0 \neq v \equiv v_- + v_0 + v_+ \in N \cap (SE_{m\tau}^- \oplus SE_{m\tau}^0 \oplus SE_{m\tau}^+)$, we have

$$0 = \phi_{m\tau}(v, v) = \phi_{m\tau}(v_-, v_-) + \phi(v_+, v_+). \tag{18}$$

If $v_- = 0$, we would have $v_+ = 0$ by (18). Thus $0 \neq v_0 = v \in N \cap SE_{m\tau}^0$. This contradicts (14). Then $P_{m\tau}^- v = v_- \neq 0$. Therefore $P_{m\tau}^-$ is injective. Since it is linear, we obtain

$$m = \dim N \leq \dim SE_{m\tau}^- = si_{m\tau}. \tag{19}$$

The proof is complete. □

Now we can give the proofs of Theorems 1 and 2. Since they are similar to that of Theorem 4.1 in a certain sense, we are very sketchy here.

Proof of Theorem 1: Let $X = SE_\tau$, $Z = \mathbf{R}^{2n}$, and

$$Y = \{x \in SE_\tau \mid x(0) = 0\}. \tag{20}$$

By the conditions (V1)–(V4), similar to the proof of Theorem 4.1, the functional f_τ satisfies the (PS) condition, (F1), and (F2) of Theorem 1.7. $f_\tau''(x)$ is Fredholm for $x \in \mathcal{K}(f_\tau)$. Thus by Theorem 1.7, for any $\tau > 0$ we obtain a critical point x of f_τ on SE_τ satisfying

$$si_\tau(x) \leq n+1, \quad f_\tau(x) > 0. \tag{21}$$

Together with (V3), this x is a non-constant τ-periodic even solution of (1). Denote the minimal period of x by τ/m for some $m \in \mathbf{N}$. Since (1) is autonomous, we have $\dot{x} \in OE_{\tau/m}\backslash\{0\}$. Thus $ov_{\tau/m}(x) \geq 1$. Applying Theorem 4 we obtain $m \leq n+1$. □

Proof of Theorem 2: The proof is similar to that of Theorem 1, except the verification of the condition (F1) in Theorem 1.7, which follows from (V5) and the definition of τ instead of (V4) as follows. For any $x \in Y$ defined by (20), since $x(0) = x(\tau) = 0$, by Poincaré–Wirtinger inequality we have

$$\|x\|_{L^2}^2 \leq \frac{\tau^2}{\pi^2}\|\dot{x}\|^2, \quad \forall x \in Y. \tag{22}$$

So for any $x \in Y$ with $\|x\|_\tau$ sufficiently small, by the Sobolev embedding theorem and (V5) we obtain

$$\begin{aligned} f_\tau(x) &\geq \int_0^\tau \left(\frac{1}{2}|\dot{x}(t)|^2 - \frac{\omega}{2}|x(t)|^2\right) dt \\ &\geq \frac{\omega}{2}\left(\frac{1}{\omega} - \frac{\tau^2}{\pi^2}\right)\|x\|^2 \qquad (23) \\ &\geq \delta, \end{aligned}$$

for some $\delta > 0$ provided $\|x\| = \rho > 0$ is sufficiently small. This proves (F1). □

In [Lon8] of 1993, the system (1) with even potentials was further studied, and the following results were obtained.

Theorem 5 1° *Suppose* V *satisfies the conditions* (V1)–(V4) *and*

(V6) V *is even, i.e.,* $V(-x) = V(x)$ *for any* $x \in \mathbf{R}^n$.

Then for every $\tau > 0$, *the system* (1) *possesses a non-constant* τ*-periodic solution with minimal period* τ *or* $\tau/3$, *which is even with respect to* $t = 0$ *and odd with respect to* $t = \tau/4$.

2° *Suppose* V *satisfies conditions* (V1)–(V3), (V5), *and* (V6). *Then for every positive* $0 < \tau < \pi/\sqrt{\omega}$, *the conclusion of* 1° *holds.*

The proof of this result is similar to those of Theorems 1 and 2. Here we consider the functional f_τ of (5.2) defined on a new symmetric function space

$$SE_\tau = \left\{x \in E_\tau \,|\, x(-t) = x(t),\ x\left(\frac{\tau}{4} + t\right) = -x\left(\frac{\tau}{4} + t\right),\ \forall t \in \mathbf{R}\right\}.$$

Then a new iteration inequality of the Morse index theory of f_τ is proved, which yields the above theorem. We omit it here and refer the readers to [Lon8] for details.

Remark 6 Theorems 1 and 2 are taken from [Lon9] in 1994 and [Lon12] in 1997. The Step 2 of the proof of Theorem 4 uses an idea of Dr. Guangcun Lu which simplified the original proof in [Lon9]. In these two papers and [Lon8] of 1993, one can find further developments on this problem. We suspect that the $\tau/3$ in Theorem 5 can be eliminated.

13.6 Subharmonics

We start from the autonomous Hamiltonian systems

$$\dot{x}(t) = JH'(x(t)), \quad x \in \mathbf{R}^{2n}. \qquad (1)$$

We call two τ-periodic solution x_1 and x_2 of (1) **geometrically distinct** if

$$x_1(\mathbf{R}) \neq x_2(\mathbf{R}). \qquad (2)$$

Theorem 1 *Let H satisfy conditions* (HC1), (R2), (R3), (R4). *For any $s > 0$, we denote by x_s the non-constant s-periodic solution of* (1) *produced by Theorem* 4.1, *which satisfies*

$$n+2-\nu(x_s,1) \le i(x_s,1) \le n+1. \tag{3}$$

Let $\mathcal{O}_s = x_s(\mathbf{R})$ be the orbit of x_s. Then this yields a one parameter family of non-constant periodic solution orbits of (1). *Denote the minimal period of x_s by $h(s)$. Thus $0 < h(s) \le s$ holds. For any two positive real numbers j and k, let*

$$p(j,k) \equiv \max\left\{\frac{j}{h(k)}, \frac{k}{h(j)}\right\}. \tag{4}$$

Then the solutions x_j and x_k of (1) *are geometrically distinct provided*

$$p(j,k) \in \mathbf{R}\backslash\mathbf{N} \quad or \quad p(j,k) > n+\frac{1}{2}. \tag{5}$$

Especially, (5) *holds if*

$$\max\left\{\frac{j}{k}, \frac{k}{j}\right\} > n+\frac{1}{2}. \tag{6}$$

Proof: Without loss of generality, we suppose $p(j,k) = \frac{k}{h(j)}$.

When $p(j,k) \in \mathbf{R}\backslash\mathbf{N}$, clearly x_k can not be any phase shift of any iteration of $x_{h(j)}$, and then is geometrically distinct from x_j.

When $p(j,k) \in \mathbf{N}$ and $p(j,k) > n+1/2$, we assume x_j and x_k are not geometrically distinct. Then x_k must be a phase shift of the $p(j,k)$-th iteration of $x_{h(j)}$. By (10.3.1) of Lemma 10.3.1 and the above (3), we then obtain

$$\begin{aligned} p(j,k) &\le 1+\frac{i(x_{h(j)},p)-i(x_{h(j)},1)}{i(x_{h(j)},1)+\nu(x_{h(j)},1)-n} \\ &= 1+\frac{i(x_k,1)-i(x_{h(j)},1)}{i(x_{h(j)},1)+\nu(x_{h(j)},1)-n} \\ &\le 1+\frac{(n+1)-(n+2-2n)}{2} \\ &= n+\frac{1}{2}. \end{aligned} \tag{7}$$

This is a contradiction and proves the theorem. □

Remark 2 After his pioneering work [Rab2], P. Rabinowitz proved in 1983 the following remarkable result by using an S^1 symmetry method and certain increasing estimates on critical values:

Theorem 3 (P. Rabinowitz [Rab6]). 1° *Suppose the conditions* (R1) *and* (R2) *hold. For any positive real numbers τ and R the autonomous Hamiltonian system* (0.1) *possesses a non-constant τ-periodic solution x such that*

$$\|x\|_C \ge R.$$

2° *Suppose the conditions* (V2) *and the following* (V1′) *hold:*

(V1′) $V \in C^1(\mathbf{R}^n, \mathbf{R})$.

For any positive real numbers τ *and* R *the autonomous second order Hamiltonian system* (5.1) *possesses a non-constant* τ*-periodic solution* x *such that*

$$\|x\|_C \geq R.$$

Note that the above Theorem 1 and Rabinowitz's result clarify the solution structure of (1) from different angles.

Next we consider the periodic boundary value problem of non-autonomous Hamiltonian systems

$$\begin{cases} \dot{x}(t) &= JH'(t,x), \\ x(m\tau) &= x(0), \end{cases} \tag{8}$$

for some $m \in \mathbf{N}$. Here we suppose $H(t,x)$ is τ-periodic in t for some $\tau > 0$. As usual, $m\tau$-periodic solutions of (8) are called **subharmonics**. For any $m\tau$-periodic solution x of (8) and any $k \in \mathbf{Z}$, the phase shift $k * x$ of x is defined by

$$(k * x)(t) = x(t + k\tau). \tag{9}$$

We call two solutions x_1 and x_2 of (8) **geometrically distinct** if

$$k * x_1 \neq h * x_2, \qquad \forall k,\ h \in \mathbf{Z}. \tag{10}$$

Theorem 4 (T. Wang and G. Fei [WaF1]). 1° *Suppose* $H(t,x)$ *satisfies the condition* (H1) *in* §4.1 *for some* $\tau > 0$ *and the following conditions:*

(H4) *There exist* $\mu > 2$ *and* $r_0 > 0$ *such that*

$$0 < \mu H(t,x) \leq H'(t,x) \cdot x, \qquad \forall (t,x) \in \mathbf{R} \times \mathbf{R}^{2n}.$$

(H5) $H(t,x) = o(|x|^2)$ *at* $x = 0$.

(H6) $H(t,x) \geq H(t,0) = 0$ *for any* $(t,x) \in \mathbf{R} \times \mathbf{R}^{2n}$.

(H7) *There exist constants* $\beta > 0$ *and* $r_1 > 0$ *such that*

$$|H'(t,x)| \leq \beta H'(t,x) \cdot x, \quad \forall t \in \mathbf{R},\ |x| \geq r_1.$$

Then for any $m \in \mathbf{N}$, *the problem* (8) *possesses a non-constant* $m\tau$*-periodic solution* x_m *which satisfies*

$$n + 1 - \nu(x_m, 1) \leq i(x_m, 1) \leq n + 1. \tag{11}$$

Moreover, x_m *and* x_{mp} *are geometrically distinct provided* $p > 2n + 1$.

2° *We further suppose the following conditions:*

(H8) $H''(t,x) \geq 0$.

(H9) *The set* $D = \{(t,x) \in \mathbf{R} \times \mathbf{R}^n \mid H''(t,x) \neq 0,\ 0 \in \sigma(H''(t,x))\}$ *is hereditarily discontinuous.*

Then for any $m \in \mathbf{N}$, *the system* (8) *possesses an* $m\tau$*-periodic solution such that* (11) *holds and* $m\tau$ *is its minimal period.*

Remark 5 Theorem 4 is due to T. Wang and G. Fei in [WaF1]. The proof of 1° of this theorem is similar to that of the above Theorem 1. The proof of the second part is similar to those of Theorem 4.1 and Corollary 4.3. Thus they are omitted here. We refer readers to [WaF1] and [Lic1] for further details.

The study on subharmonics by variational methods was started from the paper [Rab4] of P. Rabinowitz in 1980. Among many other results, [EkH3] of I. Ekeland and H. Hofer in 1987 got interesting progress. Here the geometric distinction part of Theorem 1 is new. Theorem 4 was proved by T. Wang and G. Fei via the method of [DoL1] and the Galerkin method in [FeQ2] which further generalized results of [EkH3]. We also refer readers to [Lic1] of C. Liu for further discussions on this problem.

13.7 Notes and comments

For Rabinowitz's prescribed minimal period solution conjecture, we tend to believe that one should construct counterexamples now to show that beyond Theorem 4.1 and Corollary 4.3 there is almost nothing more one can hope for.

We point out that there may be some substantial difference between first order and second order Hamiltonian systems, because estimates on the minimal period of solutions obtained by variational methods for first order Hamiltonian systems similar to those of Theorems 5.1 and 5.2 are still unknown under precisely Rabinowitz's original structure conditions.

Next we make some comments on the study of superquadratic Hamiltonian systems (0.1) and (5.1). Note that the proof of Theorem 5.3 in [Rab6] depends on the autonomous property of the systems (0.1) and (5.1) crucially, which implies that the corresponding variational functionals possess certain S^1-symmetry. A natural question is what will happen if such a symmetry is broken. There are many studies of this problem too. After significant progress on this problem obtained by P. Rabinowitz in [Rab5] of 1982 and by A. Bahri and H. Berestycki in [BaB1] of 1984 on the perturbed second order Hamiltonian systems

$$\ddot{x} + V'(x) = f(t), \quad \text{for } x \in \mathbf{R}^n, \tag{1}$$

the following result was proved in [Lon2] of 1989:

Theorem 1 *Suppose the conditions* (V1′) *in Theorem* 6.3 *and* (V2) *in Theorem* 5.1 *hold. Then given any* $\tau > 0$ *and any* $f \in L^2(\mathbf{R}/(\tau\mathbf{Z}), \mathbf{R}^n)$, *for any* $R > 0$ *the second order system* (1) *possesses a non-constant* τ*-periodic solution* x *satisfying* $\|x\|_C \geq R$.

The proof of Theorem 1 depends on an optimal increasing estimate

$$\frac{c_k}{k^2} \to +\infty, \quad \text{as } k \to +\infty,$$

of critical values $\{c_k\}$ of the functional corresponding to the autonomous system (5.1) obtained by the S^1-symmetry method, and a new S^1 action introduced in [Lon2].

Related results for the perturbed Hamiltonian system

$$\dot{x} - JH'(x) = f(t)$$

can be found in [BaB2] of A. Bahri and H. Berestycki in 1984, and [Lon1], [Lon3], and [Lon4] of the author in 1989 to 1990.

It should be pointed out that under conditions for $V(t, x)$ similar to (V1), (V2) in Theorem 5.1, and that $V(t, x)$ is τ-periodic in t, it is still unknown whether the conclusion of above Theorem 1 holds for the general non-autonomous systems

$$\ddot{x} + V'(t, x) = 0.$$

Chapter 14

Periodic Lagrangian orbits on tori

We consider the periodic solution problem for the Lagrangian system

$$\frac{d}{dt}L_{\dot{x}}(t,x,\dot{x}) - L_x(t,x,\dot{x}) = 0, \quad x \in \mathbf{R}^n, \tag{1}$$

where $L_{\dot{x}}$ and L_x denote the gradients of L with respect to $\dot{x}$ and x respectively. The following conditions on the Lagrangian function L are assumed.

(L1) $L(t,x,p) = \frac{1}{2}A(t)p \cdot p + V(t,x)$, *where* $\frac{1}{2}A(t)p \cdot p \geq \lambda |p|^2$ *for all* $(t,p) \in \mathbf{R} \times \mathbf{R}^n$ *and some fixed constant* $\lambda > 0$.

(L2) $A \in C^3(\mathbf{R}, \mathcal{L}_s(\mathbf{R}^n))$, $V \in C^3(\mathbf{R} \times \mathbf{R}^n, \mathbf{R})$, *both A and V are* 1*-periodic in all of their variables.*

Let L satisfy the conditions (L1) and (L2). The system (1) can be viewed as defined on the standard torus $T^n = \mathbf{R}^n/\mathbf{Z}^n$. We search for τ-periodic solutions of the system (1) with $\tau \in \mathbf{N}$. Solutions $x(t)$ of (1) on T^n determines a one-parameter family of diffeomorphisms $\Psi_L^t \in \mathrm{Diff}(TT^n)$ satisfying $\Psi_L^t(x(0), \dot{x}(0)) = (x(t), \dot{x}(t))$, where TT^n is the tangent bundle of T^n. Here we call the time-1-map $\Psi_L = \Psi_L^1$ the **Lagrangian diffeomorphism** of the system (1) corresponding to the Lagrangian function L. The multiplicity of periodic points of this diffeomorphism is an interesting problem considered by many mathematicians. Our aim here is to prove Theorem B in the Introduction, i.e., the following theorem proved by the author in [Lon17].

Theorem 1 *Suppose the function L satisfies the conditions* (L1) *and* (L2). *Then the corresponding Lagrangian diffeomorphism Ψ_L possesses infinitely many periodic points on TT^n produced by contractible integer periodic solutions of the system* (1).

Fix L satisfying (L1) and (L2). For any $\tau \in \mathbf{N}$, let $S_\tau = \mathbf{R}/(\tau\mathbf{Z})$ and $E_\tau = W^{1,2}(S_\tau, \mathbf{R}^n)$ with the usual inner product and the norm:

$$\langle x, y \rangle = \int_0^\tau (x \cdot y + \dot{x} \cdot \dot{y})dt, \qquad \|x\| = \langle x, x \rangle^{1/2}, \qquad \forall\, x, y \in E_\tau.$$

Define

$$f_m(x) = \int_0^{m\tau} L(t,x,\dot{x})dt, \qquad \forall\, x \in E_{m\tau},\ m \in \mathbf{N}. \tag{2}$$

By the conditions (L1) and (L2), it is well known that $f_m \in C^3(E_{m\tau}, \mathbf{R})$ satisfies the Palais–Smale condition and that critical points of f_m correspond to contractible $m\tau$-periodic solutions of (1.1) on T^n. But for distinct $h, k \in \mathbf{N}$, although two critical points $x \in E_h$ of f_h and $y \in E_k$ of f_k are different in this analytical setting, they may produce the same periodic point for Φ_L on T^n. To get over this obstacle, we consider the following analytic concepts of iterations and towers.

Let $x \in E_\tau$ with $\tau \in \mathbf{N}$. Using iteration of x defined by (8.1.2), we define the iteration map $\psi^m : E_\tau \to E_{m\tau}$ by $\psi^m(x) = x^m$. In this case, we call $\{x^m\}_{m\in\mathbf{N}} \subset \Pi_{m\in\mathbf{N}} E_{m\tau}$ the **tower** based on $x \in E_\tau$. A tower $\{x^p\}$ based on $x \in E_\tau$ is called a subtower of another tower $\{y^q\}$ based on $y \in E_\alpha$, if there exists $k \in \mathbf{N}$ and $j \in \mathbf{Z}^n$ such that $x = y^k + j$. Two towers $\{x^p\}$ based on $x \in E_\tau$ and $\{y^q\}$ based on $y \in E_\alpha$ are called T^n**-distinct**, if there exists no tower $\{z^m\}$ based on $z \in E_\beta$, $h, k \in \mathbf{N}$, and $i, j \in \mathbf{Z}^n$ such that $x = z^h + i$ and $y = z^k + j$. Note that the functional f_m defined by (2) satisfies

$$f_m(x^m + j) = m f_1(x), \qquad \forall\, x \in E_\tau,\ j \in \mathbf{Z}^n,\ \tau, m \in \mathbf{N}. \tag{3}$$

Here the solution tower is an analytic, not a geometric, concept. Using this concept, Theorem 1 can be rephrased as the following

Theorem 2 *Suppose the function L satisfies the conditions* (L1) *and* (L2). *Then the system* (1) *possesses infinitely many T^n-distinct solution towers based on integer periodic solutions in $\mathbf{R}^n$.*

Remark 3 When (L1) and (L2) hold and V is time independent, using ideas in [Lon17] and [BaK1], recently G. Lu and the author have proved in [LoL1] that the autonomous Lagrangian system (1) always possesses infinitely many T^n-distinct solution orbits based on integer periodic solutions in $\mathbf{R}^n$. Thus we suspect that (1) always possesses infinitely many contractible integer periodic solutions with mutually distinct solution orbits on T^n provided (L1) and (L2) hold.

Following ideas in [Lon17], we prove Theorem 2 indirectly by assuming that its conclusion does not hold. Then in Section 4, we find a critical point x of f_1 with $c = f_1(x)$ for some integer $\tau \in \mathbf{N}$ such that the critical module $C_n(f_1, x) \neq 0$ and by the m-th iteration this module is injectively mapped into a global homological module $H_n(E_{m\tau}, (f_m)^o_{m(c-\epsilon)})$ for $\epsilon > 0$ small enough (cf. Theorem 4.3 below). On the other hand, in Section 5 we prove that a Lagrangian version homotopy similar to the Bangert homotopy technique in [Ban1] and [BaK1] can be constructed via a new finite energy homology theory. Then we use this homotopy to show that through the above injective iteration homomorphism, a non-trivial finite energy singular chain must vanish provided m is great enough (cf. Proposition 5.2 below). Then they produce a contradiction that proves Theorem 2.

Our Theorems 1 and 2 together with the above mentioned proof are closely related to the work [BaK1] of V. Bangert and W. Klingenberg who studied the case of closed geodesics on product Riemannian manifolds including T^n. But results of [BaK1] (e.g., their Theorems 1 to 3) can not be applied to (1) directly, because the functional f_1 of (2) is strongly indefinite, all the critical points we found are non-minimal saddle points (cf. our discussion below in (4.4)), and the Morse index iteration formula of closed geodesics used in [BaK1] is not applicable to (1). Moreover, in the proof of Theorem 1 on page 382 in [BaK1], which is explained in detail in Section 2 of [Ban1] of V. Bangert, the curve shortening process for geodesics is a crucial ingredient to gain the smoothness of the initial point curves of a given singular simplex in the loop space. But it is not clear to the author whether this process can be applied to the indefinite functional f_1 and its non-minimal saddle points. In order to overcome these difficulties, we have used two new ingredients.

(A) We need the iteration theory of the index for symplectic paths in the previous chapters, especially the iteration inequality Theorem 10.1.2 which estimates the iterated index in terms of the mean index. This inequality yields the homological injectivity obtained in Theorem 4.3 mentioned above.

(B) A new finite energy singular homology and cohomology theory is introduced in this paper. Since the curve shortening process for geodesics used in [BaK1] fails in our case, our new idea here is to mount this required condition of smoothness of initial point curves to all singular simplexes in the loop space appearing in the homology theory. All such simplexes form a new homology theory, i.e., the finite energy homology theory introduced in Section 2 below. Then every generator of the nontrivial critical module $C_n(f_1, x)$ found in Theorem 4.3 inherits the required initial point curve smoothness from the finite energy property of the homology theory we chosen. This allows us to construct a Bangert type homotopy in the Lagrangian case. In the generalized Morse lemma (Theorem 1.2), the smoothness of the homeomorphism ϕ_1 is usually two orders lower than that of the given functional f_1 of (2). In our proof of Theorem 4.3, this map ϕ_1 is used to induce an isomorphism between finite energy homology modules. In order to do so, we need f_1 to be C^3 so that ϕ_1 becomes a C^1-diffeomorphism and then a finite energy map defined in Section 2 below. This request then needs the Lagrangian function L to be C^3 and explains this smoothness requirement in our condition (L2). We have also carefully chosen the definition of the critical modules in (3.11) below, to make the corresponding deformation maps become finite energy maps. The reason for this definition is explained in Example 2.7.

As preparations for our proof in Sections 4 and 5, in the following Section 2 we define finite energy homology and cohomology theory. In Section 3, we study critical modules and corresponding isomorphisms.

14.1 Critical module preparations

In this section, for the reader's convenience, we list the results contained in other research papers or books which will be used in this chapter. Here we always let X be a Hilbert space with inner product $\langle \cdot, \cdot \rangle$ and norm $\| \cdot \|$, and use notation already introduced in Sections 4.1 and 13.1.

The pseudo-gradient vector field is defined by Definition I.3.1 on page 19 of [Cha6] and Remark A.17 on page 85 of [Rab7] as follows. Let $U \subset X$ and $f \in C^1(U, \mathbf{R})$, Then $v \in T_u X$ is called a **pseudo-gradient vector** for f at $u \in U$ if there exist $\alpha > \beta > 0$ such that

$$\|v\| \le \alpha \|f'(u)\|, \quad \langle f'(u), v \rangle \ge \beta \|f'(u)\|^2. \tag{1}$$

$V : (U\backslash\mathcal{K}(f)) \to T_U X$ is a **pseudo-gradient vector field** for f on U, if V is locally Lipschitz continuous and $V(u)$ is a pseudo-gradient vector for f at any $u \in U\backslash\mathcal{K}(f)$.

The following deformation theorem is due to E. Rothe [Rot1], K.C. Chang [Cha6], and Z.Q. Wang [Wzq1].

Theorem 1 *Let $f \in C^1(X, \mathbf{R})$ satisfy the $(PS)_c$ condition for all $c \in [a, b]$ and let a be the only critical value of f in $[a, b)$. Assume that the connected components of $\mathcal{K}_a(f)$ are only isolated points. Then f_a is a strong deformation retract of $f_b\backslash\mathcal{K}_b(f)$.*

The Gromoll–Meyer pair was introduced by D. Gromoll and W. Meyer in [GrM1] of 1969. Here we use Definition I.5.1 on page 48 of [Cha6] for the Gromoll–Meyer pair as follows. Let $f \in C^2(X, \mathbf{R})$ be a function satisfying the Palais–Smale condition. Suppose that $V : (X\backslash\mathcal{K}(f)) \to TX$ is a pseudo-gradient vector field for f on X. Let p be an isolated critical point of f, $c = f(p)$. A pair of topological spaces (W, W^-) is called a **Gromoll–Meyer pair with respect to** V, if

(i) W is a closed neighborhood of p possessing the mean value property, i.e., $\forall t_1 < t_2$, $\eta(t_i) \in W$, $i = 1, 2$, implies $\eta(t) \in W$ for all $t \in [t_1, t_2]$, where $\eta(t)$ is a decreasing flow with respect to V. And there exists $\epsilon > 0$ such that $W \cap f_{c-\epsilon} = f^{-1}[c - \epsilon) \cap \mathcal{K}(f) = \emptyset$, $W \cap \mathcal{K}(f) = \{p\}$;
(ii) the set $W^- = \{x \in W \mid \eta(t, x) \notin W,\ \forall t > 0\}$ is closed in W;
(iii) W^- is a piecewise submanifold, and the flow η is transversal to W^-.

The existence and properties of the Gromoll–Meyer pair can be found on pages 48 to 50 in [Cha6].

The classical Morse lemma claims that a function at its non-degenerate critical point can be written into a quadratic function under a suitable coordinate system (cf. [Mor1]). This result was generalized by D. Gromoll and W. Meyer to functionals at degenerate critical points in [GrM1] of 1969 for smooth functionals. Here we follow the description of this theorem in [GrM1] and Theorem I.5.1 of [Cha6].

Theorem 2 *Let U be an open subset of X, $f \in C^3(U, \mathbf{R})$. Let x be the only critical point of f in U, and $A = f''(x)$ with kernel N. Suppose that 0 is either an isolated point of the spectrum $\sigma(A)$ or not in $\sigma(A)$. Then there exist an open ball $B_r(0)$ in U centered at 0 with some radius $r > 0$, a local C^2-diffeomorphism ϕ from $B_r(0)$ to an open neighborhood of x in X, and a map $h \in C^2(B_r(0) \cap N, N^\perp)$ satisfying $\phi(0) = x$ and $h(0) = 0$ such that*

$$\begin{aligned} f \circ \phi(\xi + \eta) &= \frac{1}{2}\langle A\xi, \xi\rangle + f(x + \eta + h(\eta)) \\ &\equiv \beta(\xi) + \gamma(\eta), \end{aligned} \tag{2}$$

for any $\xi + \eta \in B_r(0) \cap (N \oplus N^\perp)$, where $N^\perp$ is the orthogonal subspace of N in X. Furthermore, ϕ has the form $\phi(\eta) = x + \eta + h(\eta)$ for all $\eta \in B_r(0) \cap N$ and $(\phi^{-1})'(0)$ is the identity map. The degenerate part $\gamma \in C^3(U \cap N, \mathbf{R})$, the origin $\eta = 0$ is an isolated critical point of γ, and $\gamma(0) = f(x)$. The origin $\xi = 0$ is a *non-degenerate critical point of the quadratic function β satisfying $\beta(0) = 0$. Let $Q : X \to N$ and $Q^\perp : X \to N^\perp$ be the projectors respectively. By the implicit function theorem, the map $h : U \cap N \to N^\perp$ is uniquely defined by the equation*

$$Q^\perp f'(x + \eta + h(\eta)) = 0, \quad \forall \eta \in U \cap N. \tag{3}$$

We follow the definition given in Sections I.4 and I.5 [Cha6]. Let $f \in C^2(X, \mathbf{R})$. By H_q we denote the usual singular homology.

Definition 3 For an isolated $x \in \mathcal{K}_c(f)$, the q-th critical group of f at x with coefficient group $\mathcal{G}$ is defined by

$$C_q(f, x) = H_q(f_c \cap U, (f_c \setminus \{x\}) \cap U; \mathcal{G}), \tag{4}$$

where U is any open neighborhood of x in M such that $\mathcal{K}(f) \cap (f_c \cap U) = \{x\}$. It is proved in [Cha6] that this definition is independent of the choice of U.

Let (W, W^-) be a Gromoll–Meyer pair of f at x. One can also define

$$C_q(f, x) = H_q(W, W^-; \mathcal{G}). \tag{5}$$

Note that by Theorem I.5.1 of [Cha6], the above two definitions coincide, i.e., $H_q(f_c \cap U, (f_c \setminus \{x\}) \cap U; \mathcal{G})$ is isomorphic to $H_q(W, W^-; \mathcal{G})$ in the singular homology theory.

Using the generalized Morse lemma, the following shifting theorem was proved by Gromoll–Meyer in [GrM1] (cf. also Theorem I.5.4 of [Cha6]).

Theorem 4 *In the above generalized Morse Theorem 2, we suppose further that the Morse index of f at its isolated critical point x is j. Then*

$$C_q(f, x) \cong C_j(\beta, 0) \otimes C_{q-j}(\gamma, 0) \cong C_{q-j}(\gamma, 0). \tag{6}$$

The following example can be found on page 33 of [Cha6].

Example 5 If x is an isolated local minimum of $f \in C^2(X, \mathbf{R})$, then

$$C_q(f, x) \cong \begin{cases} \mathcal{G}, & \text{if} \quad q = 0, \\ 0, & \text{if} \quad q \neq 0. \end{cases} \tag{7}$$

In an n-dimensional manifold M, if x is an isolated local maximum of $f \in C(M, \mathbf{R})$, then

$$C_q(f, x) \cong \begin{cases} \mathcal{G}, & \text{if} \quad q = n, \\ 0, & \text{if} \quad q \neq n. \end{cases} \tag{8}$$

Let A be a bounded self-adjoint operator defined on the Hilbert space X. According to its spectral decomposition, $X = X_+ \oplus X_0 \oplus X_-$, where for $* = +, 0, -$, each X_* is an invariant subspace of X corresponding to the positive, zero, or negative spectrum of A respectively, and we denote by $P_* : X \to X_*$ the corresponding orthogonal projection. The following conditions are supposed.

(A1) $A_\pm \equiv A|_{X_\pm}$ has bounded inverse on $X_\pm$.

(A2) $p = \dim(X_- \oplus X_0) < +\infty$.

Theorem 6 (K.C. Chang, Lemma II.5.2 of [Cha6]). *Suppose conditions* (A1) *and* (A2) *hold. Let V be an n-dimensional C^2-compact manifold without boundary. Suppose that $g \in C^1(X \times V, \mathbf{R})$ has a bounded and compact g' (if $\dim X_0 = 0$, $\|g'(x, v)\| = o(\|x\|)$ for all $v \in V$), and satisfies*

$$g(P_0 x, v) \to -\infty \quad as \quad \|P_0 x\| \to +\infty, \quad if \quad \dim X_0 \neq 0. \tag{9}$$

Let

$$f(x, v) = \frac{1}{2}\langle Ax, x\rangle + g(x, v). \tag{10}$$

Then f satisfies the (PS) condition and

$$H_q(X \times V, f_a) \cong H_{q-p}(V), \tag{11}$$

for $-a > 0$ sufficiently great with $\mathcal{K}(f) \cap f_a = \emptyset$.

Theorem 7 (J.-Q. Liu, Theorem II.1.5 of [Cha6]). *Let $f \in C^1(X, \mathbf{R})$. Suppose that $\alpha \in H_k(f_b, f_a)$ is non-trivial, where $a < b$. Let*

$$c = \inf_{z \in \alpha} \sup_{x \in |z|} f(x). \tag{12}$$

Suppose f satisfies the $(PS)_c$ condition, and the critical point set $\mathcal{K}_c(f)$ of f with critical value c contains only isolated critical points. Then there exists an $x_0 \in \mathcal{K}_c(f)$ such that

$$C_k(f, x_0) \neq 0. \tag{13}$$

Theorem 8 (V. Bangert and W. Klingenberg, Lemma 1 of [BaK1]). *Let (Y, A) be a pair of topological spaces and α a singular relative p-cycle of (Y, A). Let Σ denote the set of singular simplices of α together with all their faces. Suppose to every $\sigma \in \Sigma$, $\sigma : \Delta^q \to X$, $0 \le q \le p$, there is assigned a map $P(\sigma) : \Delta^q \times [0, 1] \to X$ such that*

(i) $P(\sigma)(z, 0) = \sigma(z), \quad \forall z \in \Delta^q,$

(ii) $P(\sigma)(z, t) = \sigma(z), \quad if \quad \sigma(\Delta^q) \subset A,$

(iii) $P(\sigma)(\Delta^q \times \{1\}) \subset A,$

(iv) $P(\sigma) \circ (e_q^i \times id) = P(\sigma \circ e_q^i)$ *for $0 \le i \le q$, where e_q^i is the i-th face of Δ^q.*

Then the homotopy class $[\alpha] \in H_p(Y, A)$ vanishes.

14.2 The finite energy homology theory

In this section we introduce the finite energy homology and cohomology theories to gain certain smoothness of initial curves for singular simplexes in loop spaces.

As usual, for any integer $q \geq 0$, let Δ_q denote the standard closed q-dimensional simplex in $\mathbf{R}^q$ and F_q^i denote its i-th face. For $q \geq 0$ and $s \in [0, 1]$ denote $e(s) = (s, \ldots, s) \in \mathbf{R}^q$. Then $\hat{e} = e(1/(q+1))$ is the barycenter of Δ_q. Denote by L the straight line passing through $e(0)$ and $\hat{e}$ succesively in $\mathbf{R}^q$. According to L, we obtain an orthogonal subspace decomposition

$$\mathbf{R}^q = V_{q-1} \times L. \tag{1}$$

Thus any $w \in \Delta_q$ can be uniquely written as $w = (v, t) \in [V_{q-1} \times L] \cap \Delta_q$. For such a $w = (v, t) \in \Delta_q$, denote by $l(v)$ the intersection segment of Δ_q with the straight line passing through w and parallel to L. This segment $l(v)$ is well defined, since it is independent of the particular choice of the t-component of the point $w = (v, t)$ on $l(v)$.

Definition 1 Let M be a Hilbert manifold. For any $\phi \in C(\Delta_q, M)$, we define the **v-parametrized restriction curve** $^{\#}\phi_v : l(v, s) \to M$ of ϕ for each $w = (v, t) \in [V_{q-1} \times L] \cap \Delta_q$ by

$$^{\#}\phi_v(s) = \phi(v, s), \qquad \forall\, s \in l(v). \tag{2}$$

Definition 2 Let $k = 1$ or 2. Suppose M is a C^2-Hilbert manifold modeled on a Hilbert space X with the norm $\|\cdot\|$. Let A be a subset of M. For $q = 0$, every singular simplex $\sigma \in C(\Delta_0, A)$ is defined to be a 0-dimensional singular simplex with **finite k-energy** (k-FE 0-simplex in A for short). Next we define higher dimensional k-FE simplexes inductively. For any positive integer q, suppose m-dimensional k-FE singular simplexes have been defined for all $0 \leq m < q$. A singular simplex $\sigma \in C(\Delta_q, A)$ is a q-dimensional singular simplex in A with **finite k-energy** (k-FE q-simplex in A for short) if every $(q-1)$-dimensional face $\sigma \circ F_q^i : \Delta_{q-1} \to A$ with $0 \leq i \leq q$ of σ is a k-FE simplex, and there exists a real constant $\hat{b}(\sigma) \geq 0$ depending on σ so that for any $w = (v, t) \in \Delta_q$ it holds that $^{\#}\sigma_v \in C(l(v), A) \cap W^{1,k}(l(v), A)$ and

$$\int_{l(v)} \left\| \frac{d}{ds}\, {}^{\#}\sigma_v(s) \right\|^k ds \leq \hat{b}(\sigma). \tag{3}$$

Note that all C^1-smooth simplexes are k-FE simplexes for $k = 1$ and 2. For $\tau > 0$ let σ be a singular simplex from Δ_q to the loop space $E_\tau = W^{1,2}(S_\tau, \mathbf{R}^n)$ with $S_\tau = \mathbf{R}/(\tau\mathbf{Z})$. For $w = (v, t) \in \Delta_q$, define the initial point curve $\sigma_v^{ini} : l(v) \to \mathbf{R}^n$ of the restriction curve $^{\#}\sigma_v : l(v) \to E_\tau$ by

$$\sigma_v^{ini}(s) = {}^{\#}\sigma_v(s)(0), \qquad \forall\, s \in l(v). \tag{4}$$

Then the most important property of FE simplexes in the loop space is given in the following lemma.

Lemma 3 *For $\tau > 0$ let $\sigma : \Delta_q \to A \subset W_\tau$ be a k-FE singular simplex. Then for any $w = (v, t) \in \Delta_q$, it holds that $\sigma_v^{ini} \in C(l(v), A) \cap W^{1,k}(l(v), \mathbf{R}^n)$ and*

$$\int_{l(v)} \left| \frac{d}{ds} \sigma_v^{ini}(s) \right|^k ds \le K^k \hat{b}(\sigma), \tag{5}$$

where $K > 0$ is the Sobolev embbeding constant from the space E_τ to the space $C_\tau = C(S_\tau, \mathbf{R}^{2n})$.

Proof: By definition we have $\sigma_v^{ini} \in C(l(v), \mathbf{R}^{2n})$ and

$$\begin{aligned} \int_{l(v)} \left| \frac{d}{ds} \sigma_v^{ini}(s) \right|^k ds &\le \int_{l(v)} \left\| \frac{d}{ds} {}^{\#}\sigma_v(s) \right\|_{C_\tau}^k ds \\ &\le K^k \int_{l(v)} \left\| \frac{d}{ds} {}^{\#}\sigma_v(s) \right\|_{E_\tau}^k ds \\ &\le K^k \hat{b}(\sigma). \end{aligned}$$

Therefore $\sigma_v^{ini} \in W^{1,k}(l(v), \mathbf{R}^n)$ and (5) holds. □

Based on Definition 2, we can define new homology and cohomology theories for simplexes with finite energy as follows.

Definition 4 Let $k = 1$ or 2. Given two C^2-Hilbert manifolds M_i and their subsets A_i for $i = 0, 1$, a map $f \in C(A_0, A_1)$ is a **finite** k**-energy map** (k-FE map for short) if for every k-FE q-simplex σ in A_0, the composition map $f \circ \sigma : \Delta \to A_1$ is also a k-FE q-simplex in A_1.

Note that f is an k-FL-map if $f \in C^1(A_0, A_1)$.

Definition 5 Given C^2-Hilbert manifolds M_i and subset $B_i \subset A_i \subset M_i$ for $i = 0, 1$. Two k-FE maps $f, g : (A_0, B_0) \to (A_1, B_1)$ are k**-FE homotopic**, if there exists a k-FE map $h : A_0 \times [0, 1] \to A_1$ such that $h(\cdot, 0) = f(\cdot)$, $h(\cdot, 1) = g(\cdot)$, and $h(\cdot, s) : (A_0, B_0) \to (A_1, B_1)$ for each $s \in [0, 1]$.

For $k = 1$ or 2, let $\mathcal{A}$ be the category consisting of the following objects and morphisms:

(a) All subset pairs of any C^2-Hilbert manifolds.
(b) All k-FE maps between any two subset pairs of any two C^2-Hilbert manifolds.

Note that $\mathcal{A}$ is an admissible category in the sense of Eilenberg and Steenrod (cf. Section I.1 of [EiS1]). For any coefficient ring $\mathcal{F}$, using the usual argument as in Chapter VII of [EiS1], it is verified that, for the category $\mathcal{A}$, singular homology and cohomology theories can be defined, and the axioms for homology and cohomology theories of Eilenberg and Steenrod (cf. [EiS1]) hold. We call them the **k-finite energy** singular homology and cohomology theories, denoted by $H_*^{fe}(A, B; \mathcal{G})$ and $H^*_{fe}(A, B; \mathcal{G})$ with coefficient ring $\mathcal{G}$ respectively.

Note that based on definitions 1, 2, 4, and 5, the k-FE homology and cohomology theory can also be defined as the usual singular theory via the usual boundary operators.

For $k = 1$ or 2, we say that a pair (A, B) of subsets $B \subset A$ of a C^2-Hilbert manifold M is **k-finite**, if for any $\epsilon > 0$, $q \geq 0$, and any usual singular q-cycle $\sigma = \sum_{i=1}^k \sigma_i$ with $\sigma_i : \Delta_q \to A$, there exists a k-finite energy singular cycle $\hat{\sigma} = \sum_{i=1}^k \hat{\sigma}_i$ with $\hat{\sigma}_i : \Delta_q \to A$ such that any subsimplex η_i of σ_i and the corresponding subsimplex $\hat{\eta}_i$ of $\hat{\sigma}_i$ satisfy

$$\rho(\eta_i, \hat{\eta}_i) < \epsilon, \tag{6}$$

where $\rho(\cdot, \cdot)$ is the metric on $C(\Delta_q, M)$ induced from the metric on M, and any subsimplex η_i of σ_i satisfying $\mathrm{Im}(\eta_i) \subset B$ implies $\mathrm{Im}(\hat{\eta}_i) \subset B$ for the corresponding subsimplex $\hat{\eta}_i$ of $\hat{\sigma}_i$.

Theorem 6 *For $k = 1$ or 2, let (A, B) be a k-finite pair of subsets of any C^2-Hilbert manifold with $B \subset A$. Then for any coefficient ring $\mathcal{F}$, the usual singular homology $H_*^{sg}(A, B; \mathcal{F})$ and cohomology $H^*_{sg}(A, B; \mathcal{F})$ are isomorphic to the k-finite energy homology $H_*^{fe}(A, B; \mathcal{F})$ and cohomology $H^*_{fe}(A, B; \mathcal{F})$ as modules respectively.*

Proof: Since every finite energy cycle is a usual singular cycle, $H_q^{fe}(A, B; \mathcal{F})$ is a subgroup of $H_q^{sg}(A, B; \mathcal{F})$. On the other hand, by the 2-finite assumption on (A, B), any singular cycle can be approximated by finite energy cycles preserving the pair (A, B). Thus every cycle in $H_q^{sg}(A, B; \mathcal{F})$ can be modified to a cycle in $H_q^{fe}(A, B; \mathcal{F})$. Therefore $H_q^{sg}(A, B; \mathcal{F})$ is a subgroup of $H_q^{fe}(A, B; \mathcal{F})$. Thus the conclusion holds for 2-finite energy homology. Other cases can be proved similarly. □

Since every q-dimensional singular simplex in T^n can be approximated by a q-dimensional smooth simplex, which is an FE-simplex, for T^n the FE homology and cohomology modules are isomorphic to the usual singular homology and cohomology modules. Especially, in the FE homology, we have $H_n^{fe}(T^n; \mathcal{F}) \neq 0$ with any non-trivial coefficient ring $\mathcal{F}$.

In Section 1, we have given two ways to define the critical modules. For our finite energy homology theory, we always use (1.5) of the critical modules via Gromoll–Meyer pairs. The importance of using the Gromoll–Meyer pair in (1.5) of critical modules instead of (1.4) in our study here is that in the Gromoll–Meyer pair case we can avoid meeting curves with infinite energies, and avoid the strong deformation retractions in the proof to reach the critical point x.

Note that even if $0 \neq C_q(f_\tau, x)$ when it is defined by (1.4) in terms of the usual singular homology theory, it may still hold that $C_q(f_\tau, x) = 0$ when it is defined by (1.4) in terms of the FE homology theory.

Example 7 For $r > 0$, let $C_r = \{z \in \mathbf{R}^2 \mid |z| = r\}$ and $B_r(0) = \{z \in \mathbf{R}^2 \mid |z| < r\}$. We choose four smooth curves $\psi_i : [0, +\infty) \to \overline{B_1(0)}$ for $i = 1, 2, 3, 4$ with no mutual intersections such that $\psi_1(0) = (0, 1)$, $\psi_2(0) = (1, 0)$, $\psi_3(0) = (0, -1)$, $\psi_4(0) = (-1, 0)$, $\psi_i([0, +\infty)) \cap C_r$ contains only one point for each $r \in (0, 1]$, $\psi_i(t) \to 0$ as $t \to +\infty$ for all i, and

$$\int_t^{+\infty} |\psi_i(s)|^2 ds = +\infty, \quad \forall t \geq 0, \quad i = 1, 2, 3, 4. \tag{7}$$

Then we thicken each ψ_i to get its one open cone-shaped neighborhood Ψ_i with smooth boundaries so that all these four Ψ_is have no mutual intersections, the intersection of each Ψ_i with C_r is path connected and possesses a finite length for any $r \in (0, 1]$, and the closure of $\cup_{i=1}^4 \Psi_i$ covers $\overline{B_1(0)}$. Note that the origin is still a limit point of each Ψ_i.

These four infinitely long open cones can also be obtained in the following way. Denote by Γ_i for $i = 1, 2, 3$, and 4, the intersections of the four open quadrants in $\mathbf{R}^2$ with $\overline{B_1(0)}$. Then we fix $\cup_{i=1}^4 \Gamma_i \cap C_1$ and spin their parts in $B_1(0)$ around the origin at a variable speed, increasing to infinity as the point tends to the origin, to get four rotating open cones so that the closed disc $\overline{B_1(0)}$ is filled out by the closure of these four rotating cones, each of which has smooth boundaries, converges to the origin, and the original unit interval on each positive or negative axis becomes an infinitely long curve converging to the origin and satisfying (7). We denote the resulting rotating open cones by Ψ_i for $i = 1, 2, 3$, and 4.

Now we choose a function $f \in C^2(\overline{B_1(0)}, \mathbf{R})$ so that 0 is the only critical point of f and

$$\begin{cases} f(z) < 0, & \forall z \in \Psi_1 \cup \Psi_3, \\ f(z) > 0, & \forall z \in \Psi_2 \cup \Psi_4, \\ f(z) = 0, & \text{otherwise.} \end{cases} \tag{8}$$

Denote the usual singular homology and our finite energy homologies by H^{sg} and H^{fe} respectively. Let U be any open neighborhood of 0 in $B_1(0)$, and (W, W^-) be a Gromoll–Meyer pair with respect to the gradient field f'. Let $f_0 = \{z \in \overline{B_1(0)} \mid f(z) \leq 0\}$. Then by direct computation, we obtain that

$$\begin{cases} H_1^{sg}(f_0 \cap U, (f_0 \backslash \{0\}) \cap U; \mathbf{R}) \cong \mathbf{R}, \\ H_1^{fe}(f_0 \cap U, (f_0 \backslash \{0\}) \cap U; \mathbf{R}) \cong 0, \\ H_1^{sg}(W, W^-; \mathbf{R}) \cong \mathbf{R}, \\ H_1^{fe}(W, W^-; \mathbf{R}) \cong \mathbf{R}. \end{cases} \tag{9}$$

This result indicates the differences between the usual singular homology and our finite energy homology, and between the definitions (1.4) and (1.5) for critical modules. It shows that definition (1.5) is the right choice for FE homology to catch the useful information.

In the rest of this paper, for studies of the Lagrangian system (0.1) we use only the 2-finite energy theory and call it the finite energy (FE for short) homology and cohomology theory. In order to simplify the homology for product spaces, from now on we fix $\mathbf{R}$ to be the coefficient field for all the homology and cohomology modules in the rest of this paper. When there is no confusion, in this chapter we shall omit the super index fe in the notations of finite energy homology modules.

14.3 Critical modules and isomorphisms

In this and the next two sections, we give the proof of Theorem 0.2. Fix L satisfying (L1) and (L2). For any $\tau \in \mathbf{N}$, let $S_\tau = \mathbf{R}/(\tau\mathbf{Z})$ and $E_\tau = W^{1,2}(S_\tau, \mathbf{R}^n)$ with the usual inner product and the norm

$$\langle x, y\rangle = \int_0^\tau (x \cdot y + \dot{x} \cdot \dot{y})dt, \quad \|x\| = \langle x, x\rangle^{1/2}, \quad \forall\, x, y \in E_\tau.$$

Define

$$f_\tau(x) = \int_0^\tau L(t, x, \dot{x})dt, \quad \forall\, x \in E_\tau. \tag{1}$$

By the conditions (L1) and (L2), it is well known that $f_\tau \in C^3(E_\tau, \mathbf{R})$ satisfies the Palais–Smale condition and that critical points of f_τ correspond to contractible τ-periodic solutions of (0.1) on T^n. Note that by the 1-periodicity of the function H in (L2), the functional f_τ defined by (1) satisfies

$$f_{m\tau}(x^m + j) = mf_\tau(x), \quad \forall\, x \in E_\tau,\ j \in \mathbf{Z}^n,\ \tau, m \in \mathbf{N}. \tag{2}$$

Let $\mathcal{K}(f_\tau)$ be the critical point set of f_τ in E_τ. Then for any $x \in \mathcal{K}(f_\tau)$, the linearized system of (0.1) at x has the form

$$-(P\dot{y} + Qy)^{\cdot} + Q^T\dot{y} + Ry = 0, \tag{3}$$

where

$$\begin{cases} P(t) = L_{\dot{x},\dot{x}}(t, x(t), \dot{x}(t)), \\ Q(t) = L_{x,\dot{x}}(t, x(t), \dot{x}(t)), \\ R(t) = L_{x,x}(t, x(t), \dot{x}(t)). \end{cases} \tag{4}$$

Note that in this case, P and $R \in C(S_\tau, \mathcal{L}(\mathbf{R}^n))$ are symmetric, P is positive definite, and $Q \in C(S_\tau, \mathcal{L}(\mathbf{R}^n))$. The system (3) corresponds to the Hessian of f_τ at x given by

$$\langle f_\tau''(x)y, z\rangle = \int_0^\tau \{(P\dot{y} + Qy) \cdot \dot{z} + Q^T\dot{y} \cdot z + Ry \cdot z\}dt, \tag{5}$$

for all $y, z \in E_\tau$. We define a $2n \times 2n$ matrix function

$$B(t) \equiv B_x(t) = \begin{pmatrix} P^{-1}(t) & -P^{-1}(t)Q(t) \\ -Q(t)^T P^{-1}(t) & Q(t)^T P^{-1}(t)Q(t) - R(t) \end{pmatrix}. \tag{6}$$

Then the linear system (3) is equivalent to the linear Hamiltonian system

$$\dot{y} = JB(t)y, \qquad y \in \mathbf{R}^{2n}. \tag{7}$$

Through the fundamental solution γ_x of this system (7) with $\gamma_x(0) = I$, the iterated index of $(\dot{x}, x)$ is defined by

$$(i(x, m), \nu(x, m)) = (i_1(\gamma_x^m), \nu_1(\gamma_x^m)), \quad \forall\, m \in \mathbf{N}. \tag{8}$$

By Theorem 7.3.1, the Morse index $m^-(x)$ and the nullity $m^0(x)$ of the functional f_τ at x coincide with the corresponding index defined in Chapter 5, i.e.,

$$m^-(x) = i(x, 1), \qquad m^0(x) = \nu(x, 1). \tag{9}$$

Note that by Theorem 10.1.2,

$$\begin{aligned} \max\{0, m\hat{i}(x, 1) - n\} &\le i(x, m) \\ &\le m\hat{i}(x, 1) + n - \nu(x, m), \quad \forall\, m \in \mathbf{N}. \end{aligned} \tag{10}$$

Let $x \in E_\tau$ be an isolated critical point of f_τ in the rest of this section. As mentioned at the end of Section 2, following definition (1.5), we define the critical module $C_q(f_\tau, x)$ of f_τ at its critical point x via the FE-singular homology by

$$C_q(f_\tau, x) = H_q(W(x), W^-(x)), \tag{11}$$

where $(W(x), W^-(x))$ is a Gromoll–Meyer pair defined in Section 1 following [GrM1] and [Cha6] via some pseudo-gradient vector field of f_τ near x defined following [Rab7]. Note that the critical module $C_q(f_\tau, x)$ is independent of the choice of the Gromoll–Meyer pair and the pseudo-gradient vector field in the definition as proved in [GrM1] and [Cha6]. According to the null, negative, and positive definiteness of the quadratic form $f_\tau''(x)$, the space E_τ possesses an orthogonal decomposition

$$E_\tau = M^0(x) \oplus M^-(x) \oplus M^+(x) \equiv M^0(x) \oplus M^\perp(x). \tag{12}$$

Using the generalized Morse lemma mentioned in Section 1 following [GrM1] and [Cha6], there exist a diffeomorphism ϕ_τ which maps some open neighborhood U_τ of 0 in E_τ to $\phi_\tau(U_\tau) \subset E_\tau$ with $\phi_\tau(0) = x$, and a map $h_\tau \in C^2(U_\tau \cap M^0(x), M^\perp(x))$ such that

$$\begin{aligned} f_\tau(\phi_\tau(\eta + \xi)) &= f_\tau(x + \eta + h_\tau(\eta)) + \frac{1}{2}\langle f_\tau''(x)\xi, \xi\rangle, \\ &\equiv \gamma_\tau(\eta) + \beta_\tau(\xi), \end{aligned} \tag{13}$$

for any $\eta + \xi \in U_\tau \cap (M^0(x) \oplus M^\perp(x))$. Denote $N(x) = \phi_\tau(U_\tau \cap M^0(x))$. Since L is C^3 by (L2), the map ϕ_τ is a diffeomorphism from $U_\tau \cap M^0(x)$ to $N(x)$ and has the form $\phi_\tau(\eta) = x + \eta + h_\tau(\eta)$ for all $\eta \in U_\tau \cap M^0(x)$. The degenerate part $\gamma_\tau \in C^3(U_\tau \cap M^0(x), \mathbf{R})$, the origin $\eta = 0$ is an isolated critical point of γ_τ, and $\gamma_\tau(0) = f_\tau(x)$. The origin $\xi = 0$ is a non-degenerate critical point of the quadratic function β_τ satisfying $\beta_\tau(0) = 0$.

Lemma 1 *Let $x \in E_\tau$ have the Fourier expansion*

$$x(\tilde{t}) = \sum_{k\in\mathbf{Z}} \exp\left(\frac{2k\pi\tilde{t}}{\tau} J\right)\alpha_k, \quad \forall\, \tilde{t} \in [0, \tau]. \tag{14}$$

For $m \in \mathbf{N}$ denote $x^m \in E_{m\tau}$ by

$$x^m(\hat{t}) = \sum_{k\in\mathbf{Z}} \exp\left(\frac{2k\pi\hat{t}}{m\tau} J\right)\beta_k, \quad \forall\, \hat{t} \in [0, m\tau]. \tag{15}$$

Then for $h \in \mathbf{Z}$

$$\beta_k = \alpha_h, \qquad \text{if} \quad k = hm, \tag{16}$$

$$\beta_k = 0, \qquad \text{if} \quad k = hm + i, 1 \le i \le m-1. \tag{17}$$

Proof: The proof follows from the uniqueness of the Fourier coefficients and simple computations, and therefore is omitted. □

This lemma shows that the iteration map $\psi^m : E_\tau \to E_{m\tau}$ has the following form:

$$\psi^m\left(\sum_{k\in\mathbf{Z}} \exp\left(\frac{2k\pi t}{\tau} J\right)\alpha_k\right) = \sum_{m\in\mathbf{Z}} \exp\left(\frac{2km\pi s}{m\tau} J\right)\alpha_k, \tag{18}$$

where $\alpha_k \in \mathbf{R}^{2n}$ for all $k \in \mathbf{Z}$. Note that the left-hand side of (18) is defined for $t \in [0, \tau]$, and the right-hand side of (18) is defined for $s \in [0, m\tau]$.

It is easy to see that for all $\tau, m \in \mathbf{N}$ and $y, z \in E_\tau$,

$$\begin{cases} \psi^m(f_\tau'(z)) = f_{m\tau}'(\psi^m(z)), \\ \psi^m(f_\tau''(z)y) = f_{m\tau}''(\psi^m(z))\psi^m(y). \end{cases} \tag{19}$$

Lemma 2 *The iteration map $\psi^m : M^*(x) \to M^*(x^m)$ for $* = 0, -,$ and $+$ is linear, continuous, and injective. Moreover, if $\nu(x, m) = \nu(x, 1)$, the map $\psi^m : M^0(x) \to M^0(x^m)$ is a linear diffeomorphism. If $i(x, m) = i(x, 1)$, the map $\psi^m : M^-(x) \to M^-(x^m)$ is a linear diffeomorphism.*

Proof: This follows from the definition of ψ^m. □

For the functionals f_τ and $f_{m\tau}$ we have the following results.

Lemma 3 *Using notation of the generalized Morse lemma, suppose for some $m \in \mathbf{N}$ that $\nu(x, m) = \nu(x, 1)$.*

1° *The map $h_{m\tau} \in C^2(U_{m\tau} \cap M(x^m), M^\perp(x^m))$ defined in the generalized Morse lemma further satisfy*

$$h_{m\tau}(\psi^m(\eta)) = \psi^m(h_\tau(\eta)), \quad \forall\, \eta \in U_\tau \cap M(x). \tag{20}$$

2° *It holds that*

$$\phi_{m\tau} \circ \psi^m = \psi^m \circ \phi_\tau : U_\tau \to N(x^m). \tag{21}$$

Here we require $U_{m\tau} \cap \psi^m(E_\tau) = \psi^m(U_\tau)$.

Proof: 1° By Lemma 2, the map $\psi^m : M^0(x) \to M^0(x^m)$ is a linear diffeomorphism. Thus every element in $M^0(x^m)$ is of the form $\psi^m(\eta)$ for some $\eta \in M^0(x)$. Thus for any $\eta \in U_\tau \cap M^0(x)$,

$$\psi^m[f'_\tau(x + \eta + h_\tau(\eta))] = f'_{m\tau}(x^m + \psi^m(\eta) + \psi^m \circ h_\tau(\eta)).$$

Denote by $P^\perp_{m\tau} : E_{m\tau} \to M^0(x^m)$ the projection map. Thus

$$P^\perp_{m\tau} f'_{m\tau}(x^m + \psi^m(\eta) + \psi^m \circ h_\tau(\eta)) = 0.$$

This is equation (1.3) in the generalized Morse lemma for $f_{m\tau}$ which uniquely defines the map $h_{m\tau} : U_{m\tau} \cap M^0(x^m) \to M^\perp(x^m)$. Therefore (20) must hold.

2° Note that both sides of (21) are solutions of the equation

$$P^\#_{m\tau} f'_{m\tau}(\alpha) = 0,$$

where $P^\#_{m\tau}$ is the projection map from $E_{m\tau}$ to the orthogonal complement of $\psi^m(E_\tau)$ in $E_{m\tau}$. Thus 2° follows by an argument similar to that of 1°.

□

Corollary 4 *For $\eta \in U_\tau \cap M^0(x)$, $\xi \in U_\tau \cap M^\perp(x)$, and $m \in \mathbf{N}$,*

$$\gamma_{m\tau}(\psi^m(\eta)) = m\gamma_\tau(\eta), \quad \beta_{m\tau}(\psi^m(\xi)) = m\beta_\tau(\xi), \tag{22}$$

and $\phi_{m\tau}$ induces an isomorphism on critical modules

$$(\phi_{m\tau})_* : C_q(\gamma_{m\tau} + \beta_{m\tau}, 0) \xrightarrow{\cong} C_q(f_{m\tau}, x^m). \tag{23}$$

For any $m \in \mathbf{N}$ and $\eta \in M^0(x^m)$ near the origin, using the generalized Morse lemma we define

$$V_m(\eta) = \frac{1}{m}\psi^m \circ (\psi^m)^* \circ \gamma'_{m\tau}(\eta).$$

Then V_m is a pseudo-gradient vector field of $\gamma_{m\tau}$ near the origin.

Proposition 5 *Suppose* $\nu(x,m)=\nu(x,1)$.

1° *It holds that*

$$\psi^m \circ V_k = V_{mk} \circ \psi^m, \quad \forall\, m,k \in \mathbf{N}. \tag{24}$$

2° *For* $k,m\in\mathbf{N}$, *let* $(U_{\gamma_{k\tau}}, U^-_{\gamma_{k\tau}})$ *be a Gromoll–Meyer pair of* $\gamma_{k\tau}$ *at* 0 *in* $M^0(x^k)$ *defined by the negative flow of* V_k. *Then* $(\psi^m(U_{\gamma_{k\tau}}), \psi^m(U^-_{\gamma_{k\tau}}))$ *is a Gromoll–Meyer pair of* $\gamma_{mk\tau}$ *at* 0.

3° *Further suppose* $i(x,m)=i(x,1)$. *Then* ψ^m *induces an isomorphism on FE homology modules*

$$\psi^m_*: C_{q-i_\tau(x)}(\gamma_\tau,0) \xrightarrow{\cong} C_{q-i_{m\tau}(x^m)}(\gamma_{m\tau},0). \tag{25}$$

Proof: By the assumption $\nu(x,m)=\nu(x,1)$, the map ψ^m is a linear homeomorphism from $M^0(x)$ to $M^0(x^m)$. From Corollary 4, we obtain

$$(\psi^m)^*\gamma'_{mk\tau}(\psi^m(\eta)) = m\gamma'_{k\tau}(\eta) \quad \forall \eta\in U_{\gamma_{k\tau}}, \quad \forall\, m,k\in\mathbf{N}.$$

Then for any $\eta\in U_{\gamma_{k\tau}}$,

$$\psi^m\circ V_k(\eta) = \frac{1}{mk}\psi^m[\psi^k\circ(\psi^k)^*](\psi^m)^*\circ\gamma'_{mk\tau}(\eta) = V_{mk}\circ\psi^m(\eta).$$

Thus 1° follows. Therefore we obtain that $(\psi^m(U_{\gamma_{k\tau}}), \psi^m(U_{\gamma_{k\tau}})^-)$ is a Gromoll–Meyer pair of $\gamma_{mk\tau}$ at the origin. Other claims of 2° in the definition of the Gromoll–Meyer pair in Section 1 can be proved similarly. 3° follows from 1°–2° and Lemma 2. □

Since both β_τ and $\beta_{m\tau}$ are non-degenerate quadratic forms, by a proof similar to that of Theorem I.4.1 on pages 34 to 35 of [Cha6] we obtain

Proposition 6 *Suppose* $i(x,m)=i(x,1)$. *Then the iteration map* ψ^m *induces an isomorphism on FE homology modules*

$$\psi^m_*: C_q(\beta_\tau,0)\xrightarrow{\cong} C_q(\beta_{m\tau},0) = \begin{cases}\mathbf{R}, & \text{if } q=i(x,1),\\ 0, & \text{if } q\neq i(x,1).\end{cases} \tag{26}$$

Now we are ready to study the homomorphisms among the critical modules induced by the iteration map ψ^m.

Theorem 7 *Let* $x: S_\tau\to\mathbf{R}^n$ *be an isolated* τ*-periodic solution of* (0.1) *for some* $\tau\in\mathbf{N}$ *with* $c=f_\tau(x)$. *Suppose that for some* $m\in\mathbf{N}$, *as an* $m\tau$*-periodic solution of* (0.1), *its iteration* x^m *is isolated, and that*

$$i(x,m)=i(x,1) \quad \text{and} \quad \nu(x,m)=\nu(x,1). \tag{27}$$

Then for every $q\in\mathbf{Z}$, *the iteration map* $\psi^m: E_\tau\to E_{m\tau}$ *induces an isomorphism*

$$\psi^m_*: C_q(f_\tau,x)\xrightarrow{\cong} C_q(f_{m\tau},x^m). \tag{28}$$

Here we also note that ψ_*^m *maps generators to generators in* (28). *Moreover if similar conditions also hold for k and mk* $\in$ **N**, *then*

$$\psi_*^1 = id \quad and \quad (\psi^m)_* \circ (\psi^k)_* = (\psi^{mk})_*. \tag{29}$$

Proof: Consider the diagram

$$\begin{array}{ccc} C_q(f_\tau, x) & \overset{\psi_*^m}{\longrightarrow} & C_q(f_{m\tau}, x^m) \\ \Big\downarrow (\phi_\tau^{-1})_* & & \Big\uparrow (\phi_{m\tau})_* \\ C_q(\gamma_\tau + \beta_\tau, 0) & \overset{\psi_*^m}{\longrightarrow} & C_q(\gamma_{m\tau} + \beta_{m\tau}, 0) \end{array} . \tag{30}$$

By Lemma 3 this diagram is commutative and the iteration map on the top line induces a homomorphism on critical modules. By the shifting Theorem 1.4 with its proofs in [GrM1] and page 51 of [Cha6], and our above Propositions 5 and 6, the iteration map on the bottom line of (30) induces an isomorphism. By Corollary 4, both ϕ_τ and $\phi_{m\tau}$ induce isomorphisms. Thus the conclusion holds. In this proof, we always keep the Gromoll–Meyer pair $(U_{\gamma_{m\tau}}, U_{\gamma_{m\tau}}^-)$ of $\gamma_{m\tau}$ at the origin to be defined by the negative flow of V_m for which Proposition 5 is applicable. The proof is complete. □

Remark 8 As mentioned in Section 1, we need the C^3 smoothness of L by (L2) to guarantee the map ϕ_τ in (13) of the generalized Morse lemma to be a C^1 map, and thus it is an FE map and induces an isomorphism on FE homology modules.

14.4 Global homological injectivity

In this section we fix a Lagrangian function L satisfying (L1) and (L2). For any $k \in \mathbf{N}$, our proof from the following (2) to (4) shows that the critical point set $\mathcal{K}(f_k)$ of f_k on E_k is non-empty if it is at most finite, and thus must be non-empty. Next we prove Theorem 0.2 indirectly in Sections 4 and 5 by supposing the following assumption:

(LF) *The system* (0.1) *possesses only finitely many* T^n*-distinct solution towers based on contractible integer periodic solutions in* $\mathbf{R}^n$.

Therefore under (LF) we can assume that there exist positive integers τ and p such that (0.1) possesses only finitely many $\mathbf{Z}^n$-translation independent τ-periodic solutions $\{x_1, \ldots, x_p\}$ on $\mathbf{R}^n$ and all the other $m\tau$-periodic solutions of (0.1) on $\mathbf{R}^n$ for $m \in \mathbf{N}$ are $\mathbf{Z}^n$-translations of iterations of these x_k, i.e., the critical point set $\mathcal{K}(f_{m\tau})$ of $f_{m\tau}$ in $E_{m\tau}$ has the form

$$\mathcal{K}(f_{m\tau}) = \{x_k^m + j \mid 1 \le k \le p, \ j \in \mathbf{Z}^n\}, \tag{1}$$

and the critical value set $f_{m\tau}(\mathcal{K}(f_{m\tau})) \subset \mathbf{R}$ is a finite non-empty set.

Note that $E_{m\tau}$ possesses an orthogonal decomposition

$$E_{m\tau} = \mathbf{R}^n \oplus W_{m\tau}$$

with $W_{m\tau}$ formed by functions in $E_{m\tau}$ possessing zero mean value. Using K.C. Chang's Theorem 1.6 and our (3.2), we obtain that there exist two real numbers $c_m < d_m$ depending on $f_{m\tau}$ such that

$$c_m < f_{m\tau}(x) < d_m, \quad \forall\, x \in \mathcal{K}(f_{m\tau}), \tag{2}$$

and the following isomorphisms hold:

$$H_n((f_{m\tau})_{d_m}, (f_{m\tau})_{c_m}) \cong H_n(E_{m\tau}, (f_{m\tau})_{c_m}) \cong H_n(T^n) \neq 0, \tag{3}$$

where $(f_{m\tau})_c \equiv \{z \in E_{m\tau} \mid f_{m\tau}(z) \leq c\}$. By the isolatedness of points in $\mathcal{K}(f_{m\tau})$, J.Q. Liu's Theorem 1.7 yields that at least one critical point x of $f_{m\tau}$ satisfies

$$c_m < f_{m\tau}(x) < d_m, \quad \text{and} \quad C_n(f_{m\tau}, x) \neq 0. \tag{4}$$

Together with Example 1.5, this critical point x must be a non-minimal saddle point. From (4) and the shifting Theorem 1.4, we then obtain immediately the following result.

Lemma 1 *By the assumption* (LF), *for every* $m \in \mathbf{N}$, *there exists an* $m\tau$*-periodic solution* y_m *of the system* (0.1) *such that*

$$C_n(f_{m\tau}, y_m) \neq 0 \quad \textit{and} \quad n - \nu(y_m, 1) \leq i(y_m, 1) \leq n. \tag{5}$$

Next we consider the homological injectivity of the homomorphism induced by the iteration map ψ^m under the assumption (LF).

By (LF) and (3.2), if necessary replacing the Lagrangian function L by $L + b$ with a large enough constant $b > 0$, we always assume in the following that there exists a constant $A_0 > 0$ such that for every $m \in \mathbf{N}$ and every critical point x of $f_{m\tau}$ in $E_{m\tau}$,

$$f_{m\tau}(x) \geq A_0 > 0. \tag{6}$$

Lemma 2 *Suppose* (LF) *holds. Then there exists a constant* $A_1 \geq A_0$ *such that for every* $m \in \mathbf{N}$ *and every* $m\tau$*-periodic solution* x *of* (0.1),

$$C_{n+1}(f_{m\tau}, x) = 0, \quad \textit{whenever} \quad f_{m\tau}(x) > A_1. \tag{7}$$

Proof: By (LF) we obtain (1). Note that $x_k^m + j = (x_k + j)^m$. By the shifting Theorem 1.4 and the generalized Morse Theorem 1.2, (3.2), Corollary 3.4, and the Künneth formula (cf. [Spa1]), we obtain

$$\begin{aligned} C_{n+1}(f_{m\tau}, x_k^m + j) &\cong C_{n+1-i_{m\tau}(x_k^m)}(\gamma_{m\tau}, 0) \otimes C_{i_{m\tau}(x_k^m)}(\beta_{m\tau}, 0) \\ &\cong C_{n+1-i_{m\tau}(x_k^m)}(\gamma_{m\tau}, 0) \otimes \mathbf{R} \\ &\cong C_{n+1-i_{m\tau}(x_k^m)}(\gamma_{m\tau}, 0). \end{aligned} \tag{8}$$

Note that $0 \leq \nu_{m\tau}(x_k^m + j) \leq 2n$ always holds. If $\hat{i}(x_k, 1) > 0$, then for sufficiently large $m \in \mathbf{N}$, the right-hand side of (8) must be killed because of the iteration Theorem 10.1.2. If $\hat{i}(x_k, 1) = 0$, then for any $m \in \mathbf{N}$, the right-hand side of (8) is still killed because of the iteration Theorem 10.1.2. This proves the lemma. □

By (LF), (3.2), and Lemma 1, we obtain an infinite subsequence Q of $\{2^m \mid m \in \{0\} \cup \mathbf{N}\}$, some $k \in \mathbf{N}$ and an x in $\{x_1, \ldots, x_p\}$ such that for every $m \in Q$ it holds that $C_n(f_{k\tau}, x^k) \neq 0$, $i(x, mk) = i(x, k)$, and $\nu(x, mk) = \nu(x, k)$. To simplify notation, without loss of generality, from now on we rename $k\tau$ by τ and suppose $1 \in Q$, $x \in \{x_1, \ldots, x_p\}$, and for any $m \in Q$,

$$\begin{cases} C_n(f_\tau, x) \neq 0, & \\ i(x, m) = i(x, 1), & \nu(x, m) = \nu(x, 1). \end{cases} \tag{9}$$

Thus by Theorem 3.7, the iteration map ψ^m induces isomorphisms

$$0 \neq C_n(f_\tau, x) \xrightarrow{\cong} C_n(f_{m\tau}, x^m), \qquad \forall\, m \in Q.$$

Let $c = f_\tau(x)$. By (1), $\mathcal{K}(f_{m\tau}))$ contains only finitely many points. By (3.2), the deformation Theorem 1.1, where we also refer to the idea of the proof of Theorem I.4.2 on pages 35 to 36 of [Cha6], there exists a sufficiently small $\epsilon > 0$ independent of $m \in \mathbf{N}$ such that for all $m \in \mathbf{N}$

$$f_{m\tau}(\mathcal{K}(f_{m\tau})) \cap [m(c - 3\epsilon), m(c + 3\epsilon)] = \{mc\},$$

and the inclusion map

$$\begin{aligned} h_2 \circ h_1 : (W(x^m), W(x^m)^-) &\xrightarrow{h_1} ((f_{m\tau})_{m(c+2\epsilon)}, (f_{m\tau})_{m(c-2\epsilon)}) \\ &\xrightarrow{h_2} ((f_{m\tau})_{m(c+2\epsilon)}, (f_{m\tau})^\circ_{m(c-\epsilon)}) \end{aligned} \tag{10}$$

induces a monomorphism on homology modules

$$(h_2 \circ h_1)_* : C_n(f_{m\tau}, x^m) \to H_n((f_{m\tau})_{m(c+2\epsilon)}, (f_{m\tau})^\circ_{m(c-\epsilon)}), \tag{11}$$

where B° denotes the interior of B.

If $H_{n+1}(E_{m\tau}, (f_{m\tau})_{m(c+2\epsilon)}) \neq 0$, using (LF) and J.Q. Liu's Theorem 1.7 as in the above discussions, we would obtain a critical point z of $f_{m\tau}$ such that $C_{n+1}(f_{m\tau}, z) \neq 0$. When m is sufficiently large, this violates Lemma 2. Thus there exists $m_0 > 0$ such that

$$H_{n+1}(E_{m\tau}, (f_{m\tau})_{m(c+2\epsilon)}) = 0, \quad \forall\, m \in Q(m_0). \tag{12}$$

Here and in the rest of this chapter we denote $Q(k) = \{m \in Q \mid m \geq k\}$. Then the exact sequence of FE homology modules for the triple

$$(E_{m\tau},\ (f_{m\tau})_{m(c+2\epsilon)},\ (f_{m\tau})^\circ_{m(c-\epsilon)})$$

shows that the inclusion map

$$h_3 : ((f_{m\tau})_{m(c+2\epsilon)}, (f_{m\tau,s})^{\circ}_{m(c-\epsilon)}) \to (E_{m\tau}, (f_{m\tau})^{\circ}_{m(c-\epsilon)}) \tag{13}$$

induces a monomorphism on homology modules

$$(h_3)_* : H_n((f_{m\tau})_{m(c+2\epsilon)}, (f_{m\tau})^{\circ}_{m(c-\epsilon)}) \to H_n(E_{m\tau}, (f_{m\tau})^{\circ}_{m(c-\epsilon)}). \tag{14}$$

Summarizing our discussion now, we obtain

Theorem 3 *Suppose conditions* (L1) *and* (L2) *and the assumption* (LF) *hold. Then there exist a* τ*-periodic solution of* (0.1) *with* $\tau \in \mathbf{N}$*, an integer* $m_0 > 0$*, an infinite integer set* Q *containing* 1*, and a small* $\epsilon > 0$ *such that, for* $j_{m\tau} = h_3 \circ h_2 \circ h_1$ *with* h_1 *to* h_3 *defined by* (10) *and* (13) *and all* $m \in Q(m_0) = \{k \in Q \mid k \geq m_0\}$*, we obtain the diagram*

$$\begin{aligned} 0 \neq C_n(f_\tau, x) &\xrightarrow{\psi_*^m} C_n(f_{m\tau}, x^m) \\ &\xrightarrow{(j_{m\tau})_*} H_n(E_{m\tau}, (f_{m\tau})^{\circ}_{m(c-\epsilon)}) \equiv \mathcal{H}_m, \end{aligned} \tag{15}$$

where ψ_*^m *is an isomorphism, and* $(j_{m\tau})_*$ *is a monomorphism among the FE homology modules. Especially let* $[\sigma]$ *be a generator of* $C_n(f_\tau, x)$*. Then going through the diagram* (15),

$$(j_{m\tau})_* \circ \psi_*^m[\sigma] \neq 0, \qquad \text{in} \quad \mathcal{H}_m. \tag{16}$$

14.5 Global homological vanishing

In this section we give a Lagrangian system version of V. Bangert's global deformation method introduced in [Ban1] for closed geodesics, and construct global homotopies as in the proofs of Theorems 1 to 3 of [BaK1] to show that the left-hand side of (4.16) must vanish in $\mathcal{H}_m$ for all large enough $m \in Q$. Then this yields a contradiction to Theorem 4.3 and proves our Theorem 0.2. We carry out the details of this proof in three steps.

Step 1 For a given topological space V, let $\sigma_i : [a_i, b_i] \to V$, for $i = 1$ and 2, be two paths defined on bounded intervals with $b_1 = a_2$ and $\sigma_1(b_1) = \sigma_2(a_2)$. As usual we define new paths $\sigma_1^{-1} : [a_1, b_1] \to V$ and $\sigma_2 * \sigma_1 : [a_1, b_2] \to V$ by

$$\sigma_1^{-1}(s) = \sigma(b_1 + a_1 - s), \qquad \text{for } a_1 \leq s \leq b_1,$$

$$\sigma_2 * \sigma_1(s) = \begin{cases} \sigma_1(s), & \text{for } a_1 \leq s \leq b_1, \\ \sigma_2(s), & \text{for } a_2 \leq s \leq b_2. \end{cases}$$

Proposition 1 *For* $\tau \in \mathbf{N}$*,* $c \in \mathbf{R}$*,* $\epsilon > 0$*, and* $q \geq 0$*, let*

$$\eta : (\Delta_q, \partial\Delta_q) \to (E_\tau, (f_\tau)^{\circ}_{c-\epsilon}) \tag{1}$$

be an FE q-simplex. Then there exists an integer $m(\eta) > 0$ *such that for every integer* $m \geq m(\eta)$*, the q-simplex*

$$\eta^m \equiv \psi^m(\eta) : (\Delta_q, \partial\Delta_q) \to (E_{m\tau}, (f_{m\tau})^{\circ}_{m(c-\epsilon)}) \tag{2}$$

is FE homotopic to an FE singular q-simplex η_m *satisfying*

$$\eta_m : (\Delta_q, \partial\Delta_q) \to ((f_{m\tau})^{\circ}_{m(c-\epsilon)}, (f_{m\tau})^{\circ}_{m(c-\epsilon)}), \tag{3}$$

$$\eta^m = \eta_m \qquad \text{on} \quad \partial\Delta_q, \tag{4}$$

and the homotopy fixes $\eta^m|_{\partial\Delta_q}$.

Proof: We carry out the proof in two steps.

(A) *The definition of a new path* $\alpha_m : [a, b] \to E_{m\tau}$ *with* $\alpha_m(a) = \psi^m(\alpha(a))$ *and* $\alpha_m(b) = \psi^m(\alpha(b))$ *for a given FE-path* $\alpha : [a, b] \to E_\tau$.

Fix the integer $m \geq 3$. Define the initial point curve β_α of α by

$$\beta_\alpha(t) = \alpha(t)(0), \qquad \text{for } a \leq t \leq b. \tag{5}$$

For $0 \leq s \leq (b-a)/m$ and $1 \leq k \leq m-2$, following [Ban1] we define

$$\begin{aligned}
\alpha_m(a+s) &= (\beta_\alpha|_{[a,a+ms]})^{-1} * \alpha(a+ms) * (\beta_\alpha|_{[a,a+ms]}) * \alpha(a)^{m-1},\\
\alpha_m(a+k(b-a)/m+s) &= (\beta_\alpha)^{-1} * \alpha(b)^k * (\beta_\alpha|_{[a+ms,b]})\\
&\quad * \alpha(a+ms) * (\beta_\alpha|_{[a,a+ms]}) * \alpha(a)^{m-k-1},\\
\alpha_m(b-(b-a)/m+s) &= (\beta_\alpha|_{[a+ms,b]})^{-1} * \alpha(b)^{m-1}\\
&\quad * (\beta_\alpha|_{[a+ms,b]}) * \alpha(a+ms).
\end{aligned}$$

For each $r \in [a, b]$, we reparametrize the loop $\alpha_m(r)$ as follows: in the above formula, each fixed part $\alpha(a)$ or $\alpha(b)$ spends the original time τ, the moving part $\alpha(a+ms)$ and the pieces of the initial point curve β_α share the time τ proportionally to their original parametrizations. This defines a new path $\alpha_m : [a, b] \to E_{m\tau}$ with $\alpha_m(a) = \psi^m(\alpha(a)) = \alpha(a)^m$ and $\alpha_m(b) = \psi^m(\alpha(b)) = \alpha(b)^m$.

Let $\beta_{\alpha,m} : [a, b] \to E_\tau$ be the path defined via replacing all the terms of powers of $\alpha(a)$ and $\alpha(b)$ by the identity matrix in the formula (5) of α_m. Let

$$\begin{cases}
M_0(\alpha) = \max\{f_\tau(\alpha(a)), f_\tau(\alpha(b))\},\\
M_1(\alpha) = \max_{a \leq s \leq b} |f_\tau(\alpha(s))|,\\
M_2(\alpha) = \int_a^b |\mathcal{F}(\beta_\alpha)(t))| dt,
\end{cases} \tag{6}$$

where

$$\mathcal{F}(\beta_\alpha)(t) = \frac{1}{2}A(\beta_\alpha(t))\dot{\beta}_\alpha(t) \cdot \dot{\beta}_\alpha(t) + V(t, \beta_\alpha(t))$$

for t belonging to the domain of β_α. Clearly $M_0(\alpha)$ and $M_1(\alpha)$ are finite real numbers. Since α is an FE path in the loop space E_τ, by Lemma 2.3 and the continuity of α we obtain $M_2(\alpha)$ is also finite and depends only on α.

Then by direct computations for $a \le s \le b$ and $k = [m(s-a)/(b-a)]$, we obtain

$$f_{m\tau}(\alpha_m(s)) = (m-k-1)f_\tau(\alpha(a)) + kf_\tau(\alpha(b)) + f_\tau(\beta_{\alpha,m}(s)).$$

Then

$$f_{m\tau}(\alpha_m(s)) \le (m-1)M_0(\alpha) + M_1(\alpha) + 2M_2(\alpha), \quad \forall s \in [a,b], \tag{7}$$

$$\limsup_{m\to+\infty} \max_{a\le s\le b} \frac{1}{m} f_{m\tau}(\alpha_m(s)) = M_0(\alpha). \tag{8}$$

(B) *The definition of a new FE simplex $\eta_m : \Delta_q \to E_\tau$ and the homotopy from η^m to η_m for the given η.*

As in Section 3, for $q \ge 0$ and $t, s \in [0,1]$ denote $e(t,s) = (t, \dots, t, s) \in \mathbf{R}^q \times [0,1]$. Then the point $\hat{e}(s) = e(1/(q+1), s)$ is the barycenter of $\Delta_q \times \{s\} \subset \mathbf{R}^q \times \{s\}$. Define

$$\Delta_q(s) = e((1-s)/(q+1), s) + s\Delta_q \subset \Delta_q, \quad \text{for } 0 \le s \le 1.$$

Denote by $L(s)$ the straight line passing through $e(0,s)$ and $\hat{e}(s)$ successively in $\mathbf{R}^q \times \{s\}$. According to $L(s)$, we obtain a new orthogonal subspace decomposition of $\mathbf{R}^q \times \{s\}$ as $V_{q-1,s} \times L(s)$. Thus any $w \in \Delta_q \times \{s\}$ can be uniquely written as $w = (v,t) \in V_{q-1,s} \times L(s)$. For such a $w = (v,t) \in \Delta_q(s)$, denote by $l(v,s)$ the intersection segment of $\Delta_q(s)$ with the straight line passing through w and parallel to $L(s)$. This segment $l(v,s)$ is well defined, since it is independent of the particular choice of the t-component in $w = (v,t) \in \Delta_q(s)$.

For any $s \in [0,1]$ and $\phi : \Delta_q(s) \to E_\tau$, similar to (2.2) we define a v-parametrized curve ${}^{\#}\phi_v : l(v,s) \to E_\tau$ for each $w = (v,t) \in [V_{q-1,s} \times L(s)] \cap \Delta_q(s)$ by

$${}^{\#}\phi_v(t) = \phi(v,t), \qquad \forall\, t \in l(v,s). \tag{9}$$

Now for the given FE simplex map $\eta : (\Delta_q, \partial\Delta_q) \to (E_\tau, (f_\tau)^\circ_{c-\epsilon})$, its m-th iteration is given by $\eta^m : (\Delta_q, \partial\Delta_q) \to (E_{m\tau}, (f_{m\tau})^\circ_{m(c-\epsilon)})$. Since η is an FE q-simplex in the loop space E_τ, by Lemma 2.3 we can apply the above Step A to the initial value curve $\eta_v^{ini} : l(v,s) \to \mathbf{R}^{2n}$ of each ${}^{\#}\eta_v : l(v,s) \to E_\tau$. Thus using (9) we can define

$$\eta_m(w) = ({}^{\#}\eta_v)_m(t), \quad \forall\, w \in \Delta_q, \tag{10}$$

where (v, t) is the unique $V_{q-1,1} \times L(1)$ decomposition of $(w, 1) \in \Delta_q(1) = I_q \times \{1\}$. For any $(w, s) \in \Delta_q \times [0, 1]$, writing $(w, s) = (v, t, s) \in [V_{q-1,s} \times L(s)] \times [0, 1]$, since η is FE, using Lemma 2.3 we define

$$h(w, s) = \begin{cases} \eta^m(w), & \text{if } (w, s) \in (\Delta_q \times \{s\}) \backslash \Delta_q(s), \\ ({}^{\#}\eta_v)_m(t), & \text{if } (w, s) = (v, t, s) \in \Delta_q(s). \end{cases} \tag{11}$$

Then by this definition, h is an FE continuous map, and satisfies $h(\cdot, 0) = \eta^m$, $h(\cdot, 1) = \eta_m$, and $h(w, s) = \eta^m(w)$ whenever $w \in \partial\Delta_q$ and $0 \le s \le 1$. Thus h is an FE-homotopy from η^m to η_m which fixes $\eta^m|_{\partial\Delta_q}$. The proof is complete. □

Step 2 For any FE singular cycle σ, let $\Sigma(\sigma)$ be the set of all simplexes together with all their faces contained in the singular chain σ. Then $\Sigma(\sigma)$ is a finite set.

Proposition 2 *Suppose L satisfies* (L1) *and* (L2). *Suppose* (4.15) *holds via the composition $(j_{m\tau})_* \circ \psi_*^m$ of the homomorphism $(j_{m\tau})_*$ and the isomorphism ψ_*^m for some τ-periodic solution x of* (0.1) *on $\mathbf{R}^n$ with $\tau \in \mathbf{N}$ obtained via* (4.3) *to* (4.4), *all $m \ge m_1$ with some integer $m_1 > 0$, and some $\epsilon > 0$. Let $\sigma \in C_n(f_\tau, x)$ be any FE singular chain which is mapped into $\mathcal{H}_m$ in* (4.15) *for all $m \ge m_1$. Then there exists a sufficiently large integer $m(\sigma) \ge m_1$ such that for every integer $m \in Q(m(\sigma))$ and for every $\mu \in \Sigma(\sigma)$ with $\mu : \Delta_r \to E_\tau$ and $0 \le r \le n$, there exists a homotopy $P(\mu^m) : \Delta_r \times [0, 1] \to E_{m\tau,t}$ such that the properties* (i) *to* (iv) *in Bangert and Klingenberg's Theorem* 1.8 *hold for $(X, A) = (E_{m\tau}, (f_{m\tau})^\circ_{m(c-\epsilon)})$.*

Proof: For any integer $p \ge 0$, define $\Sigma_p(\sigma) = \{\mu \in \Sigma(\sigma) \mid \dim \mu \le p\}$. By induction on p we will determine the integer $m(\sigma)$ and construct the mentioned homotopy $P(\mu^m)$ for $m \in Q(m(\sigma))$ and each $\mu \in \Sigma_p(\sigma)$.

When $p = 0$, by our discussion following (4.4), x^m in (4.15) as a critical point of $f_{m\tau}$ is a non-minimum saddle point. Thus we can connect every $\mu \in \Sigma_0(\sigma)$ to a point in $(f_{m\tau})^\circ_{m(c-\epsilon)}$ by smooth paths in $E_{m\tau}$. Using such paths the homotopy $P(\mu^m)$ can be defined for all $m \ge m_1$. We set $m_0(\sigma) = m_1$.

When $p \ge 1$, by induction, there exists an integer $m_{p-1}(\sigma) \ge 1$ such that the homotopy $P(\mu^m)$ satisfying the properties (i) to (iv) in Theorem 1.8 is defined for every $m \in Q(m_{p-1}(\sigma))$ and every $\mu \in \Sigma_{p-1}(\sigma)$. By reparametrization we may further assume

$$P(\mu^m)(\cdot, s) = P(\mu^m)\left(\cdot, \frac{1}{2}\right), \qquad \forall s \in \left[\frac{1}{2}, 1\right],$$
$$m \in Q(m_{p-1}(\sigma)), \ \mu \in \Sigma_{p-1}(\sigma).$$

Let $q = m_{p-1}(\sigma)$. Fix $\mu \in \Sigma_p(\sigma)$ with $\mu^q(\Delta_p) \not\subset (f_{q\tau})^\circ_{q(c-\epsilon)}$. By the property (iv), the homotopies $P(\mu^q \circ e_p^i)$, $0 \le i \le p$ induce a map $R'_{q,\mu} : \partial\Delta_p \times [0, 1] \to E_{q\tau}$ such that $R'_{q,\mu} \circ (e_p^i \times id) = P(\mu^q \circ e_p^i)$. Especially it holds that $R'_{q,\mu}(w, 0) = \mu^q(w)$ and $R'_{q,\mu}(w, \frac{1}{2}) \in (f_{q\tau})^\circ_{q(c-\epsilon)}$ for all $w \in \partial\Delta_p$. Let $R_{q,\mu} : \Delta_p \times [0, 1] \to E_{q\tau}$ be the homotopy extension of the map $R'_{q,\mu}$ such that $R_{q,\mu}(\cdot, 0) = \mu^q$. Then

$$R_{q,\mu}(\cdot, 1/2) : (\Delta_p, \partial\Delta_p) \to (E_{q\tau}, (f_{q\tau})^\circ_{q(c-\epsilon)}).$$

By Proposition 1, there exists a sufficiently large integer $\hat{m}_p(\mu) \in Q(q)$ for the above fixed μ such that for every integer $k \in Q(\hat{m}_p(\mu))$ there exists a homotopy $F^k : \Delta_p \times [\frac{1}{2}, 1] \to E_{kq\tau}$ with $F_s^k = h^k(\cdot, s)$ for $s \in [\frac{1}{2}, 1]$ such that

$$\begin{aligned} F^k_{\frac{1}{2}} &= \psi^k \circ R_{q,\mu}\left(\cdot, \frac{1}{2}\right), \\ F^k_s &= \psi^k \circ R_{q,\mu}(\cdot, s), \quad \text{for} \quad \frac{1}{2} \le s \le 1, \quad \text{on} \quad \partial\Delta_p, \\ F^k_1 &= (R_{q,\mu}(\cdot, 1))_k : \Delta_p \to (f_{q\tau k})^{\circ}_{qk(c-\epsilon)}. \end{aligned}$$

Since $\Sigma_p(\sigma)$ is a finite set, let $m_p(\sigma) = \max\{\hat{m}_p(\mu)^2 \mid \mu \in \Sigma_p(\sigma)\}$. Then for every $m \in Q(m_p(\sigma))$, we define the homotopy functor $P(\mu^m)$ for every $\mu \in \Sigma_p(\sigma)$ as follows:

1° If $\dim \mu < p$, then $Q(m_p(\mu)) \subset Q(m_{p-1}(\mu))$ and $P(\mu^m)$ is already defined.

2° If $\dim \mu = p$ and $\mu(\Delta_p) \subset (f_\tau)^{\circ}_{c-\epsilon}$, define $P(\mu^m)(\cdot, s) = \mu^m(\cdot)$ for $0 \le s \le 1$.

3° If $\dim \mu = p$ and $\mu(\Delta_p) \not\subset (f_\tau)^{\circ}_{c-\epsilon}$, by the induction assumption, and the definition of $m \in Q(m_p(\sigma))$, there exists a unique integer $k \in \mathbf{N}$ such that $m = km_{p-1}(\sigma)$. Then $k \ge \hat{m}_p(\mu)$ by the definition of $m_p(\sigma)$. We define

$$P(\mu^m)(\cdot, s) = \begin{cases} \psi^k \circ R_{q,\mu}(\cdot, s), & 0 \le s \le \frac{1}{2}, \\ F^k_s(\cdot), & \frac{1}{2} \le s \le 1. \end{cases}$$

Now one can verify that the map $P(\mu^m)$ satisfies the properties (i) to (iv) of Theorem 1.8.

By the finiteness of the set $\Sigma(\sigma)$ and induction on p, we can define $m(\sigma) \equiv m_n(\sigma)$. Then by induction, the homotopy $P(\mu^m)$ is defined for all $\mu \in \Sigma(\sigma)$ and $m \in Q(m_n(\sigma))$ and possesses properties (i) to (iv) in Theorem 1.8 as claimed. This completes the proof. □

Step 3 Now we can give

Proof of Theorem 0.2: Suppose the Lagrangian function L satisfies the conditions (L1) and (L2), and the assumption (LF) holds. We obtain a τ-periodic solution x of (0.1) on $\mathbf{R}^n$ for some $\tau \in \mathbf{N}$, $m_0 > 0$, a small $\epsilon > 0$, and a singular cycle $[\sigma] \ne 0$ in $C_n(f_\tau, x)$ such that (4.15) and (4.16) in Theorem 4.3 hold for all $m \in Q(m_0)$. But on the other hand, we can take a sufficiently great integer $m(\sigma) \ge m_0$ such that Proposition 2 holds for all $m \in Q(m(\sigma))$. Thus by Proposition 2, the same homomorphism in (4.16) sends $[\sigma]$ to the zero element of the module $\mathcal{H}_m$ in (4.15) for all $m \in Q(m(\sigma))$. This contradicts (4.16), and then (LF) can not hold. The proof is complete.

14.6 Notes and comments

The study of integer periodic solutions of the system (0.1) under the conditions (L1) and (L2) possesses a very long history. The most standard model is the multiple pendulum type system with periodic forcing terms (cf. [CLZ1]). We refer to [Rab2] and [FoW1] for further references. Note that in Theorems 1 and 2 we have no non-degeneracy restriction on any solutions of (1). This is rather different from the known results for Lagrangian system (1) as mentioned in [FoW1].

Our Theorem 0.1 can be viewed as a confirming answer to the Lagrangian system analogue of C. Conley's conjecture on periodic points of Hamiltonian maps mentioned on pages 1304 to 1305 of [SaZ2]:

C. Conley's conjecture: On the torus T^{2n} with standard symplectic form, any Hamiltonian diffeomorphism (i.e., the time 1-map defined by the flow of some 1-periodic Hamiltonian system) possesses infinitely many periodic points which are produced by contractible 1-periodic solutions of that Hamiltonian system.

So far this conjecture is still open. Partial results on this conjecture can be found in [CoZ3], [CoZ4], [LZe1], [SaZ1], and [SaZ2]. Among them the following significant progress was made by D. Salamon and E. Zehnder in [SaZ2] of 1992:

Theorem 1 (Theorem A of [SaZ2]). *Suppose the symplectic form ω and the first Chern class c_1 vanish on $\pi_2(M)$ of a compact symplectic manifold (M, ω). Then any Hamiltonian diffeomorphism on (M, ω) possesses infinitely many periodic points which are produced by contractible 1-periodic solutions of that Hamiltonian system, provided every contractible Hamiltonian solution has at least one Floquet multiplier not equal to* 1.

Here I would also like to remind readers of the following related conjecture of H. Hofer and E. Zehnder proposed on Page 263 of their celebrated book [HoZ3] in 1994: *Every Hamiltonian diffeomorphism on a compact symplectic manifold (M, ω) possessing more fixed points than necessarily required by the V. Arnold conjecture possesses always infinitely many periodic points.*

Note that the famous V. Arnold conjecture claims that *the number of fixed points of every Hamiltonian diffeomorphism on a compact symplectic manifold (M, ω) is at least cuplength$(M) + 1$. This number is at least the Betti number of M, provided all the fixed points are non-degenerate.*

In 1978, Y. Eliashberg proved this conjecture for Riemannian surfaces. In 1983, C. Conley and E. Zehnder proved the Arnold conjecture for T^{2n} in their celebrated paper [CoZ1]. After that, a great number of research works have been published on this conjecture, including the great contributions [Gro1] of M. Gromov and [Flo2] of A. Floer. Other contributions can be found in [For1], [Hof3], [HoS1], [Mar1], [RuT1]. For more details on these contributions we refer to the excellent books [HoZ3] of

H. Hofer and E. Zehnder in 1994, [McS1] and [McS2] of D. McDuff and D. Salamon in 1994, 1996, and 1998. Note that in the recent celebrated research works [FuO1] of K. Fukaya and K. Ono, [LiT1] of G. Liu and G. Tian, [Rua2] of Y. Ruan, the non-degenerate Arnold conjecture was completely proved. But the degenerate Arnold conjecture is still open for general compact symplectic manifolds. Further results on this degenerate conjecture can be found in [LeO1], [Sch2] and the references therein.

Chapter 15

Closed characteristics on convex hypersurfaces

Let Σ be a C^2-compact hypersurface in $\mathbf{R}^{2n}$, bounding a strictly convex compact set C with non-empty interior, and with a non-vanishing Gaussian curvature. Without loss of generality, we assume $0 \in C$. We denote the set of all such hypersurfaces in $\mathbf{R}^{2n}$ by $\mathcal{H}(2n)$. We denote by $\mathcal{H}_s(2n)$ the set of elements in $\mathcal{H}(2n)$ which are symmetric with respect to the origin, i.e., $x \in \Sigma$ implies $-x \in \Sigma$.

For $x \in \Sigma$, let $N_\Sigma(x)$ be the outward unit normal vector at x of Σ. We consider the dynamics problem of finding $\tau > 0$ and a C^1 curve $x : [0, \tau] \to \mathbf{R}^{2n}$ such that

$$\begin{cases} \dot{x}(t) = JN_\Sigma(x(t)), \quad x(t) \in \Sigma, \quad \forall t \in \mathbf{R}, \\ x(\tau) = x(0), \end{cases} \tag{1}$$

where J is the standard symplectic matrix on $\mathbf{R}^{2n}$ given in Section 1.1. A solution (τ, x) of the problem (1) is called a **closed characteristic** on Σ. Two closed characteristics (τ, x) and (σ, y) are **geometrically distinct**, if $x(\mathbf{R}) \neq y(\mathbf{R})$. We denote by $\mathcal{J}(\Sigma)$ and $\tilde{\mathcal{J}}(\Sigma)$ the set of all closed characteristics (τ, x) on Σ with τ being the minimal period of x and the set of all geometrically distinct ones respectively. For $(\tau, x) \in \mathcal{J}(\Sigma)$, we denote by $[(\tau, x)]$ the set of all elements in $\mathcal{J}(\Sigma)$ which are geometrically the same as (τ, x). ${}^{\#}A$ denotes the total number of elements in a set A.

To cast the given energy problem (1) into a Hamiltonian version, we follow Section V.3 of I. Ekeland's celebrated book [Eke3]. Fix a $\Sigma \in \mathcal{H}(2n)$ bounding a convex set C. Then the origin is in the interior of C. Let $j_\Sigma : \mathbf{R}^{2n} \to [0, +\infty)$ be the gauge function of Σ defined by

$$\begin{cases} j_\Sigma(x) = \inf\{\lambda > 0 \mid \frac{x}{\lambda} \in C\}, \quad \text{for} \quad x \neq 0, \\ j_\Sigma(0) = 0. \end{cases} \tag{2}$$

Fix a constant α satisfying $1 < \alpha < 2$ in this chapter. As usual we define the Hamiltonian function $H_\alpha : \mathbf{R}^{2n} \to [0, +\infty)$ by

$$H_\alpha(x) = j_\Sigma(x)^\alpha, \qquad \forall x \in \mathbf{R}^{2n}. \tag{3}$$

Then $H_\alpha \in C^1(\mathbf{R}^{2n}, \mathbf{R}) \cap C^2(\mathbf{R}^{2n} \setminus \{0\}, \mathbf{R})$ is convex and $\Sigma = H_\alpha^{-1}(1)$. The problem (1) is equivalent to the following given energy problem of the Hamiltonian system:

$$\dot{x}(t) = JH'_\alpha(x(t)), \quad \forall t \in \mathbf{R}, \tag{4}$$

$$x(\tau) = x(0), \tag{5}$$

$$H_\alpha(x(t)) = 1, \quad \forall t \in \mathbf{R}. \tag{6}$$

Denote by $\mathcal{J}(\Sigma, \alpha)$ the set of all solutions (τ, x) of the problem (4)–(6) where τ is the minimal period of x. Note that elements in $\mathcal{J}(\Sigma)$ and $\mathcal{J}(\Sigma, \alpha)$ are in one to one correspondence with each other.

Let $(\tau, x) \in \mathcal{J}(\Sigma, \alpha)$. We call the fundamental solution $\gamma_x : [0, \tau] \to \mathrm{Sp}(2n)$ with $\gamma_x(0) = I$ of the linearized Hamiltonian system

$$\dot{y}(t) = JH_\alpha''(x(t))y(t), \quad \forall t \in \mathbf{R}, \tag{7}$$

the **associated symplectic path** of (τ, x). The eigenvalues of $\gamma_x(\tau)$ are called **Floquet multipliers** of (τ, x). By Proposition I.6.13 of [Eke3], the Floquet multipliers with their multiplicity and Krein type numbers of $(\tau, x) \in \mathcal{J}(\Sigma)$ do not depend on the particular choice of the Hamiltonian function in (4). As usual a $(\tau, x) \in \mathcal{J}(\Sigma)$ is **elliptic, hyperbolic**, or **non-degenerate**, if so is $\gamma_x(\tau)$ as in Definitions 1.8.1 and 5.0.1 with $\omega = 1$. It is well known that these concepts are independent of the choice of $\alpha > 1$.

Note that $(\tau, x) \in \mathcal{J}(\Sigma, \alpha)$ is called **linearly stable**, if there exists a constant $b > 0$ such that

$$\|M^m\|_{\mathcal{L}(\mathbf{R}^{2n-2})} \le b, \quad \forall m \in \mathbf{N},$$

for any M given by the decomposition (2.2.2) below of $\gamma_x(\tau)$. By this definition, that $(\tau, x) \in \mathcal{J}(\Sigma, \alpha)$ is linearly stable implies that it is elliptic.

In this chapter we study the existence, multiplicity, and stability of closed characteristics on any $\Sigma \in \mathcal{H}(2n)$. The main results of this chapter are Theorems C and D on the existence and multiplicity, and the stability given in the Introduction of this book.

Note that a typical example of $\Sigma \in \mathcal{H}_s(2n)$ is the ellipsoid $\mathcal{E}_n(r)$ defined as follows. Let $r = (r_1, \ldots, r_n)$ with $r_k > 0$ for $1 \le k \le n$. Define

$$\mathcal{E}_n(r) = \left\{ x = (x_1, \ldots, x_n) \in \mathbf{R}^{2n} \mid \frac{1}{2} \sum_{k=1}^{n} \frac{|x_k|^2}{r_k^2} = 1 \right\}. \tag{8}$$

If r_j/r_k is irrational whenever $j \ne k$, this $\mathcal{E}_n(r)$ is called a **weakly non-resonant ellipsoid**. In this case, the corresponding Hamiltonian system is linear, and all the solutions can be computed explicitly. Thus it is easy to verify that ${}^{\#}\tilde{\mathcal{J}}(\mathcal{E}_n(r)) = n$, that all the closed characteristics on $\mathcal{E}_n(r)$ are elliptic and non-degenerate, and that all of them have irrational mean indices greater than 2 (cf. §I.7 of [Eke3]).

For the reader's convenience, next we briefly describe the main ideas in the proofs of the main results in this section, i.e., Theorems C and D in the Introduction of this book.

A Ideas of the proof of the multiplicity results (0.0.16)–(0.0.18) in Theorem C

Fix a $\Sigma \in \mathcal{H}(2n)$. Our study is carried out in the following four steps.

⟨1⟩ Using the gauge function $j_\Sigma(x)$ of Σ, the problem of finding closed characteristics on Σ is transformed to the above given energy problem (4)–(6) with Hamiltonian function H_α for some $\alpha \in (1, 2)$. Then this given energy problem is further transformed to a given period problem (2.1) below of finding 1-periodic solutions of the Hamiltonian system (4). By the dual action principle of F. Clarke and I. Ekeland (cf. Section V.3 of [Eke3]), solutions of the 1-periodic boundary value problem correspond to critical points of the dual action functional $f_\alpha : E_\alpha \to \mathbf{R}$ defined by (2.3) below, and f_α possesses an increasing infinite sequence of negative critical values $\{c_k\}$ with $c_k \to 0$ as $k \to +\infty$. These c_k are obtained by the Fadell–Rabinowitz S^1-cohomological index method as in Section V.3 of [Eke3]. Note that here each infinite variationally visible (see Definition 3.3 below) $[(\tau, x)] \in \tilde{\mathcal{J}}(\Sigma)$ and all of its iterations $\{x^m\}$ yield a strictly increasing infinite subsequence of $\{c_k\}$, and the corresponding critical points of f_α are given by $\{u_m^x\}_{m\in\mathbf{N}}$ of (2.10) below.

Denote the index interval of the m-th iteration of (τ, x) by

$$\mathcal{I}_m(\tau, x) = [i(x, m), i(x, m) + \nu(x, m) - 1].$$

An important result of I. Ekeland and H. Hofer (cf. [EkH2]) is that for every $k \in \mathbf{N}$, there exists a $[(\tau, x)] \in \tilde{\mathcal{J}}(\Sigma)$ and an $m \in \mathbf{N}$ such that

$$f'_\alpha(u_m^x) = 0, \qquad f_\alpha(u_m^x) = c_k, \tag{9}$$

$$2k - 2 + n \in \mathcal{I}_m(\tau, x). \tag{10}$$

Especially this implies

$$2\mathbf{N} - 2 + n \subset \bigcup_{[(\tau,x)]\in\tilde{\mathcal{J}}(\Sigma)} \bigcup_{m\in\mathbf{N}} \mathcal{I}_m(\tau, x). \tag{11}$$

Thus instead of studying $^\#\tilde{\mathcal{J}}(\Sigma)$ from corresponding subsequences in the whole sequence $\{c_k\}$, we study how their index interval sequences cover the set $2\mathbf{N} - 2 + n$.

⟨2⟩ Suppose $\tilde{\mathcal{J}}(\Sigma) < +\infty$ and denote

$$\tilde{\mathcal{J}}(\Sigma) = \{[(\tau_j, x_j)] \mid j = 1, \ldots, q\}. \tag{12}$$

The precise iteration formulae established in Chapter 8 allow us to classify the iterated index sequences into 8 different iteration patterns as in [Lon18] for symplectic paths in Sp(4) as fundamental solutions of closed characteristics on Σ when $n = 2$. Indices of iterations of paths in Sp(6) are also classified and studied similarly. The most important phenomenon we observed in these studies is that there are infinitely many chances of iteration times $m_1, \ldots, m_p$ and subsets $\mathcal{E}$ of $2\mathbf{N} - 2 + n$ such that $\cup_{j=1}^q \mathcal{I}_{m_j}(\tau_j, x_j)$ covers $\mathcal{E}$ and $^\#\mathcal{E}$ yields a lower bound for the number of symplectic paths under study.

In order to effectively catch the set $\mathcal{E}$s, we enlarge the closed index interval $\mathcal{I}_{m+1}(\tau, x)$ to a bigger open interval

$$\mathcal{G}_m(\tau, x) = (i(x, m) + \nu(x, m) - 1, i(x, m + 2)),$$

which we call the index jump of the m-th iteration of (τ, x). Here by our study of iterated indices, the closed index interval $\mathcal{I}_m(\tau, x)$ of (τ, x) are mutually disjoint and strictly increasing (cf. Lemma 3.2 below). Thus

$$\mathcal{I}_{m+1}(\tau, x) \subset \mathcal{G}_m(\tau, x), \tag{13}$$

$$\mathcal{I}_j(\tau, x) \cap \mathcal{G}_m(\tau, x) = \emptyset, \quad \forall j \neq m+1, \tag{14}$$

and $\mathcal{G}_m(\tau, x)$ is the largest interval possessing these two properties. Then we observe that there exist infinitely many $\{m_1, \dots, m_q\}$ and non-empty interval $\mathcal{I}$ such that

$$\mathcal{I} \subset \bigcap_{j=1}^{q} \mathcal{G}_{2m_j-1}(\tau_j, x_j). \tag{15}$$

Thus the set $\mathcal{E}$ is captured by $\mathcal{G}_{2m_j-1}(\tau_j, x_j)$ in the following way:

$$\begin{aligned} &(2\mathbf{N} - 2 + n) \cap \bigcap_{j=1}^{q} \mathcal{G}_{2m_j-1}(\tau_j, x_j) \\ &\supset (2\mathbf{N} - 2 + n) \cap \mathcal{I} \equiv \mathcal{E}. \end{aligned} \tag{16}$$

Now we noticed that the following crucial estimate holds:

$$q \geq {}^{\#}\mathcal{E} \equiv p. \tag{17}$$

In fact, assume $p > q$. Denote $\mathcal{E} = \{2k - 2 + n \mid k_0 \leq k \leq k_0 + p - 1\}$ for some $k_0 \in \mathbf{N}$. Then by (9), (10), (13), (14), and (16), there are critical points $u_{2m_j}^{x_j}$ of f_α such that

$$c_k = f_\alpha(u_{2m_j}^{x_j}), \qquad \forall k_0 \leq k \leq k_0 + p - 1, \ 1 \leq j \leq q. \tag{18}$$

Because $p > q$, there exist at least two $k \neq l$ in $\{k_0, \dots, k_0 + p - 1\}$ and one $j \in \{1, \dots, q\}$ such that

$$c_k = f_\alpha(u_{2m_j}^{x_j}) = c_l. \tag{19}$$

Then by a Ljusternik–Schnirelman type argument of Ekeland et al. in terms of the Fadell–Rabinowitz index (cf. (3.3) below), (19) implies ${}^{\#}\tilde{\mathcal{J}}(\Sigma) = +\infty$. This contradiction proves the estimate (17).

In Section 3 below, (17) is realized by showing the existence of an injection map Ψ : $k \mapsto ([(\tau_j, x_j)], m)$ defined by Lemma 3.5 below, where $([(\tau_j, x_j)], m)$ corresponds to the critical point $u_m^{x_j}$ of f_α such that (9) and (10) hold, and Ψ induces an injection map $\tilde{\Psi}$ from $(2\mathbf{N} - 2 + n) \cap \bigcap_{j=1}^{q} \mathcal{G}_{2m_j-1}(\tau_j, x_j)$ to $\tilde{\mathcal{J}}(\Sigma)$.

⟨3⟩ Because of (17), precise values of the iterated indices appearing in (16) become very crucial. To further our study, we need to find precisely the iteration time m_j and the interval $\mathcal{I}$ in (15) simultaneously and to make ${}^{\#}\mathcal{E}$ as large as possible. This is realized by applying the common index jump Theorems 11.2.1 and 11.2.2, and Corollary 11.2.4 to associated symplectic paths γ_{x_j} of (τ_j, x_j). By this method, we obtain an integer N and iteration times $m_1, \ldots, m_q$ such that (15) holds with $\mathcal{I} = [2N - \kappa_1, 2N + \kappa_2]$ for some constants κ_1 and $\kappa_2 > 0$ depending only on Σ.

⟨4⟩ To further estimate ${}^{\#}\mathcal{E}$ in (17), we define a new invariant $\varrho_n(\Sigma)$ of $\Sigma \in \mathcal{H}(2n)$ by the initial indices $(i(x_j, 1), \nu(x_j, 1))$ and splitting numbers $S^+(x_j)$ for $1 \le j \le q$ (cf. Definition 4.1 below). By our understanding of these indices and splitting numbers for closed characteristics, we further obtain:

$$\begin{aligned} {}^{\#}\mathcal{E} &\ge \varrho_n(\Sigma) \\ &\ge \left[\frac{n}{2}\right] + 1. \end{aligned} \tag{20}$$

This proves (0.0.16) and (0.0.17) of Theorem C in the Introduction. When every (τ_j, x_j) is non-degenerate, by the definition of $\varrho_n(\Sigma)$ and the first inequality in (20), we obtain (0.0.18), i.e., $\varrho_n(\Sigma) \ge n$.

Ideas in this proof are illustrated in Example 3.7 below for three closed characteristics by $\mathcal{I} = [33, 38]$ and $\mathcal{E} = \{33, 35, 37\}$.

B Ideas of the proof of the multiplicity result (0.0.19) in Theorem C for symmetric hypersurfaces

Fix a $\Sigma \in \mathcal{H}_s(2n)$ and suppose ${}^{\#}\tilde{\mathcal{J}}(\Sigma) < +\infty$. The main idea of our proof is the following.

⟨1⟩ Let $\mathcal{O}(x) = x(\mathbf{R})$ for any closed characteristic $(\tau, x) \in \mathcal{J}(\Sigma)$. Lemma 6.1 below shows that every $(\tau, x) \in \mathcal{J}(\Sigma)$ is either **symmetric**, i.e., $\mathcal{O}(x) = \mathcal{O}(-x)$, or **asymmetric**, i.e., $\mathcal{O}(x) \cap \mathcal{O}(-x) = \emptyset$. Moreover, any symmetric (τ, x) satisfies

$$x\left(t + \frac{\tau}{2}\right) = -x(t), \qquad \forall t \in \mathbf{R}, \tag{21}$$

and any asymmetric (τ, x) satisfies

$$(i(x, m), \nu(x, m)) = (i(-x, m), \nu(-x, m)), \quad \forall m \in \mathbf{N}. \tag{22}$$

Suppose ${}^{\#}\tilde{\mathcal{J}}(\Sigma) < +\infty$. Denote the numbers of symmetric and asymmetric elements in $\tilde{\mathcal{J}}(\Sigma)$ by p and q. Write

$$\begin{aligned} \tilde{\mathcal{J}}(\Sigma) = &\{[(\tau_j, x_j)] \mid j = 1, \ldots, p\} \\ &\cup \{[(\tau_k, x_k)], [(\tau_k, -x_k)] \mid k = p+1, \ldots, p+q\}. \end{aligned} \tag{23}$$

⟨2⟩ Applying the common index jump Theorem 11.2.1 to associated symplectic paths of

$$(\tau_1, x_1), \ldots, (\tau_{p+q}, x_{p+q}), (2\tau_{p+1}, x_{p+1}^2), \ldots, (2\tau_{p+q}, x_{p+q}^2),$$

we obtain an integer N and iteration times $m_1, \ldots, m_{p+2q}$ with the rather precise information on the Maslov-type indices of iterations of (τ_j, x_j) as listed in (6.8)–(6.15) below. By Lemma 3.5 below, we obtain an injection map $\Psi : \mathbf{N} \to \tilde{\mathcal{J}}(\Sigma) \times \mathbf{N}$ given by

$$\Psi(N - s + 1) = ([(\tau_{k(s)}, x_{k(s)})], m(s)), \qquad \text{for } s = 1, \ldots, n, \tag{24}$$

so that

$$2(N - s + 1) - 2 + n \in \mathcal{I}_{m(s)}(\tau_{k(s)}, x_{k(s)}), \qquad \text{for } s = 1, \ldots, n.$$

⟨3⟩ Let

$$S_1 = \{s \in \{1, \ldots, n\} \,|, k(s) \le q_1\}, \quad S_2 = \{1, \ldots, n\} \backslash S_1. \tag{25}$$

Then we prove

$$^{\#}S_1 \le p, \quad \text{and} \quad ^{\#}S_2 \le 2q. \tag{26}$$

Then this implies

$$^{\#}\tilde{\mathcal{J}}(\Sigma) = p + 2q \ge {}^{\#}S_1 + {}^{\#}S_2 = n.$$

⟨4⟩ To prove the first estimate in (26), using the property (21) of symmetric orbit (τ, x) and precise index information of iterations of (τ_j, x_j) obtained in the step ⟨2⟩, we conclude that the Maslov-type index, nullity, and splitting numbers of $x|_{[0,\tau]}$ are the same as those of $y \equiv (x|_{[0,\tau/2]})^2$. Therefore

$$i(x, 1) + 2S^+(x) - \nu(x, 1) = i(y, 2) + 2S^+(y^2) - \nu(y, 2) \ge n.$$

By this estimate, we get that the integer $m(s)$ in (24) is uniquely determined by $k(s)$, provided $k(s) \le p$. Then the injection map Ψ of (24) induces an injection map from S_1 to $\{[(\tau_j, x_j)] \,|\, 1 \le j \le p\}$ and yields the first estimate in (26).

⟨5⟩ To prove the second estimate in (26), using the property (22) of asymmetric orbit (τ, x) and precise index information of iterations of (τ_j, x_j) obtained in step ⟨2⟩, we conclude that the integer $m(s)$ in (24) possesses at most two choices determined by the $k(s)$ there, provided $p < k(s) \le p + q$. Then the injection map Ψ of (24) induces a map from S_2 to $\Gamma \equiv \{[(\tau_j, x_j)] \,|\, p < j \le p + q\}$ such that any element in Γ is the image of at most two numbers in S_2. This yields the second estimate in (26), and completes the proof of (0.0.19) of Theorem C.

C Ideas of the proof of the stability result in Theorem D

For the existence of at least one elliptic closed characteristic on Σ when $^{\#}\tilde{\mathcal{J}}(\Sigma) < +\infty$, we use the common index jump Theorems 11.2.1 and 11.2.2, and the iteration inequality Theorem 10.2.4, and conclude that the elliptic height $e(x_j) = 2n$ for at least one $j \in \{1, \ldots, q\}$. Note that the first elliptic closed characteristic corresponds to the vertex point χ of the cube $[0, 1]^n$ in (11.1.22) found in Step 2 of the proof of the common selection Theorem 11.1.1. For the proof of the second conclusion of Theorem D, our method is to find out two such vertices, and show that they correspond to two elliptic closed characteristics.

This chapter is organized as follows. In Sections 1 to 3, we study basic properties of closed characteristics. In Section 4 we study the existence and multiplicity problem and prove results (0.0.16)–(0.0.18) in Theorem C of the Introduction of this book for general hypersurfaces. In Section 5, we study the stability problem and prove Theorem D in the Introduction. In Section 6, we study the multiplicity result (0.0.19) in Theorem C of the Introduction for symmetric hypersurfaces, and prove a beautiful result [DDE1] of G.F. Dell'Antonio, B. D'Onofio, and I. Ekeland, which shows that any $\Sigma \in \mathcal{H}_s(2n)$ carries at least one elliptic closed characteristic. In the last section of Notes and Comments, we give more comments on the history and further considerations for this closed characteristic problem.

15.1 Index theorem for dual action principle

The aim in this section is to prove the relation theorem of the Ekeland index theory for convex Hamiltonian systems and the index theory of symplectic paths defined in Chapter 5.

For any τ and $s > 0$, we consider the linear Hamiltonian system

$$\begin{cases} \dot{y}(t) = JB(t)y(t), \\ y(s\tau) = y(0). \end{cases} \tag{1}$$

Here we suppose the coefficient B to satisfy the following conditions:

(B1) $B \in C(S_\tau, \mathcal{L}_s(\mathbf{R}^{2n}))$ *with* $S_\tau = \mathbf{R}/(\tau\mathbf{Z})$,

(B2) $B(t)$ *is positive definite for all* $t \in \mathbf{R}$.

Fix $\alpha \in (1, 2)$. For any τ and $s > 0$, following §I.4 of I. Ekeland's book [Eke3], we define

$$L_s = \left\{ u \in L^2((0, s\tau), \mathbf{R}^{2n}) \mid \int_0^{s\tau} u dt = 0 \right\}, \tag{2}$$

with the usual L^2-norm

$$\|u\|_s = \left(\int_0^s |u|^2 dt \right)^{1/2}. \tag{3}$$

Then L_s becomes a Hilbert space. The Clarke–Ekeland dual action functional $Q_s : L_s \to \mathbf{R}$ corresponding to problem (1) is defined by

$$Q_s(u) = \frac{1}{2} \int_0^{s\tau} \{Ju \cdot \Pi_s u + B^{-1}(t) Ju \cdot Ju\} dt, \tag{4}$$

where $\Pi_s u$ is defined by $\frac{d}{dt}\Pi_s u = u$ and $\int_0^{s\tau} \Pi u dt = 0$. Then by the study in [Eke3], L_s possesses a Q_s orthogonal decomposition

$$L_s = L_s^+ \oplus L_s^0 \oplus L_s^-,$$

such that Q_s is positive, null, and negative definite on L_s^+, L_s^0, and L_s^- respectively, and

$$i_s^E(B) \equiv \dim L_s^- < +\infty, \tag{5}$$

$$0 \le \nu_s^E(B) \equiv \dim L_s^0 \le 2n. \tag{6}$$

$(i_1^E(B), \nu_1^E(B))$ are the **Ekeland index and nullity** of (1) with $s = 1$ introduced by I. Ekeland in [Eke1] of 1984.

The following theorem is the main result in this section.

Theorem 1 (V. Brousseau [Bro1], Y. Long [Lon14]). *Suppose* (B1) *and* (B2) *hold, and* $\alpha \in (1, 2)$. *Then*

$$i(B, 1) = i_1^E(B) + n, \quad \nu(B, 1) = \nu_1^E(B). \tag{7}$$

In order to prove this theorem, we need the following characterization of Ekeland index and nullity, whose proof can be found in Chapter 1 of [Eke3].

Proposition 2 *Suppose conditions* (B1) *and* (B2) *hold. Then*

1° *(Theorem* 1.4.4 *of* [Eke3]*) it holds that*

$$\nu_s^E(B) = \nu(B_s, 1), \tag{8}$$

where B_s is the restriction of B to $[0, s]$.

2° *(Theorem* 1.4.6 *of* [Eke3]*) it holds that*

$$i_1^E(B) = \sum_{0<s<1} \nu_s^E(B_s). \tag{9}$$

Denote the fundamental solution of system (1) by γ. Then we have the following similar result.

Proposition 3 *Suppose conditions* (B1) *and* (B2) *hold. Then*

$$i(\gamma, 1) = \sum_{0<s<1} \nu(\gamma(s\tau), 1) + n. \tag{10}$$

Proof: Set $\tau = 1$. Denote $\gamma_s(t) = \gamma(st)$ for $s, t \in [0, 1]$. By (8) and (9), there exist only finitely many times $\{s_j \mid j = 1, \ldots, k\} \subset (0, 1)$ such that $\gamma(s_j) \in \mathrm{Sp}(2n)_1^0$, i.e., $\nu(\gamma_j, 1) > 0$, where $\gamma_j = \gamma_{s_j}$, for $|\epsilon| > 0$ so small that

$$\nu(B|_{[0,s_j+s]}, 1) = 0, \quad \forall 0 < |s| \le \epsilon, \ 1 \le j \le k.$$

Denote $\gamma_{j+\epsilon}(t) = \gamma((s_j + \epsilon)t)$ for $t \in [0, 1]$.

We claim that for $\epsilon > 0$ sufficiently small,

$$\gamma_{j+\epsilon} \sim_1 (\gamma_j)_1. \tag{11}$$

where $(\gamma_j)_1$ is the perturbation path of γ_j defined by (5.4.6).

In fact, we first define a new path $\beta \in \mathcal{P}_1(2n)$ by

$$\beta(t) = \begin{cases} \gamma((s_j + \epsilon)t), & \text{for } 0 \le t \le \theta, \\ \gamma(s_j)PR(t\theta^{-1} - 1)^{\diamond n}P^{-1}, & \text{for } \theta \le t \le 1, \end{cases} \tag{12}$$

where $\theta = s_j/(s_j + \epsilon)$, $P \in \mathrm{Sp}(2n)$ is given by the normal form Theorem 1.7.3 for $M = \gamma(s_j)$. Then for $t \in [\theta, 1]$ we have

$$\begin{aligned} A(t) &= -J\dot{\beta}(t)\beta(t)^{-1} \\ &= -\theta^{-1}J\gamma(s_j)PJ(\gamma(s_j)P)^{-1} \\ &= -\theta^{-1}JJ((\gamma(s_j)P)^{-0})^T(\gamma(s_j)P)^{-1} \\ &= \theta^{-1}((\gamma(s_j)P)^{-1})^T(\gamma(s_j)P)^{-1}. \end{aligned}$$

Thus $A(t)$ is positive definite for any $t \in [\theta, 1]$. Let

$$D_s(t) = (1 - s)B_{s_j+\epsilon}(t) + sA(t), \quad \forall t \in [\theta, 1],\ s \in [0, 1].$$

Then $D_s(t)$ is positive definite, and the fundamental solution $\beta_s(\cdot)$ of the system (1) with B replaced by D_s satisfying $\beta_s(\theta) = \gamma(s_j)$ yields a homotopy $\delta : [0, 1] \times [\theta, 1] \to \mathrm{Sp}(2n)$ from $\gamma_{s_j+\epsilon}$ to β by

$$\delta(s, t) = \beta_s(t), \quad \text{for } \theta \le t \le 1.$$

We extend $\delta(s, t)$ by

$$\delta(s, t) = \gamma((s_j + \epsilon)t), \quad \text{for } 0 \le t \le \theta,\ 0 \le s \le 1.$$

By Proposition 2, for sufficiently small $\epsilon > 0$ this yields $\nu(\delta(s, \cdot), 1) = 0$ for all $s \in [0, 1]$. Thus

$$\gamma_{s_j+\epsilon} \sim_1 \beta. \tag{13}$$

Now applying rotations in the definition of $(\gamma_j)_1$ in (5.4.6) first, and then plugging a parameter s into other rotation parts in $\beta|_{[\theta,0]}$ yields the second homotopy

$$\beta \sim_1 (\gamma_j)_1. \tag{14}$$

Thus claim (11) holds.

Especially this argument implies

$$i(\gamma_s, 1) = n, \quad \forall 0 < s < s_1, \tag{15}$$

where $\gamma_s(t) = \gamma(st)$ for $0 \le t \le 1$. Note that (15) can also be obtained by Proposition 13.2.3.

Similarly, we can prove

$$\gamma_{s_j-\epsilon} \sim_1 (\gamma_j)_{-1}. \tag{16}$$

for sufficiently small $\epsilon > 0$ and $1 \le j \le k$, where $(\gamma_j)_{-1}$ is defined by (5.4.6).

Therefore together with Theorem 5.4.1, we obtain

$$i(\gamma_{j+\epsilon}, 1) - i(\gamma_{j-\epsilon}, 1) = \nu(\gamma_j, 1). \tag{17}$$

Now summing (17) up for $j = 1, \dots, k$, together with (15) we obtain

$$\begin{aligned} i(\gamma, 1) &= i((\gamma)_{-1}, 1) \\ &= i(\gamma_{k+\epsilon}, 1) \\ &= \sum_{j=1}^{k} (i(\gamma_{j+\epsilon}, 1) - i(\gamma_{j-\epsilon}, 1)) + i(\gamma_{1-\epsilon}, 1) \\ &= \sum_{j=1}^{k} \nu(\gamma_j, 1) + n. \end{aligned}$$

Thus (10) holds. □

Now we can give

Proof of Theorem 1: (7) follows from (8), (9), and (10).

As a corollary of Theorem 1, we obtain

Corollary 4 *It holds that*

$$i(x, 1) \ge n, \quad \forall (\tau, x) \in \mathcal{J}(\Sigma, \alpha). \tag{18}$$

This is also a special case of Proposition 13.2.3.

Remark 5 For $\tau > 0$ denote by $\mathcal{E}_\tau(2n)$ the set of $\gamma \in \hat{\mathcal{P}}_\tau(2n)$ which are fundamental solutions of the linear systems (1) with negative definite $B(t)$ for all $t \in [0, \tau]$. In his works [Eke1]–[Eke3], I. Ekeland systematically studied such Hamiltonian systems and defined his index function theory via the dual action principle. For any $\gamma \in \mathcal{E}_\tau(2n)$, denote the Ekeland index function by $(j_\omega(\gamma), n_\omega(\gamma))$ for all $\omega \in \mathbf{U}$ via Definition I.5.3 of [Eke3]. Note that $(j_1(\gamma), n_1(\gamma))$ is the Ekeland index defined by Definition I.4.3 of [Eke3]. It should be pointed out that $(j_1(\gamma), n_1(\gamma))$ here is not the index (i_1^E, ν_1^E) defined by above (5) and (6), because of the different "J" used in these two cases. The following result gives the relation between our index function theory and Ekeland's. For the proof of this theorem, we refer the reader to [Lon16].

Theorem 6 *For any* $\gamma \in \mathcal{E}_\tau(2n)$,

$$\nu_\omega(\gamma) = n_\omega(\gamma), \quad \forall\, \omega \in \mathbf{U}, \tag{19}$$

$$i_1(\gamma) + \nu_1(\gamma) = -j_1(\gamma) - n, \tag{20}$$

$$i_\omega(\gamma) + \nu_\omega(\gamma) = -j_\omega(\gamma), \quad \forall\, \omega \in \mathbf{U}\backslash\{1\}. \tag{21}$$

Remark 7 Theorem 1 was proved by V. Brousseau in [Bro1] of 1990 for the non-degenerate case, and by the author in [Lon14] of 1998 for the degenerate case. The proof given here is different from those in [Bro1] and [Lon14] and is based on I. Ekeland's Theorems I.4.4 and I.4.6 in [Eke3] and properties of our index, especially Theorem 5.4.1.

15.2 Variational properties

In this section we study the basic variational properties of closed characteristics.

To solve the given energy problem (0.4)–(0.6), we consider a fixed period problem with $\alpha \in (1, 2)$ and H_α given by (0.3),

$$\begin{cases} \dot z(t) = JH'_\alpha(z(t)), & \forall t \in \mathbf{R}, \\ z(1) = z(0). \end{cases} \tag{1}$$

Let

$$E_\alpha = \left\{ u \in L^{\alpha/(\alpha-1)}(\mathbf{R}/\mathbf{Z}), \mathbf{R}^{2n}) \mid \int_0^1 u dt = 0 \right\}, \tag{2}$$

with the usual $L^{\alpha/(\alpha-1)}$ norm. The Clarke–Ekeland dual action functional $f_\alpha : E_\alpha \to \mathbf{R}$ is defined by

$$f_\alpha(u) = \int_0^1 \left\{ \frac{1}{2} Ju \cdot \Pi u + H^*_\alpha(-Ju) \right\} dt, \tag{3}$$

where Πu is defined by $\frac{d}{dt}\Pi u = u$ and $\int_0^1 \Pi u dt = 0$. The Fenchel conjugate $H^*_\alpha : \mathbf{R}^{2n} \to \mathbf{R}$ is defined by

$$H^*_\alpha(x) = \sup_{y \in \mathbf{R}^{2n}} \{x \cdot y - H_\alpha(y)\}. \tag{4}$$

Then $f_\alpha \in C^2(E_\alpha, \mathbf{R})$. Suppose $u \in E_\alpha \backslash \{0\}$ is a critical point of f_α. By [Eke3], there exists $\xi_u \in \mathbf{R}^{2n}$ such that $z_u(t) = \Pi u(t) + \xi_u$ is a 1-periodic solution of the problem (0.4).

In [Eke1]–[Eke3], I. Ekeland proved that the Hessian of f_α at its critical point u possesses finite Morse index and nullity, which are the so-called Ekeland index and nullity, and denoted by $i_1^E(u)$ and $\nu_1^E(u)$ respectively.

Let $h = H_\alpha(z_u(t))$ and $1/m$ be the minimal period of z_u for some $m \in \mathbf{N}$. Define

$$x_u(t) = h^{-1/\alpha} z_u(h^{(2-\alpha)/\alpha} t) \quad \text{and} \quad \tau = \frac{1}{m} h^{(\alpha-2)/\alpha}. \tag{5}$$

Then $x_u(t) \in \Sigma$ for all $t \in \mathbf{R}$ and $(\tau, x_u) \in \mathcal{J}(\Sigma, \alpha)$. Note that the period 1 of z_u corresponds to the period $m\tau$ of the solution $(m\tau, x_u^m)$ of (0.4)–(0.6) with minimal period τ.

The following propositions relate the Ekeland index to the index given by Definitions 5.4.3 and 8.0.1.

Proposition 1 *For u and x_u defined above,*

$$i(x_u, 1) = i_1^E(u) + n, \qquad \text{and} \quad \nu_1(x_u) = \nu_1^E(u). \tag{6}$$

Proof: This follows from Theorem 1.1 directly. □

Proposition 2 *For z_u, x_u, τ and m defined above,*

$$i(x_u, m) = i(z_u, 1), \qquad \nu(x_u, m) = \nu(z_u, 1). \tag{7}$$

Proof: Let $\psi : [0, +\infty) \to \mathrm{Sp}(2n)$ be the associated symplectic path of z_u, i.e., the fundamental solution of (1.1) with $B(t) = H_\alpha''(z_u(t))$. Thus by definition, $i(z_u, 1) = i(\psi, 1)$.

Define

$$\gamma(t) = \psi(h^{(2-\alpha)/\alpha} t), \qquad \forall t \in [0, +\infty). \tag{8}$$

Then using the positive homogeneity of H_α'', we obtain that $\gamma : [0, +\infty) \to \mathrm{Sp}(2n)$ is the fundamental solution of the system (1.1) with coefficient $B(t) = H_\alpha''(x_u(t))$, i.e., γ coincides with the associated symplectic path of x_u. Thus by (5) and (8), $\gamma(m\tau) = \psi(1)$. This implies $\nu(x_u, m) = \nu(z_u, 1)$. Since $\gamma|_{[0,m\tau]}$ is only a rescaling of $\psi|_{[0,1]}$, they are geometrically the same path in $\mathrm{Sp}(2n)$. This yields $i(x_u, m) = i(\gamma|_{[0,m\tau]}, 1) = i(\psi|_{[0,1]}, 1) = i(z_u, 1)$, and completes the proof. □

On the other hand, every solution $(\tau, x) \in \mathcal{J}(\Sigma, \alpha)$ gives rise to a sequence $\{z_m^x\}_{m\in\mathbf{N}}$ of solutions of the problem (1), and a sequence $\{u_m^x\}_{m\in\mathbf{N}}$ of critical points of f_α defined by

$$z_m^x(t) = (m\tau)^{-1/(2-\alpha)} x(m\tau t), \tag{9}$$

$$u_m^x(t) = (m\tau)^{(\alpha-1)/(2-\alpha)} \dot{x}(m\tau t). \tag{10}$$

For every $m \in \mathbf{N}$,

$$f_\alpha(u_m^x) = -\left(1 - \frac{\alpha}{2}\right)\left(\frac{2m}{\alpha}A(\tau, x)\right)^{-\alpha/(2-\alpha)}, \tag{11}$$

where

$$A(\tau, x) = \frac{1}{2}\int_0^\tau (-J\dot{x} \cdot x)dt. \tag{12}$$

Lemma 3 (Lemma I.7.3 of [Eke3]). *For any $(\tau, x) \in \mathcal{J}(\Sigma, \alpha)$, denote by γ_x the associated symplectic path of x. Then*

$$\gamma_x(\tau)\dot{x}(0) = \dot{x}(0), \tag{13}$$

$$\gamma_x(\tau)x(0) = \tau(\alpha - 2)\dot{x}(0) + x(0). \tag{14}$$

Proof: For the reader's convenience, the proof is given here.

Let $x_h(t) = h^{1/\alpha}x(th^{(\alpha-2)/\alpha})$ and $\tau_h = \tau h^{(2-\alpha)/\alpha}$ for $h > 0$. Then (τ_h, x_h) is a solution of (0.4)–(0.5) with

$$H_\alpha(x_h(t)) = h, \tag{15}$$

and it holds that

$$x_h(\tau_h) = h^{1/\alpha}x(\tau) = h^{1/\alpha}x(0) = x_h(0). \tag{16}$$

Let $\gamma_h(t) = \gamma(th^{(\alpha-2)/\alpha})$ for $t \geq 0$. Then γ_h is the associated symplectic path of x_h, i.e., the fundamental solution of the linear system (0.7) with x replaced by x_h. Especially,

$$\gamma_h(\tau_h)\left(\frac{d}{dh}x_h\right)(0) = \frac{d}{dh}x_h(\tau_h). \tag{17}$$

Differentiating (15) with respect to h yields

$$\dot{x}_h(\tau_h)\frac{d}{dh}\tau_h + \frac{d}{dh}x_h(\tau_h) = \frac{d}{dh}x_h(0). \tag{18}$$

Plugging (17) into (18) yields

$$\dot{x}_h(\tau_h)\frac{d}{dh}\tau_h + \gamma_h(\tau_h)\frac{d}{dh}x_h(0) = \frac{d}{dh}x_h(0). \tag{19}$$

Note that

$$\dot{x}_h(\tau_h) = h^{(\alpha-1)/\alpha}\dot{x}(\tau), \quad \frac{d}{dh}\tau_h = \frac{2-\alpha}{\alpha}\tau h^{2(\alpha-1)/\alpha},$$

and by the definition of x_h we have

$$\frac{d}{dh}x_h(0) = \frac{1}{\alpha}h^{(1-\alpha)/\alpha}x(0).$$

Together with $\gamma_h(\tau_h) = \gamma(\tau)$, we obtain from (19),

$$\begin{aligned} &h^{(\alpha-1)/\alpha}\dot{x}(\tau)\frac{2-\alpha}{\alpha}\tau h^{2(\alpha-1)/\alpha} + \gamma(\tau)\frac{1}{\alpha}h^{(1-\alpha)/\alpha}x(0) \\ &= \frac{1}{\alpha}h^{(1-\alpha)/\alpha}x(0). \end{aligned} \tag{20}$$

Setting $h = 1$ in (20) yields

$$\tau(2-\alpha)\dot{x}(\tau) + \gamma(\tau)x(0) = x(0). \tag{21}$$

This implies (14). □

Lemma 4 *For any* $(\tau, x) \in \mathcal{J}(\Sigma, \alpha)$, *there exist* $P \in \mathrm{Sp}(2n)$ *and* $M \in \mathrm{Sp}(2n-2)$ *such that*

$$\gamma_x(\tau) = P^{-1}(N_1(1,1)\diamond M)P, \tag{22}$$

where $N_1(1,1)$ *is the normal form defined by* (1.4.1).

Proof: Fix $(\tau, x) \in \mathcal{J}(\Sigma, \alpha)$. Let γ_x be the associated symplectic path of x and

$$\xi_1 = \tau(\alpha-2)\dot{x}(0), \quad \xi_2 = x(0).$$

We carry out the proof in three steps.

Step 1 Since $x = x(t)$ is a solution of (0.4), we have $x(0) \in \Sigma$ and $\dot{x}(0) = JH'_\alpha(x(0))$. Since Σ is convex and C^2, by a direct computation and the fact that $1 < \alpha < 2$, we obtain

$$\xi_1^* J\xi_2 = \tau(\alpha-2)H'_\alpha(x(0)) \cdot x(0) < 0. \tag{23}$$

Note that from Lemma 3, the span of ξ_1 and ξ_2 form an invariant subspace of $\gamma_x(\tau)$. By (23), a rescaling of ξ_1 and ξ_2 forms a symplectic basis of this invariant subspace. Since $\gamma_x(\tau)$ is symplectic, (22) then results. In the following we use the argument in Section 1.4 to give a different proof.

Step 2 Now suppose $\{\xi_1, \xi_2, \ldots, \xi_p\}$ form a Jordan block of $\gamma_x(\tau)$ belonging to the eigenvalue 1, i.e., setting $\xi_0 = 0$,

$$\gamma_x(\tau)\xi_i = \xi_i + \xi_{i-1}, \qquad \forall 1 \le i \le p. \tag{24}$$

For $1 \le i, j \le p$ by (24) we have

$$\xi_i^* J\xi_j = (M\xi_i)^* J(M\xi_j) = \xi_i^* J\xi_j + \xi_{i-1}^* J\xi_j + \xi_i^* J\xi_{j-1} + \xi_{i-1}^* J\xi_{j-1}.$$

This yields

$$\xi_{i-1}^* J\xi_{j-1} + \xi_{i-1}^* J\xi_j + \xi_i^* J\xi_{j-1} = 0, \quad \forall 1 \le i, j \le p, \tag{25}$$

$$\xi_i^* J\xi_j = 0, \quad \forall 1 \le i \le p - j,\ 1 \le j \le \left[\frac{p}{2}\right], \tag{26}$$

where $[a] = \max\{m \in \mathbf{Z} \,|\, m \le a\}$ for any $a \in \mathbf{R}$, and (26) follows from (25) by induction.

Thus from (23) and (26) we must have $p = 2$, i.e., ξ_1 and ξ_2 form a Jordan block of $\gamma_x(\tau)$ belonging to the eigenvalue 1.

Step 3 Define

$$\delta_1 = \frac{1}{\sqrt{|\xi_1^* J\xi_2|}}\xi_1, \quad \delta_2 = \frac{1}{\sqrt{|\xi_1^* J\xi_2|}}\xi_2. \tag{27}$$

Then

$$\delta_1^* J\delta_2 = -1, \qquad F \equiv \mathrm{span}\{\delta_1, \delta_2\} = \mathrm{span}\{\xi_1, \xi_2\}, \tag{28}$$

i.e., $\{\delta_1, \delta_2\}$ form a symplectic base for F. Denote by K the $2n \times 2$ matrix formed by δ_1 and δ_2 as the first and the second columns. From (21), (27), and (28), we obtain $\gamma_x(\tau)K = KN_1(1, 1)$. Now we can extend K to a matrix $P \in \mathrm{Sp}(2n)$ such that δ_1 and δ_2 form the first and the $(n + 1)$-st columns of P and for some $M \in \mathrm{Sp}(2n - 2)$ such that (22) holds. This proves the lemma. □

Remark 5 Lemma 4 holds for $\alpha = 2$ or $\alpha > 2$ with $N_1(1, 1)$ in (22) being replaced by I_2 or $N_1(1, -1)$ respectively. Note also that Lemmas 3 and 4 hold for star-shape hypersurfaces too.

Proposition 2 and Lemma 4 were proved in [Lon14] of 1998.

15.3 Critical orbits and index jumps

Next we consider critical points produced by the Fadell–Rabinowitz cohomological index theory.

For a principal $U(1)$-bundle $E \to B$, the Fadell–Rabinowitz index of E is defined to be

$$\sup\{k \,|\, c_1(E)^{k-1} \neq 0\},$$

where $c_1(E) \in H^2(B, \mathbf{Q})$ is the first rational Chern class. For a $U(1)$-space, i.e., a topological space X with a $U(1)$-action, the Fadell–Rabinowitz index is defined to be the index of the bundle

$$X \times S^\infty \to X \times_{U(1)} S^\infty,$$

where $S^\infty \to CP^\infty$ is the universal $U(1)$-bundle. For the details of the Fadell–Rabinowitz index theory, we refer readers to their original paper [FaR1] and Section V.3 of I. Ekeland's book [Eke3]. A more analytical approach can be found in [Ben1] of V. Benci.

Following §V.3 of [Eke3], we denote by "ind" the Fadell–Rabinowitz S^1-action cohomology index theory for S^1-invariant subsets of E_α. Using the index theory defined in Chapter 5, the results of Ekeland et al contained in (V.3.21), (V.3.22), Proposition V.3.3, and Theorem V.3.4 in Section V.3 of [Eke3] can be rephrased as follows.

Theorem 1 *For*

$$[f_\alpha]_c \equiv \{u \in E_\alpha \mid f_\alpha(u) \le c\},$$

define

$$c_k = \inf\{c < 0 \mid \mathrm{ind}([f_\alpha]_c) \ge k\}, \qquad \forall k \in \mathbf{N}. \tag{1}$$

Then $\{c_k\}$ are critical values of f.

They satisfy

$$-\infty < c_1 = \inf_{u \in E_\alpha} f_\alpha(u) \le c_2 \le \cdots \le c_k \le c_{k+1} \le \cdots < 0, \tag{2}$$

$$^{\#}\tilde{\mathcal{J}}(\Sigma) = +\infty, \quad \text{if} \quad c_k = c_{k+1} \quad \text{for some} \quad k \in \mathbf{N}. \tag{3}$$

For any given $k \in \mathbf{N}$, there exists $(\tau, x) \in \mathcal{J}(\Sigma, \alpha)$ and $m \in \mathbf{N}$ such that for u_m^x defined by (2.10),

$$f'_\alpha(u_m^x) = 0, \quad f_\alpha(u_m^x) = c_k, \tag{4}$$

$$i(x, m) \le 2k - 2 + n \le i(x, m) + \nu(x, m) - 1. \tag{5}$$

By our study in the previous sections, we have

Lemma 2 *Fix $\Sigma \in \mathcal{H}(2n)$ and $\alpha \in (1, 2)$. For any $(\tau, x) \in \mathcal{J}(\Sigma, \alpha)$ and $m \in \mathbf{N}$,*

$$i(x, m+1) - i(x, m) \ge 2, \tag{6}$$

$$\begin{aligned} i(x, m+1) - i(x, m) - \nu(x, m) &\ge i(x, 1) - \frac{e(\gamma_x(\tau))}{2} + 1 \\ &\ge i(x, 1) - n + 1, \quad \forall m \in \mathbf{N}, \end{aligned} \tag{7}$$

$$\begin{aligned} i(x, m+1) + \nu(x, m+1) - 1 &\geq i(x, m+1) \\ &> i(x, m) + \nu(x, m) - 1, \end{aligned} \tag{8}$$

$$\nu(x, m) \geq \nu(x, 1) \geq 1, \tag{9}$$

$$\hat{i}(x, 1) > 2. \tag{10}$$

Proof: By Theorem 10.2.4 and (1.18) we obtain (6), (7), and (8). Since the system (0.4) is autonomous, (9) follows. (10) follows from Corollary 8.3.2, (1.18), and Lemma 2.4. □

Based upon (4) and (5), we introduce the following definitions.

Definition 3 For any $\Sigma \in \mathcal{H}(2n)$ and $\alpha \in (1, 2)$, $(\tau, x) \in \mathcal{J}(\Sigma, \alpha)$ is (m, k)-**variationally visible**, if there exist some m and $k \in \mathbf{N}$ such that (4) and (5) hold for u_m^x defined by (2.10). We call $(\tau, x) \in \mathcal{J}(\Sigma, \alpha)$ **infinite variationally visible**, if there exist infinitely many (m, k) such that (τ, x) is (m, k)-variationally visible. We denote by $\mathcal{V}(\Sigma, \alpha)$ (or $\mathcal{V}_\infty(\Sigma, \alpha)$) the subset of $\tilde{\mathcal{J}}(\Sigma)$ in which a representative $(\tau, x) \in \mathcal{J}(\Sigma, \alpha)$ of each $[(\tau, x)]$ is variationally visible (or infinite variationally visible).

Suggested by (4) and (5), we introduce index intervals and index covers.

Definition 4 For any $(\tau, x) \in \mathcal{J}(\Sigma, \alpha)$, we define the ***m*-th index interval** of (τ, x) by the closed interval

$$\mathcal{I}_m(\tau, x) = [i(x, m), i(x, m) + \nu(x, m) - 1]. \tag{11}$$

We call the set

$$\mathcal{I}(\tau, x) = \bigcup_{m \geq 1} \mathcal{I}_m(\tau, x), \tag{12}$$

the **index cover set** of (τ, x).

Note that the inequalities (7) and (1.18) especially imply that all index intervals of (τ, x) are mutually disjoint. (7) especially introduces a way to estimate the ellipticity of (τ, x) in terms of its iterated indices for some $m \in \mathbf{N}$.

Now (5) can be restated as

$$2\mathbf{N} - 2 + n \subset \bigcup_{[(\tau,x)] \in \tilde{\mathcal{J}}(\Sigma,\alpha)} \mathcal{I}(\tau, x). \tag{13}$$

We call integers in the sequence $2\mathbf{N} - 2 + n$ **effective integers**.

Now in the following we suppose

$$^{\#}\tilde{\mathcal{J}}(\Sigma) < +\infty. \tag{14}$$

Then by (3) no equality in (2) can hold, i.e., we must have

$$-\infty < c_1 < c_2 < \cdots < c_k < c_{k+1} < \cdots < 0. \tag{15}$$

Here we have used the method of Fadell–Rabinowitz S^1-cohomology index theory (cf. [FaR1]) via Theorem V.3.4 of [Eke3]. Note that here each c_k corresponds uniquely to an effective integer $2k-2+n$. We observe that in this case there is a one to one correspondence between the effective integers and index intervals of all closed characteristic $[(\tau,x)]$ in $\mathcal{V}_\infty(\Sigma,\alpha)$. In other words, under condition (14), an injective map $\Psi : \mathbf{N} \to \mathcal{V}_\infty(\Sigma,\alpha) \times \mathbf{N}$ can be defined by (4), (5), and (15), which makes precise the correspondence between effective integers and iterations of closed characteristics.

Lemma 5 (Y. Long and C. Zhu [LZu3]). *Suppose $^{\#}\tilde{\mathcal{J}}(\Sigma,\alpha) < +\infty$. Then there exist an integer $K \geq 0$ and an injection map $\Psi : \mathbf{N}+K \to \mathcal{V}_\infty(\Sigma,\alpha) \times \mathbf{N}$ such that*

(i) *for any $k \in \mathbf{N}+K$, $(\tau,x) \in \mathcal{J}(\Sigma,\alpha)$ and $m \in \mathbf{N}$ satisfying $\Psi(k) = ([(\tau,x)],m)$, (4) and (5) hold, and*

(ii) *for any $k_j \in \mathbf{N}+K$, $k_1 < k_2$, $(\tau_j,x_j) \in \mathcal{J}(\Sigma,\alpha)$ satisfying $\Psi(k_j) = ([(\tau_j,x_j)],m_j)$ with $j=1,2$,*

$$\hat{i}(x_1,m_1) < \hat{i}(x_2,m_2). \tag{16}$$

Proof: (i) By Theorem V.3.4 of [Eke3], for each $k \in \mathbf{N}$, there is a $(\tau,x) \in \mathcal{J}(\Sigma,\alpha)$ such that (τ,x) is (m,k)-variationally visible for some $m \in \mathbf{N}$, i.e., $[(\tau,x)] \in \mathcal{V}(\Sigma,\alpha)$. We define a map $\Psi_1 : \mathbf{N} \to \mathcal{V}(\Sigma,\alpha) \times \mathbf{N}$ by $\Psi_1(k) = ([(\tau,x)],m)$. Since $\tilde{\mathcal{J}}(\Sigma) < +\infty$, (15) holds by Theorem V.3.4 of [Eke3], i.e., $c_j < c_k < 0$ whenever $j < k$. Thus if $\Psi_1(j) = \Psi_1(k) = ([\tau,x)],m)$ for some $j < k$, by (4), we obtain

$$c_j = f(u_m^x) = c_k.$$

This contradiction proves that Ψ_1 is injective.

Since $^{\#}\tilde{\mathcal{J}}(\Sigma) < +\infty$, there exists an integer $K \geq 0$ such that all critical values c_{k+K} with $k \in \mathbf{N}$ come from iterations of elements in $\mathcal{V}_\infty(\Sigma,\alpha)$. Thus $\Psi_1(k+K) \in \mathcal{V}_\infty(\Sigma,\alpha) \times \mathbf{N}$ for any $k \in \mathbf{N}$. We define

$$\Psi(k) = \Psi_1(k), \quad \forall k \in \mathbf{N}+K. \tag{17}$$

Then Ψ is injective, and (4) and (5) hold.

(ii) By Theorem V.3.11 of [Eke3], we have $\gamma \equiv \gamma^-(\Sigma) = \gamma^+(\Sigma) > 0$, where $\gamma^\pm(\Sigma)$ are defined by (V.3.62)–(V.3.65) of [Eke3] as follows:

$$\begin{aligned}
\gamma^+(\Sigma) &= C_\alpha^{-1} \limsup_{k\to\infty}((-c_k)^{(2-\alpha)/\alpha}k)^{-1}, \\
\gamma^-(\Sigma) &= C_\alpha^{-1} \liminf_{k\to\infty}((-c_k)^{(2-\alpha)/\alpha}k)^{-1},
\end{aligned}$$

with $C_\alpha = \frac{4}{\alpha}(1-\frac{\alpha}{2})^{\frac{\alpha-2}{\alpha}}$. Thus we have

$$\lim_{k\to+\infty} k|c_k|^{\frac{2-\alpha}{\alpha}} C_\alpha = \frac{1}{\gamma}. \tag{18}$$

For $k \in \mathbf{N}+K$, let $([(\tau,x)],m) = \Psi(k)$ for some $[(x,\tau)] \in \mathcal{V}_\infty(\Sigma,\alpha)$ and $m \in \mathbf{N}$. Then (x,τ) is (m,k)-variationally visible. So by (V.3.45) of [Eke3],

$$\begin{aligned} c_k &= f(u_m^x), \\ |c_k|^{\frac{2-\alpha}{\alpha}} &= 2(C_\alpha m A(\tau,x))^{-1} = 2(C_\alpha A(m\tau, x^m))^{-1}, \end{aligned}$$

where $A(\tau,x)$ is defined by (2.12). Note that by Lemma V.3.12 of [Eke3] and (9.2.25),

$$\frac{\hat{i}(x,m)}{A(m\tau,x^m)} = \frac{\hat{i}(x,1)}{A(\tau,x)} = \frac{1}{\gamma}.$$

This implies

$$|c_k|^{\frac{2-\alpha}{\alpha}}\hat{i}(x,m) = \frac{2}{\gamma C_\alpha}.$$

Since $c_k < c_{k+1} < 0$ for $k \in \mathbf{N}$, we get our results. □

From our observations on the weakly non-resonant ellipsoid and studies of the case of $\mathcal{H}(4)$ in [Lon18] as well as our study of $\mathcal{H}(6)$, we noticed that in order to maximize the effect of the Fadell–Rabinowitz S^1-index theory, instead of the index interval $\mathcal{I}_m(\tau,x)$, we should consider the largest open interval which contains $\mathcal{I}_m(\tau,x)$, possesses no part of any other index interval of (τ,x), and still can be used as the target of the map Ψ. This leads to our introduction of the index jump of (τ,x) using Definition 11.2.3.

Definition 6 (Y. Long and C. Zhu [LZu3]). For $\Sigma \in \mathcal{H}(2n)$ and $\alpha \in (1,2)$, we define the m**-th index jump** $\mathcal{G}_m(\tau,x)$ of $(\tau,x) \in \mathcal{J}(\Sigma,\alpha)$ by the open interval

$$\mathcal{G}_m(\tau,x) = \mathcal{G}_m(\gamma_x) = (i(x,m)+\nu(x,m)-1, i(x,m+2)). \tag{19}$$

When (14) holds, we have $\mathcal{V}_\infty(\Sigma,\alpha) \neq \emptyset$ and we write

$$\mathcal{V}_\infty(\Sigma,\alpha) = \{[(\tau_1,x_1)],\ldots,[(\tau_q,x_q)]\}. \tag{20}$$

By property (10), we observe that the common index jump Theorems 11.2.1, 11.2.2, and Corollary 11.2.4 can be applied to yield infinitely many chances that the index jumps of all the q closed characteristics contain common intervals, i.e., there exist infinitely many $(N,m_1,\ldots,m_q) \in \mathbf{N}^{q+1}$ such that

$$[2N-\kappa_1, 2N+\kappa_2] \subset \bigcap_{j=1}^{q} \mathcal{G}_{2m_j-1}(\tau_j,x_j), \tag{21}$$

where

$$\kappa_1 \equiv \kappa_1(\Sigma, \alpha) = \min_{1 \le j \le q} (i(x_j, 1) + 2S^+(x_j) - \nu(x_j, 1)), \tag{22}$$

$$\kappa_2 \equiv \kappa_2(\Sigma, \alpha) = \min_{1 \le j \le q} (i(x_j, 1) - 1), \tag{23}$$

and $S^+(x_j) = S^+_{\gamma_x(\tau_j)}(1)$. Note that here by (1.18), Lemma 2.4, Theorems 9.1.7, and 9.1.10 on the splitting numbers, we obtain

$$\kappa_1 \ge n + 2 - 2n = 2 - n, \tag{24}$$

$$\kappa_2 \ge n - 1. \tag{25}$$

Thus especially we have

$$\kappa_1 + \kappa_2 \ge 1. \tag{26}$$

So the left interval in (21) is non-empty. We shall get a better estimate on κ_1 later in Section 4 below to estimate the size of the interval $[2N - \kappa_1, 2N + \kappa_2]$. This important phenomenon will be used in our proofs of the main theorems on the existence of multiple closed characteristics on any $\Sigma \in \mathcal{H}(2n)$ in this chapter.

Example 7 We study the index jumps of closed characteristics on the weakly non-resonant ellipsoid $\mathcal{E}_3(r)$ in $\mathbf{R}^6$ defined by (0.8) as an example, where we set $r = (r_1, r_2, r_3)$ with

$$0 < r_1 = \frac{1}{2\pi} < r_2 = \frac{1}{5} < r_3 = \frac{1}{\sqrt{11}}. \tag{27}$$

This example suggests our general observations on the common index jump theorem.

In this case the corresponding Hamiltonian function is quadratic and is given by

$$H(y) = \frac{1}{2} \sum_{j=1}^{3} \frac{|y_j|^2}{r_j}, \quad \forall y = (y_1, y_2, y_3) \in (\mathbf{R}^2)^3.$$

Then solving explicitly the linear Hamiltonian system

$$\dot{y} = JH'(y), \tag{28}$$

we obtain precisely three two-parameter families of periodic solutions of (28). By requiring $H(y) = 1$, we obtain precisely three elements in $\tilde{\mathcal{J}}(\Sigma)$, which we denote by $[(\tau_j, x_j)]$ for $j = 1, 2$, and 3 respectively according to the action

$$A(\tau_j, x_j) = \frac{1}{2} \int_0^{\tau_j} J\dot{x}_j \cdot x_j dt = -2\pi r_j. \tag{29}$$

Note that all of them are strongly elliptic, so their nullities are always 2. When we choose $\alpha \in (1, 2)$, their nullities are always 1, and their indices are the same as those when $\alpha = 2$ by Proposition I.7.5 of [Eke3]. Denote the indices and the splitting numbers of these three closed characteristics by $(i(x_j, m), \nu(x_j, m))$ and $S^+(x_j)$ respectively for $m \in \mathbf{N}$ and $j = 1, 2, 3$. Then for these m and j when $\alpha \in (1, 2)$,

$$\begin{aligned}
i(x_1, m) &= 2m + 1 + 2\left[\frac{m\sqrt{11}}{2\pi}\right] + 2\left[\frac{5m}{2\pi}\right],\\
i(x_2, m) &= 2m + 1 + 2\left[\frac{m\sqrt{11}}{5}\right] + 2\left[\frac{2m\pi}{5}\right],\\
i(x_3, m) &= 2m + 1 + 2\left[\frac{5m}{\sqrt{11}}\right] + 2\left[\frac{2m\pi}{\sqrt{11}}\right],\\
\nu(x_j, m) &= 1,\\
S^+(x_j) &= 1.
\end{aligned}$$

Thus we obtain $\kappa_1 = 3$ and $\kappa_2 = 2$. By these iteration formulae, we obtain

$$\begin{aligned}
&2\mathbf{N} - 2 + 3 = \cup_{j=1}^{3} \mathcal{I}(\tau_j, x_j),\\
&\mathcal{I}(\tau_j, x_j) \cap \mathcal{I}(\tau_k, x_k) = \emptyset, \qquad \forall 1 \le j < k \le 3.
\end{aligned}$$

Therefore all these three closed characteristics are infinite variationally visible. The invariant $\varrho_3(\Sigma)$ for $\Sigma = \mathcal{E}_3(r)$ defined by (15.4.1) below is given by

$$\varrho_3(\mathcal{E}_3(r)) = \left[\frac{1}{2}(i(x_1, 1) + 2S^+(x_1) - \nu(x_1, 1) + 3)\right] = 3.$$

From the above iteration formulae of x_j, we can list the sequences of indices as follows, for example, up to $m = 10$:

m	1	2	3	4	5	6	7	8	9	10
$i(x_1, m)$	3	9	13	19	21	27	31	37	41	45
$i(x_2, m)$	5	11	15	23	29	33	39	47	51	57
$i(x_3, m)$	7	17	25	35	43	53	61	71	79	87.

Then we get

$$\begin{aligned}
\{33, 35, 37\} &\subset [33, 38]\\
&= [2N - \kappa_1, 2N + \kappa_2]\\
&\subset (31, 41) \cap (29, 39) \cap (25, 43)\\
&= \bigcap_{j=1}^{3} \mathcal{G}_{2m_j - 1}(\tau_j, x_j),
\end{aligned}$$

where $m_1 = 4$, $m_2 = 3$, $m_3 = 2$, and $N = 18$. This illustrates the common index jump property. Here we notice that

$$\begin{aligned} &{}^{\#}((2\mathrm{N}-2+n)\cap[2N-\kappa_1, 2N+\kappa_2]) = {}^{\#}\{33, 35, 37\} \\ &\quad = 3 = {}^{\#}\tilde{\mathcal{J}}(\Sigma) = \varrho_3(\mathcal{E}_3(r)). \end{aligned}$$

In the next section, we shall use numbers of effective integers in the interval $[2N - \kappa_1, 2N + \kappa_2]$ to estimate ${}^{\#}\tilde{\mathcal{J}}(\Sigma)$ from below as in (22) of the Introduction of this book.

Remark 8 Definitions 3 and 4 were introduced in [Lon14] of 1998. Lemma 5 and Definition 6 were proved or introduced by C. Zhu and the author in the recent [LZu3]. The explanation of the Fadell-Rabinowitz cohomological index at the beginning of this section was given by J. Mather as an appendix of [LZu3].

15.4 Existence and multiplicity

In this section, we study the existence and multiplicity of closed characteristics on any $\Sigma \in \mathcal{H}(2n)$. We introduce the invariant $\varrho_n : \mathcal{H}(2n) \to \mathbf{N} \cup \{+\infty\}$ first.

Definition 1 (Y. Long and C. Zhu [LZu3]). For $\alpha \in (1, 2)$, we define a map $\varrho_n : \mathcal{H}(2n) \to \mathbf{N} \cup \{+\infty\}$ by

$$\varrho_n(\Sigma) = \begin{cases} +\infty, & \text{if} \quad {}^{\#}\mathcal{V}(\Sigma, \alpha) = +\infty, \\ \min\{[\frac{i(x,1)+2S^+(x)-\nu(x,1)+n}{2}] & | \quad [(\tau, x)] \in \mathcal{V}_\infty(\Sigma, \alpha)\}, \\ & \text{if} \quad {}^{\#}\mathcal{V}(\Sigma, \alpha) < +\infty. \end{cases} \tag{1}$$

where $(i(x, 1), \nu(x, 1))$ is the index of (τ, x) defined in Chapter 5, $S^+(x) = S_M^+(1)$ is the splitting number of (τ, x) with $M = \gamma_x(\tau)$, and $\mathcal{V}(\Sigma, \alpha)$ and $\mathcal{V}_\infty(\Sigma, \alpha)$ are given by Definition 3.3.

Note that when ${}^{\#}\tilde{\mathcal{J}}(\Sigma) < +\infty$, the set $\mathcal{V}_\infty(\Sigma, \alpha)$ is non-empty and finite. Our next lemma shows that $\varrho_n(\Sigma)$ does not depend on the choice of $\alpha \in (1, 2)$ and is shape invariant, i.e., is independent of dilation of Σ.

Lemma 2 *Fix $\Sigma \in \mathcal{H}(2n)$ and $\alpha \in (1, 2)$.*

(i) *$\varrho_n(\Sigma)$ defined by (1) does not depend on the choice of $\alpha \in (1, 2)$.*
(ii) *For any $s > 0$ and $\Sigma \in \mathcal{H}(2n)$, it holds that $\varrho_n(s\Sigma) = \varrho_n(\Sigma)$.*

Proof: (i) follows from Proposition I.7.5 of [Eke3] and a similar proof for the index functions and splitting numbers.

(ii) By the definition of the gauge function j_Σ of Σ for any $s > 0$ we have

$$\begin{aligned} j_{s\Sigma}(x) &= s^{-1} j_\Sigma(x), \\ H_{s,\alpha}(x) &\equiv j_{s\Sigma}(x)^\alpha = s^{-\alpha} H_\alpha(x), \end{aligned}$$

where H_α is defined by (0.3). Let $(\tau, x) \in \mathcal{J}(\Sigma, \alpha)$. Then τ is the minimal period of x and $H_\alpha(x(t)) = 1$. Set $y(t) = sx(s^{-2}t)$. Then $s^2\tau$ is the minimal period of y and $H_{s,\alpha}(y(t)) = 1$. Since x is a solution of (0.4)–(0.6), we have

$$\begin{aligned} \dot{y}(t) &= s^{-1}JH'_\alpha(x(s^{-2}t)) \\ &= s^{-\alpha}JH'_\alpha(sx(s^{-2}t)) \\ &= JH'_{s,\alpha}(y(t)). \end{aligned}$$

So we have $(s^2\tau, y) \in \mathcal{J}(s\Sigma, \alpha)$. So the map from $\mathcal{J}(\Sigma, \alpha)$ to $\mathcal{J}(s\Sigma, \alpha)$ defined by $(\tau, x) \mapsto (s^2\tau, sx(s^{-2}t))$ is a bijection.

Let γ_x be the associated symplectic path of (τ, x). Then we have

$$\begin{aligned} \frac{d}{dt}\gamma_x(s^{-2}t) &= s^{-2}JH''_\alpha(x(s^{-2}t))\gamma_x(s^{-2}t)) \\ &= s^{-\alpha}JH''_\alpha(sx(s^{-2}t))\gamma_x(s^{-2}t)) \\ &= JH''_{s,\alpha}(y(t))\gamma_x(s^{-2}t)). \end{aligned}$$

So the associated symplectic path of $(s^2\tau, y)$ is $\gamma_y(t) = \gamma_x(s^{-2}t)$. Hence our lemma follows. □

The following is our main existence result on the closed characteristics, which implies the result (0.0.16) and (0.0.17) in Theorem C in the Introduction of the book.

In the following, when ${}^\#\tilde{\mathcal{J}}(\Sigma) < +\infty$, we always denote the elements in $\mathcal{V}_\infty(\Sigma, \alpha)$ by

$$\mathcal{V}_\infty(\Sigma, \alpha) = \{[(\tau_j, x_j)] \mid j = 1, \ldots, q\}, \tag{2}$$

for some integer q, where $(\tau_j, x_j) \in \mathcal{J}(\Sigma, \alpha)$ for $j = 1, \ldots, q$. Note that by (3.13), we always have

$$q \geq 1.$$

Theorem 3 (Y. Long and C. Zhu [LZu3]). *Let $\Sigma \in \mathcal{H}(2n)$ and $1 < \alpha < 2$. Suppose ${}^\#\tilde{\mathcal{J}}(\Sigma) < +\infty$. Then*

$$\min\{{}^\#\mathcal{V}_\infty(\Sigma, \alpha), n\} \geq \varrho_n(\Sigma) \geq \left[\frac{n}{2}\right] + 1. \tag{3}$$

Proof: Because ${}^\#\tilde{\mathcal{J}}(\Sigma) < +\infty$, by Lemma 3.5 there exist an integer $K \geq 0$ and an injection map $\Psi : \mathbf{N} + K \to \mathcal{V}_\infty(\Sigma, \alpha) \times \mathbf{N}$ such that

$$i(x, m) \leq 2k - 2 + n \leq i(x, m) + \nu(x, m) - 1 \tag{4}$$

holds for any $k \in \mathbf{N} + K$, $(\tau, x) \in \mathcal{J}(\Sigma, \alpha)$ and $\Psi(k) = ([(\tau, x)], m)$.

By Lemma 2.4 and (3.10), we can apply the common index jump Theorems 11.2.1 and 11.2.2, and obtain infinitely many $(N, m_1, \dots, m_q) \in \mathbf{N}^{q+1}$ such that

$$\bigcap_{j=1}^{q} \mathcal{G}_{2m_j-1}(\tau_j, x_j) \supset [2N - \kappa_1, 2N + \kappa_2], \tag{5}$$

$$\bigcup_{j=1}^{q} \mathcal{I}_{2m_j}(\tau_j, x_j) \subset [2N - n, 2N - 2 + n], \tag{6}$$

where $\kappa_1 = \kappa_1(\Sigma, \alpha)$ and $\kappa_2 = \kappa_2(\Sigma, \alpha)$ are defined by (3.22) and (3.23) respectively. We continue the proof in the following three claims.

Claim 1 *It holds that*

$$q \geq {}^{\#}((2\mathbf{N} - 2 + n) \cap [2N - \kappa_1, 2N + \kappa_2]) \geq \varrho_n(\Sigma). \tag{7}$$

In fact, by the definition of the interval $[2N - \kappa_1, 2N + \kappa_2]$, the injection map Ψ defined by Lemma 3.5 induces another injection map Ψ' from the set $(2\mathbf{N} - 2 + n) \cap [2N - \kappa_1, 2N + \kappa_2]$ to $\mathcal{V}_\infty(\Sigma, \alpha) \times \mathbf{N}$. But by (4) and the definition of the index jump $\mathcal{G}_{2m_j-1}(\tau_j, x_j)$, no two iterations of one element $[(\tau_j, x_j)]$ in $\mathcal{V}_\infty(\Sigma, \alpha)$ can be the image points of Ψ' simultaneously. Thus we obtain the left inequality in (7).

To prove the right inequality in (7), we set

$$\Psi(N - s + 1) = ([(\tau_{j(s)}, x_{j(s)}], m(s)), \qquad \forall s = 1, \dots, \varrho_n(\Sigma), \tag{8}$$

where $j(s) \in \{1, \dots, q\}$ and $m(s) \in \mathbf{N}$. Then by the definition of the map Ψ in Lemma 3.5, we obtain

$$i(x_{j(s)}, m(s)) \leq 2N - 2s + n \leq i(x_{j(s)}, m(s)) + \nu(x_{j(s)}, m(s)) - 1. \tag{9}$$

Note that by the definitions (3.22) of κ_1 and (1) of $\varrho_n(\Sigma)$ we have

$$\varrho_n(\Sigma) = \left[\frac{\kappa_1 + n}{2}\right]. \tag{10}$$

Together with (3.25), we obtain

$$\begin{aligned} 2N - \kappa_1 &\leq 2N - 2\left[\frac{\kappa_1 + n}{2}\right] + n \\ &\leq 2N - 2s + n \\ &\leq 2N - 2 + n \\ &\leq 2N + \kappa_2, \end{aligned} \tag{11}$$

for $s = 1, \dots, \varrho_n(\Sigma)$. Thus the right inequality of (7) holds. This proves Claim 1.

Note that the only index interval $\mathcal{I}_m(\tau_{j(s)}, x_{j(s)})$ contained in the index jump $\mathcal{G}_{2m_{j(s)}-1}(\tau_{j(s)}, x_{j(s)})$ is the one with $m = 2m_{j(s)}$. Therefore we must have

$$m(s) = 2m_{j(s)}, \qquad \forall s = 1, \ldots, \varrho_n(\Sigma). \tag{12}$$

Thus by reordering the elements in $\mathcal{V}_\infty(\Sigma, \alpha)$ and Claim 1,

$$\Psi(N - s + 1) = ([(\tau_s, x_s)], 2m_s), \qquad \forall s = 1, \ldots, \varrho_n(\Sigma). \tag{13}$$

Next we derive this fact on $m(s)$ and $j(s)$ and give the proof of Claim 1 and the above (13) slightly differently by our iteration inequalities obtained before.

From Definition 3.6 of the index jump, (5), (6), and (9), we obtain

$$\begin{aligned} & i(x_{j(s)}, 2m_{j(s)} - 1) + \nu(x_{j(s)}, 2m_{j(s)} - 1) - 1 \\ & < 2N - \kappa_1 \\ & \leq 2N - 2s + n \\ & \leq i(x_{j(s)}, m(s)) + \nu(x_{j(s)}, m(s)) - 1, \end{aligned} \tag{14}$$

and

$$\begin{aligned} i(x_{j(s)}, m(s)) & \leq 2N - 2s + n \\ & < 2N + \kappa_2 \\ & < i(x_{j(s)}, 2m_{j(s)} + 1). \end{aligned} \tag{15}$$

Comparing (14), (15), and (3.8) of Lemma 3.2, we obtain

$$2m_{j(s)} - 1 < m(s) < 2m_{j(s)} + 1.$$

Hence $m(s) = 2m_{j(s)}$, and then (13) holds. Note that (7) and then Claim 1 follows also from (13) and the injectivity of the map Ψ.

Claim 2 $\varrho_n(\Sigma) \leq n$.

In fact, assuming $\varrho_n(\Sigma) > n$ we prove the claim indirectly. By Claim 1 and (13) with $s = n + 1$ we obtain

$$\begin{aligned} q & \geq n + 1, \\ \Psi(N - n) & = ([(\tau_{n+1}, x_{n+1})], 2m_{n+1}). \end{aligned}$$

Then by (3.5) and (6) we have

$$i(x_{n+1}, 2m_{n+1}) \leq 2(N - n) - 2 + n < 2N - n \leq i(x_{n+1}, 2m_{n+1}). \tag{16}$$

This contradiction proves our Claim 2.

Claim 3 $\varrho_n(\Sigma) \geq [\frac{n}{2}] + 1$.

In fact, let $(\tau, x) \in \mathcal{J}(\Sigma, \alpha)$ and γ_x be its associated symplectic path. In the following, we estimate $i(x, 1) + 2S^+(x) - \nu(x, 1) + n$.

By Lemma 2.4, (2.22) holds for some $P \in \mathrm{Sp}(2n)$ and $M \in \mathrm{Sp}(2n-2)$. By 1° and 2° of Theorem 9.1.9, the splitting numbers are constant on each homotopy component and symplectic additive. Thus by (2.22), we get

$$\begin{aligned} &2S^+(x) - \nu(x, 1) \\ &\quad = 2S^+_{N_1(1,1)}(1) - \nu_1(N_1(1, 1)) + 2S^+_M(1) - \nu_1(M). \end{aligned} \tag{17}$$

By Theorem 9.1.7, we obtain

$$S^+_{N_1(1,a)}(1) = \begin{cases} 1, & \text{if } a \geq 0, \\ 0, & \text{if } a < 0. \end{cases} \tag{18}$$

Thus

$$2S^+_{N_1(1,a)}(1) - \nu_1(N_1(1, a)) = a, \quad \text{for } a = \pm 1, 0. \tag{19}$$

By Theorem 1.8.10, we obtain

$$N_1(1, 1)^{\diamond p_-} \diamond I_2^{\diamond p_0} \diamond N_1(1, -1)^{\diamond p_+} \diamond G \in \Omega^0(M), \tag{20}$$

for some nonnegative integers p_-, p_0, and p_+, and some symplectic matrix G satisfying $1 \notin \sigma(G)$. By (19) and (20), we then obtain

$$2S^+_M(1) - \nu_1(M) \geq p_- - p_+ \geq -p_+ \geq 1 - n. \tag{21}$$

From (1.18), (17), (19) with $a = 1$, and (21), we obtain

$$\begin{aligned} i(x, 1) + 2S^+(x) - \nu(x, 1) + n &\geq n + 1 + (1 - n) + n \\ &= n + 2. \end{aligned} \tag{22}$$

So Claim 3 holds.

The proof of Theorem 3 is complete. □

Next we study the non-degenerate case. The following result implies the corresponding result (0.0.18) in Theorem C in the Introduction of the book.

Theorem 4 (Y. Long and C. Zhu [LZu3]). *Fix $\Sigma \in \mathcal{H}(2n)$ and $\alpha \in (1, 2)$. Suppose every $(\tau, x) \in \mathcal{J}(\Sigma, \alpha)$ satisfies*

$$i(x, 1) + 2S^+(x) - \nu(x, 1) \geq n. \tag{23}$$

Then

$$\varrho_n(\Sigma) \geq n. \tag{24}$$

In particular, when every $(\tau, x) \in \mathcal{J}(\Sigma)$ is non-degenerate, (23) *must hold, and then*

$$^{\#}\tilde{\mathcal{J}}(\Sigma) \geq n. \tag{25}$$

Proof: By the definition of $\varrho_n(\Sigma)$ and (23), we obtain (24).

If $(\tau, x) \in \mathcal{J}(\Sigma, \alpha)$ is non-degenerate, we must have $1 \notin \sigma(M)$ in (17). Thus by (17) and (19), we obtain $2S^+(x) - \nu(x, 1) = 1$. Together with (1.18), we obtain (23). So (24) holds. By (3) of Theorem 3, this implies (25). □

Corollary 5 *Let $\Sigma \in \mathcal{H}(2n)$ and $\alpha \in (1, 2)$. Assume $^{\#}\tilde{\mathcal{J}}(\Sigma) < +\infty$. Then there exists an element $[(\tau, x)] \in \mathcal{V}_\infty(\Sigma, \alpha)$ with $i(x, 1) = n$.*

Proof: Using (3), the definition of the map Ψ in Lemma 3.5, and notation in the proof of Theorem 3, we can denote $([(\tau_j, x_j)], m) = \Psi(N + 1)$ for some $1 \le j \le q$ and some $m \in \mathbf{N}$. Note that (13) is not applicable here. By the definition of Ψ and (3.5), we obtain

$$i(x_j, m) \le 2N + n \le i(x_j, m) + \nu(x_j, m) - 1. \tag{26}$$

On the other hand, (5) and (6) hold for the $2m_j$-th iteration of x_j. Thus by (3.25), and the right end point properties of (5) and (6), we obtain

$$\begin{aligned} i(x_j, 2m_j) + \nu(x_j, 2m_j) - 1 &\le 2N - 2 + n \\ &< 2N + n \\ &\le 2N + \kappa_2 + 1 \\ &\le i(x_j, 2m_j + 1). \end{aligned} \tag{27}$$

Hence by Lemma 3.2, comparing (26) and (27) we obtain

$$2m_j < m \le 2m_j + 1.$$

Thus $m = 2m_j + 1$. By (26) and (27) again we obtain

$$i(x_j, 2m_j + 1) = 2N + n.$$

Now by (11.2.5) of Theorem 11.2.1, we obtain

$$i(x_j, 1) = i(x_j, 2m_j + 1) - 2N = n.$$

This completes the proof. □

Remark 6 Although the global minimal point (τ, x) of f_α on E_α satisfies $i(x, 1) = n$, it is still not clear to us now whether $[(\tau, x)] \in \mathcal{V}_\infty(\Sigma, \alpha)$.

15.5 Stability results

In this section, we study the ellipticity of closed characteristics, and prove Theorem D in the Introduction of the book.

For $\Sigma \in \mathcal{H}(2n)$, $\alpha \in (1,2)$, and $(\tau, x) \in \mathcal{J}(\Sigma, \alpha)$, we define the **elliptic height** $e(x)$ of (τ, x) to be the elliptic height $e(\gamma_x(\tau))$, where γ_x is the associated symplectic path of (τ, x).

We start from the following theorem which implies the first part of Theorem D mentioned in the Introduction under condition (0.0.20) there.

Theorem 1 (Y. Long and C. Zhu [LZu3]). *Let* $\Sigma \in \mathcal{H}(2n)$ *and* $\alpha \in (1,2)$. *Suppose* $^{\#}\tilde{\mathcal{J}}(\Sigma) < +\infty$. *We use notation introduced in the proof of Theorem* 4.3. *By reordering elements in* $\mathcal{V}_\infty(\Sigma, \alpha)$, (4.13) *holds, i.e.,*

$$\Psi(N - s + 1) = ([(\tau_s, x_s)], 2m_s), \qquad \forall s = 1, \ldots, \varrho_n(\Sigma). \tag{1}$$

Then we have

$$e(x_s) \ge 2|n - 2s + 1| + 2, \quad \forall s = 1, \ldots, \varrho_n(\Sigma). \tag{2}$$

Especially $[(\tau_1, x_1)]$ *is an elliptic element in* $\mathcal{V}_\infty(\Sigma, \alpha)$.

Proof: By the definition of the injection map Ψ in Lemma 3.5, for $s = 1, \ldots, \varrho_n(\Sigma)$ we obtain

$$\begin{aligned} i(x_s, 2m_s) &\le 2N - 2s + n \\ &\le i(x_s, 2m_s) + \nu(x_s, 2m_s) - 1. \end{aligned} \tag{3}$$

By the first inequality of (10.2.22) in Theorem 10.2.4 and (11.2.5) of the common index jump Theorem 11.2.1, we have

$$\begin{aligned} \frac{e(x_s)}{2} &\ge i(x_s, 2m_s) + \nu(x_s, 2m_s) \\ &\quad - i(x_s, 2m_s + 1) + i(x_s, 1) + 1 \\ &= i(x_s, 2m_s) + \nu(x_s, 2m_s) - 2N + 1 \\ &\ge 2N - 2s + n + 1 - 2N + 1 \\ &= (n - 2s + 1) + 1, \end{aligned} \tag{4}$$

where the second inequality in (4) follows from the second inequality in (3).

On the other hand, by the second inequality of (10.2.22) in Theorem 10.2.4, we obtain

$$\begin{aligned} \frac{e(x_s)}{2} &\ge i(x_s, 2m_s + 1) + \nu(x_s, 2m_s + 1) \\ &\quad - i(x_s, 2m_s) - i(x_s, 1) - \nu(x_s, 1). \end{aligned}$$

By (11.2.3) and (11.2.5) of the common index jump Theorem 11.2.1 and the first inequality in (3), we then obtain

$$\begin{aligned} \frac{e(x_s)}{2} &\ge 2N - (2N - 2s + n) \\ &= (2s - n - 1) + 1. \end{aligned} \tag{5}$$

Now (4) and (5) yield (2). □

Our following theorem implies the second part of Theorem D in the Introduction on the multiplicity of closed characteristics with irrational mean indices. Note that by Theorem 4.3, $\varrho_n(\Sigma) - 1 \geq [\frac{n}{2}]$ always holds.

Theorem 2 (Y. Long and C. Zhu [LZu3]). *Let $\Sigma \in \mathcal{H}(2n)$ with $n \geq 2$ and $\alpha \in (1, 2)$. Suppose $^{\#}\tilde{\mathcal{J}}(\Sigma) < +\infty$. Then there exist at least $\varrho_n(\Sigma) - 1$ elements in $\mathcal{V}_\infty(\Sigma, \alpha)$ such that each $[(\tau, x)]$ of these elements possesses irrational mean index, i.e.,*

$$\hat{i}(x, 1) \in \mathbf{R}\backslash\mathbf{Q}. \tag{6}$$

Especially such a closed characteristic $[(\tau, x)]$ must possess at least an irrational rotator on U *in the sense of Theorem* 9.3.1, *i.e., in* (9.3.2) *there is a term $E(\frac{m\theta}{2\pi})$ with $S_M^-(e^{\sqrt{-1}\theta}) > 0$ and θ is an irrational multiple of 2π.*

Proof: Since $n \geq 2$, we have $\varrho_n(\Sigma) - 1 \geq 1$. We use the notation (4.2) of $\mathcal{V}_\infty(\Sigma, \alpha)$. By (11.2.13) and Corollary 9.3.2, the D_j defined by (11.1.2) according to Step 1 of the proof of Theorem 11.2.1 satisfies

$$D_j = \hat{i}(x_j, 1), \quad \forall j = 1, \ldots, q. \tag{7}$$

By the proof of Theorem 4.3, we obtain $(N, m_1, \ldots, m_q) \in \mathbf{N}^{q+1}$ such that the injection map Ψ of Lemma 3.5 satisfies (1). Thus by Lemma 3.5 and (9.2.25), we have

$$\begin{aligned} 2m_k D_k - 2m_j D_j &= 2m_k\hat{i}(x_k, 1) - 2m_j\hat{i}(x_j, 1) \\ &= \hat{i}(x_k, 2m_k) - \hat{i}(x_j, 2m_j) \\ &< 0, \qquad \text{if } 1 \leq j < k \leq \varrho_n(\Sigma). \end{aligned} \tag{8}$$

Now it suffices to prove that among the first $\varrho_n(\Sigma)$ of D_j, at most one of them is rational. We prove this claim indirectly by assuming

$$D_j \quad \text{and} \quad D_k \in \mathbf{Q}, \tag{9}$$

for some j and k satisfying $1 \leq j < k \leq \varrho_n(\Sigma)$. Then in our choice of $(N, m_1, \ldots, m_q)$ in the proof of Theorem 11.1.1, we require that $N \in \mathbf{N}$ further satisfies

$$\frac{N}{MD_i} \in \mathbf{N}, \quad \text{for } i = j, \text{ and } k. \tag{10}$$

Then the closure of the set $\{\{Nv\} \mid N \in \mathbf{N}$ satisfies (10)$\}$ is still a closed additive subgroup of $\mathcal{T}^h$ for some $h \in \mathbf{N}$. Thus (11.1.22) holds in the j-th and k-th coordinates for infinitely many N, if we choose χ to be a vertex of $[0, 1]^h$ with $\chi_j = \chi_k = 0$. This implies

$$\left\| \left\{ \frac{N}{MD_i} \right\} \right\| = 0, \quad \text{for } i = j, k.$$

Therefore by the definition (11.1.10) of m_i, we obtain

$$m_i = \left[\frac{N}{MD_i}\right]M = \frac{N}{D_i}, \quad \text{for } i = j, k$$

This yields

$$m_j D_j = N = m_k D_k, \tag{11}$$

which contradicts (8) and completes the proof of the theorem. □

Our following theorem studies the existence of multiple elliptic orbits and implies the third part of Theorem D in the Introduction.

Theorem 3 (Y. Long and C. Zhu [LZu3]). *Let* $\Sigma \in \mathcal{H}(2n)$ *with* $n \geq 2$ *and* $\alpha \in (1, 2)$. *Suppose* ${}^{\#}\tilde{\mathcal{J}}(\Sigma) < +\infty$ *and*

$${}^{\#}\mathcal{V}_\infty(\Sigma, \alpha) \leq 2\varrho_n(\Sigma) - 2. \tag{12}$$

Then there exist at least two elliptic elements in $\mathcal{V}_\infty(\Sigma, \alpha)$.

In particular by the right-hand side inequality of (4.3), *there are at least two elliptic elements in* $\mathcal{V}_\infty(\Sigma, \alpha)$, *provided*

$${}^{\#}\mathcal{V}(\Sigma, \alpha) \leq 2\left[\frac{n}{2}\right]. \tag{13}$$

Proof: We use the notation (4.2) of $\mathcal{V}_\infty(\Sigma, \alpha)$. Let $v \in \mathbf{R}^k$ be the vector given by (11.1.21) according to quantities in (11.2.14)–(11.2.15) in the proof of Theorem 11.2.1. Let $A(v)$ be the set given by Theorem 11.1.2 according to this v.

We use notations introduced in the proof of Theorem 2. Note that by Theorem 4.3 and $n \geq 2$, we have $q \geq \varrho_n(\Sigma) \geq 2$. By Theorem 2, $D_j \in \mathbf{R}\backslash\mathbf{Q}$ holds for at least $[n/2]$ integers in $\{1, \ldots, q\}$. Therefore by Theorem 11.1.2, we obtain

$$\dim A(v) \geq 1, \quad 0 \notin A(v). \tag{14}$$

We continue our proof in two steps.

Step 1 For a given $a \in A(v)$, we define $\chi \equiv \chi(a) = (\psi(a_1), \ldots, \psi(a_k))$ by (11.1.26). Let $(N, m_1, \ldots, m_q) \in \mathbf{N}^{q+1}$ be given in the proof for Theorem 4.3 via an application of Theorems 11.1.1, 11.1.2, and 11.2.1 with this $\chi = \chi(a)$ used in their proofs. By the proof of Theorem 4.3, there exists a subset $L(a, N)$ of $\{1, \ldots, q\}$ such that

(i) $1 \in L(a, N)$ and ${}^{\#}L(a, N) = \varrho_n(\Sigma)$,
(ii) $\Psi(N) = ([(\tau_1, x_1), 2m_1)]$ and x_1 is elliptic,
(iii) for each $s = 2, \ldots, \varrho_n(\Sigma)$, there is a $j(s) \in L(a, N)\backslash\{1\}$ such that $\Psi(N - s + 1) = ([\tau_{j(s)}, x_{j(s)}], 2m_{j(s)})$.

By Lemma 3.5, we have

$$m_1 D_1 - m_j D_j > 0, \quad \forall j \in L(a, N)\backslash\{1\}. \tag{15}$$

Now assume that there exists only one elliptic element in $\mathcal{V}_\infty(\Sigma, \alpha)$. Thus we always get the same elliptic element $[(\tau_1, x_1)] \in \mathcal{V}_\infty(\Sigma, \alpha)$ for all $a \in A(v)$.

By (14), we have $-a \in A(v)$. By the definition (11.1.26), we then have $\chi(-a) \neq \chi(a)$. Thus by the proof of Theorem 4.3 for this vertex $\chi(-a)$ we get $(\tilde{N}, \tilde{m_1}, \ldots, \tilde{m_q}) \in \mathbf{N}^{q+1}$ and another subset $L(-a, \tilde{N})$ of $\{1, \ldots, q\}$ such that Theorems 4.3, 4.4, 1, and 2, especially the above (i)–(iii) still hold correspondingly.

By the assumption $q \leq 2\varrho_n(\Sigma) - 2$ of (12), we have

$$1 \leq {}^{\#}((L(a, N) \cap L(-a, \tilde{N}))\backslash\{1\}) \leq \varrho_n(\Sigma). \tag{16}$$

Let

$$t_0 = \frac{\delta_1}{6(|a| + 1)(M\Lambda + 1)}, \quad \Lambda = \max_{1 \leq j \leq q} D_j. \tag{17}$$

Note that in the proof of Theorem 11.1.1 we can further require $N \in \mathbf{N}$ such that the vector $\{Nv\} - \chi(a)$ defined by (11.1.22) and (11.1.21) are located in a sufficiently small neighborhood inside the open ball in V with radius $\delta_1/(6M\Lambda + 1)$ and centered at at_0, i.e.,

$$\{Nv\} - \chi(a) \in V, \tag{18}$$

$$|\{Nv\} - \chi(a) - at_0| < \frac{\delta_1}{6M\Lambda + 1}, \tag{19}$$

where $V = T_0 \pi^{-1}(\overline{\{\{Nv\} \mid N \in \mathbf{N}\}})$ is defined in Theorem 11.1.2. Under this requirement, we still have (11.1.22):

$$\left|\left\{\frac{N}{MD_i}\right\} - \chi_i\right| < |a_i t_0| + \frac{\delta_1}{6M\Lambda + 1} \leq \frac{\delta_1}{3M\Lambda}, \tag{20}$$

$$\left|\left\{\frac{N\alpha_{i,j}}{D_i}\right\} - \chi_{i,j}\right| < |a_{i,j} t_0| + \frac{\delta_1}{6M\Lambda + 1} \leq \frac{\delta_1}{3}, \tag{21}$$

for infinitely many $N \in \mathbf{N}$. Here $\chi_i = \psi(a_i)$ and $\chi_{i,j} = \psi(a_{i,j})$, the function ψ is defined in Theorem 11.1.2.

Claim: $a_1 D_1 - a_j D_j = 0$ *for all* $j \in (L(a, N) \cap L(-a, \tilde{N}))\backslash\{1\}$.

Assuming the claim does not hold, we prove it by contradiction. In fact, we can further require $N \in \mathbf{N}$ so that

$$\left|\left\{\frac{N}{MD_k}\right\} - \chi_k - a_k t_0\right| < \frac{t_0}{3\Lambda} \min_{a_i D_i - a_j D_j \neq 0} |a_i D_i - a_j D_j|,$$

for all $1 \le k \le q$. By our choice of $(N, m_1, \ldots, m_q)$, we have

$$\begin{aligned}
&m_1 D_1 - m_j D_j \\
&= M\left(\left[\frac{N}{MD_1}\right] + \chi_1\right) D_1 - M\left(\left[\frac{N}{MD_j}\right] + \chi_j\right) D_j \\
&= M\left(\left(\chi_1 - \left\{\frac{N}{MD_1}\right\}\right) D_1 - \left(\chi_j - \left\{\frac{N}{MD_j}\right\}\right) D_j\right) \qquad (22)\\
&= -Mt_0(a_1 D_1 - a_j D_j) + M\left(\left(\chi_1 - \left\{\frac{N}{MD_1}\right\} + a_1 t_0\right) D_1\right.\\
&\quad \left. - \left(\chi_j - \left\{\frac{N}{MD_j}\right\} + a_j t_0\right) D_j\right).
\end{aligned}$$

For any $j \in (L(a, N) \cap L(-a, \tilde{N}))\setminus\{1\}$, by (15) and (22) we have

$$a_1 D_1 \le a_j D_j. \tag{23}$$

For any $j > 1$ satisfying $j \in L(a, N) \cap L(-a, \tilde{N})$, we apply the above argument to a and $-a$ respectively and obtain $a_1 D_1 \le a_j D_j$ and $-a_1 D_1 \le -a_j D_j$. Therefore $a_1 D_1 = a_j D_j$, and the claim is proved.

Step 2 Set

$$V_j = \{a \in V \mid a_1 D_1 = a_j D_j\}, \quad j = 2, \ldots, q, \tag{24}$$

$$B(v) = A(v) \Big\backslash \bigcup_{V_j \neq V,\ j=2,\ldots,q} V_j. \tag{25}$$

Since $\dim A(v) \ge 1$, by Theorem 11.1.2, $A(v)$ is obtained from V by deleting finitely many proper linear subspaces of V, and so is $B(v)$. Hence $B(v)$ is non-empty.

Now we choose an $a \in B(v)$ and $j \in L(a, N) \cap L(-a, \tilde{N})\setminus\{1\}$. By the claim in Step 1, we have $a_1 D_1 = a_j D_j$. By the definitions of $a \in B(v)$ and j we have $V_j = V$.

By (18), the vector $\{Nv\} - \chi$ defined by (11.1.22) belongs to V, and thus belongs to V_j. Then by the definition (24) of V_j, this implies

$$0 = (\{Nv_1\} - \chi_1)D_1 - (\{Nv_j\} - \chi_j)D_j. \tag{26}$$

By the definition (11.1.21) of v, we then obtain

$$0 = \left(\left\{\frac{N}{MD_1}\right\} - \chi_1\right) D_1 - \left(\left\{\frac{N}{MD_j}\right\} - \chi_j\right) D_j.$$

By the second equality in (22), this implies $m_1 D_1 = m_j D_j$. Then it contradicts (15) and completes the proof. □

Corollary 4 *Suppose* $\Sigma \in \mathcal{H}(4)$ *with* ${}^{\#}\tilde{\mathcal{J}}(\Sigma) = 2$. *Then both the two closed characteristics on* Σ *are elliptic and belong to* $\mathcal{V}_\infty(\Sigma, \alpha)$, *one of them possesses an irrational mean index and therefore is strongly elliptic.*

Remark 5 Corollary 4 was proved in [Lon18] by using the precise iteration formulae in Chapter 8. All the other results in Sections 4 and 5 were proved by C. Zhu and the author in [LZu3].

15.6 Symmetric hypersurfaces

Suppose $\Sigma \in \mathcal{H}_s(2n)$, i.e., $\Sigma \in \mathcal{H}(2n)$ is symmetric with respect to the origin. We study the multiplicity and stability of closed characteristics on Σ.

Our first lemma describes a geometric property of closed characteristics on $\Sigma \in \mathcal{H}_s(2n)$. Let $\mathcal{O}(x) = \{x(t) \mid t \in \mathbf{R}\}$. We denote by H_α the Hamiltonian function corresponding to Σ defined by (0.3).

Lemma 1 *For* $\Sigma \in \mathcal{H}_s(2n)$, $\alpha \in (1, 2)$, *and* $(\tau, x) \in \mathcal{J}(\Sigma, \alpha)$, *we have* $(\tau, -x) \in \mathcal{J}(\Sigma, \alpha)$. *If* $\mathcal{O}(x) \cap \mathcal{O}(-x) \neq \emptyset$, *it must hold that*

$$x(t) = -x\left(t + \frac{\tau}{2}\right), \qquad \forall t \in \mathbf{R}. \tag{1}$$

Proof: Since $\Sigma = -\Sigma$,

$$H_\alpha(x) = H_\alpha(-x), \tag{2}$$

$$H'_\alpha(x) = -H'_\alpha(-x), \tag{3}$$

$$H''_\alpha(x) = H''_\alpha(-x). \tag{4}$$

So for any $(\tau, x) \in \mathcal{J}(\Sigma, \alpha)$, it holds that $(\tau, -x) \in \mathcal{J}(\Sigma, \alpha)$.

If $\mathcal{O}(x) \cap \mathcal{O}(-x) \neq \emptyset$, there exist s_1 and $s_2 \in [0, \tau]$ such that $x(s_1) = -x(s_2)$. By the fact $x(\tau) = x(0) \neq 0$ and $x(t) \neq 0$ for any $t \in \mathbf{R}$, we have $s_2 - s_1 \neq 0$, and $\pm\tau$. Since $x(s_1 + t)$ and $-x(s_2 + t)$ satisfy the same Hamiltonian system

$$\dot{y} = JH'_\alpha(y),$$

by the uniqueness theorem for initial value problems of ordinary differential equations, we obtain $x(s_1 + t) = -x(s_2 + t)$ and hence

$$x(t) = x(2(s_2 - s_1) + t), \qquad \forall t \in \mathbf{R}.$$

Since τ is the minimal period of x, we have $2(s_2 - s_1) = \pm\tau$. Therefore (1) holds. □

From $x(t) = -x(t + \frac{\tau}{2})$ we obtain

$$H''_\alpha(x(t)) = H''_\alpha\left(x\left(t + \frac{\tau}{2}\right)\right).$$

Then the associated symplectic path γ_x of (τ, x) satisfies

$$\gamma_x\left(t + \frac{\tau}{2}\right) = \gamma_x(t)\gamma_x\left(\frac{\tau}{2}\right), \quad \forall t \in \left[0, \frac{\tau}{2}\right]. \tag{5}$$

From this lemma, the orbits of closed characteristics are classified into two families: symmetric ones satisfying $\mathcal{O}(x) \cap \mathcal{O}(-x) \neq \emptyset$, and non-symmetric ones satisfying $\mathcal{O}(x) \cap \mathcal{O}(-x) = \emptyset$. Now we can give the proof of the third part of Theorem C in the Introduction of this book.

Theorem 2 (C. Liu, Y. Long, and C. Zhu [LLZ1]). *For any $\Sigma \in \mathcal{H}_s(2n)$,*

$$^{\#}\tilde{\mathcal{J}}(\Sigma) \geq n. \tag{6}$$

Proof: It suffices to consider the case $^{\#}\tilde{\mathcal{J}}(\Sigma) < +\infty$. By (2) and (3), we have $(\tau, -x) \in \mathcal{J}(\Sigma, \alpha)$ if $(\tau, x) \in \mathcal{J}(\Sigma, \alpha)$. Note that we have

$$(i(x_k, m), \nu(x_k, m)) = (i(-x_k, m), \nu(-x_k, m)), \qquad \forall m \in \mathbf{N}.$$

By Lemma 1, we can denote the elements in $\tilde{\mathcal{J}}(\Sigma, \alpha)$ by

$$\{[(\tau_j, x_j)] | j = 1, \ldots, p\} \cup \{[(\tau_k, x_k)], [(\tau_k, -x_k)] | k = p+1, \ldots, p+q\}, \tag{7}$$

where $\mathcal{O}(x_j) = \mathcal{O}(-x_j)$ for $j = 1, \ldots, p$, and $\mathcal{O}(x_k) \cap \mathcal{O}(-x_k) = \emptyset$ for $k = p+1, \ldots, p+q$.

By Lemma 3.5, we get an injection map $\Psi = \Psi(\Sigma, \alpha) : \mathbf{N} \to \tilde{\mathcal{J}}(\Sigma, \alpha) \times \mathbf{N}$. Note that (τ_k, x_k) and $(\tau_k, -x_k)$ have the same index intervals. Thus by Lemma 3.5 we can further require that

$$\mathrm{Im}(\Psi) \subset \{[(\tau_k, x_k)] \mid k = 1, \ldots, p+q\} \times \mathbf{N}.$$

Set $i(k, m) = i(x_k, m)$ and $\nu(k, m) = \nu(x_k, m)$. By (3.10) we have $\hat{i}(x_k, 1) > 2$ for $k = 1, \ldots, p+q$. Applying the common index jump Theorem 11.2.1 to the associated symplectic paths of

$$(\tau_1, x_1), \ldots, (\tau_{p+q}, x_{p+q}), (2\tau_{p+1}, x_{q+1}^2), \ldots, (2\tau_{p+q}, x_{p+q}^2),$$

we get infinitely many $(N, m_1, \ldots, m_{p+2q}) \in \mathbf{N}^{p+2q+1}$ such that

$$i(k, 2m_k + 1) = 2N + i(k, 1), \tag{8}$$

$$\begin{aligned} &i(k, 2m_k - 1) + \nu(k, 2m_k - 1) \\ &\quad = 2N - (i(k, 1) + 2S^+(x_k) - \nu(k, 1)), \end{aligned} \tag{9}$$

$$i(k, 2m_k) \geq 2N - n, \tag{10}$$

$$i(k, 2m_k) + \nu(k, 2m_k) \leq 2N + n \tag{11}$$

for $k = 1, \ldots, p + q$, and

$$i(k, 4m_{k+q} + 2) = 2N + i(k, 2), \tag{12}$$

$$\begin{aligned} &i(k, 4m_{k+q} - 2) + \nu(k, 4m_{k+q} - 2), \\ &= 2N - (i(k, 2) + 2S^+(x_k^2) - \nu(k, 2)), \end{aligned} \tag{13}$$

$$i(k, 4m_{k+q}) \geq 2N - n, \tag{14}$$

$$i(k, 4m_{k+q}) + \nu(k, 4m_{k+q}) \leq 2N + n \tag{15}$$

for $k = p + 1, \ldots, p + q$.

Denote

$$([(\tau_{k(s)}, x_{k(s)})], m(s)) = \Psi(N - s + 1), \qquad \forall s = 1, \ldots, n, \tag{16}$$

where $k(s) \in \{1, \ldots, p + q\}$ and $m(s) \in \mathbf{N}$. Let

$$S_1 = \{s \in \{1, \ldots, n\} \mid k(s) \leq p\}, \qquad S_2 = \{1, \ldots, n\} \backslash S_1. \tag{17}$$

We continue our proof in two steps and study the symmetric and non-symmetric orbits separately. By our following Claims 1 and 3, we then obtain upper bounds for the numbers of integers contained in the sets S_1 and S_2 respectively, which yields our Theorem 2.

Step 1 Study on symmetric orbits.

We need

Claim 1 $^{\#}S_1 \leq p$.

Since S_1 corresponds to those symmetric characteristics, our proof of Claim 1 depends on the index increasing estimates contained in the following two lemmas.

Lemma 3 *Let $\gamma \in \mathcal{P}_\tau(2n)$ be a symplectic path and extend it to* $[0, +\infty]$ *by* (8.0.4). *Let $M = \gamma(\tau)$. If $i(\gamma, 1) \geq n$, we have*

$$i(\gamma, 2) + 2S^+_{M^2}(1) - \nu(\gamma, 2) \geq n. \tag{18}$$

Proof: By Proposition 9.1.11, Theorem 9.1.7, and Lemma 1.8.14, we obtain

$$\begin{aligned} &\sum_{\theta \in (0,\pi)} S^-_M(e^{\sqrt{-1}\theta}) + (\nu_1(M) - S^-_M(1)) + (\nu_{-1}(M) - S^-_M(-1)) \\ &\leq \sum_{\theta \in (0,\pi)} S^-_M(e^{\sqrt{-1}\theta}) + p(1) + p(-1) \\ &\leq \sum_{\theta \in (0,\pi)} q(e^{\sqrt{-1}\theta}) + m(1) + m(-1) \\ &\leq n, \end{aligned} \tag{19}$$

where we denote by $(p(\omega), q(\omega))$ the Krein type of $\omega \in \sigma(M) \cap \mathbf{U}$, and $2m(\pm 1)$ the total multiplicity of $\pm 1 \in \sigma(M)$.

Note that $S_M^+(\pm 1) = S_M^-(\pm 1)$ and $i(\gamma, 1) \geq n$. By Theorem 9.2.1, Corollary 9.2.4, (9.3.3), and (19), we have

$$\begin{aligned}
& i(\gamma, 2) + 2S_{M^2}^+(1) - \nu(\gamma, 2) \\
& = i_1(\gamma) + i_{-1}(\gamma) + 2S_M^+(1) + 2S_M^+(-1) - \nu_1(\gamma) - \nu_{-1}(\gamma) \\
& = 2i_1(\gamma) + \left[S_M^+(1) + \sum_{\theta \in (0,\pi)} (S_M^+(e^{\sqrt{-1}\theta}) - S_M^-(e^{\sqrt{-1}\theta}) - S_M^-(-1) \right] \\
& \quad + 2S_M^+(1) + 2S_M^+(-1) - \nu_1(\gamma) - \nu_{-1}(\gamma) \\
& = 2i_1(\gamma) + 2S_M^+(1) + \sum_{\theta \in (0,\pi)} (S_M^+(e^{\sqrt{-1}\theta}) - S_M^-(e^{\sqrt{-1}\theta})) \\
& \quad - (\nu_1(M) - S_M^-(1)) - (\nu_{-1}(M) - S_M^-(-1)) \\
& \geq 2i_1(\gamma) - \left[\sum_{\theta \in (0,\pi)} S_M^-(e^{\sqrt{-1}\theta}) \right. \\
& \quad \left. + (\nu_1(M) - S_M^-(1)) + (\nu_{-1}(M) - S_M^-(-1)) \right] \\
& \geq 2n - n \\
& = n.
\end{aligned} \tag{20}$$

This proves (18). □

Lemma 4 *For any symmetric closed characteristic* $[(\tau, x)] \in \tilde{\mathcal{J}}(\Sigma, \alpha)$,

$$i(x, 1) + 2S^+(x) - \nu(x, 1) \geq n. \tag{21}$$

Proof: By Lemma 1, (τ, x) satisfies (1), i.e., it is $\tau/2$-antiperiodic. Therefore $H_\alpha''(x(t))$ is $\tau/2$-periodic by (4). Let $\psi = \gamma_x|_{[0,\tau/2]}$, $M = \gamma_x(\tau)$, and $G = \gamma_x(\tau/2)$. Then we have $\gamma_x|_{[0,\tau]} = \psi^2$, $M = G^2$, and

$$\begin{aligned}
i(x, 1) + 2S^+(x) - \nu(x, 1) &= i(\gamma_x, 1) + 2S_M^+(1) - \nu(\gamma_x, 1) \\
&= i(\psi, 2) + 2S_{G^2}^+(1) - \nu(\psi, 2) \\
&\geq n,
\end{aligned} \tag{22}$$

where in the last step we used Lemma 3. □

Now we can give

The proof of Claim 1: In fact, by the definition (17) of S_1, (7) and (16), $\Psi(N-s+1) = ([(\tau_{k(s)}, x_{k(s)})], m(s))$ must be symmetric when $k(s) \leq p$. We further claim that $m(s)$ is uniquely determined by $k(s)$ if $k(s) \leq p$. More precisely we have

Claim 2 $m(s) = 2m_{k(s)}$ *when* $k(s) \le p$.

In fact, by the definition of Ψ and (3.5) we have

$$\begin{aligned} i(k(s), m(s)) &\le 2(N - s + 1) - 2 + n \\ &= 2N - 2s + n \\ &\le i(k(s), m(s)) + \nu(k(s), m(s)) - 1. \end{aligned} \tag{23}$$

So from $i(k(s), 1) \ge n$ we have

$$\begin{aligned} i(k(s), m(s)) &\le 2N - 2s + n \\ &< 2N + n \\ &\le 2N + i(k(s), 1) \\ &= i(k(s), 2m_{k(s)} + 1), \end{aligned} \tag{24}$$

for every $s = 1, \ldots, n$, where we have used (8) in the last equality. Note that the proofs of (23) and (24) do not depend on the condition $k(s) \le p$.

By (9), Lemma 4, and the right-hand side of (23), if $k(s) \le p$ we have

$$\begin{aligned} &i(k(s), 2m_{k(s)} - 1) + \nu(k(s), 2m_{k(s)} - 1) \\ &= 2N - (i(k(s), 1) + 2S^+(x_{k(s)}) - \nu(k(s), 1)) \\ &\le 2N - n \\ &\le 2N - 2s + n \\ &\le i(k(s), m(s)) + \nu(k(s), m(s)) - 1. \end{aligned} \tag{25}$$

Thus by Lemma 3.2, (24), and (25) we have

$$2m_{k(s)} - 1 < m(s) < 2m_{k(s)} + 1, \quad \text{for } k(s) \le p.$$

This proves our Claim 2.

By Claim 2, we obtain especially:

$$\Psi(N - s + 1) = ([(\tau_{k(s)}, x_{k(s)})], 2m_{k(s)}), \qquad \forall s \in S_1.$$

Then by the injectivity of the map Ψ, it induces another injection

$$\Psi_1 : N - S_1 + 1 \to \{1, \ldots, p\}.$$

Therefore we obtain Claim 1.

Step 2 Study on non-symmetric orbits.

We need

Claim 3 ${}^{\#}S_2 \le 2q$.

In fact, by the definition (17) of S_2, (7) and (16), $\Psi(N-s+1) = ([(\tau_{k(s)}, x_{k(s)})], m(s))$ must be non-symmetric when $s \in S_2$, i.e., $p < k(s) \le p+q$. Note that in this case, $[(\tau_{k(s)}, x_{k(s)})]$ and $[(\tau_{k(s)}, -x_{k(s)})]$ are different closed characteristics possessing the same iterated index sequences and non-intersecting orbits. Our following claim studies a special property of such characteristics.

Claim 4 *$m_k = 2m_{k+q}$ for $p < k \le p+q$.*

In fact, by (10), (13), and Lemma 3, for $k = p+1, \ldots, p+q$ we have

$$\begin{aligned} i(k, 2m_k) &\ge 2N - n \\ &\ge 2N - (i(k,2) + 2S^+(x_k^2) - \nu(k,2)) \\ &= i(k, 4m_{k+q} - 2) + \nu(k, 4m_{k+q} - 2) \\ &> i(k, 4m_{k+q} - 2). \end{aligned}$$

By (11), (12), (1.18) and Lemma 3.2, we have

$$\begin{aligned} i(k, 2m_k) &< i(k, 2m_k) + \nu(k, 2m_k) \\ &\le 2N + n \\ &\le 2N + i(k,2) \\ &= i(k, 4m_{k+q} + 2). \end{aligned}$$

By (3.6) of Lemma 3.2, we then obtain

$$4m_{k+q} - 2 < 2m_k < 4m_{k+q} + 2.$$

Therefore Claim 4 holds.

To continue our proof of Claim 3, by Claim 4 we obtain

$$m_{k(s)} = 2m_{k(s)+q}, \tag{26}$$

if $p < k(s) \le p+q$, i.e., $s \in S_2$. By (26), (13), Lemma 3, and the right-hand side of (23), we then have

$$\begin{aligned} & i(k(s), 2m_{k(s)} - 2) + \nu(k(s), 2m_{k(s)} - 2) \\ &= i(k(s), 4m_{k(s)+q} - 2) + \nu(k(s), 4m_{k(s)+q} - 2) \\ &= 2N - (i(k(s),2) + 2S^+(x_{k(s)}^2) - \nu(k(s),2)) \\ &\le 2N - n \\ &\le 2N - 2s + n \\ &\le i(k(s), m(s)) + \nu(k(s), m(s)) - 1. \end{aligned} \tag{27}$$

Thus by (3.6) and (3.8) of Lemma 3.2, (27), and (24), we obtain

$$2m_{k(s)} - 2 < m(s) < 2m_{k(s)} + 1, \quad \text{for } p < k(s) \le p+q.$$

Hence

$$m(s) \in \{2m_{k(s)} - 1, 2m_{k(s)}\}, \qquad \text{if } p < k(s) \le p + q. \tag{28}$$

Especially this yields that for any s_0 and $s \in S_2$, if $k(s) = k(s_0)$, then

$$m(s) \in \{2m_{k(s)} - 1, 2m_{k(s)}\} = \{2m_{k(s_0)} - 1, 2m_{k(s_0)}\}.$$

Thus by the injectivity of the map Ψ we obtain

$$^{\#}\{s \in S_2 \mid k(s) = k(s_0)\} \le 2. \tag{29}$$

This proves Claim 3.

Step 3 Completion of the proof of Theorem 2.

Now by Claims 1 and 3 we obtain

$$\begin{aligned} {}^{\#}\tilde{\mathcal{J}}(\Sigma) &= {}^{\#}\tilde{\mathcal{J}}(\Sigma, \alpha) \\ &= p + 2q \\ &\ge {}^{\#}S_1 + {}^{\#}S_2 \\ &= n. \end{aligned} \tag{30}$$

This completes the proof. □

Next we present a stability theorem for closed characteristics on $\Sigma \in \mathcal{H}_s(2n)$.

Theorem 5 (G.F. Dell'Antonio, B. D'Onofrio, and I. Ekeland [DDE1]). *For any $\Sigma \in \mathcal{H}_s(2n)$, there exists at least one elliptic closed characteristic on Σ.*

Proof: The proof is similar to those in the previous sections. We are rather sketchy here.

Fix $\Sigma \in \mathcal{H}_s(2n)$. To solve the given energy problem (0.4)–(0.6), we consider the fixed period problem with $\alpha \in (1, 2)$ and H_α given by (0.3),

$$\begin{cases} \dot{x}(t) = JH'_\alpha(x(t)), & \forall t \in \mathbf{R}, \\ x(1/2) = -x(0). \end{cases} \tag{31}$$

For $1 < \alpha < 2$, let

$$E_\alpha = \left\{ u \in L^{\alpha/(\alpha-1)}\left(\mathbf{R}\Big/\left(\frac{1}{2}\mathbf{Z}\right), \mathbf{R}^{2n}\right) \mid u(t + 1/2) = -u(t) \right\}. \tag{32}$$

The Clarke–Ekeland dual action functional $f_\alpha : E_\alpha \to \mathbf{R}$ is defined by

$$f_\alpha(u) = \int_0^{1/2} \left\{ \frac{1}{2} Ju \cdot \Pi u + H^*_\alpha(-Ju) \right\} dt. \tag{33}$$

Then we obtain a global minimal point u of f_α on E_α. This u corresponds to a solution x of the problem (0.4)–(0.6).

The linearized Hamiltonian system of (31) yields

$$\begin{cases} \dot{y}(t) = JH''_\alpha(x(t))y(t), & y \in \mathbf{R}^{2n}, \\ y(1/2) = -y(0). \end{cases} \tag{34}$$

Denote the fundamental solution of (34) by γ. Then $\gamma \in \mathcal{P}_{1/2}(2n)$ and the singular hypersurface in Sp$(2n)$ for the definition of the index corresponding to (34) is given by $\omega = -1$. Then using our notation we obtain for this global minimum point x that

$$i_{-1}(\gamma) = 0. \tag{35}$$

On the other hand, by the convexity of H_α and our discussion in Section 1, we have

$$i_1(\gamma|_{[0,\epsilon]}) \geq n, \tag{36}$$

for any sufficiently small $\epsilon > 0$. By (35), (36) and formula (9.3.3), the total number of eigenvalues of $\gamma(1/2)$ on the upper semi-circle in $\mathbf{U}$ is at least n. Thus the solution x is elliptic. □

Remark 6 Theorem 2 was proved by C. Liu, C. Zhu, and the author in the recent paper [LLZ1]. Here the proof is rewritten to make it easier to understand. Theorem 5 was proved by G.F. Dell'Antonio, B. D'Onofrio, and I. Ekeland in [DDE1] of 1992.

15.7 Notes and comments

Since $\Sigma \in \mathcal{H}(2n)$ is compact smooth and orientable, the flow of the ordinary differential equation system (0.1) on Σ is always defined all the time. But it is completely non-trivial to say more about the periodic boundary value problem (0.1). We give next a brief historical survey on the multiplicity and stability problems of (0.1). We also give a brief introduction to the Weinstein conjecture which is closely related to the material of this chapter. Then we propose some open problems in this area.

I Existence and multiplicity

As mentioned on page 235 of [Eke3], the most famous long-standing conjecture on (0.1) is

Conjecture 1 $^\#\tilde{\mathcal{J}}(\Sigma) \geq n$ for any $\Sigma \in \mathcal{H}(2n)$.

It seems that the first study of this problem was started by A.M. Liapunov in [Lia1] of 1892 and J. Horn in [Hor1] of 1903. It is surprising enough that more than one hundred years ago, they were able to prove the following great result for (0.1) in the local sense: *Suppose* $H : \mathbf{R}^{2n} \to \mathbf{R}$ *is analytic,* $\sigma(JH''(0)) = \{\pm\sqrt{-1}\omega_1, \ldots, \pm\sqrt{-1}\omega_n\}$ *are purely imaginary, and satisfy* $\frac{\omega_i}{\omega_j} \notin \mathbf{Z}$ *for all* $i \neq j$. *Then there exists* $\epsilon_0 > 0$ *so small that*

$$^{\#}\tilde{\mathcal{J}}(H^{-1}(\epsilon)) \geq n, \quad \forall 0 < \epsilon < \epsilon_0. \tag{1}$$

This deep result was greatly improved by A. Weinstein in [Wei2] of 1973. He was able to prove that for $H \in C^2(\mathbf{R}^{2n}, \mathbf{R})$, if $H''(0)$ is positive definite, then there exists $\epsilon_0 > 0$ small such that (1) still holds. Weinstein's theorem then was further generalized by J. Moser in [Mos5] of 1976 to more general systems and by T. Bartsch in [Bar1] of 1997 to non-convex cases for the local problem.

In the full generality of the Conjecture 1, i.e., the problem (0.1) in the global sense, an exciting breakthrough was made by P. Rabinowitz in [Rab2] (for Σ being star-shaped) and A. Weinstein in [Wei5] of 1978 independently. They proved

$$^{\#}\tilde{\mathcal{J}}(\Sigma) \geq 1, \quad \forall \Sigma \in \mathcal{H}(2n). \tag{2}$$

Here in his pioneering work, P. Rabinowitz first introduced the variational method to study the problem (0.1), which stimulated great interest and much research on this problem.

Among the great number of papers on this problem, I. Ekeland introduced his Morse-type index theory for convex Hamiltonian systems, and made profound studies on Conjecture 1. I. Ekeland-L. Lassoued in [ELa1] and I. Ekeland-H.Hofer in [EkH2] of 1987, and A. Szulkin in [Szu1] of 1988 proved the multiplicity result

$$^{\#}\tilde{\mathcal{J}}(\Sigma) \geq 2, \qquad \forall \Sigma \in \mathcal{H}(2n) \text{ with } n \geq 2. \tag{3}$$

Because it is so difficult to completely answer Conjecture 1, it is natural to study the multiplicity of closed characteristics on special subclasses of $\mathcal{H}(2n)$ under additional conditions. Following studies for closed geodesics on Riemannian manifolds, the pinching condition is introduced to the study of problem (0.1). We say $\Sigma \in \mathcal{H}(2n)$ is b-pinched, if there exist two balls B_r and B_R with radius r and R respectively satisfying

$$B_r \subset C \subset B_R \quad \text{and} \quad R/r < b,$$

where C is the convex domain bounded by Σ. In [EkL1] of 1980, I. Ekeland and J.M. Lasry proved

$$^{\#}\tilde{\mathcal{J}}(\Sigma) \geq n, \quad \text{if } \Sigma \in \mathcal{H}(2n) \text{ is } \sqrt{2}\text{–pinched}. \tag{4}$$

A different proof of this result can be found in [Hof1] of H. Hofer of 1982. In [AmM2] of 1982, A. Ambrosseti and G. Mancini further proved

$$^{\#}\tilde{\mathcal{J}}(\Sigma) \geq \left[\frac{n}{k-1}\right], \quad \text{if } \Sigma \in \mathcal{H}(2n) \text{ is } \sqrt{k}\text{–pinched}, \tag{5}$$

where as usual $[a] = \max\{k \in \mathbf{Z} \mid k \leq a\}$ for any $a \in \mathbf{R}$.

When Σ is symmetric to the origin, a result which is slightly better than (4) was obtained by M. Girardi in [Gir1] of 1984, where he proved

$$^{\#}\tilde{\mathcal{J}}(\Sigma) \geq n, \quad \text{if } \Sigma \in \mathcal{H}_s(2n) \text{ is } \sqrt{3}\text{–pinched.} \tag{6}$$

In [BLMR1] of 1985, H. Berestycki, J.M. Lasry, G. Mancini, and B. Ruf generalized (4) to (star-shaped hypersurfaces under) pinching conditions in terms of ellipsoids.

Using a theorem in [Fra1] of J. Franks, in [HWZ1] of 1998, H. Hofer, K. Wysocki, and E. Zehnder further improved (3) when $n = 2$ to

$$^{\#}\tilde{\mathcal{J}}(\Sigma) = 2 \text{ or } +\infty, \quad \forall \Sigma \in \mathcal{H}(4). \tag{7}$$

Another interesting result proved by I. Ekeland in [Eke2] of 1986 and [Eke3] of 1990 is the following: $^{\#}\tilde{\mathcal{J}}(\Sigma) = +\infty$ holds generically for $\Sigma \in \mathcal{H}(2n)$ in C^p-topology with $3 \leq p \leq +\infty$.

Then Theorems 4.3 and 4.4 were proved in [LZu3] of Y. Long and C. Zhu, and Theorem 6.2 by C. Liu, Y. Long, and C. Zhu in [LLZ1] recently.

Note that one can construct a path $\gamma \in \mathcal{P}_1(6)$ (cf. the first manuscript of [LZu3], Nankai Inst. of Math. Preprint No. 1999-M-002) such that γ is the fundamental solution of a linear Hamiltonian system (1.1) with $s = 1$ satisfying (B1) and (B2), $\gamma(1) = N_1(1, 1)\diamond N_1(1, -1)^{\diamond 2}$, $i(\gamma, 1) = 3$. Especially $S^+_{\gamma(1)}(1) = 1$ and

$$i(\gamma, 1) + 2S^+_{\gamma(1)}(1) - \nu(\gamma, 1) + 3 = 5 < 6.$$

This linear example is not a counterexample to the above Conjecture 1 on the nonlinear problem, but it may suggest that the lower bound $[\frac{n}{2}] + 1$ of $\varrho_n(\Sigma)$ in (0.0.17) of Theorem C in the Introduction of the book as well as (4.3) of the above Theorem 4.3 may not be further improved in general.

II Stability

We are interested in the ellipticity of closed characteristic (τ, x) on a given $\Sigma \in \mathcal{H}(2n)$. On this stability problem, the following is the most important long-standing conjecture (cf. p.235 of [Eke3]):

Conjecture 2 There exists an elliptic closed characteristic on any $\Sigma \in \mathcal{H}(2n)$.

The first result on this topic in the global sense seems to be that proved by I. Ekeland in [Eke2] of 1986:

There exists at least one elliptic element in $\mathcal{J}(\Sigma, \alpha)$, if $\Sigma \in \mathcal{H}(2n)$ is $\sqrt{2}$-pinched.

In [Eke2] of I. Ekeland in 1986 and [Lon14] of the author in 1998, the following alternative result for any $\Sigma \in \mathcal{H}(2n)$ was proved: either there exist infinitely many geometrically distinct hyperbolic closed characteristics $[(\tau_j, x_j)]$ for $j \in \mathbf{N}$ with their minimal periods $\tau_j \to +\infty$, or there exists a non-hyperbolic closed characteristic on Σ. Especially, this implies that there exists at least one non-hyperbolic closed characteristic on Σ, if $^{\#}\tilde{\mathcal{J}}(\Sigma) < +\infty$.

Note that the above remarkable Theorem 6.5 of G.F. Dell'Antonio, B. D'Onofrio, and I. Ekeland in [DDE1] of 1992 solved Conjecture 2 whenever Σ is symmetric.

Considering the above mentioned result (7) of Hofer-Wysocki-Zehnder, in the recent [Lon18] based upon the precise iteration formulae developed there, the author proved that both the closed characteristics are elliptic if $\Sigma \in \mathcal{H}(4)$ and $^{\#}\tilde{\mathcal{J}}(\Sigma) = 2$.

Then in the recent [LZu3] of Y. Long and C. Zhu, Theorems 5.1–5.3, especially Theorem D in the Introduction were further proved.

III The Weinstein conjecture

Based on the results of [Wei5] and [Rab2], A. Weinstein proposed the following conjecture in [Wei6]: *On every contact hypersurface Σ in a symplectic manifold (M, ω), there exists at least one closed characteristic provided $H^1(M) = 0$.* Here a hypersurface $\Sigma \subset M$ is of contact type, if it is compact, orientable, and there exists a 1-form α on Σ such that $d\alpha = j^*\omega$, and $\alpha(\xi) \neq 0$ for $0 \neq \xi \in \mathcal{L}_\Sigma$, where $j : \Sigma \to M$ is the inclusion map, and $\mathcal{L}_\Sigma$ is the canonical bundle of Σ defined by

$$\mathcal{L}_\Sigma = \{(x, \xi) \in T_x\Sigma \mid \omega_x(\xi, \eta) = 0, \forall \eta \in T_x\Sigma\}.$$

A closed characteristic of Σ is an embedded circle $P \subset \Sigma$ satisfying

$$TP = \mathcal{L}_\Sigma|_P.$$

A compact smooth hypersurface Σ in $\mathbf{R}^{2n}$ is **star-shaped**, if there exists one point O located inside the open set enclosed by Σ such that the ray from O to any point $x \in \Sigma$ intersects Σ at the point x only and transversally. Note that every hypersurface in $\mathcal{H}(2n)$ is star-shaped. Every compact smooth star-shaped hypersurface in $\mathbf{R}^{2n}$ is of contact type.

Note that in [Vit1] of 1987, C. Viterbo proved the Weinstein conjecture for every contact hypersurface Σ in $\mathbf{R}^{2n}$ without the assumption $H^1(\Sigma) = 0$. Then H. Hofer and E. Zehnder gave a rather simple proof for this result in [HoZ1] of 1987. Further results on this conjecture can be found in [FHV1], [HoV1], [HoZ2], [Mar1], and [LiT2].

For any compact smooth star-shaped hypersurface Σ in $\mathbf{R}^{2n}$, as we have mentioned, P. Rabinowitz in [Rab2] of 1978 proved the existence of at least one closed characteristic, i.e., $^{\#}\tilde{\mathcal{J}}(\Sigma) \geq 1$. In [BLMR1] of 1985, H. Berestycki, J.-P. Lasry, G. Mancini, and B. Ruf proved $^{\#}\tilde{\mathcal{J}}(\Sigma) \geq n$ provided Σ is suitably pinched by two ellipsoids. Note that very recently in [HuL1], Xijun Hu and the author proved the following result: *For any smooth compact star-shaped hypersurface Σ in $\mathbf{R}^{2n}$, suppose every closed characteristic $(\tau, x) \in \mathcal{J}(\Sigma)$ is completely non-degenerate, i.e., $\nu(x, m) = 0$ for all $m \in \mathbf{N}$. Then,*

$$^{\#}\tilde{\mathcal{J}}(\Sigma) \geq 2, \qquad \text{if } n \geq 2.$$

In [LLo2] of 1999, C. Liu and the author obtained some partial results for non-hyperbolic closed characteristics on star-shaped hypersurfaces in $\mathbf{R}^{2n}$.

To the author's knowledge, no multiplicity and stability results are known so far for closed characteristics on general contact type hypersurfaces.

IV Open problems

The aim of the study on the problem (0.1) is threefold. The first aim is to completely understand the structure of the set of closed characteristics on any given hypersurface $\Sigma \in \mathcal{H}(2n)$, and then to further understand the characteristic flow on Σ. The second aim is to understand the structure of the set $\mathcal{H}(2n)$. The third is to use it as the most typical example in the study of the closed orbits of the Reeb vector field on any general contact manifold.

Based on our Theorem C in the Introduction, as well as Theorems 4.3, 4.4 and 6.2 in this chapter, earlier results mentioned above, and other observations, we suspect that the following conjecture holds:

Conjecture A

$$\{{}^{\#}\tilde{\mathcal{J}}(\Sigma) \mid \Sigma \in \mathcal{H}(2n)\} = \left\{k \in \mathbf{N} \mid \left[\frac{n}{2}\right] + 1 \leq k \leq n\right\} \cup \{+\infty\}.$$

Here we would also like to remind the readers of the following reduced version of Conjecture 2 based on our Theorem 5.1:

Conjecture B There exists at least an elliptic closed characteristic on any $\Sigma \in \mathcal{H}(2n)$ with ${}^{\#}\tilde{\mathcal{J}}(\Sigma) = +\infty$.

More generally, it is interesting to know the global topological structure of the set $\tilde{\mathcal{J}}(\Sigma)$ for any $\Sigma \in \mathcal{H}(2n)$.

It is an interesting problem to clarify the relation between the contact (or corresponding symplectic) invariants of the given contact manifold and the number of periodic orbits of the Reeb vector field on it. Such a relation should include results on the case of hypersurfaces in $\mathcal{H}(2n)$, especially our Theorems 4.3, 4.4, and 6.2 in Sections 4 and 6 as special cases. It is interesting to know the relation between our shape invariant $\varrho_n(\Sigma)$ and the possible topological invariants. For this problem, we refer the readers to the recent exciting work [LiT2] of G. Liu and G. Tian.

As pointed out in [Lon18], the study of the stability of closed characteristics is related to the following conjectures on primes in number theory.

Given any $p \in N$, we define an integer pair set

$$Y(p) = \{(pn - 1, pn + 1) \mid n \in \mathbf{N}\}.$$

Conjecture C For any $p \in \mathbf{N}$, the set $Y(p)$ contains infinitely many prime number pairs.

Note that this is a slight general version of the conjecture on twin primes.

Given any $\varphi \in [0, 1]\backslash\mathbf{Q}$, we define an integer set

$$Z(\varphi) = \mathbf{N} \setminus \{2n + 2[n\varphi] + 1 \mid n \in \mathbf{N}\}.$$

Conjecture D For any $\varphi \in [0, 1]\backslash\mathbf{Q}$, the set $Z(\varphi)$ contains infinitely many prime numbers.

It is even more interesting if one can develop dynamical system studies to these conjectures on prime numbers.

Bibliography

[Abb1] A. Abbondanolo, On the Morse index of Lagrangian systems. Preprint, 2001

[AbM1] R. Abraham and J.E. Marsden, Foundations of Mechanics, 2nd ed. Benjamin/ Cummings *Publ. Com.* London. 1978

[Ada1] R.A. Adams, Sobolev Spaces. Academic Press, 1975

[Ama1] H. Amann, Saddle points and multiple solutions of differential equations. *Math. Z.* 169(1979), 122–166

[AmZ1] H. Amann and E. Zehnder, Nontrivial solutions for a class of non-resonance problems and applications to nonlinear differential equations. *Ann. Scuola Norm. Sup. Pisa. Cl.Sci. Series* 4. 7(1980), 539–603

[AmZ2] ———, Periodic solutions of asymptotically linear Hamiltonian systems. *Manuscripta Math.* 32(1980), 149–189

[ABL1] A. Ambrosetti, V. Benci and Y. Long, A note on the existence of multiple brake orbits. *Nonlinear Anal. TMA.* 21(1993), 643–649

[AmC1] A. Ambroseti and V. Coti Zelati, Solutions with minimal period for Hamiltonian systems in a potential well. *Ann. IHP. Anal. Nonlineaire.* 4 (1987), 242–275

[AmC2] A. Ambrosetti and V. Coti Zelati, Periodic Solutions of Singular Lagrangian Systems. Birkhäuser, Basel, 1993

[AmM1] A. Ambrosetti and G. Mancini, Solutions of minimal period for a class of convex Hamiltonian systems. *Math. Ann.* 255(1981), 405–421

[AmM2] A. Ambrosetti and G. Mancini, On a theorem by Ekeland and Lasry concerning the number of periodic Hamiltonian trajectories. *J. Diff. Equa.* 43 (1982), 249–256

[AmR1] A. Ambrosetti and P. Rabinowitz, Dual variational methods in critical point theory and applications. *J. Funct. Anal.* 14(1973), 349–381

[AnL1] T. An and Y. Long, Index theories of second order Hamiltonian systems. *Nonlinear Anal. TMA.* 34(1998), 585–592

[AnL2] ———, The classification of exponential paths in Sp($2n$). *Advances in Math. (China).* 27(1998), No. 3, 209–213

[Arn1] V.I. Arnold, On a characteristic class entering into the quantization condition. *Funkt. Anal. Prilozh.* 1(1967), 1–14. (Russian). *Funct. Anal. Appl.* 1(1967), 1–13. (English transl.)

[Arn2] ———, Mathematical Methods of Classical Mechanics. Springer. Berlin, 1978

[AKN1] V.I. Arnold, V.V. Kozlov and A.I. Neishtadt, Mathematical Aspects of Classical and Celestial Mechanics. Dynamical Systems III. Springer. Berlin, 1988

[AuE1] J.-P. Aubin and I. Ekeland, Applied Nonlinear Analysis. John Wiley & Sons. New York, 1984

[BaB1] A. Bahri and H. Berestycki, Existence of forced oscillations for some nonlinear differential equations. *Comm. Pure Appl. Math.* 37(1984), 403–442

[BaB2] ———, Forced vibrations of superquadratic Hamiltonian systems. *Acta Math.* 152(1984), 143–197

[BaL1] A. Bahri and P.L. Lions, Morse index of some min-max critical points. I. Applications to multiplicity results. *Comm. Pure Appl. Math.* 41 (1988), 1027–1037

[BaR1] A. Bahri and P. Rabinowitz, Periodic solutions of Hamiltonian systems of the 3-body type. *Ann. Inst. H. Poincaré Anal. non linéaire.* 8(1991), 561–649

[BTZ1] W. Ballmann, G. Thobergsson and W. Ziller, Closed geodesics on positively curved manifolds. *Ann. of Math.* 116(1982), 213–247

[Ban1] V. Bangert, Closed geodesics on complete surfaces. *Math. Ann.* 251(1980), 83–96

[Ban2] V. Bangert, Geodetische Linien auf Riemannschen Mannigfaltigkeiten. *Jber. D. Dt. Math.-Verein.* 87(1985), 39–66

[BaK1] V. Bangert and W. Klingenberg, Homology generated by iterated closed geodesics. *Topology* 22(1983), 379–388

[Bar1] T. Bartsch, A generalization of the Weinstein-Moser theorems on periodic orbits of a Hamiltonian system near an equilibrium. *Ann. Inst. H. Poincaré Anal. non linéaire.* 14(1997), 691–718

[BaW1] T. Bartsch and Z.-Q. Wang, Periodic solutions of even Hamiltonian systems on the torus. *Math. Z.* 224(1997), 65–76

[Ben1] V. Benci, A geometric index for the group S^1 and some applications to the periodic solutions of ordinary differential equations. *Comm. Pure Appl. Math.* 34(1981), 393–432

[Ben2] ———, On the critical point theory for indefinite functionals in the presence of symmetries. *Trans. Amer. Math. Soc.* 274(1982), 533–572

[Ben3] ———, Normal modes of a Lagrangian system constrained in a potential well. *Ann. Inst. H. Poincaré Anal. non linéaire.* 1(1984), 362–379

[BeR1] V. Benci and P. Rabinowitz, Critical point theorems for indefinite functionals. *Inv. Math.* 52(1979), 241–273

[BLMR1] H. Berestycki, J.M. Lasry, G. Mancini and B. Ruf, Existence of multiple periodic orbits on star-shaped Hamiltonian surfaces. *Comm. Pure Appl. Math.* 38(1985), 252–290

[Ber1] M. Berger, Nonlinearity and Functional Analysis. Academic Press. New York, 1977

[Bir1] G.D. Birkhoff, Dynamical Systems. *Colloquium Publ. Amer. Math. Soc.* 9(1927)

[BiL1] G.D. Birkhoff, and D.C. Lewis, On the periodic motions near a given periodic motion of a dynamical system. *Annali di Math.* 12(1933), 117–133

[Bol1] S. Bolotin, Libration motions of natural dynamical systems. *Moscow Univ. Math. Bull.* 33(1978), 49–53

[Bot1] R. Bott, On the iteration of closed geodesics and the Sturm intersection theory. *Comm. Pure Appl. Math.* 9(1956), 171–206

[Bot2] ———, Lectures on Morse theory, old and new. *Bull. Amer. Math. Soc.* 7(1982), 331–358

[Bro1] V. Brousseau, L'index d'un systéme hamiltonien linéaire. *C. R. Acad. Sc. Paris,* t.303. Série 1(1986), 351–354

[BuC1] N. Burgoyne and R. Cushman, Normal forms for real linear Hamiltonian systems with purely imaginary eigenvalues. *Celest. Mech.* 8(1973), 435–443

[CLM1] S.E. Cappell, R. Lee and E.Y. Miller, On the Maslov index. *Comm. Pure Appl. Math.* 47(1994), 121–186

[Cha1] K.C. Chang, Solutions of asymptotically linear operator equations via Morse theory. *Comm. Pure Appl. Math.* 34(1981), 693–712

[Cha2] ———, A variant mountain pass lemma. *Sci. Sinica. Ser.* A 26(1983), 1241–1255

[Cha3] ———, Critical Point Theory and its Applications. *Shanghai Sci. Tech. Press. Shanghai.*, 1986 (in Chinese)

[Cha4] ———, On the periodic nonlinearity and multiplicity of solutions. *Nonlinear Anal. TMA.* 13(1989), 527–537

[Cha5] K.C. Chang, Applications of homology theory to some problems in differential equations. *Proc. Symp. in Pure Math.* F. Browder ed. *Amer. Math. Soc.* 45(1986), 253–262

[Cha6] ———, Infinite Dimensional Morse Theory and Multiple Solution Problems. Birkhäuser. Basel, 1993

[ChJ1] K.C. Chang and M.-Y. Jiang, The Lagrangian intersections for $(\mathbf{C}P^n, \mathbf{R}P^n)$. *Manuscripta Math.* 68(1990), 89–100

[CGL1] K.C. Chang, M.Z. Guo, and Y.Q. Lin, Lectures on Functional Analysis. Chinese Ed. Peking Univ. Press. Beijing. I. 1987, II. 1990

[ChL1] K.C. Chang and J.-Q. Liu, A strong resonance problem. *Chinese Ann. Math.* 11B(1990), 191–210

[CLL1] K.C. Chang, J.-Q. Liu and M. Liu, Nontrivial periodic solutions for strong resonance Hamiltonian systems. *Ann. Inst. H. Poincaré Anal. non linéaire.* 14(1997), 103–117

[CLZ1] K.C. Chang, Y. Long and E. Zehnder, Forced oscillations for the triple pendulum, *Analysis et cetera*, P. Rabinowitz and E. Zehnder ed. Academic Press. New York, 1990

[CLo1] G. Chen and Y. Long, Periodic Solutions of Second Order Nonlinear Hamilton Systems with Zero Mean Value superquadratic Potentials. *Chinese Ann. Math.* 19A (1998), No. 4, 525–532 (Chinese ed.). *Chinese J. Cont. Math.* 19(1998), 333–342

[Che1] W.Y. Chen, Nonlinear Functional Analysis. Chinese ed. Gansu People's Press. Lanzhou, 1982

[ChS1] C.-Q. Cheng and Y.-S. Sun, Regular and Stochastic Motions in Hamiltonian Systems. *Publ. House. Shanghai.*, 1996 (in Chinese)

[ClE1] F. Clarke and I. Ekeland, Hamiltonian trajectories having prescribed minimal period. *Comm. Pure Appl. Math.* 33(1980), 103–116

[ClE2] ———, Nonlinear oscillations and boundary value problems for Hamiltonian systems. *Arch. for Rat. Mech. and Anal.* 78(1982), 315–333

[CoL1] E. Coddington and N. Levinson, Theory of Ordinary Differential Equations. McGraw-Hill. New York, 1955

[Con1] C. Conley, Isolated Invariant Sets and the Morse Index. *CBMS. Regional Reports.* Series 38. *Amer. Math. Soc.*, 1978

[CoZ1] C. Conley and E. Zehnder, The Birkhoff-Lewis fixed point theorem and a conjecture of V.I. Arnold. *Invent. Math.* 73(1983), 231–249

[CoZ2] ———, Morse-type index theory for flows and periodic solutions for Hamiltonian equations. *Comm. Pure Appl. Math.* 37(1984), 207–253

[CoZ3] ———, Subharmonic solutions and Morse theory. *Physica* 124A(1984), 649–658

[CoZ4] ———, A global fixed point theorem for symplectic maps and subharmonic solutions of Hamiltonian equations on tori. *Proc. Symp. in Pure Math.* F. Browder ed. *Amer. Math. Soc.* 45(1986), 283–300

[CoR1] V. Coti Zelati and P. Rabinowitz, Homoclinic orbits for second order Hamiltonian systems possessing surperquadratic potentials. *J. Amer. Math. Soc.* 4(1991), 693–727

[CrW1] C. Croke and A. Weinstein, Closed curves on convex hypersurfaces and periods of nonlinear oscillations. *Invent. Math.* 64(1981), 199–202

[Dan1] E.N. Dancer, Degenerate critical points, homotopy indices and Morse inequalities. *J. Reine Angew. Math.* 50(1984), 1–22

[Del1] G. Dell'Antonio, Variational calculus and stability of periodic solutions of a class of Hamiltonian systems. *SISSA Ref. 185/92/FM* Oct., 1992

[DDE1] G. Dell'Antonio, B. D'Onofrio and I. Ekeland, Les systém hamiltoniens convexes et pairs ne sont pas ergodiques en general. *C. R. Acad. Sci. Paris*. Series I. 315(1992), 1413–1415

[DoE1] B. D'Onofrio and I. Ekeland, Hamiltonian systems with elliptic periodic orbits. *Nonlinear Anal. TMA*. 14(1990), 11–21

[Den1] S. Deng, Minimal period solutions of a class of Hamiltonian equation systems. *Acta Math. Sinica*. 27(1984), 664–675

[Din1] W.-Y. Ding, A generalization of the Poincaré-Birkhoff theorem. *Proc. Amer. Math. Soc.* 88(1983), 341–346

[DoL1] D. Dong and Y. Long, The iteration formula of the Maslov-type index theory with applications to nonlinear Hamiltonian systems. *Trans. Amer. Math. Soc.* 349(1997), 2619–2661

[Dui1] J.J. Duistermaat, On the Morse index in variational calculus. *Adv. in Math.* 21(1976), 173–195

[Edw1] H.M. Edwards, A generalized Sturm theorem. *Ann. of Math.* 80(1964), 22–57

[EiS1] S. Eilenberg and N. Steenrod, Foundations of Algebraic Topology. *Princeton Univ. Press. Princeton.*, 1952

[Eke1] I. Ekeland, Une théorie de Morse pour les systémes hamiltoniens convexes. *Ann. IHP. Anal. non Linéaire*. 1(1984), 19–78

[Eke2] I. Ekeland, An index theory for periodic solutions of convex Hamiltonian systems. *Proc. Symp. in Pure Math.* 45(1986), 395–423

[Eke3] ———, Convexity Methods in Hamiltonian Mechanics. Springer-Verlag. Berlin, 1990

[EkH1] I. Ekeland and H. Hofer, Periodic solutions with prescribed period for convex autonomous Hamiltonian systems. *Invent. Math.* 81(1985), 155–188

[EkH2] ———, Convex Hamiltonian energy surfaces and their closed trajectories. *Comm. Math. Phys.* 113(1987), 419–467

[EkH3] ———, Subharmonics of convex nonautonomous Hamiltonian systems. *Comm. Pure Appl. Math.* 40(1987), 1–37

[EkH4] ———, Symplectic topology and Hamiltonian dynamics. *Math. Z.* 200(1990), 355–378. (II) 203(1990), 553–567

[EkL1] I. Ekeland and J.M. Lasry, On the number of periodic trajectories for a Hamiltonian flow on a convex energy surface. *Ann. of Math.* 112(1980), 283–319

[ELa1] I. Ekeland and L. Lassoued, Multiplicité des trajectoires fermées d'un systéme hamiltonien sur une hypersurface d'energie convexe. *Ann. IHP. Anal. non Linéaire*. 4(1987), 1–29

[Eli1] Y. Eliashberg, Contact 3-manifolds twenty years since J. Martinet's work. *Ann. Inst. Fourier (Grenoble)*. 42(1992), 165–192

[FHR1] E. Fadell, S. Husseini and P. Rabinowitz, Borsuk-Ulam theorems for arbitrary S^2 actions and applications. *Trans. Amer. Math. Soc.* 274(1982), 345–360

[FaR1] E. Fadell and P. Rabinowitz, Generalized cohomological theories for Lie group actions with an application to bifurcation questions for Hamiltonian systems. *Invent. Math.* 45(1978), 139–174

[Fei1] G. Fei, Relative Morse index and its applications to the Hamiltonian systems in the presence of symmetry. *J. Diff. Equa.* 122(1995), 302–315

[FeQ1] G. Fei and Q. Qiu, Minimal period solutions of nonlinear Hamiltonian systems. *Nonlinear Analysis, T.M.A.* 27(1996), 821–839

[FeQ2] G. Fei and Q. Qiu, Periodic solutions of asymptotically linear Hamiltonian systems. *Chinese Ann. of Math.* 18B(1997), no. 3, 359–372

[FKW1] G. Fei, S.-K. Kim, and T. Wang, Minimal period estimates of periodic solutinos for superquadratic Hamiltonian systems. *J. Math. Anal. Appl.* 238(1999), 216–233

[Fel1] P. Felmer, Periodic solutions of spatially periodic Hamiltonian systems. CMS Report No. 90-3, *Univ.* of Wisconsin, 1990

[Flo1] A. Floer, Morse theory for Lagrangian intersection theory. *J. Diff. Geom.* 28(1988), 513–547

[Flo2] ———, Symplectic fixed points and holomorphic spheres. *Comm. Math. Phys.* 120(1989), 575–611

[FHV1] A. Floer, H. Hofer and C. Viterbo, The Weinstein conjecture in $P \times \mathbf{C}^l$. *Math. Z.* 203(1990), 355–378

[FoW1] A. Fonda and M. Willem, Subharmonic osillations of forced pendulum-type equations. *J. Diff. Equa.* 81(1989), 215–220

[For1] B, Fortune, A symplectic fixed point theorem for $\mathbf{CP}^n$. *Invent. Math.* 81(1985), 29–46

[Fra1] J. Franks, Generalizations of the Poincaré-Birkhoff theorem. *Ann. of Math.* 128(1988), 131–151

[FuO1] K. Fukaya and K. Ono, Arnold conjecture and Gromov-Witten invariant for general symplectic manifolds. *Topology*. 38(1999), 933–1048

[GaC1] J.M. Gamboudo and P. Le Calvez, Infinité d'orbits péridiques pour les difféomorphismes conservatifs de l'anneau. Preprint, 1991

[GeL1] I.M. Gelfand and V.B. Lidskii, On the structure of the regions of stability of linear canonical systems of differential equations with periodic coefficients. *Usp. Mat. Nauk.* 10(1955), 3–40 (Russian); *Transl. Amer. Math. Soc.* 8(1958), No. 2, 143–181

[Gho1] N. Ghoussoub, Location, multiplicity and Morse indices of min-max critical points. *J. reine angew Math.* 417(1991), 27–76

[Gho2] ———, Duality and Perturbation Methods in Critical Point Theory. Cambridge Univ. Press, 1993

[GiT1] D. Gilbarg and N.S. Trudinger, Elliptic Partial Differential Equations of Second Order. 2nd ed. Springer. Berlin, 1983

[Gin1] V. Ginzburg, An embedding $S^{2n-1} \to \mathbf{R}^{2n}$, $2n-1 \geq 7$, whose Hamiltonian flow has no periodic trajectories. *Inter. Math. Res. Notices.* 2(1995), 84–98

[Gir1] M. Girardi, Multiple orbits for Hamiltonian systems on starshaped energy surfaces with symmetry. *Ann. IHP. Analyse non linéaire.* 1(1984), 285–294

[GiM1] M. Girardi and M. Matzeu, Some results on solutions of minimal period to superquadratic Hamiltonian equations. *Nonlinear Anal. TMA.* 7(1983), 475–482

[GiM2] ———, Solutions of minimal period for a class of nonconvex Hamiltonian systems and applications to the fixed energy problem. *Nonlinear Anal. TMA.* 10(1986), 371–382

[GiM3] ———, Periodic solutions of convex autonomous Hamiltonian systems with a quadratic growth at the origin and superquadratic at infinity. *Ann. Mat. pura ed appl.* 147(1987), 21–72

[GiM4] ———, Dual Morse index estimates for periodic solutions of Hamiltonian systems in some nonconvex superquadratic case. *Nonlinear Anal. TMA.* 17(1991), 481–497

[GeG1] K. Geba and A. Granas, Infinite dimensional cohomology theories. *J. Math. Pure Appl.* 52(1973), 145–270

[GrM1] D. Gromoll and W. Meyer, On differentiable functions with isolated critical points. *Topology.* 8(1969), 361–369

[GrM2] ———, Periodic geodesics on compact Riemannian manifolds. *J. Diff. Geom.* 3(1969), 493–510

[Gro1] M. Gromov, Pseudoholomorphic curves in symplectic manifolds. *Invent. Math.* 82(1985), 307–347

[Guo1] D.J. Guo, Nonlinear Functional Analysis. Chinese ed. Shandong *Sci. Tech.* Press. Jinan, 1985

[HaL1] J. Han and Y. Long, Normal forms of symplectic matrices (II). *Acta Sci. Nat. Univ. Nankai.* 32(1999), No. 3, 30–41

[HLP1] G.H. Hardy, J.E. Littlewood and G. Polya, Inequality. 2nd ed. Cambridge Univ. Press. Cambridge, 1952

[Hin1] N. Hingston, Equivariant Morse theory and closed geodesics. *J. Diff.Geom.* 19(1984), 85–116

[Hir1] M. Hirsch, Differential Topology. Springer. New York, 1976

[Hof1] H. Hofer, A new proof for a result of Ekeland and Lasry concerning the number of periodic Hamiltonian trajectories on a prescribed energy surface. *Boll. U. M. I.* 16(1982), 931–942

[Hof2] ———, A geometric description of the neighborhood of a critical point given by the mountain pass theorem. *J. London Math. Soc.* 31(1985), 566–570

[Hof3] ———, Lagrangian embeddings and critical point theory. *Ann. Inst. H. Poincaré. Anal. non linéaire.* 2(1985), 407–462

[Hof4] ———, On the topological properties of symplectic maps. *Proc. Roy. Soc. Edinburgh.* 115A(1990), 25–38

[HoV1] H. Hofer and C. Viterbo, The Weinstein conjecture in the presence of J-holomorphic spheres. *Comm. Pure Appl. Math.* XLV(1992), 583–622

[HoS1] H. Hofer and D. Salamon, Floer homology and Novikov rings. In *The Floer Memorial Volume.* H. Hofer, C. Taubes, A. Weinstein, E. Zehnder ed. *Progress in Math.* 133(1995), 483–524. Birkhäuser, Basel

[HWZ1] H. Hofer, K. Wysocki and E. Zehnder, The dynamics on three-dimensional strictly convex energy surfaces. *Ann. of Math.* 148(1998), 197–289

[HWZ2] ———, Finite energy foliations of tight three-spheres and Hamiltonian dynamics. FIM. ETH-Zürich Preprint, 2000

[HWZ3] ———, Pseudoholomorphic curves and dynamics in three dimensions. Preprint, 2000

[HoZ1] H. Hofer and E. Zehnder, Periodic solutions on hypersurfaces and a result by C. Viterbo. *Invent. Math.* 90(1987), 1–9

[HoZ2] ———, A new capacity for symplectic manifolds. *Analysis et cetera.* P. Rabinowitz and E. Zehnder ed. 1990, 405–428

[HoZ3] ———, Symplectic Invariants and Hamiltonian Dynamics. Birkhäuser. Basel, 1994

[Hom1] L. Hörmander, The Analysis of Linear Partial Differential Operators, I–IV. Springer. Berlin, 1983

[Hom2] ———, Symplectic classification of quadratic forms and general Mahler formulas. *Math. Z.* 219(1995), 413–449

[Hor1] V.J. Horn, Beiträge zur Theorie der kleinen Schwingungen. *Zeit. Math. Phys.* 48(1903), 400–434

[HuL1] X. Hu and Y. Long, Multiplicity of closed characteristics on non-degenerate star-shaped hypersurfaces in $\mathbf{R}^{2n}$. *Nankai Inst. of Math. Nankai Univ.* Preprint, 2001

[Jia1] M.-Y. Jiang, Hofer-Zehnder symplectic capacity for two dimensional manifolds. *Proc. Royal Soc. Edinb.* 123A(1993), 945–950

[Jia2] ———, Periodic solutions of Hamiltonian systems on hypersurfaces in a torus. *Manuscr. Math.* 85(1994), 307–321

[Kat1] T. Kato, Perturbation Theory for Linear Operators. Springer. Berlin, 1995

[Kli1] W. Klingenberg, Lectures on Closed Geodesics. Springer. Berlin, 1978

[Kli2] ———, Riemannian Geometry. 2nd ed. Walter de Gruyter. Berlin, 1995

[Kra1] M.A. Krasnoselski, Topological Methods in the Theory of Nonlinear Integral Equations. Macmillan. New York, 1964

[LaM1] A. Laub and K. Meyer, Canonical forms for symplectic and Hamiltonian matrices. Celestial Mech. 9 (1974), 213–238

[LaS1] A. Lazer and S. Solimini, Nontrivial solutions of operator equations and Morse indices of critical points of min-max type. *Nonlinear Anal. TMA.* 12(1988), 761–775

[LeO1] H.V. Le and K. Ono, Cup-length estimate for symplectic fixed points. In *Contact and Symplectic Gemetry*. C.B. Thomas ed. (1996) Cambridge Univ. Press. Cambridge.

[Lef1] S. Lefschetz, Algebraic Topology. *Amer. Math. Soc.* New York, 1942

[Lev1] M. Levi, Stability of linear Hamiltonian systems with epriodic coefficients. IBM T.J. Watson Research Center. Research Report No. RC6610 (1977)

[LiL1] S.J. Li and J.-Q. Liu, Morse theory and asymptotic linear Hamiltonian systems. *J. Diff. Equa.* 78(1989), 53–73

[Lia1] A. Liapunov, Probléme général de la stabilité du mouvement. Russian edition (1892), *Ann. Fac. Sci. Toulouse* 9(1907), 203–474

[LiM] P. Libermann and C.-M. Marle, Symplectic Geometry and Analytical Mechanics. D. Reidel. Dordrecht, 1987

[Lic1] C. Liu, Subharmonic solutions of Hamiltonian systems. *Nankai Inst. of Math.* Preprint, 1997. *Nonlinear Anal. TMA.* to appear

[LLo1] C. Liu and Y. Long, An optimal increasing estimate for iterated Maslov-type indices. *Chinese Sci. Bull.* 42(1997), 2275–2277

[LLo2] ———, Hyperbolic characteristics on star-shaped hypersurfaces. *Ann. Inst. H. Poincaré. Anal. non linéaire.* 16(1999), 725–746

[LLo3] ———, Iteration inequalities of the Maslov-type index theory with applications. *J. Diff. Equa.* 165(2000), 355–376

[LLo4] ———, Iterated index formulae for closed geodesics with applications. Nankai Inst. Math. Preprint, 2001. *Science in China* to appear

[LiT1] G. Liu and G. Tian, Floer homology and Arnold conjecture. Preprint, 1996, 1997

[LiT2] ———, Weinstein conjecture and GW invariants. Preprint, 1997, 1999

[Liu1] J.-Q. Liu, A generalized saddle point theorem. *J. Diff. Equa.* 82(1989), 372–385

[LiW1] J.-Q. Liu and Z.-Q. Wang, Remarks on subharmonics with minimal periods of Hamiltonian systems. *Nonlinear Anal. TMA.* 20(1993), 803–821

[LLZ1] C. Liu, Y. Long and C. Zhu, Multiplicity of closed characteristics on symmetric convex hypersurfaces in $\mathbf{R}^{2n}$. *Nankai Inst. Math.* Preprint No. 1999-M-003. Math. Ann. to appear

[Lon1] Y. Long, On the density of the range for some nonlinear operators. *Ann. Inst. H. Poincaré. Anal. non linéaire.* 6(1989), 139–151

[Lon2] ———, Multiple solutions of perturbed superquadratic second order Hamiltonian systems. *Trans. Amer. Math. Soc.* 311(1989), 749–780

[Lon3] ———, Periodic solutions of perturbed superquadratic Hamiltonian systems. *Ann. Scuola Norm. Sup. Pisa.* Series 4. 17(1990), 35–77

[Lon4] ———, Periodic solutions of superquadratic Hamiltonian systems with bounded forcing terms. *Math. Z.* 203(1990), 453–467

[Lon5] ———, Maslov-type index, degenerate critical points, and asymptotically linear Hamiltonian systems. *Science in China (Scientia Sinica).* Series A. 7(1990), 673–682. (Chinese edition), 33(1990), 1409–1419. (English edition)

[Lon6] ———, The structure of the singular symplectic matrix set. *Science in China (Scientia Sinica).* Series A. 5(1991), 457–465. (Chinese edition), 34(1991), 897–907. (English edition)

[Lon7] ———, Index Theory of Hamiltonian Systems with Applications. Science Press. Beijing, 1993. (In Chinese)

[Lon8] ———, The minimal period problem for classical Hamiltonian systems with even potentials. *Ann. Inst. H. Poincare. Anal. non linéaire.* 10(1993), 605–626

[Lon9] ———, The minimal period problem of periodic solutions for autonomous superquadratic second order Hamiltonian systems. *J. Diff. Equa.* 111(1994), 147–174

[Lon10] ———, Nonlinear oscillations for classical Hamiltonian systems with bi-even subquadratic potentials. *Nonlinear Analysis, TMA.* 24(1995), 1665–1671

[Lon11] ———, Geodesics in the compactly supported Hamiltonian diffeomorphism group. *Math. Z.* 220(1995), 279–294

[Lon12] ———, On the minimal period for periodic solutions of nonlinear Hamiltonian systems. *Chinese Ann. of Math.* 18B(1997), 481–484

[Lon13] ———, A Maslov-type index theory for symplectic paths. *Top. Meth. Nonl. Anal.* 10(1997), 47–78

[Lon14] ———, Hyperbolic closed characteristics on compact convex smooth hypersurfaces. *J. Diff. Equa.* 150(1998), 227–249

[Lon15] ———, The topological structures of ω-subsets of symplectic groups. *Acta Math. Sinica.* English Series. 15(1999), 255–268

[Lon16] Y. Long, Bott formula of the Maslov-type index theory. *Pacific J. Math.* 187(1999), 113–149

[Lon17] ———, Multiple periodic points of the Poincaré map of Lagrangian systems on tori. *Math. Z.* 233(2000), No. 3, 443–470

[Lon18] ———, Precise iteration formulae of the Maslov-type index theory and ellipticity of closed characteristics. *Advances in Math.* 154(2000), 76–131

[Lon19] ———, The Maslov-type index and its iteration theory with applications to Hamiltonian systems. *Third School on Nonlinear Analysis and Applications to Differential Equations.* (10.12-30, 1998). ICTP Lecture Notes. SMR 1071/2. *Minimax Theory, Morse Theory and the Applications to Differential Equations.* H. Brezis, S. Li, J.-Q. Liu, P. Rabinowitz ed. International Press. to appear

[Lon20] ———, Dynamics on compact convex hypersurfaces in $\mathbf{R}^{2n}$. *Progress in Nonlinear Analysis.* Proceedings of the Second International Conference on Nonlinear Analysis at Tianjin. K.C. Chang and Y. Long ed. June 1999. World Sci., 2000, 271–288

[LoA1] Y. Long and T. An, Indexing the domains of instability for Hamiltonian systems. Nankai Inst. of Math. Preprint, 1996. (Revised 1997). *NoDEA*. 5(1998), 461–478

[LoD1] Y. Long and D. Dong, Normal forms of symplectic matrices. Nankai Inst. of Math. Preprint, 1995. *Acta Math.Sinica.* 16(2000), 237–260

[LLu1] Y. Long and G. Lu, Infinitely many periodic solution orbits of the autonomous Lagrangian systems on tori, 1999 Submitted

[LoX1] Y. Long and X. Xu, Periodic solutions of a class of nonautonomous Hamiltonian systems. *Nonlinear Anal. T.M.A.* 41(2000), 455–463

[LZe1] Y. Long and E. Zehnder, Morse theory for forced oscillations of asymptotically linear Hamiltonian systems. In *Stoc. Proc. Phys. and Geom.*, S. Albeverio et al. ed. World Sci. 1990, 528–563

[LZe2] ———, Remarks on isoenergetic periodic orbits in tori. *Advances in dynamical systems and quantum physics.* (Capri, 1993), World Sci., 1995, 326–344

[LZa1] Y. Long and S. Zhang, Geometric characterizations for variational minimization solutions of the 3-body problem. *Chinese Sci. Bull.* (Announcement) 44(1999), No. 18, 1653–1655. *Acta Math.Sinica.* 16(2000), 579–592

[Lza2] ———, Geometric characterizations for variational minimization solutions of the 3-body problem with fixed energy. *J. Diff. Equa.* 160(2000), 422–438

[LZu1] Y. Long and C. Zhu, Maslov-type index theory for symplectic paths and spectral flow (II). *Chinese Ann. of Math.* 21B(2000), No. 1, 89–108

[LZu2] ———, The differentiable structure of ω-subsets of symplectic groups. *Science in China.* Series A. 30(2000), 680–689

[LZu3] ———, Closed characteristics on compact convex hypersurfaces in $\mathbf{R}^{2n}$. Nankai Inst. Math. Preprint No. 1999-M-002. Revised March 2001. *Annals of Math.* to appear

[Mar1] R. Ma, A remark on the Weinstein conjecture in $M \times R^{2n}$. *Nonlinear Analysis and Microlocal Analysis*, edited by K.C. Chang, Y.-M. Huang, and T.-T. Li, World Sci., 1992, 176–184

[MaF1] V.P. Maslov and M.V. Fedoryuk, The Quasiclassical Approximation for Equations of Quantum Mechanics. Nauka, Moscow, 1976 (in Russian). English transl. Reidel Publ. Comp. Dordrecht, 1981

[Mat1] J. Mather, Existence of quasiperiodic orbits for twist homeomorphisms of the annulus. *Topology* 21(1982), 457–467

[MFo1] J. Mather and G. Forni, Action Minimizing Orbits in Hamiltonian Systems. *Lect. In Math.* 1589. Springer. Berlin, 1994

[MaW1] J. Mawhin and M. Willem, Critical Point Theory and Hamiltonian systems. Springer. Berlin, 1989

[McS1] D. McDuff and D. Salamon, J-holomorphic Curves and Quantum Cohomology. *Amer. Math. Soc.* Providence, 1994

[McS2] ———, Introduction to Symplectic Topology. Oxford Univ. Press, 1994. Second ed., 1998

[MeH1] K.R. Meyer and G.R. Hall, Introduction to Hamiltonian Dynamical Systems and the N-body Problem. Springer. Berlin, 1992

[MSS1] A.S. Mishchenko, V.E. Shatalov and B.Yu. Sternin, Lagrangian Manifolds and the Maslov Operator. Nauka. Moscow, 1978. (Russian). English transl. Springer. Berlin, 1990

[Mor1] M. Morse, The Calculus of Variations in the Large. *Amer. Math. Soc. Colloquium Publ.* 18(1934)

[Mos1] J.K. Moser, New aspects in the theory of stability of Hamiltonian systems. *Comm. Pure Appl. Math.* 11(1958), 81–114

[Mos2] ———, On invariant curves of area preserving mappings of the annulus. *Nachr. Acad. Wiss. Göttingen Math. phys. Kl.*, 1962, 1–20

[Mos3] ———, On the volume elements on a manifold. *Trans. Amer. Math. Soc.* 120(1965), 286–294

[Mos4] ———, Stable and Random Motions in Dynamical Systems. Princeton Univ. Press, 1973

[Mos5] ———, Periodic orbits near an equilibrium and a theorem of A. Weinstein. *Comm. Pure Appl. Math.* 29(1976), 727–747

[Mos6] ———, Proof of a generalized form of a fixed point theorem due to G.D. Birkhoff. *Geometry and Topology. Lecture Notes in Math.* 597(1977), 464–494

[Mos7] ———, Minimal solutions of variational problms on a torus. *Ann. Inst. H. Poincare. Anal. non linéaire.* 3(1986), 229–272

[MoZ1] J.K. Moser and E. Zehnder, Notes on Dynamical Systems. Courant Inst., 1980

[Ni1] W.-M. Ni, Some minimax principles and their applications in nonlinear elliptic equations. *J. d'Analyse Math.* 37(1980), 248–275

[Nir1] L. Nirenberg, Variational and topological methods in nonlinear problems. *Bull. Amer. Math. Soc.* New Series 4(1981), 267–302

[Poi1] H. Poincaré, Les Méthodes nouvelles de la Mecanique celeste. (3 vol.) Cauthier Villars. Paris, 1892–1899. New Methods of Celestial Mechanics. (English transl.) *Amer. Inst. Phys.*, 1993

[Rab1] P.H. Rabinowitz, Variational methods for nonlinear eigenvalue problems. Eigenvalues of Nonlinear Problems. (G. Prodi ed.) *CIME. Cremonese. Roma.* 1975, 141–195

[Rab2] ———, Periodic solutions of Hamiltonian systems. *Comm. Pure Appl. Math.* 31(1978), 157–184

[Rab3] ———, Periodic solutions of Hamiltonian systems on a prescribed energy surface. *J. Diff. Equa.* 33(1979), 336–352

[Rab4] ———, On subharmonic solutions of Hamiltonian systems. *Comm. Pure Appl. Math.* 33(1980), 609–633

[Rab5] ———, Multiple critical points of perturbed symmetric functionals. *Trans. Amer. Math. Soc.* 272(1982), 753–770

[Rab6] ———, Periodic solutions of Large norm of Hamiltonian systems. *J. Diff. Equa.* 50(1983), 33–48

[Rab7] P. Rabinowitz, Minimax methods in critical point theory with applications to differential equations. *CBMS Regional Conf. Ser. in Math.* No. 65. *Amer. Math. Soc.*, 1986

[Rab8] ———, A variational approach to multibump solutions of differential equations. *Contemp. Math.* 198(1996), 31–43

[Rad1] H.-B. Rademacher, On the average indices of closed geodesics. *J. Diff. Geom.* 29(1989), 65–83

[RiN1] F. Riesz and B. Sz.-Nagy, Leçon d'analyse fonctionnelle. 4th ed. Akadémiai Kiadó. Budapest. 1965. Chinese ed. 1980

[RoS1] J. Robbin and D. Salamon, The Maslov index for paths. *Topology* 32(1993), 827–844

[RoS2] ———, The spectral flow and the Maslov index. *Bull. London Math. Soc.* 27(1995), 1–33

[Rot1] E. Rothe, Morse theory in Hilbert space. *Rocky Mountain J. Math.* 3(1973), 251–274

[Rua1] Y. Ruan, Symplectic topology on algebraic 3-folds. *J. Diff. Geom.* 39(1994), 215–227

[Rua2] ———, Virtual neighborhood and pseudo-holomorphic curves. *Turkish J. Math.* 23(1999), 161–231

[RuT1] Y. Ruan and G. Tian, A mathematical theory of quantum cohomology. *J. Diff. Geom.* 42(1995)

[RuT2] Y. Ruan and G. Tian, Bott-type symplectic Floer cohomology and its multiplication structures. *Math. Rev. Let.* 2(1995), 203–219

[SaZ1] D. Salamon and E. Zehnder, Floer homology, the Maslov index and periodic orbits of Hamiltonian equations. *Analysis, et cetera.* ed. by P. Rabinowitz and E. Zehnder. Academic Press, 1990, 573–600

[SaZ2] ———, Morse theory for periodic solutions of Hamiltonian systems and the Maslov index. *Comm. Pure and Appl. Math.* 45(1992), 1303–1360

[Stz1] J.T. Schwartz, Nonlinear Functional Analysis. Gordon and Breach *Sci. Publ.* New York, 1969

[Sch1] M. Schwarz, Morse Homology. Birkhauser. Basel, 1993

[Sch2] M. Schwarz, A quantum cup-length estimate for symplecic fixed points. *Invent. Math.* 133(1998), 353–397

[Sei1] H. Seifert, Periodischer Bewegungen mechanischen systeme. *Math. Z.* 51(1948), 197–216

[Ser1] E. Séré, Existence of infinitely many homoclinic orbits in Hamiltonian systems. *Math. Z.* 209(1992), 27–42

[SiM1] C.L. Siegel and J.K. Moser, Lectures on Celestial Mechanics. Springer. Berlin, 1971

[Sol1] S. Solimini, Morse index estimates in min-max theorems. *Manus. Math.* 63(1989), 421–453

[Spa1] E.H. Spanier, Algebraic Topology. New York. McGraw-Hill, 1966

[Str1] M. Struwe, Infinitely many critical points for functionals which are not even and applications to nonlinear BVP. *Manuscripta Math.* 32(1982), 753–770

[Str2] ———, Existence of periodic solutions of Hamiltonian systems on almost every energy surfaces. *Bol. Soc. Brasil Mat.* 20(1990), 49–58

[Str3] ———, Variational Methods. Springer. Berlin, 1990

[Szu1] A. Szulkin, Morse theory and existence of periodic solutions of convex Hamiltonian systems. *Bull. Soc. Math. France.* 116(1988), 171–197

[Szu2] ———, Cohomology and Morse theory for strongly indefinite functionals. *Math. Z.* 209(1992), 375–418

[Tan1] K. Tanaka, Non-collision solutions for a second order singular Hamiltonian system with weak force. *Ann. Inst. H. Poincaré Anal. non linéaire.* 10(1993), 215–238

[Tia1] G. Tian, On the mountain pass theorem. *Kexue Tongbao.* (Chinese Sci. Bull.) 20(1984), 1150–1154

[Tia2] ———, Gauge theory and calibrated geometry, I. *Annals of Math.* to appear

[Tia3] ———, Constructing symplectic invariants. Preprint, 1999

[Vit1] C. Viterbo, A proof of the Weinstein conjecture in $\mathbf{R}^{2n}$. *Ann. Inst. H. Poincaré Anal. non linéaire.* 4(1987), 337–357

[Vit2] ———, Indice de Morse des points critiques obtenus par minimax. *Ann. Inst. H. Poincaré Anal. non linéaire.* 5(1988), 221–226

[Vit3] ———, Equivariant Morse theory for starshaped Hamiltonian systems. *Trans. Amer. Math. Soc.* 311(1989), 621–655

[Vit4] C. Viterbo, A new obstruction to embedding Lagrangian tori. *Invent. Math.* 100(1990), 301–320

[Wan1] J.-H. Wang, Three Body Problems. Beijing. Science Press, 1961, (in Chinese)

[WaF1] T. Wang and G. Fei, Subharmonics for superquadratic Hamiltonian systems via the iteration method of the Maslov-type index theory. Preprint, 1996

[Wzq1] Z.-Q. Wang, A note on the deformation theorem. *Acta Math. Sinica.* 30(1987), 106–110

[Wzq2] Z.-Q. Wang, On a superlinear elliptic equation. *Ann. Inst. H. Poincaré. Anal. non linéaire.* 8(1991), 43–58

[Wei1] A. Weinstein, Lagrangian submanifolds and Hamiltonian systems. *Ann. of Math.* 108(1978), 377–410

[Wei2] ———, Normal modes for nonlinear Hamiltonian systems. *Invent. Math.* 20(1973), 45–57

[Wei3] ———, Lectures on Symplectic Manifolds. *CBMS regional Reports* No. 29. *Amer. Math. Soc.*, 1977

[Wei4] ———, Bifurcations and Hamilton's principle. *Math. Z.* 159(1978), 235–248

[Wei5] ———, Periodic orbits for convex Hamiltonian systems. *Ann. of Math.* 108(1978), 507–518

[Wei6] ———, On the hypothesis of Rabinowitz's periodic orbit theorems. *J. Diff. Equa.* 33(1979), 353–358

[Wey1] H. Weyl, The Classical Groups. 2nd ed. Princeton Univ. Press, 1953

[Wil1] J. Williamson, On the normal forms of linear canonical transformations in dynamics. *Amer. J. Math.* 59(1937), 599–617

[Wim1] H. Wimmer, Normal forms of symplectic pencils and the discrete-time algebraic Riccati equation. *Linear Alg. Appl.* 147(1991), 411–440

[Win1] A. Wintner, The Analytical Foundation of Celestial Mechanics. Princeton Univ. Press. Princeton, 1947

[Xia1] Z.-H. Xia, The existence of noncollision singularities in Newtonian systems. *Ann. of Math.* 135(1992), 411–468

[Xia2] ———, Existence of invariant tori in volume preserving diffeomorphisms. *Egord. Th. Dynam. Sys.* 12(1992), 621–631

[YaS1] V. Yakubovich and V. Starzhinskii, Linear Differential Equations with Periodic Coefficients. New York, John Wiley & Sons, 1975

[Yi1] Zh.-H. Yi, Introduction on Celestial Mechanics. Nanjing. Nanjing Univ. Press, 1993. (in Chinese)

[YiS1] Zh.-H. Yi and Y.-S. Sun, Perturbation Theory. Beijing. Science Press, 1981 (in Chinese)

[Zeh1] E. Zehnder, Homoclinic points near elliptic fixed points. *Comm. Pure Appl. Math.* 26(1973), 131–182

[Zeh2] ———, A Poincaré-Birkhoff-type result in higher dimensions. Travaux eu Cours 25. Hermann. Paris, 1987, 119–146

[Zeh3] E. Zehnder, Remarks on periodic solutions on hypersurfaces. In *Periodic Solutions of Hamiltonian Systems and Related Topics.* NATO ASI Series. P. Rabinowitz et al. ed. (1987), 267–279

[Zei1] E. Zeidler, Nonlinear Functional Analysis and its Applications. III. Variational Methods and Optimization. Springer. Berlin, 1985

[Zha1] S. Zhang and Q. Zhou, Symmetric periodic noncollision solutions for N-body type problems. *Acta Math. Sinica.* New Series 11(1995), 37–43

[Zha2] S. Zhang, Multiple closed orbits of fixed energy for N-body type problems with gravitational potentials. *J. Math. Anal. Appl.* 208(1997), 462–475

[ZhL1] C. Zhu and Y. Long, Maslov-type index theory for symplectic paths and spectral flow (I). *Chinese Ann. of Math.* 208(1999), 413–424

Index

MIX
Papier aus verantwortungsvollen Quellen
Paper from responsible sources
FSC® C105338

If you have any concerns about our products,
you can contact us on
ProductSafety@springernature.com

In case Publisher is established outside the EU,
the EU authorized representative is:
Springer Nature Customer Service Center GmbH
Europaplatz 3, 69115 Heidelberg, Germany

Printed by Libri Plureos GmbH
in Hamburg, Germany